SKELETAL BIOLOGY AND MEDICINE, PART A

Aspects of Bone Morphogenesis and Remodeling

ANNALS OF THE NEW YORK ACADEMY OF SCIENCES

Volume 1116

SKELETAL BIOLOGY AND MEDICINE, PART A

Aspects of Bone Morphogenesis and Remodeling

Edited by Mone Zaidi

Published by Blackwell Publishing on behalf of the New York Academy of Sciences
Boston, Massachusetts
2007

Library of Congress Cataloging-in-Publication Data

Skeletal biology and medicine/edited by Mone Zaidi
    p.; cm. – (Annals of the New York Academy of Sciences, ISSN
0077-8923; v. 1116)
  Includes bibliographical references.
  ISBN-13: 978-1-57331-684-2 (paper: alk. paper)
  ISBN-10: 1-57331-684-9 (paper: alk. paper)
  1. Bones–Growth.
  2. Bones–Pathophysiology. I. Zaidi, Mone. II. New York Academy of
Sciences. III. Series.
  [DNLM: 1. Bone Development–physiology. 2. Bone
Diseases–physiopathology. 3. Hormones. 4. Intercellular Signaling
Peptides and Protiens. W1 AN626YL v. 1116 2007/WE 200 S62676
2007]

  QP88.2.S525 2007
  612.7′5–dc22

                                     2007034590

The *Annals of the New York Academy of Sciences* (ISSN: 0077-8923 [print]; ISSN: 1749-6632 [online]) is published 28 times a year on behalf of the New York Academy of Sciences by Blackwell Publishing with offices at 350 Main St., Malden, MA 02148 USA; 9600 Garsington Road, Oxford, OX4 2ZG UK; and 600 North Bridge Rd, #05-01 Parkview Square, 18878 Singapore.

Information for subscribers: For new orders, renewals, sample copy requests, claims, changes of address and all other subscription correspondence please contact the Journals Department at your nearest Blackwell office (address details listed above). UK office phone: +44 (0)1865 778315, fax +44 (0)1865 471775; US office phone: 1-800-835-6770 (toll free US) or 1-781-388-8599; fax: 1-781-388-8232; Asia office phone: +65 6511 8000, fax; +44 (0)1865 471775, Email: customerservices@blackwellpublishing.com

Subscription rates:
Institutional Premium   The Americas: $4043   Rest of World: £2246
The Premium institutional price also includes online access to full-text articles from 1997 to present, where available. For other pricing options or more information about online access to Blackwell Publishing journals, including access information and terms and conditions, please visit www.blackwellpublishing. com/nyas
*Customers in Canada should add 6% GST or provide evidence of entitlement to exemption.
**Customer in the UK or EU: add the appropriate rate for VAT EC for non-registered customers in countries where this is applicable. If you are registered for VAT please supply your registration number.

Mailing: The *Annals of the New York Academy of Sciences* is mailed Standard Rate. Mailing to rest of world by International Mail Express (IMEX). Canadian mail is sent by Canadian publications mail agreement number 40573520. **Postmaster:** Send all address changes to *Annals of the New York Academy of Sciences*, Blackwell Publishing Inc., Journals Subscription Department, 350 Main St., Malden, MA 02148-5020.

Membership information: Members may order copies of *Annals* volumes directly from the Academy by visiting www.nyas.org/annals, emailing membership@nyas.org, faxing 212-298-3650, or calling 800-843-6927 (US only), or 212-298-8640 (International). For more information on becoming a member of the New York Academy of Sciences, please visit www.nyas.org/membership. Claims and inquiries on member orders should be directed to the Academy at email: membership@nyas.org or Tel: 212-298-8640 (International) or 800-843-6927 (US only).

Annals are available to subscribers online at the New York Academy of Sciences and also at Blackwell Synergy. Visit www.blackwell-synergy.com or www.annalsnyas.org to search the articles and register for table of contents e-mail alerts. Access to full text and PDF downloads of *Annals* articles are available to nonmembers and subscribers on a pay-per-view basis at www.blackwell-synergy.com and www.annalsnyas.org.

The paper used in this publication meets the minimum requirements of the National Standard for Information Sciences Permanence of Paper for Printed Library Materials, ANSI Z39.48-1984.

ISSN: 0077-8923 (print); 1749-6632 (online)
ISBN-10: 1-57331-684-9 (alk. paper); ISBN-13: 978-1-57331-684-2 (alk. paper)

A catalogue record for this title is available from the British Library.

ANNALS OF THE NEW YORK ACADEMY OF SCIENCES

Volume 1116
November 2007

SKELETAL BIOLOGY AND MEDICINE, PART A

Aspects of Bone Morphogenesis and Remodeling

Editor
MONE ZAIDI

This volume is the result of a conference entitled **2nd Conference on Skeletal Biology and Medicine**, held on April 25–28, 2007 in New York City, and sponsored by Mount Sinai School of Medicine and the New York Academy of Sciences.

CONTENTS

Financial assistance was received from:

- Amgen, Inc.
- Genzyme Corporation
- Eli Lilly & Company
- Merck Research Laboratories
- Mitzy and Warren Eisenberg Family Foundation
- National Institute of Arthritis and Musculoskeletal and Skin Diseases - NIH
- Novartis Pharmaceuticals
- NPS Pharmaceuticals
- Procter & Gamble Alliance for Better Bone Health
- Roche Pharmaceuticals
- Sanofi-Aventis
- Diane Wolf Foundation
- Erving & Joyce Wolf Foundation
- Wyeth Pharmaceuticals

The Second Conference on Skeletal Medicine and Biology

Overview and Some Comments

The skeleton is the target not only of aging and hormone deprivation but is also adversely affected by a variety of systemic diseases, including diabetes, as well as drugs used in the management of such diseases and exemplified by the glitazones. The skeleton can now be seen as an endocrine organ that regulates insulin release and resistance.[1] Skeletal medicine is thus emerging as its own discipline encompassing fields as broadly disparate as aging, endocrinology, pharmacology, developmental biology, cell and molecular biology, radiology, rheumatology, obstetrics and gynecology, primary care, radiology, cancer medicine, and hematology. The 2nd Conference on Skeletal Biology and Medicine, held in New York City between April 25 and 28, 2007, cosponsored by the New York Academy of Sciences (NYAS) and Mount Sinai School of Medicine,[2] provided a continuing platform for the sequestration of elements from many such traditional disciplines focused toward understanding diseases of the human skeleton.

The meeting had three goals. The first was to allow the latest research from any area,[3] as it related to bones and joints, both clinical and basic, to be discussed in a free, trusting, and interactive environment. Our desire was to provide the opportunity for unpublished scientific data to be shared and discussed openly. As with the larger Gordon Conferences and Keystone Symposia, all speakers and session chairs, totaling 87 in number, were invited; all other registrants were preselected. However, unlike Gordon and Keystone, this conference, albeit small, was not particularly theme specific and therefore purposefully displayed current investigation from laboratories worldwide.

A second goal, carried over seamlessly from the 2005 conference, sought to allow junior investigators, including postdoctoral fellows and students, to participate in constructive unencumbered dialogue with peers and mentors. It was anticipated that this dialogue would not only allow their exposure (in the form of a crash course) to top-notch science in skeletal biology, but also pave the way for networking with the very best researchers, without the time constraints experienced in larger forums, such as the American Society for Bone and Mineral Research. A new development, however, was the inclusion of a young investigator session. Four abstracts, selected for oral presentations, were presented by postdoctoral fellows and students, namely Morten Karsdel (Nordic Bioscience, Denmark), Christa Maes (Harvard Medical School), Ritu Saxena (University of Alabama at Birmingham), and Emi Shimuzu (Robert Wood Johnson Medical School). The session was chaired by Allen Spiegel,

Ann. N.Y. Acad. Sci. 1116: xi–xiii (2007). © 2007 New York Academy of Sciences.
doi: 10.1196/annals.1402.083

MD, formerly director of the National Institute of Diabetes and Digestive and Kidney Diseases, and currently dean of the Albert Einstein College of Medicine in New York City.

A third, more interesting goal actually emerged as the organization of this conference progressed over the past year. It became obvious that, because of their interdisciplinary and translational tenor, discovery groups from pharmaceutical houses and biotechnology companies have much to offer. The conference had thus become a forum whereby academic physicians and scientists were presented with the unrivalled opportunity of bouncing new ideas off their industry counterparts.

Skeletal Biology and Medicine, Part A: Aspects of Bone Morphogenesis and Remodeling is a compendium of part of the enduring material—authoritative reviews, provocative original articles, and short commentaries—that has resulted from the 2nd NYAS Conference on Skeletal Biology of Medicine, thanks to its contributors. The volume comprises three sections: Part I provides insights into skeletal development and includes novel ideas about the role of the perichondrium in skeletogenesis, current studies on joint formation, and the human consequences of dysregulated morphogenic signaling. Part II contains a traditional series of papers, both reviews and original research articles, on bone cells—osteoblasts, osteoclasts, and osteocytes—as they relate to mechanisms underlying their formation, fate, and function. Part III focuses on new information on the role of hormones, both steroid and peptide, as well as cytokines and the central nervous system in regulating skeletal remodeling. I sincerely hope that these two volumes go beyond the past one in providing a sophisticated audience that is dedicated to bone disease investigation with a collection of articles from the latest research.

ACKNOWLEDGMENTS

My personal gratitude goes first to the speakers, moderators, chairs, and attendees, who, despite their busy schedules, found time to come to New York, committing themselves to make the 2nd NYAS Conference on Skeletal Medicine and Biology a successful forum for purposeful scientific exchange among scientists of varied disciplines, among mentors and mentees, and between academia and industry. I wish also to acknowledge my two co-chairs, Steve Teitelbaum and Gerard Karsenty, as well as my colleagues at Mount Sinai, Baljit Moonga, Li Sun, Guozhe Yang, Jameel Iqbal, and my worthy assistant, Mary Jo Sweeney, without whose commitment and assistance, the conference could not have been realized. In this regard, I must also gratefully acknowledge the assistance, cooperation, and hard work of Renee Wilkerson and her staff at the New York Academy of Sciences, who toiled for three months to make this meeting a success; and to the direction, and particularly the forbearance, of Stacie Bloom, PhD, in her new position as director of programs at the New York Academy of Sciences. The editorial staff who put this volume together—namely Ralph Brown, Steve Bohall, Kirk Jensen (executive editor), and Linda Mehta (acquisitions editor), and my assistant Mary Jo—indeed deserve special mention. The CME office at Mount Sinai, under the leadership of Alfie Truchan, did an outstanding job in ensuring that credits were passed along appropriately.

Without industrial sponsors, meetings such as these would simply not be possible. For their generosity, we, as a bone community, remain grateful to Amgen, Genzyme Corporation, Eli Lilly and Company, Merck Research Laboratories, Novartis Pharmaceuticals, NPS Pharmaceuticals, the Procter and Gamble Alliance for Better Bone Health (includes sanofi-aventis Pharmaceuticals), Roche Pharmaceuticals, and Wyeth Pharmaceuticals. The highest level of appreciation and gratitude goes to the following private donors for their generosity and graciousness: the Erving and Joyce Wolf Foundation, the Diane Wolf Foundation, and the Mitzy and Warren Eisenberg Family Foundation; as well as to the National Institute of Arthritis and Musculoskeletal and Skin Diseases, whose commitment to scientific interchange remains unparalleled. Finally, on a personal note, I am grateful to my wife Meenakshi, our children Neeha and Samir, and my father Professor Sibte Zaidi, for their resilience and support during the long hours of my involvement in organizing this meeting.

—MONE ZAIDI
Mount Sinai School of Medicine
New York, New York

REFERENCES

1. LEE, N.K., H. SOWA, E. HINOI, *et al.* 2007. Endocrine regulation of energy metabolism by the skeleton. Cell **130:** 456–469.
2. E-briefing, New York Academy of Sciences. www.nyas.org.
3. M. ZAIDI. 2007. Skeletal remodeling in health and disease. Nature Medicine **13:** 791–801.

Intracellular Protein Degradation From a Vague Idea through the Lysosome and the Ubiquitin-Proteasome System and on to Human Diseases and Drug Targeting

Nobel Lecture, December 8, 2004[a]

AARON CIECHANOVER

Cancer and Vascular Biology Research Center, Faculty of Medicine, Technion, Israel Institute of Technology, Haifa, Israel

ABSTRACT: Between the 1950s and 1980s, scientists were focusing mostly on how the genetic code is transcribed to RNA and translated to proteins, but how proteins are degraded has remained a neglected research area. With the discovery of the lysosome by Christian de Duve, it was assumed that cellular proteins are degraded within this organelle. Yet, several independent lines of experimental evidence strongly suggested that intracellular proteolysis is largely nonlysosomal, but the mechanisms involved had remained obscure. The discovery of the ubiquitin–proteasome system resolved this enigma. We now recognize that ubiquitin- and proteasome-mediated degradation of intracellular proteins is involved in the regulation of a broad array of cellular processes, such as cell cycle and division, regulation of transcription factors, and assurance of the cellular quality control. Not surprisingly, aberrations in the system have been implicated in the pathogenesis of many human diseases, malignancies, and neurodegenerative disorders among them, which led subsequently to an increasing effort to develop mechanism-based drugs; one is already in use.

KEYWORDS: ubiquitin–proteasome system; proteolysis; ubiquitin system

Address for correspondence: Aaron Ciechanover, Vascular and Tumor Biology Research Center, The Rappaport Faculty of Medicine and Research Institute, Technion–Israel Institute of Technology, Efron Street, Bat Galim, Haifa 31096, Israel. Voice: +972-4-829-5427; fax: +972-4-852-1193. c_tzachy@netvision.net.il

[a]This chapter is a reprint, with minor revisions, of the Nobel Lecture given by Aaron Ciechanover in 2004, and with permission from the Nobel Foundation 2004.

Ann. N.Y. Acad. Sci. 1116: 1–28 (2007). © 2007 New York Academy of Sciences.
doi: 10.1196/annals.1402.078

INTRODUCTION

The concept of protein turnover is hardly 60 years old. Earlier, body proteins were viewed as essentially stable constituents that were subject to only minor wear and tear: dietary proteins were believed to function primarily as energy-providing fuel, traversing metabolic pathways that were completely distinct from those of the structural and functional proteins of the body. The problem was hard to approach experimentally, as research tools were not available. Important research reagents that were lacking at that time were stable isotopes. While radioactive isotopes were developed earlier by George de Hevesy (de Hevesy G., Chemistry 1943. In: *Nobel Lectures in Chemistry 1942–1962*. World Scientific 1999. pp. 5–41), they were mostly unstable and could not be used to follow metabolic pathways. The concept that the body structural proteins are static and the dietary proteins are used only as a fuel was challenged by Rudolf Schoenheimer from Columbia University in New York City. Schoenheimer escaped from Nazi Germany and joined the Department of Biochemistry in Columbia University founded by Hans T. Clarke.[1–3] There he met Harold Urey who worked in the Department of Chemistry and discovered deuterium, the heavy isotope of hydrogen, a discovery that enabled him to prepare heavy water, D_2O. David Rittenberg who had recently received his Ph.D. in Urey's laboratory, joined Schoenheimer, and together they entertained the idea of *employing a stable isotope as a label in organic compounds, destined for experiments in intermediary metabolism, which should be biochemically indistinguishable from their natural analog.*[1] Urey later succeeded in enriching nitrogen with ^{15}N, which provided Schoenheimer and Rittenberg with a tag for amino acids, resulting in a series of studies on protein dynamics. They discovered that following administration of ^{15}N-labeled tyrosine to rat, only ~50% was recovered in the urine, *while most of the remainder is deposited in tissue proteins, an equivalent of protein nitrogen is excreted.*[4] They further discovered that from the half that was incorporated into body proteins *only a fraction was attached to the original carbon chain, namely to tyrosine, while the bulk was distributed over other nitrogenous groups of the proteins,*[4] mostly as an NH_2 group in other amino acids. These experiments demonstrated unequivocally that the body structural proteins are in a dynamic state of synthesis and degradation, and that even individual amino acids are in a state of dynamic interconversion. Similar results were obtained using ^{15}N-labeled leucine.[5] This series of findings shattered the paradigm in the field at that time that: (i) ingested proteins are completely metabolized and the products are excreted, and (ii) that body structural proteins are stable and static. Schoenheimer was invited to deliver the prestigious Harvey Lecture (1937) and Edward K. Dunham Lecture (1941; at Harvard University) where he presented his revolutionary findings. After his untimely tragic death in 1941, his lecture notes were edited by Hans Clarke, David Rittenberg, and Sarah Ratner, and were published in a small book by Harvard University Press. The editors called the book the

Dynamic State of Body Constituents,[6] adopting the title of Schoenheimer's presentation. In the book, the new hypothesis is clearly presented: *the simile of the combustion engine pictured the steady state flow of fuel into a fixed system, and the conversion of this fuel into waste products. The new results imply that not only the fuel, but the structural materials are in a steady state of flux. The classical picture must thus be replaced by one which takes account of the dynamic state of body structure.* However, the idea that body proteins are turning over was not accepted easily and was challenged as late as the mid 1950s. For example, Hogness and colleagues studied the kinetics of galactosidase in *E. coli* and summarized their findings[7]: *to sum up: there seems to be no conclusive evidence that the protein molecules within the cells of mammalian tissues are in a dynamic state. Moreover, our experiments have shown that the proteins of growing E. coli are static. Therefore it seems necessary to conclude that the synthesis and maintenance of proteins within growing cells is not necessarily or inherently associated with a dynamic state.* While the experimental study involved the bacterial β-galactosidase, the conclusions were broader, including also the authors' hypothesis on mammalian proteins. The use of the term *dynamic state* was not incidental, as they challenged directly Schoenheimer's studies, using his own term. It should be noted, however, that Schoenheimer's result related, as we know now, to the entire population of body proteins, extracellular and intracellular alike. It is now clear that these two classes of proteins are targeted by two different mechanisms (see below). The complex catabolic pathways, along with the lack of recognition of the importance of the process, made progress in the field slow.

Now, after more than six decades of research in the field and with the discovery of the lysosome and later the complex ubiquitin–proteasome system with its numerous tributaries, it is clear that the area has been revolutionized. While the lysosome is involved mostly (but not solely) in targeting extracellular proteins, the ubiquitin–proteasome system degrades intracellular proteins. We now realize that intracellular proteins are turning over extensively, that the process is specific, and that the stability of many proteins is regulated individually and can vary under different pathophysiological conditions. From a scavenger, unregulated and nonspecific end process, it has become clear that proteolysis of cellular proteins is a highly complex, temporally controlled, and tightly regulated process that plays major roles in a broad array of basic pathways. Among these processes are cell cycle, development, differentiation, regulation of transcription, antigen presentation, signal transduction, receptor-mediated endocytosis, quality control, and modulation of diverse metabolic pathways. Subsequently, this development has changed the paradigm that regulation of cellular processes occurs mostly at the transcriptional and translational levels, and has set regulated protein degradation in an equally important position. With the multitude of substrates targeted and processes involved, it is not surprising that aberrations in the pathway have been implicated in the pathogenesis of many diseases, among them certain malignancies, neurodegeneration, and

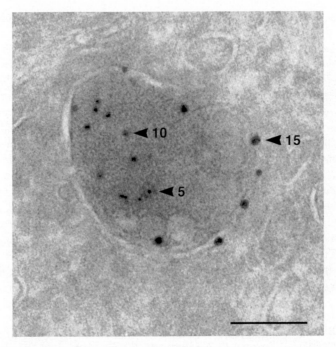

FIGURE 1. The lysosome: ultrathin cryosection of a rat PC12 cell that had been loaded for 1 h with bovine serum albumin (BSA)-gold (5 nm particles) and immunolabeled for the lysosomal enzyme cathepsin B (10-nm particles) and the lysosomal membrane protein LAMP1 (15 nm particles). Lysosomes are recognized also by their typical dense content and multiple internal membranes. Bar, 100 nm. Courtesy of Viola Oorschot and Judith Klumperman, Department of Cell Biology, University Medical Center, Utrecht, the Netherlands.

disorders of the immune and inflammatory system. As a result, the system has become a platform for drug targeting, and mechanism-based drugs are currently developed, one of them is already on the market.

THE LYSOSOME AND INTRACELLULAR PROTEIN DEGRADATION: FACTS

In the mid1950s, Christian de Duve discovered the lysosome (see, for example, Refs. 8 and 9 and FIG. 1). The lysosome was first recognized biochemically in the rat liver as a vacuolar structure that contains various hydrolytic enzymes, which function optimally at an acidic pH. It is surrounded by a membrane that endows the contained enzymes with latency that is required to protect the cellular contents from their action (see below). The definition of the lysosome has been broadened over the years. This is because it has been recognized that the digestive process is dynamic and involves numerous stages of lysosomal

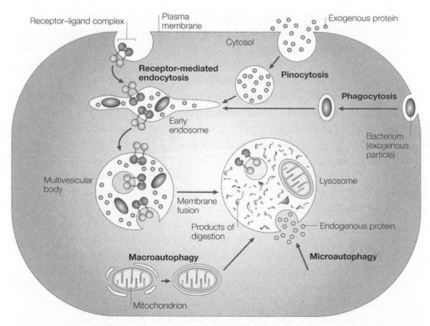

FIGURE 2. The four digestive processes mediated by the lysosome: (i) specific receptor-mediated endocytosis, (ii) pinocytosis (nonspecific engulfment of droplets containing extracellular fluid), (iii) phagocytosis (of extracellular particles), and (iv) autophagy (micro- and macro-; of intracellular proteins and organelles) (with permission from Nature Publishing Group. Published originally in Ref. 83).

maturation together with the digestion of both exogenous proteins (which are targeted to the lysosome through receptor-mediated endocytosis and pinocytosis) and exogenous particles (which are targeted via phagocytosis; the two processes are known as heterophagy), as well as digestion of endogenous proteins and cellular organelles (which are targeted via micro- and macroautophagy, in many cases, under stress; see FIG. 2). The lysosomal–vacuolar system as we currently recognize it is a discontinuous and heterogeneous digestive system that also includes structures that are devoid of hydrolases—for example, early endosomes that contain endocytosed receptor–ligand complexes and pinocytosed–phagocytosed extracellular contents. On the other extreme it includes the residual bodies—the end products of the completed digestive processes of heterophagy and autophagy. In between these extremes one can observe: primary/nascent lysosomes that have not been engaged yet in any proteolytic process; early autophagic vacuoles that might contain intracellular organelles; intermediate/late endosomes and phagocytic vacuoles (heterophagic vacuoles) that contain extracellular contents/particles; and multivesicular bodies (MVBs), which are the transition vacuoles between endosomes/phagocytic vacuoles and the digestive lysosomes.

The discovery of the lysosome along with independent experiments that were carried out at the same time and that have further strengthened the notion that cellular proteins are indeed in a constant state of synthesis and degradation (see, for example, Ref. 10), led scientists to feel, for the first time, that they have at hand an organelle that can potentially mediate degradation of intracellular proteins. The separation of the proteases from their substrates by a membrane provided an explanation for controlled degradation, and the only problem left to be explained (at the time) was how the substrates are translocated into the lysosomal lumen where they are exposed to the activity of the lysosomal proteases and degraded. An important discovery in this respect was the unraveling of the basic mechanism of action of the lysosome—autophagy (reviewed in Ref. 11). Under basal metabolic conditions, portions of the cytoplasm, which contain a fraction of the entire population of cellular (cytosolic) proteins, are segregated within a membrane-bound compartment, and are then fused to a primary nascent lysosome, and their contents digested. This process was denoted microautophagy. Under more extreme conditions, starvation for example, mitochondria, endoplasmic reticulum membranes, glycogen bodies, and other cytoplasmic entities, can also be engulfed by a process called macroautophagy (see, for example, Ref. 12; the different modes of action of the lysosome in digesting extra- and intracellular proteins particles and organelles, are shown in FIG. 2).

THE LYSOSOME AND INTRACELLULAR PROTEIN DEGRADATION

However, over a period of more than two decades, between the mid 1950s and the late 1970s, it has become gradually more and more difficult to explain several aspects of intracellular protein degradation based on the known mechanisms of lysosomal activity: accumulating lines of independent experimental evidence indicated that the degradation of at least certain classes of cellular proteins must be nonlysosomal. Yet, in the absence of any alternative, researchers came with different explanations, some more substantiated and others less, to defend the lysosomal hypothesis.

First was the gradual discovery, coming from several laboratories, that different proteins vary in their stability, and their half-life times can span a range of three orders of magnitude, from a few minutes to many days. Thus, the $t_{1/2}$ of ornitihine decarboxylase (ODC) is $\sim$10 min, while that of glucose-6-phosphate dehydrogenase (G6PD) is 15 h (for review articles, see, for example, Refs. 13, 14). Also, rates of degradation of many proteins were shown to change under different physiological conditions, such as availability of nutrients or hormones. It was conceptually difficult to reconcile the findings of distinct half-lives of different proteins with the mechanism of action of the lysosome, where the microautophagic vesicle contains the entire cohort of cellular (cytosolic) proteins that are therefore expected to be degraded at the same rate. Similarly, changing pathophysiological conditions, such as starvation or resupplementation of nutrients, were expected to affect the stability of all cellular proteins to the same extent. Clearly, this was not the case.

Another source of concern about the lysosome as the organelle in which intracellular proteins are degraded were the findings that both specific and general inhibitors of lysosomal proteases have different effects on different classes of proteins, making it clear that distinct groups of proteins are targeted by different proteolytic machineries. Thus, the degradation of endocytosed–pinocytosed extracellular proteins was significantly inhibited by lysosomal inhibitors, a partial effect was observed on the degradation of long-lived cellular proteins, and almost no effect was observed on the degradation of short-lived and abnormal/mutated proteins.

Finally, the thermodynamically paradoxical observation that the degradation of cellular proteins requires metabolic energy, and more importantly, the emerging evidence that the proteolytic machinery uses the energy directly, were in contrast with the known mode of action of lysosomal proteases that under the appropriate acidic conditions, and similar to all known proteases, degrade proteins in an exergonic manner.

The assumption that the degradation of intracellular proteins is mediated by the lysosome was nevertheless logical. Proteolysis results from direct interaction between the target substrates and proteases, and therefore it was clear that active proteases cannot coexist free and active in the cytosol along with their substrates, which would have resulted in destruction of the cell. Thus, it was recognized that any suggested proteolytic machinery that mediates degradation of intracellular proteins must also be equipped with a mechanism that separates— physically or virtually—between the proteases and their substrates, and enables them to associate only when needed. The lysosomal membrane provided this separating/fencing mechanism. Obviously, nobody could have predicted that a new mode of posttranslational modification—ubiquitination—could function as a chemical fence and as a proteolysis signal, and that untagged proteins will remain protected. Thus, while the structure of the lysosome as a distinct organelle could explain the separation necessary between the proteases and their substrates, and autophagy could explain the mechanism of entry of cytosolic proteins into the lysosomal lumen, major problems have remained unsolved. As noted, important among them were: (i) the various half-lives of cellular proteins, (ii) the energy requirement for intracellular proteolysis, and (iii) the distinct response of different populations of proteins to lysosomal inhibitors. Thus, according to one model, it was proposed that different proteins have different sensitivities to lysosomal proteases, since their half-lives *in vivo* were in correlation with their sensitivity to the action of lysosomal proteases *in vitro*.[15] To explain an extremely long half-life for a protein that was nevertheless sensitive to lysosomal proteases, or alterations in the stability of a single protein under various physiological states, it was suggested that although all cellular proteins are engulfed into the lysosome, only the short-lived proteins are degraded, whereas the long-lived proteins exit back into the cytosol: *to account for differences in half-life among cell components or of a single component in various physiological states, it was necessary to include in the model the possibility of an exit of native components back to the extralysosomal compartment.*[16] According to a different model, selectivity is determined by the binding affinity of the different proteins for the lysosomal membrane, which controls their entry rates into the lysosome, and subsequently their degradation rates.[17] For a selected group of proteins, such as the gluconeogenetic enzymes phosphoenol-pyruvate carboxykinase (PEPCK) and fructose-1,6-biphosphatase, it was suggested, though not firmly substantiated, that their degradation in the yeast vacuole is regulated by glucose via a mechanism called catabolite inactivation that possibly involves their phosphorylation. However, this regulated mechanism for vacuolar degradation was limited only to a small and specific group of proteins (see, for example, Ref. 18; reviewed in Ref. 19). More recent studies have shown that at least for stress-induced macroautophagy, a general sequence of amino acids, KFFERQ, which was identified in numerous cellular proteins, directs, via binding to a specific receptor and in cooperation with cytosolic and lysosomal chaperones, the regulated entry of many cytosolic proteins into the lysosomal lumen. While further corroboration of this hypothesis is still required, it can explain the mass, stress-induced, entry into the lysosome of a large population of proteins that contain a homologous sequence, but not the targeting for degradation of a specific protein under defined conditions (reviewed in Refs. 20, 21). The energy requirement for protein degradation was described as indirect, and necessary, for example, for protein transport across the lysosomal membrane[22] and/or for the activity of the H^+ pump and the maintenance of the low acidic intralysosomal pH that is necessary for optimal activity of the proteases.[23] We now know that both mechanisms require energy. Thus, despite the strained explanations for different stabilities and energy requirement, the lysosome still appeared as the most logical organelle/system that mediates degradation of intracellular proteins. Christian de Duve's view on the subject as summarized in a review article he published in the mid 1960s, saying: *Just as extracellular digestion is successfully carried out by the concerted action of enzymes with limited individual capacities, so, we believe, is intracellular digestion*[24] still appeared to be valid even toward the mid 1970s. The problem of different sensitivities of distinct protein groups to lysosomal inhibitors has remained unsolved, and may have served as an important trigger in the future quest for a nonlysosomal proteolytic system that may be involved in at least certain aspects of intracellular protein degradation.

Progress in identifying the elusive, nonlysosomal proteolytic system(s) was hampered by the lack of a cell-free preparation that could faithfully replicate the cellular proteolytic events—degrading proteins in a specific and energy-requiring mode and at neutral pH. An important breakthrough was made by Rabinovitz and Fisher who found that rabbit reticulocytes degrade abnormal, amino acid analogue-containing hemoglobin.[25] Their experiments modeled known disease states, the hemoglobinopathies. In these diseases, abnormal mutated hemoglobin chains (such as sickle cell hemoglobin) or excess of unassembled normal hemoglobin chains (generated in thalassemias, diseases in which the pairing chain is not synthesized at all or is mutated and rapidly degraded. Consequently, the hemoglobin complex ($\alpha_2\beta_2$) is not assembled and the free chains or partially assembled complexes) are rapidly degraded in the reticulocyte.[26,27] Reticulocytes are terminally differentiating red blood cells that do not contain lysosomes. Therefore, it was postulated that the degradation of hemoglobin in these cells is mediated by a nonlysosomal machinery. Etlinger and Goldberg[28] were the first to isolate and characterize a cell-free proteolytic preparation from reticulocytes. The crude extract selectively degraded abnormal hemoglobin, required ATP hydrolysis, and acted optimally at a neutral pH, which further corroborated the assumption that the proteolytic activity was nonlysosomal. A similar system was isolated and characterized later by our group.[29] Additional studies by the group led subsequently to resolution, characterization, and purification of the major enzymatic components from this extract, and to the discovery of the ubiquitin signaling system (see below).

THE LYSOSOME HYPOTHESIS IS CHALLENGED

As mentioned above, the unraveled mechanism(s) of action of the lysosome could explain only partially, and at times not satisfactorily, several key emerging characteristics of intracellular protein degradation. Among them were: (i) the heterogeneous stability of individual proteins, (ii) the effect of nutrients and hormones on their degradation, and (iii) the dependence of intracellular proteolysis on metabolic energy. The differential effect of selective inhibitors on the degradation of different classes of cellular proteins could not be explained at all.

The development of methods to monitor protein dynamics in cells together with the discovery of specific and general lysosomal inhibitors have resulted in the identification of different classes of cellular proteins, long- and short-lived, and the findings of the differential effects of the inhibitors on these groups (see, for example, Refs. 30, 31). An elegant experiment in this respect was carried out by Brian Poole and his colleagues[32] in the Rockefeller University. Poole was studying the effect of lysosomotropic agents—weak bases, such as ammonium chloride and chloroquine—that accumulate in the lysosome and dissipate its low acidic pH by neutralizing the H^+ ions: the lysosomal proteases are inactive in neutral pH. It was assumed that this mechanism also underlies the antimalarial activity of chloroquine and similar drugs that inhibit the activity parasite's lysosome, paralyzing its ability to digest the host's hemoglobin during the intraerythrocytic stage of its life cycle. Poole and his colleagues labeled metabolically endogenous proteins in living macrophages with ^{3}H-leucine and fed them with broken macrophages that had been labeled metabolically with ^{14}C-leucine prior to their disruption. They assumed, apparently correctly, that the dead macrophages debris and proteins will be phagocytosed by the live macrophages and targeted to the lysosome for degradation. They monitored the effect of lysosomotropic agents on the degradation of these two protein populations, the endogenous and the exogenous. In particular, they studied the effect of the weak bases chloroquine and ammonium chloride, and the acid ionophore X537A, which dissipates the H$^+$ gradient across the lysosomal membrane. They found that these drugs inhibited specifically the degradation of extracellular proteins, but not of intracellular proteins.[32] Poole summarized beautifully (I would say also poetically) these experiments and explicitly predicted the existence of a nonlysosomal proteolytic system that degrades intracellular proteins: *Some of the macrophages labeled with tritium were permitted to endocytise the dead macrophages labeled with ^{14}C. The cells were then washed and replaced in fresh medium. In this way we were able to measure in the same cells the digestion of macrophage proteins from two sources. The exogenous proteins will be broken down in the lysosomes, while the endogenous proteins will be broken down wherever it is that endogenous proteins are broken down during protein turnover.*[33]

The requirement for metabolic energy for the degradation of both prokaryotic[34] and eukaryotic[10,35] proteins was difficult to interpret. Proteolysis is an exergonic process and the thermodynamically paradoxical energy requirement for intracellular proteolysis made researchers believe that energy cannot be consumed directly by proteases or the proteolytic process per se, and is used indirectly. As Simpson summarized his findings[10]: *the data can also be interpreted by postulating that the release of amino acids from protein is itself directly dependent on energy supply. A somewhat similar hypothesis, based on studies on autolysis in tissue minces, has recently been advanced, but the supporting data are very difficult to interpret. However, the fact that protein hydrolysis as catalyzed by the familiar proteases and peptidases occurs exergonically, together with the consideration that autolysis in excised organs or tissue minces continues for weeks, long after phosphorylation or oxidation ceased, renders improbable the hypothesis of the direct energy dependence of the reactions leading to protein breakdown.* Being cautious however, and probably unsure about a single interpretation and conclusion, Simpson still left a narrow slit opened for a proteolytic process that requires energy in a direct manner: *However, the results do not exclude the existence of two (or more) mechanisms of protein breakdown, one hydrolytic, the other energy-requiring.* Since any proteolytic process must be at one point or another hydrolytic, the statement that makes a distinction between a hydrolytic process and an energy requiring (yet non-hydrolytic one), is not clear. Judging the statement from a historical perspective and knowing the mechanism of action of the ubiquitin system, where energy is required also in the prehydrolytic step (ubiquitin conjugation), Simpson may have thought of a two-step mechanism: an initial conditioning step that is energy requiring yet not hydrolytic, followed by a second hydrolytic step that is energy-independent. However, he did not provide us with a clear explanation. Clearly, he could not have imagined that even the second step will be energy requiring (as we know is proteasomal activity). At the end of this clearly understandable and apparently difficult deliberation, he left us with a vague explanation linking protein degradation to protein synthesis, a process that was known already at the time to require metabolic energy: the fact that a supply of energy seems to be necessary for both the incorporation and the release of amino acids from protein might well mean that the two processes are interrelated. Additional data suggestive of such a view are available from other types of experiments. Early investigations on nitrogen balance by Benedict, Folin, Gamble, Smith, and others point to the fact that the rate of protein catabolism varies with the dietary protein level. Since the protein level of the diet would be expected to exert a direct influence on synthesis rather than breakdown, the altered catabolic rate could well be caused by a change in the rate of synthesis.[10] With the discovery of lysosomes in eukaryotic cells it could be argued that energy is required for the transport of substrates into the lysosome or for maintenance of the low intralysosomal pH, for example (see above). The observation by Hershko and Tomkins that the activity of tyrosine aminotransferase (TAT) was stabilized following depletion of ATP[36] indicated that energy may be required at an early stage of the proteolytic process, most probably before proteolysis occurs. Yet, it did not provide a clue as for the mechanism involved: the energy could be used, for example, for specific modification of TAT, for example, phosphorylation, which would sensitize it to degradation by the lysosome or by a yet unknown proteolytic mechanism, or for a modification that activates its putative protease. It could also be used for a more general lysosomal mechanism, one that involves transport of TAT as well as other proteins into the lysosome or dissipation of the intralysosomal acidic pH, for example. The energy inhibitors used by Hershko and Tomkins abolished almost completely degradation of the entire population of cell proteins, confirming previous studies (e.g., Ref. 10) and suggesting a general role for energy in protein catabolism. In that respect energy inhibitors had an effect that was distinct from that of protein synthesis inhibitors (e.g., cycloheximide), which affected only enhanced degradation of TAT (induced by steroid hormone depletion), but not basal degradation.[36] This finding ruled out, at least partially, a tight linkage between protein synthesis and degradation. In bacteria, which lack lysosomes, an argument involving energy requirement for lysosomal degradation could not have been proposed, but other indirect effects of ATP hydrolysis could have affected proteolysis in E. coli, such as phosphorylation of substrates and/or proteolytic enzymes, or maintenance of the energized membrane state. According to this model, proteins could become susceptible to proteolysis by changing their conformation, for example, following association with the cell membrane that maintains a local, energy-dependent gradient of a certain ion. However, such an effect was ruled out,[37] and since there was no evidence for a phosphorylation-mediated mechanism (although, similar to the state of affairs in eukaryotes, the proteolytic machinery in prokaryotes had not been identified at that time), it seemed that at least in bacteria, energy is required directly for the proteolytic process. In any event, the requirement for metabolic energy for protein degradation in both prokaryotes and eukaryotes, a process that is exergonic thermodynamically, strongly indicated that proteolysis is a highly regulated process, and that a similar principle/mechanism has been preserved along evolution of the two kingdoms. Implying from the possible direct requirement for ATP in degradation of proteins in bacteria, it was not too unlikely to assume a similar direct mechanism in the degradation of cellular proteins in eukaryotes. Supporting this notion was the description of the cell-free proteolytic system in reticulocytes,[28,29] a cell that lacks lysosomes, which suggested that energy is probably required directly for the proteolytic process, although here too, the underlying mechanisms had remained enigmatic at the time. The description of the cell-free system paved, however, the road for the later detailed dissection of the underlying mechanisms involved and to the discovery of the ubiquitin system.

THE UBIQUITIN–PROTEASOME SYSTEM

The cell-free proteolytic system from reticulocytes[28,29] turned out to be an important and rich source for the purification and characterization of the enzymes that are involved in the ubiquitin–proteasome system. Initial fractionation of the crude reticulocyte cell extract on the anion-exchange resin diet-hylaminoethyl cellulose yielded two fractions that were both required to reconstitute the energy-dependent proteolytic activity found in the crude extract: the unadsorbed, flow through material was denoted fraction I, and the high salt eluate of the adsorbed proteins was denoted fraction II (TABLE 1).[38]

This finding was an important observation and a methodological lesson for future dissection of the system. For one, it suggested that the system is

TABLE 1. Resolution of the ATP-dependent proteolytic activity from crude reticulocyte extract into two essentially required complementing activities

	Degradation of [^{3}H]globin (%)	
Fraction	–ATP	+ATP
Lysate	1.5	1.0
Fraction I	0.0	0.0
Fraction II	1.5	2.7
Fraction I and fraction II	1.6	10.6

Adapted from Ciechanover *et al*; with permission from Elsevier/Biochem. Biophys. Res. Commun.[38]

not composed of a single "classical" protease that has evolved evolutionarily to acquire energy-dependence (although energy-dependent multicomponent–complex proteases, such as the mammalian 26S proteasome (see below) and the prokaryotic Clp family of proteases have been described later), but that it is made of at least two components. This finding of a two-component, energy-dependent protease that had no precedent, left us with no paradigm to follow. In an attempt to explain this finding, we suggested, for example, that the two fractions could represent an inhibited protease and its activator. Second, learning from this reconstitution experiment and the essential dependence between the two active components, we continued to reconstitute activity from resolved fractions whenever we encountered a loss of activity along further purification steps. This biochemical "complementation" approach resulted in the discovery of additional enzymes of the system, all required to be present in the reaction mixture in order to catalyze the multistep proteolysis of the target substrate. We chose first to purify the active component from fraction I (see accompanying biography). It was found to be a small, ~8.5 kDa heat-stable protein that was designated ATP-dependent proteolysis factor 1, APF-1. APF-1 was later identified as ubiquitin (see below; I am using the term APF-1 to the point in which it was identified as ubiquitin and then change the terminology accordingly). In retrospect, the decision to start the purification efforts with fraction I turned out to be strategically important, as fraction I contained only one single protein—APF-1—that was necessary to stimulate proteolysis of the model substrate we used at the time, while fraction II, as we learnt later contained many more components. Starting the purification efforts with fraction II would have resulted in a much more complicated route. Later studies showed that fraction I contains additional components necessary for the degradation of other substrates, but these were not necessary for the reconstitution of the system at that time. This choice enabled us not only to purify APF-1, but also to quickly decipher its mechanism of action. A critically important finding that paved the road for future developments in the field was that multiple moieties of APF-1 are covalently conjugated to the target substrate when incubated in the presence of fraction II, and that the modification requires ATP (Refs. 39, 40; FIGS. 3 and 4). It was also found that the modification is reversible, and APF-1 can be removed from the substrate or its degradation products and reutilized.[40]

The discovery that APF-1 is covalently conjugated to protein substrates and stimulates their proteolysis in the presence of ATP and crude fraction II, led in 1980 to the proposal of a model according to which protein substrate modification by multiple moieties of APF-1 targets it for degradation by a downstream, at that time an yet unidentified, protease that cannot recognize the unmodified substrate; following degradation, reusable APF-1 is released.[40] It was also shown that APF-1 can be released from the intact substrate.[40] This can occur, for example, if the substrate was marked erroneously. It was further predicted that the release is mediated by specific protease(s) and not via rever-

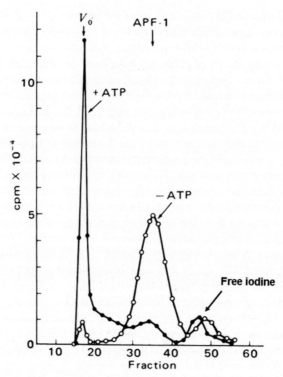

FIGURE 3. APF-1 is shifted to a high molecular mass "compound(s)" following incubation in ATP-containing crude reticulocyte cell extract. ^{125}I-labeled APF-1 was incubated with reticulocyte crude fraction II in the absence (*open circles*) or presence (*closed circles*) of ATP, and the reaction mixtures were resolved via gel filtration chromatography. Shown is the radioactivity measured in each fraction. As can be seen, following addition of ATP, APF-1 becomes associated with some component(s) in fraction II, which could be an enzyme(s) or a substrate(s) of the system (with permission from Proceedings of the National Academy of the USA; published originally in Ref. 39).

sal of the conjugation reaction. Amino acid analysis of APF-1, along with its known molecular mass and other general characteristics raised the suspicion that APF-1 is ubiquitin,[41] a known protein of previously unknown function. Contributing to this suspicion was the previously described conjugate between ubiquitin and histone H2A that raised the possibility that a similar, though obviously not an identical, conjugate is generated between APF-1 and its target substrate (see below). Indeed, in a series of biochemical and functional experiments, Wilkinson and colleagues confirmed unequivocally that APF-1 is ubiquitin.[42] Ubiquitin is a small, heat-stable, and highly evolutionarily conserved protein of 76 residues. It was first purified during the isolation of thymopoietins[43] and was subsequently found to be ubiquitously expressed in all kingdoms of living cells, including prokaryotes.[44] Interestingly, it was initially

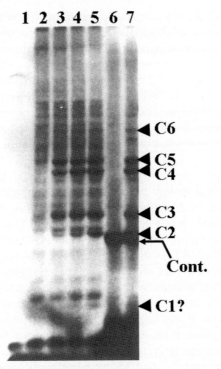

FIGURE 4. Multiple molecules of APF-1/ubiquitin are conjugated to the proteolytic substrate, probably signaling it for degradation. To interpret the data described in the experiment depicted in Figure 3 and to test the hypothesis that APF-1 is conjugated to the target proteolytic substrate, ^{125}I-APF-1 was incubated along with crude fraction II (FIG. 3 and text) in the absence (lane 1) or presence (lanes 2–5) of ATP and in the absence (lanes 1,2) or presence (lanes 3–5) of increasing concentrations of unlabeled lysozyme. Reaction mixtures resolved in lanes 6 and 7 were incubated in the absence (lane 6) or presence (lane 7) of ATP, and included unlabeled APF-1 and ^{125}I-labeled lysozyme. C1–C6 denote specific APF-1-lysozyme adducts in which the number of APF-1 moieties bound to the lysozyme moiety of the adduct is probably increasing, from 1 to 6. Reactions mixtures were resolved via sodium dodecyl sulfate–polyacrylamide gel electrophoresis (SDS-PAGE) and visualized following exposure to an X-ray film (autoradiography) (with permission from Proceedings of the National Academy of the USA; published originally in Ref. 40).

found to induce lymphocyte differentiation, a characteristic that was attributed to the stimulation of adenylate cyclase.[44,45] Accordingly, it was named UBIP for ubiquitous immunopoietic polypeptide.[44] However, later studies showed that ubiquitin is not involved in the immune response,[46] and that it was a contaminating endotoxin in the preparation that was responsible for the adenylate cyclase and the T cell-differentiating activities. Furthermore, the unraveling of the sequence of several eubacteria and archaebacteria genomes, as well as earlier biochemical analyses in these organisms (unpublished) showed that ubiquitin is restricted only to eukaryotes. The finding of ubiquitin in bacteria[44] was probably due to contamination of the bacterial extract with yeast ubiquitin derived from the yeast extract in which the bacteria were grown. While in retrospect the name ubiquitin is a misnomer as it is restricted to eukaryotes and is not ubiquitous as was previously thought, from historical reasons it has still remained the name of the protein. Accordingly, and in order to avoid confusion, we suggest that names of other novel enzymes and components of the ubiquitin system, but of other systems as well, should remain as were first coined by their discoverers.

An important development in the ubiquitin research field was the discovery that a single ubiquitin moiety can be covalently conjugated to histones (see above), and particularly to histones H2A and H2B. While the function of these adducts has remained elusive until recently, their structure was unraveled in the mid 1970s. The structure of the ubiquitin conjugate with H2A (uH2A; was also designated protein A24) was deciphered by Goldknopf and Busch[47,48] and by Hunt and Dayhoff,[49] who found that the two proteins are linked through a fork-like, branched isopeptide bond between the carboxyterminal glycine of ubiquitin (Gly^{76}) and the -NH_2 group of 119 internal lysine (Lys^{119}) of the histone molecule. It was suggested that the isopeptide bond found in the histone–ubiquitin adduct is identical to the bond that was found between ubiquitin and the target proteolytic substrate,[50] and between the ubiquitin moieties in the polyubiquitin chain[51,52] that is synthesized on the substrate and that functions, most probably, as a proteolysis recognition signal for the downstream 26S proteasome. In this particular polyubiquitin chain the linkage is between Gly^{76} of one ubiquitin moiety and internal Lys^{48} of the previously conjugated ubiquitin moiety. It appears that only Lys^{48}-based polyubiquitin chain is recognized by the 26S proteasome and serves as a proteolytic signal. In recent years it has been shown that the first ubiquitin moiety can also be attached in a linear mode to the N-terminal residue of the proteolytic target substrate.[53] However, even in this case, the subsequent ubiquitin moieties are generating a Lys^{48}-based polyubiquitin chain on the first linearly fused moiety. N-terminal ubiquitination is clearly required for targeting naturally occurring lysine-less proteins for degradation. Yet, several lysine-containing proteins have also been described that traverse this pathway, the muscle-specific transcription factor MyoD, for example. In these proteins the internal lysine residues are probably not accessible to their cognate ligases. Other types of polyubiquitin chains have also been described that are not involved in targeting the conjugated substrates for proteolysis. Thus, a Lys^{63}-based polyubiquitin chain has been described that is probably necessary to activate transcription factors (reviewed recently in Ref. 54). Interestingly, the role of monoubiquitination of histones has also been identified recently. It is involved in the regulation of transcription, probably via modulation of the structure of the nucleosomes (for recent reviews, see, for example, Refs. 55, 56).

The identification of APF-1 as ubiquitin, and the discovery that a high-energy isopeptide bond, similar to the one that links ubiquitin to histone H2A, links it also to the target proteolytic substrate, resolved at that time the enigma of the energy requirement for intracellular proteolysis (see however below) and paved the road to the untangling of the complex mechanism of isopeptide bond formation. This process turned out to be similar to that of peptide bond formation that is catalyzed by tRNA synthetase or during nonribosomal synthesis of short peptides.[57] In these processes, that occurs during protein synthesis a single amino acid is activated, whereas during activation of ubiquitin, the c-terminal Gly^{76} residue of the protein is activated, but otherwise the three processes appear to be almost identical. Using the unraveled mechanism of ubiquitin activation and immobilized ubiquitin as a "covalent" affinity bait, the three enzymes that are involved in the cascade reaction of ubiquitin conjugation were purified by Ciechanover, Hershko, and their colleagues. These enzymes are: (i) E1, the ubiquitin-activating enzyme, (ii) E2, the ubiquitin–carrier protein, and (iii) E3, the ubiquitin–protein ligase.[58,59] The discovery of the E3 component, which is the specific substrate-binding component of the system, pointed to a possible solution to the problem of the heterogeneous stabilities of different proteins—they might be specifically recognized and targeted by different ligases.

Within a short period, the ubiquitin tagging hypothesis received substantial support. For example, Chin and colleagues injected into HeLa cells labeled ubiquitin and hemoglobin and denatured the injected hemoglobin by oxidizing it with phenylhydrazine. They found that ubiquitin conjugation to globin is markedly enhanced by denaturation of hemoglobin, and the level of globin–ubiquitin conjugates was proportional to the rate of hemoglobin degradation.[60] Hershko and colleagues observed a similar correlation for abnormal, amino acid analogue-containing short-lived proteins.[61] A previously isolated cell cycle arrest mutant that loses the ubiquitin–histone H2A adduct at the permissive temperature,[62] was found by Finley, Ciechanover, and Varshavsky to harbor a thermolabile E1.[63] Following heat inactivation, the cells fail to degrade normal short-lived proteins.[64] Although the cells did not provide direct evidence for substrate ubiquitination as a destruction signal, they still provided a strong and direct linkage between ubiquitin conjugation and intracellular protein degradation.

At this point, the only missing link was the identification of the downstream protease that would specifically recognize ubiquitinated substrates. Tanaka and colleagues identified a second ATP-requiring step in the reticulocyte proteolytic system, which was independent of ubiquitination,[65] and Hershko and colleagues demonstrated that energy is required for conjugate degradation.[66] An important advance in the field was a discovery by Hough and colleagues who partially purified and characterized a high-molecular mass alkaline protease that degraded, in an ATP-dependent mode, ubiquitin adducts of lysozyme but not untagged lysozyme.[67] This protease, which was later called the 26S proteasome (see below), provided all the necessary criteria for being the specific proteolytic arm of the ubiquitin system. This finding was confirmed, and the protease was further characterized by Waxman and colleagues who found that it is an unusually large, ~1.5 MDa enzyme, unlike any other known protease.[68] A further advance in the field was the discovery[69] that a smaller, neutral multisubunit 20S protease complex that was discovered together with the larger 26S complex, is similar to a "multicatalytic proteinase complex" (MCP) that was described earlier by Wilk and Orlowski in bovine pituitary gland.[70] This 20S protease is ATP-independent and has various catalytic activities, cleaving on the carboxy-terminal side of hydrophobic, basic, and acidic residues. Hough and colleagues raised the possibility—although they did not show it experimentally—that this 20S protease can be a part of the larger 26S protease that degrades ubiquitin adducts.[69] Later studies showed that indeed, the 20S complex is the core catalytic particle of the larger 26S complex.[71,72] However, a strong evidence that the active "mushroom"-shaped 26S protease is generated through the assembly of two distinct subcomplexes—the catalytic 20S cylinder-like MCP and an additional 19S ball-shaped subcomplex (that was predicted to have a regulatory role)—was provided only in the early 1990s by Hoffman and colleagues[73]: these researchers mixed the two purified particles and generated the active 26S enzyme.

The proteasome is a large, 26S, multicatalytic protease that degrades polyubiquitinated proteins to small peptides. It is composed of two subcomplexes: a 20S core particle (CP) that carries the catalytic activity, and a regulatory 19S regulatory particle (RP) (for the structure of the 26S proteasome, see FIG. 5; for a scheme describing the ubiquitin system, see FIG. 6).

CONCLUDING REMARKS

The evolvement of proteolysis as a centrally important regulatory mechanism is a remarkable example for the development of a novel biological concept and the accompanying battles to change paradigms. The five decades journey between the early 1940s and early 1990s began with fierce discussions on whether cellular proteins are static as had been thought for a long time, or are in a "dynamic state" of synthesis and degradation. The discovery of protein dynamics was followed by the discovery of the lysosome, which was believed—between the mid 1950s and mid 1970s—to be the organelle within which intracellular proteins are destroyed. Independent lines of experimental evidence gradually eroded the lysosomal hypothesis and resulted in the evolvement of new concept, that the bulk of intracellular proteins are degraded—under basal metabolic conditions—via a nonlysosomal machinery. This resulted in the discovery of the ubiquitin system in the late 1970s and early 1980s.

With the identification of the reactions and enzymes that are involved in the ubiquitin–proteasome cascade, a new era in the protein degradation field

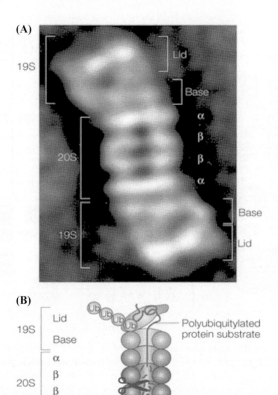

FIGURE 5. The Proteasome. The proteasome is a large, 26S, multicatalytic protease that degrades polyubiquitinated proteins to small peptides. It is composed of two subcomplexes: a 20S core particle (CP) that carries the catalytic activity, and a regulatory 19S regulatory particle (RP). The 20S CP is a barrel-shaped structure composed of four stacked rings, two identical outer rings and two identical inner rings. The eukaryotic and rings are composed each of seven distinct subunits, giving the 20S complex the general structure of 1-71-71-71-7. The catalytic sites are localized to some of the subunits. Each extremity of the 20S barrel can be capped by a 19S RP, each composed of 17 distinct subunits, 9 in a "base" subcomplex, and 8 in a "lid" subcomplex. One important function of the 19S RP is to recognize ubiquitinated proteins and other potential substrates of the proteasome. Several ubiquitin-binding subunits of the 19S RP have been identified, however, their biological roles and mode of action have not been discerned. A second function of the 19S RP is to open an orifice in the ring that will allow entry of the substrate into the proteolytic chamber. Also, since a folded protein would not be able to fit through the narrow proteasomal channel, it is assumed that the 19S particle unfolds substrates and inserts them into the 20S CP. Both the channel opening function and the unfolding of the substrate require metabolic energy, and indeed, the 19S RP "base" contains six different ATPase subunits. Following degradation of the substrate, short peptides derived from the substrate are released, as well as reusable ubiquitin (with permission from Nature Publishing Group. Published originally in Ref. 83). **(A)** Electron microscopy image of the 26S proteasome from the yeast *S. cerevisiae*. **(B)** Schematic representation of the structure and function of the 26S proteasome.

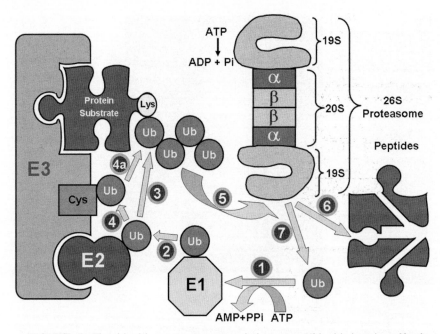

FIGURE 6. The ubiquitin–proteasome proteolytic system. Ubiquitin is activated by the ubiquitin-activating enzyme, E1 (1) followed by its transfer to a ubiquitin-carrier protein (ubiquitin-conjugating enzyme, UBC), E2 (2). E2 transfers the activated ubiquitin moieties to the protein substrate that is bound specifically to a unique ubiquitin ligase E3. The transfer is either direct ((3) in the case of RING finger ligases and possibly some other smaller groups of E3s, such as U-Box ligases), or via an additional thiol-ester intermediate on the ligase ((4, 4a) in the case of HECT domain ligases). Successive conjugation of ubiquitin moieties to one another generates a polyubiquitin chain that serves as the binding (5) and degradation signal for the downstream 26S proteasome. The substrate is degraded to short peptides (6), and free and reusable ubiquitin is released by deubiquitinating enzymes (DUBs) (7).

began in the late 1980s and early 1990s. Studies that showed that the system is involved in targeting key regulatory proteins—such as light-regulated proteins in plants, transcriptional factors, cell cycle regulators, and tumor suppressors and promoters—started to emerge (see for example Refs. 74–78). These studies were accompanied by functional analysis of the system in the yeast *Saccharomyces cerevisial* (carried out initially mostly by Varshavsky and colleagues), taking advantage of the power of genetics. They were then followed by numerous studies on the underlying mechanisms involved in the degradation of specific proteins, each with its own unique mode of recognition and regulation. The unraveling of the human genome revealed the existence of hundreds of distinct E3s, attesting to the complexity and the high specificity and selectivity of the system. Two important advances in the field were the discovery of the nonproteolytic functions of ubiquitin, such as activation

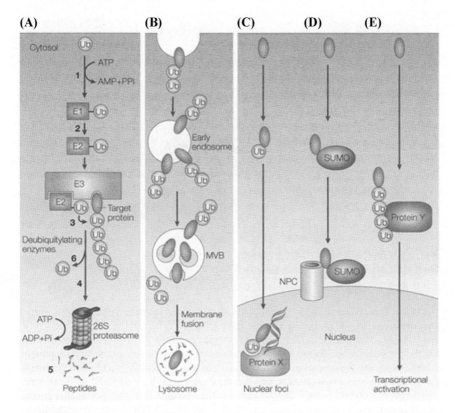

FIGURE 7. Some of the different functions of modification by ubiquitin and ubiquitin-like proteins. (**A**) Proteasomal-dependent degradation of cellular proteins (see Fig. 6). (**B**) Mono⁻ or oligoubiquitination targets membrane proteins to degradation in the lysosome/vacuole. (**C**) Monoubiquitination, or (**D**) a single modification by a ubiquitin-like protein (UBL), SUMO, for example, can target proteins to different subcellular destinations, such as nuclear foci or the nuclear pore complex (NPC). Modification by UBLs can serve also other, nonproteolytic functions, such as protecting proteins from ubiquitination or activation of E3 complexes (not shown). (**E**) Generation of a Lys[63]-based polyubiquitin chain can activate transcriptional regulators, directly or indirectly via recruitment of other proteins (protein Y, for example; shown), or activation of upstream components, such as kinases. Ub denotes ubiquitin. (With permission from Nature Publishing Group. Published originally in Ref. 83).

of transcription and routing of proteins to the lysosome/vacuole, and the discovery of modification by ubiquitin-like proteins (UBLs), which is involved in numerous nonproteolytic functions such as directing proteins to their subcellular destination, protecting proteins from ubiquitination, and controlling entire processes, such as autophagy (see, for example, Ref. 79) (for some of the different roles of modifications by ubiquitin and UBLs, see Fig. 7). All

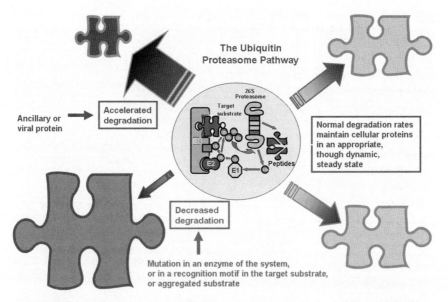

FIGURE 8. Aberrations in the ubiquitin–proteasome system and pathogenesis of human diseases. Normal degradation of cellular proteins maintains them in a steady-state level, though, this level may change under various physiological conditions (upper- and lower-right side). When degradation is accelerated due to an increase in the level or activity of an E3 (Skp2 in the case of p27, for example), or overexpression of an ancillary protein that generates a complex with the protein substrate and targets it for degradation (the human papillomavirus E6 oncoprotein that associates with p53 and targets it for degradation by the E6-AP ligase, or the cytomegalovirus-encoded ER proteins US2 and US11 that target MHC class I molecules for endoplasmic reticulum-associated degradation, ERAD), the steady-state level of the protein decreases (*upper-left side*). A mutation in a ubiquitin ligase or in a substrate's scaffold/binding protein (such as occurs in Adenomatous Polyposis Coli—APC, or in E6-AP (Angelmans' Syndrome)) or in the substrate's recognition motif (such as occurs in β-catenin or in ENaC) will result in decreased degradation and accumulation of the target substrate (*lower-left side*).

these studies have led to the emerging realization that this novel mode of post-translational modification/covalent conjugation plays key roles in regulating a broad array of cellular processes—among them cell cycle and division, growth and differentiation, activation and silencing of transcription, apoptosis, the immune and inflammatory response, signal transduction, receptor-mediated endocytosis, various metabolic pathways, and the cell quality control—all that through proteolytic and nonproteolytic mechanisms. The discovery that ubiquitin modification plays a role in routing proteins to the lysosome/vacuole, and that modification by specific and unique UBLs and modification system controls autophagy closed an exciting historical cycle, since it demonstrated that the two apparently distinct proteolytic systems—the ubiquitin/proteasome and the lysosome—communicate with one another. With the many processes and innumerable substrates targeted by the ubiquitin pathway, it is not surprising to find that aberrations in the system underlie, directly or indirectly, the pathogenesis of many diseases. While inactivation of a major enzyme, such as E1 is obviously lethal, mutations in enzymes or in recognition motifs in substrates that do not affect vital pathways or that affect the involved process only partially, may result in a broad array of phenotypes. Similarly, acquired changes in the activity of the system can also evolve into certain pathologies. The pathological states associated with the ubiquitin system can be classified into two groups: (*a*) those that result from loss of function—mutation in a ubiquitin system enzyme or in the recognition motif in the target substrate, which results in stabilization of certain proteins, and (*b*) those that result from gain of function—abnormal or accelerated degradation of the protein target (for a general scheme describing aberrations in the ubiquitin system that result in disease states, see FIG. 8). Studies that employ targeted inactivation of genes coding for specific ubiquitin system enzymes and substrates in animals can provide a more systematic view into the broad spectrum of pathologies that may result from aberrations in ubiquitination, ubiquitin-mediated proteolysis, and modification by UBLs. Better understanding of the processes and identification of the components involved in the degradation of key regulators will lead to the development of mechanism-based drugs that will target specifically only the involved proteins. While the first drug, a specific proteasome inhibitor is already on the market,[80] it appears that one important hallmark of the new era we are entering now will be the discovery of novel drugs based on targeting of specific processes, such as inhibiting aberrant Mdm2- or E6-AP-mediated accelerated targeting of the tumor suppressor p53, which will result in regaining its lost function.

Many review articles have been published on different aspects of the ubiquitin system. The purpose of this article was to bring to the reader several milestones along the historical pathway of the evolvement of the ubiquitin system. For additional reading on the ubiquitin system the reader is referred to the many reviews written on the system, among them, for example, are Refs. 81, 82. Some parts of this review, including several figures, are based on another recently published review article (Ref. 83).

ACKNOWLEDGMENTS

Research in the laboratory of Aaron Ciechanover has been supported along the years by grants from the US–Israel Binational Science Foundation (BSF), the Israel Science Foundation (ISF) founded by the Israeli National Academy of Humanities, Arts, and Sciences, the German–Israeli Foundation (GIF) for Scientific Research and Development, the Israel Cancer Research Fund (ICRF) USA, the Deutsche–Israeli Cooperation Program (DIP), the European Commission (EC), the Israel Cancer Society (ICS), the Prostate Cancer Foundation

(PCF)—Israel, the Foundation for Promotion of Research in the Technion and various research grants administered by the Vice President of the Technion for Research. Infrastructural equipment for the laboratory of A.C. and for the Cancer and Vascular Biology Research Center has been purchased with the support of the Wolfson Charitable Fund—Center of Excellence for Studies on Turnover of Cellular Proteins and its Implications to Human Diseases.

REFERENCES

1. CLARKE, H.T. 1958. Impressions of an organic chemist in biochemistry. Annu. Rev. Biochem. **27**: 1–14.
2. KENNEDY, E.P. 2001. Hitler's gift and the era of biosynthesis. J. Biol. Chem. **276**: 42619–42631.
3. SIMONI, R.D., R.L. HILL & M. VAUGHAN. 2002. The use of isotope tracers to study intermediary metabolism: Rudolf Schoenheimer. J. Biol. Chem. **277**(issue 43): e1-e3 (available on-line at: http:/www.jbc.org).
4. SCHOENHEIMER, R., S. RATNER & D. RITTENBERG. 1939. Studies in protein metabolism. VII: The metabolism of tyrosine. J. Biol. Chem. **127**: 333–344.
5. RATNER, S., D. RITTENBERG, A.S. KESTON & R. SCHOENHEIMER. 1940. Studies in protein metabolism. XIV: The chemical interaction of dietary glycine and body proteins in rats. J. Biol. Chem. **134**: 665–676.
6. SCHOENHEIMER, R. 1942. The Dynamic State of Body Constituents. Harvard University Press. Cambridge, Massachusetts.
7. HOGNESS, D.S., M. COHN & J. MONOD. 1955. Studies on the induced synthesis of β-galactosidase in *Escherichia coli*: The kinetics and mechanism of sulfur incorporation. Biochim. Biophys. Acta **16**: 99–116.
8. DE DUVE, C., R. GIANETTO, F. APPELMANS & R. WATTIAUX. 1953. Enzymic content of the mitochondria fraction. Nature (London) **172**: 1143–1144.
9. GIANETTO, R. & C. DE DUVE. 1955. Tissue fractionation studies 4. Comparative study of the binding of acid phosphatase, -glucoronidase and cathepsin by rat liver particles. Biochem. J. **59**: 433–438.
10. SIMPSON, M.V. 1953. The release of labeled amino acids from proteins in liver slices. J. Biol. Chem. **201**: 143–154.
11. MORTIMORE, G.E. & A.R. POSO. 1987. Intracellular protein catabolism and its control during nutrient deprivation and supply. Annu. Rev. Nutr. **7**: 539–564.
12. ASHFORD, T.P. & K.R. PORTER. 1962. Cytoplasmic components in hepatic cell lysosomes. J. Cell Biol. **12**: 198–202.
13. SCHIMKE, R.T. & D. DOYLE. 1970. Control of enzyme levels in animal tissues. Annual Rev. Biochem. **39**: 929–976.
14. GOLDBERG, A.L. & A.C. St. JOHN. 1976. Intracellular protein degradation in mammalian and bacterial cells: Part 2. Annu. Rev. Biochem. **45**: 747–803.
15. SEGAL, H.L., J.R. WINKLER & M.P. MIYAGI. 1974. Relationship between degradation rates of proteins in vivo and their susceptibility to lysosomal proteases. J. Biol. Chem. **249**: 6364–6365.
16. HAIDER, M. & H.L. SEGAL. 1972. Some characteristics of the alanine-aminotransferase and arginase-inactivating system of lysosomes. Arch. Biochem. Biophys. **148**: 228–237.

17. DEAN, R.T. 1977. Lysosomes and protein degradation. Acta Biol. Med. Ger. **36:** 1815–1820.
18. MÜLER, M., H. MÜLER & H. HOLZER. 1981. Immunochemical studies on catabolite inactivation of phosphoenolpyruvate carboxykinase in Saccharomyces cerevisiae. J. Biol. Chem. **256:** 723–727.
19. HOLZER, H. 1989. Proteolytic catabolite inactivation in Saccharomyces cerevisiae. Revis. Biol. Celular **21:** 305–319.
20. MAJESKI, A.E. & J.F. DICE. 2004. Mechanisms of chaperone-mediated autophagy. Intl. J. Biochem. Cell Biol. **36:** 2435–2444.
21. CUERVO, A.M. & J.F. DICE. 1998. Lysosomes, a meeting point of proteins, chaperones, and proteases. J. Mol. Med. **76:** 6–12.
22. HAYASHI, M., Y. HIROI & Y. NATORI. 1973. Effect of ATP on protein degradation in rat liver lysosomes. Nat. New Biol. **242:** 163–166.
23. SCHNEIDER, D.L. 1981. ATP-dependent acidification of intact and disrupted lysosomes: Evidence for an ATP-driven proton pump. J. Biol. Chem. **256:** 3858–3864.
24. DE DUVE, C. & R. WATTIAUX. 1966. Functions of lysosomes. Annu. Rev. Physiol. **28:** 435–492.
25. RABINOVITZ, M. & J.M. FISHER. 1964. Characteristics of the inhibition of hemoglobin synthesis in rabbit reticulocytes by threo–amino–chlorobutyric acid. Biochim. Biophys. Acta **91:** 313–322.
26. CARRELL, R.W. & H. LEHMANN. 1969. The unstable hemoglobin hemolytic anaemias. Semin. Hematol. **6:** 116–132.
27. HUEHNS, E.R. & A.J. BELLINGHAM. 1969. Diseases of function and stability of hemoglobin. Br. J. Hematol. **17:** 1–10.
28. ETLINGER, J.D. & A.L. GOLDBERG. 1977. A soluble ATP-dependent proteolytic system responsible for the degradation of abnormal proteins in reticulocytes. Proc. Natl. Acad. Sci. USA **74:** 54–58.
29. HERSHKO, A., H. HELLER, D. GANOTH & A. CIECHANOVER. 1978. Mode of degradation of abnormal globin chains in rabbit reticulocytes. In: Protein Turnover and Lysosome Function. H.L. Segal & D.J. Doyle, Eds.: 149–169. Academic Press. New York, NY.
30. KNOWLES, S.E. & F.J. BALLARD. 1976. Selective control of the degradation of normal and aberrant proteins in Reuber H35 hepatoma cells. Biochem. J. **156:** 609–617.
31. NEFF, N.T., G.N. DE MARTINO & A.L. GOLDBERG. 1979. The effect of protease inhibitors and decreased temperature on the degradation of different classes of proteins in cultured hepatocytes. J. Cell Physiol. **101:** 439–457.
32. POOLE, B., S. OHKUMA & M.J. WARBURTON. 1977. The accumulation of weakly basic substances in lysosomes and the inhibition of intracellular protein degradation. Acta Biol. Med. Germ. **36:** 1777–1788.
33. POOLE, B., S. OHKUMA & M.J. WARBURTON. 1978. Some aspects of the intracellular breakdown of exogenous and endogenous proteins. In: Protein Turnover and Lysosome Function. H.L. Segal & D.J. Doyle, Eds.: 43–58. Academic Press. New York, NY.
34. MANDELSTAM, J. 1958. Turnover of proteins in growing and non-growing populations of *Escherichia coli*. Biochem. J. **69:** 110–119.
35. STEINBERG, D. & M. VAUGHAN. 1956. Observations on intracellular protein catabolism studied *in vitro*. Arch. Biochem. Biophys. **65:** 93–105.

36. HERSHKO, A. & G.M. TOMKINS. 1971. Studies on the degradation of tyrosine amino-transferase in hepatoma cells in culture: influence of the composition of the medium and adenosine triphosphate dependence. J. Biol. Chem. **246:** 710–714.

37. GOLDBERG, A.L., J.D. KOWIT & J.D. ETLINGER. 1976. Studies on the selectivity and mechanisms of intracellular protein degradation. In: Proteolysis and Physiological Regulation. D.W. Ribbons & K. Brew, Eds.: 313–337. Academic Press. New York, NY.

38. CIECHANOVER, A., Y. HOD & A. HERSHKO. 1978. A heat-stable polypeptide component of an ATP-dependent proteolytic system from reticulocytes. Biochem. Biophys. Res. Common. **81:** 1100–1105.

39. CIECHANOVER, A., H. HELLER, S. ELIAS, et al. 1980. ATP-dependent conjugation of reticulocyte proteins with the polypeptide required for protein degradation. Proc. Natl. Acad. Sci. USA **77:** 1365–1368.

40. HERSHKO, A., A. CIECHANOVER, H. HELLER, et al. 1980. Proposed role of ATP in protein breakdown: conjugation of proteins with multiple chains of the polypeptide of ATP-dependent proteolysis. Proc. Natl. Acad. Sci. USA **77:** 1783–1786.

41. CIECHANOVER, A., S. ELIAS, H. HELLER, et al. 1980. Characterization of the heat-stable polypeptide of the ATP-dependent proteolytic system from reticulocytes. J. Biol. Chem. **255:** 7525–7528.

42. WILKINSON, K.D., M.K. URBAN & A.L. HAAS. 1980. Ubiquitin is the ATP-dependent Proteolysis Factor I of rabbit reticulocytes. J. Biol. Chem. **255:** 7529–7532.

43. GOLDSTEIN, G. 1974. Isolation of bovine thymin, a polypeptide hormone of the thymus. Nature (London) **247:** 11–14.

44. GOLDSTEIN, G., M. SCHEID, U. HAMMERLING, et al. 1975. Isolation of a polypeptide that has lymphocyte-differentiating properties and is probably represented universally in living cells. Proc. Natl. Acad. Sci. USA **72:** 11–15.

45. SCHLESSINGER, D.H., G. GOLDSTEIN & H.D. NIALL. 1975. The complete amino acid sequence of ubiquitin, an adenylate cyclase stimulating polypeptide probably universal in living cells. Biochemistry **14:** 2214–2218.

46. LOW, T.L.K. & A.L. GOLDSTEIN. 1979. The chemistry and biology of thymosin: Amino acid analysis of thymosin 1 and polypeptide 1. J. Biol. Chem. **254:** 987–995.

47. GOLDKNOPF, I.L. & H. BUSCH. 1975. Remarkable similarities of peptide finger-prints of histone 2A and non-histone chromosomal protein A24. Biochem. Biophys. Res. Commun. **65:** 951–955.

48. GOLDKNOPF, I.L. & H. BUSCH. 1977. Isopeptide linkage between non-histone and histone 2A polypeptides of chromosome conjugate-protein A24. Proc. Natl. Acad. Sci. USA **74:** 864–868.

49. HUNT, L.T. & M.O. DAYHOFF. 1977. Amino-terminal sequence identity of ubiquitin and the non-histone component of nuclear protein A24. Biochim. Biophys. Res. Commun. **74:** 650–655.

50. HERSHKO, A., A. CIECHANOVER & I.A. ROSE. 1981. Identification of the active amino acid residue of the polypeptide of ATP-dependent protein breakdown. J. Biol. Chem. **256:** 1525–1528.

51. HERSHKO, A. & H. HELLER. 1985. Occurrence of a polyubiquitin structure in ubiquitin-protein conjugates. Biochem. Biophys. Res. Common. **128:** 1079–1086.

52. CHAU, V., J.W. TOBIAS, A. BACHMAIR, et al. 1989. A multiubiquitin chain is confined to specific lysine in a targeted short lived protein. Science **243:** 1576–1583.

53. CIECHANOVER, A. & R. BEN-SAADON. 2004. N-terminal ubiquitination: More protein substrates join in. Trends Cell Biol. **14:** 103–106.
54. MURATANI, M. & W.P. TANSEY. 2003. How the ubiquitin-proteasome system controls transcription. Nat. Rev. Mol. Cell Biol. **4:** 192–201.
55. ZHANG, Y. 2003. Transcriptional regulation by histone ubiquitination and deubiquitination. Genes Dev. **17:** 2733–2740.
56. OSLEY, M.A. 2004. H2B ubiquitylation: the end is in sight. Biochim. Biophys. Acta **1677:** 74–78.
57. LIPMAN, F. 1971. Attempts to map a process evolution of peptide biosynthesis. Science **173:** 875–884.
58. CIECHANOVER, A., S. ELIAS, H. HELLER & A. HERSHKO. 1982. Covalent affinity purification of ubiquitin-activating enzyme. J. Biol. Chem. **257:** 2537–2542.
59. HERSHKO, A., H. HELLER, S. ELIAS & A. CIECHANOVER. 1983. Components of ubiquitin-protein ligase system: resolution, affinity purification and role in protein breakdown. J. Biol. Chem. **258:** 8206–8214.
60. CHIN, D.T., L. KUEHL & M. RECHSTEINER. 1982. Conjugation of ubiquitin to denatured hemoglobin is proportional to the rate of hemoglobin degradation in HeLa cells. Proc. Natl. Acad. Sci. USA **79:** 5857–5861.
61. HERSHKO, A., E. EYTAN, A. CIECHANOVER & A.L. HAAS. 1982. Immunochemical analysis of the turnover of ubiquitin-protein conjugates in intact cells: relationship to the breakdown of abnormal proteins. J. Biol. Chem. **257:** 13964–13970.
62. MATSUMOTO, Y., H. YASUDA, T. MARUNOUCHI & M. YAMADA. 1983. Decrease in uH2A (protein A24) of a mouse temperature-sensitive mutant. FEBS Lett. **151:** 139–142.
63. FINLEY, D., A. CIECHANOVER & A. VARSHAVSKY. 1984. Thermolability of ubiquitin-activating enzyme from the mammalian cell cycle mutant ts85. Cell **37:** 43–55.
64. CIECHANOVER, A., FINLEY, D. & A. VARSHAVSKY. 1984. Ubiquitin dependence of selective protein degradation demonstrated in the mammalian cell cycle mutant ts85. Cell **37:** 57–66.
65. TANAKA, K., L. WAXMAN & A.L. GOLDBERG. 1983. ATP serves two distinct roles in protein degradation in reticulocytes, one requiring and one independent of ubiquitin. J. Cell Biol. **96:** 1580–1585.
66. HERSHKO, A., E. LESHINSKY, D. GANOTH & H. HELLER. 1984. ATP-dependent degradation of ubiquitin-protein conjugates. Proc. Natl. Acad. Sci. USA **81:** 1619–1623.
67. HOUGH, R., G. PRATT & M. RECHSTEINER. 1986. Ubiquitin-lysozyme conjugates: identification and characterization of an ATP-dependent protease from rabbit reticulocyte lysates. J. Biol. Chem. **261:** 2400–2408.
68. WAXMAN, L., J. FAGAN & A.L. GOLDBERG. 1987. Demonstration of two distinct high molecular weight proteases in rabbit reticulocytes, one of which degrades ubiquitin conjugates. J. Biol. Chem. **262:** 2451–2457.
69. HOUGH, R., G. PRATT & M. RECHSTEINER. 1987. Purification of two high molecular weight proteases from rabbit reticulocyte lysate. J. Biol. Chem. **262:** 8303–8313.
70. WILK, S. & M. ORLOWSKI. 1980. Cation-sensitive neutral endopeptidase: Isolation and specificity of the bovine pituitary enzyme. J. Neurochem. **35:** 1172–1182.
71. EYTAN, E., D. GANOTH, T. ARMON & A. HERSHKO. 1989. ATP-dependent incorporation of 20S protease into the 26S complex that degrades proteins conjugated to ubiquitin. Proc. Natl. Acad. Sci. USA **86:** 7751–7755.

72. DRISCOLL, J. & A.L. GOLDBERG. 1990. The proteasome (multicatalytic protease) is a component of the 1500-kDa proteolytic complex which degrades ubiquitin-conjugated proteins. J. Biol. Chem. **265:** 4789–4792.

73. HOFFMAN, L., G. PRATT & M. RECHSTEINER. 1992. Multiple forms of the 20S multi-catalytic and the 26S ubiquitin/ATP-dependent proteases from rabbit reticulocyte lysate. J. Biol. Chem. **267:** 22362–22368.

74. SHANKLIN, J., M. JABEN & R.D. VIERSTRA. 1987. Red light-induced formation of ubiquitin-phytochrome conjugates: identification of possible intermediates of phytochrome degradation. Proc. Natl. Acad. Sci. USA **84:** 359–363.

75. HOCHSTRASSER, M. & A. VARSHAVSKY. 1990. *In vivo* degradation of a transcriptional regulator: the yeast 2 repressor. Cell **61:** 697–708.

76. SCHEFFNER, M., B.A. WERNESS, J.M. HUIBREGTSE, *et al.* 1990. The E6 oncoprotein encoded by human papillomavirus types 16 and 18 promotes the degradation of p53. Cell **63:** 1129–1136.

77. GLOTZER, M., A.W. MURRAY & M.W. KIRSCHNER. 1991. Cyclin is degraded by the ubiquitin pathway. Nature **349:** 132–138.

78. CIECHANOVER, A., J.A. DI GIUSEPPE, B. BERCOVICH, *et al.* 1991. Degradation of nuclear oncoproteins by the ubiquitin system in vitro. Proc. Natl. Acad. Sci. USA **88:** 139–143.

79. MIZUSHIMA, N., T. NODA, T. YOSHIMORI, *et al.* 1998. A protein conjugation system essential for autophagy. Nature **395:** 395–398.

80. ADAMS, J. 2003. Potential for proteasome inhibition in the treatment of cancer. Drug Discov. Today **8:** 307–315.

81. GLICKMAN, M.H. & A. CIECHANOVER. 2002. The ubiquitin-proteasome pathway: destruction for the sake of construction. Physiol. Rev. **82:** 373–428.

82. PICKART, C.M. & R.E. COHEN. 2004. Proteasomes and their kin: proteases in the machine age. Nat. Rev. Mol. Cell Biol. **5:** 177–187.

83. CIECHANOVER, A. 2005. From the lysosome to ubiquitin and the proteasome. Nat. Rev. Mol. Cell Biol. **6:** 79–86.

Multiplicity of BMP Signaling in Skeletal Development

XIANGWEI WU,[a,b] WEIBIN SHI,[a] AND XU CAO[a]

[a]Department, of Pathology, University of Alabama at Birmingham, Birmingham, Alabama, USA

[b]Department of General Surgery, First Affiliated Hospital, School of Medicine Shihezi University, Shihezi, Xinjiang, China

ABSTRACT: Skeletal development and homeostasis are complex processes involving systemic regulatory factors and local factors present in the bone microenvironment. Among these factors, bone morphogenetic proteins (BMPs) represent the major potent bone morphogens, which coordinate with other factors to exert their multiple functions by regulating cell condensation, as well as inducing chondrogenesis and osteogenesis. This article reviews the current status of BMP signaling in skeletal development, with focus on regulation at the level of the receptors. Crosstalk of the BMP pathway with other major signaling pathways is also summarized.

KEYWORDS: bone morphogenetic protein; transforming growth factor; skeletogenesis; chondrocyte; osteoblast

INTRODUCTION

Skeletal development and homeostasis are determined by systemic hormones and local factors. Bone morphogenetic proteins (BMPs), members of the transforming growth factor-β (TGF-β) superfamily, were originally identified for their ability to induce ectopic bone formation.[1] More than 20 BMP-related proteins have now been identified, and can be classified into multiple subgroups based on their structure and function.[2,3] They play important roles in determining the fate of mesenchymal cells by stimulating their differentiation into cells of the osteoblastic lineage, and by inhibiting their differentiation into myoblastic lineage cells. Their effects on skeletogenesis include increased osteoclastogenesis, a process tightly coordinated with osteoblastogenesis via a set of coupling factors.

BMPs bind to and activate specific receptors, and induce cell signaling by phosphorylating cytoplasmic receptor-regulated Smads (R-Smads, Smad1, 5,

Address for correspondence: Xu Cao, 1670 University Blvd, VH G003, Birmingham, AL 35294-0019. Voice: (205) 934-0162; fax: (205) 934-1775.
cao@uab.edu

and 8). The activated R-Smads form heterodimers with the common partner Smad (Co-Smad, Smad4) and, after nuclear translocation, recruit distinct transcription cofactors and regulate transcription. The mechanisms governing the classical BMP signaling pathway have been well characterized. In this review, we will focus on recent advances in mechanistic research in BMP-mediated skeletal development, that is, the regulatory signaling mechanism that enables the cross-talk of BMPs with other signaling pathways. In particular, the receptor-level regulatory mechanism of BMP signaling will be extensively discussed.

BMP AND SKELETAL DEVELOPMENT

BMPS

BMPs, accounting for most of the TGF-β superfamily of peptides, are secreted as precursor proteins with a hydrophobic stretch of about 50–100 amino acids. The mature BMP monomer is derived from the carboxyterminal region of the precursor and is obtained by proteolyticcleavage.[4] BMPs display extensive conservation among species, containing seven characteristic cysteine knot domains, six of which build a cystin knot, while the seventh is used for dimerization with a second monomer.[5,6] According to the homology in their sequences, BMPs can be further subdivided into three subgroups: the BMP2/4 group, including BMP2, BMP4, and the *Drosophila decapentaplegic* (*dpp*) gene product; the osteogenic protein 1 (OP1) group, including BMP5, BMP6, OP1 (BMP7), BMP8 (OP2) and the *Drosophila gbb-60A* gene product; and the growth-differentiation factor 5 (GDF5) group, including GDF5, GDF6 (BMP13), and GDF7 (BMP12).[2] Members of the BMP family have distinct spatiotemporal expression patterns. Moreover, the biological activities of BMPs are not identical among members, since they bind to their receptors with different affinities and in different combinations.

BMP1 is unrelated to other BMPs and does not regulate the growth and differentiation of skeletal cells. It is a protease that cleaves procollagen fibrils, as well as chordin. Chordin can bind to and antagonize the actions of BMP2/4.[7] BMP2 and BMP4 are the most extensively studied BMPs with potent osteogenic effects. Both of them are very important in embryonic development. Mice deficient in BMP2 are not viable because of defects in the amnion/chorion and cardiac development. The BMP4-null mutation is lethal between 6.5 and 9.5 days of gestation because of the lack of mesodermal differentiation and patterning defects,[8] which manifests the importance of these two proteins. BMP3 (osteogenin) appears to antagonize the osteogenic effects of BMP2 in stromal cells. Neither BMP3 nor the closely related GDF10 seems to play a role in skeletal development because embryos and newborn mice from *bmp3* and *gdf10*-null mutations do not display an apparent skeletal phenotype.[9,10]

Unexpectedly, *bmp3*-null mice display an increase in bone mineral density and in trabecular bone volume.[9,11] BMP3 does not prevent BMP2 binding to its receptors. Instead, it likely acts via an activin-mediated pathway. Thus far, two naturally occurring loss-of-function mutations have been assigned to BMP3: *bmp5* (short ear) and *gdf5* (brachypodism). The short-eared mouse, apart from the short-ear phenotype, develops multiple cartilage and skeletal abnormalities affecting both the skull and axial skeleton.[12] Mice with GDF5 mutations display brachypodism and exhibit short limbs and joint fusions.[13] In humans, mutations in cartilage-derived morphogenetic protein 1 (CDMP1), the human homolog of GDF5, cause autosomal recessive chondrodysplasias and display a phenotype similar to that of the brachypodism defect in mice. Mutations in BMP6 only result in minor sternal defects.[14] Conversely, disruption of its close relative, BMP7 (osteogenin protein 1), leads to multiple skeletal defects, lack of eye and glomerular development, and subsequent renal failure and neonatal death.[15–17] Targeted disruptions of BMP8 (OP2) result in an extraskeletal defect in spermatogenesis, leading to infertility.[18] Mice with a targeted disruption of BMP10 die in mid-gestation due to heart failure resulting from a thinned, poorly developed myocardium.[19] Mutations in *bmp11* (*gdf11*) cause defects in anterior–posterior patterning of the axial skeleton.[20] Regulatory mechanisms governed by these ligands have been extensively reviewed elsewhere.[21]

BMP Receptors

Regulation at the Molecular Level of the Receptors

Signaling by BMP proteins is initiated by binding of BMPs to different sets of receptor complexes,[22–25] which determine the specificity of the intracellular signals. Thus far, three type II receptors have been described for BMP signaling: BMP type II receptor (BMPRII); activin type IIA receptor (ActRIIA); and activin type IIB receptors (ActRIIB). Three type I receptors have been identified: activin receptor-like kinase (ALK)2; ALK3 (BMPRIA); and ALK6 (BMPRIB). Of the three type I receptors for BMPs, ALK2 is less similar than ALK3 and ALK6. BMP2 and BMP4 preferentially bind to ALK3 and ALK6, whereas members of the OP1 group bind to ALK2 and ALK6. Members of the GDF5 group bind efficiently to ALK6, but not to other receptors.[2] In addition, Müllerian inhibiting substance (MIS), also known as anti-Müllerian hormone (AMH) and a member of the large TGF-β superfamily, has been demonstrated to bind to the unique receptor complex consisting of the BMP type I receptor and AMH type II receptor to elicit BMP-like signaling.[21,26]

Normally, type I and type II BMP receptors are present on the cell surface as either homomeric or heteromeric complexes, even before BMP stimulation.[27] BMP2 stimulation of the cells leads to rearrangement of receptor complexes at the cell surface. It appears that binding of BMP2 to preformed heteromeric receptor complexes results in activation of the Smad pathway, whereas formation of heteromeric receptor complexes induced by BMP2 activates a Smad-independent pathway, resulting in the induction of alkaline phosphatase activity via p38 MAP kinase.[27] Generally, the type-I receptors are the high-affinity binding receptors, whereas the type II receptors bind BMPs alone with low affinity.[24,28,29] One exception is GDF-9, which binds BMPRII with high affinity and ActRII with low affinity. Although there may be no direct interaction between type I and type II receptors, the type I receptor affinity for its ligands can be increased in the presence of the type II receptor.[30] Coexpression of BMPRII influences the aggregation and distribution of ALK3 in the COS7 and A431 cell lines. A kinase-inactive BMPRII blocks rearrangement of ALK3. It seems that specific structural organization of the BMP receptors prior to BMP2 binding is a key prerequisite for activation of distinct signaling pathways at the cell surface.[31]

It has recently been reported that, while wild-type cells transduce BMP signals via BMPRII, BMPRII-deficient cells transduce BMP signals via ActRIIA in conjunction with a set of type I receptors distinct from those used by BMPR2. Disruption of *bmprII* leads to diminished signaling by BMP2 and BMP4 and augmented signaling by BMP6 and BMP7. The observation that wild-type cells do not transduce BMP signals through ActRIIA, despite its expression, indicates that expression of BMPR2 suppresses BMP signaling via ActRIIA.[32]

Dampening of BMP receptor signaling can be influenced by many factors, such as pseudo receptor BMP and activin membrane-bound inhibitor (BAMBI).[33] Moreover, in the TGF-β signaling pathway, it has been determined that Smad7 acts as an adaptor protein in the formation of the PP1 holoenzyme that targets TβRI for dephosphorylation. SARA (Smad anchor for receptor activation) enhances the recruitment PP1c to the Smad7-GADD34 complex by controlling the specific subcellular localization of PP1c.[34,35] A similar mechanism has also been characterized in BMP signaling. An FYVE domain protein, Endofin (endosome-associated FYVE-domain protein), like SARA in the TGF-β signaling pathway, preferentially recruits Smad1 to the receptor complex for its phosphorylation. Endofin regulates Smad1 signaling by modulation of its phosphorylation and nuclear translocation upon BMP activation. Endofin also has a functional PP1c binding domain, which recruits PP1c to facilitate dephosphorylation of the type I BMP receptor; mutation of this domain enhances BMP signaling.[36]

Receptors and Bone Development

In mice, ALK2 is expressed primarily in the extraembryonic visceral endoderm before gastrulation, and later in both embryonic and extraembryonic cells during gastrulation.[37] Disruption of the *alk2* gene causes defective visceral endoderm. An absence of the BMP type I receptor, Alk2, in neural crest cells leads to craniofacial defects in mice. Constitutively active (CA) ALK2 misexpression *in vitro* enhances chondrocyte maturation and induces Ihh, while CA ALK2 viral infection in stage 19–23 limbs results in cartilage expansion with joint fusion. A recent study demonstrated that ALK2 can be integrated into the Ihh/PTHrP signaling loop as a BMP type I receptor that regulates chondrocyte maturation and skeletal development.[38] In addition to the binding affinity of BMPs for type IA and IB receptors being different, the level of receptor expression in different cells is variable. Moreover, *BmprIA* and *BmprIB* are expressed in different, though broadly overlapping, patterns in developing appendicular, axial, and craniofacial elements. ALK3 is expressed at low levels throughout the limb bud mesenchyme during chondrogenesis and, at a later stage, it regulates chondrocyte differentiation. Conversely, ALK6 is expressed in prechondrogenic cells and is required to initiate cartilage formation. After initiation of chondrocyte differentiation, ALK6 expression decreases.[39] Functionally, expression of constitutively active (CA) forms of ALK3 and ALK6 promotes chondrogenesis and osteogenic differentiation, whereas expression of dominant negative (DN) ALK6, but not DN ALK3, blocks these events.[40–44] In preosteoblast 2T3 cell cultures, ALK3 directs adipocytic differentiation, while ALK6 promotes osteoblastic differentiation.[40,45] These findings further suggest that, in addition to the mode of receptor oligodimerization, activation of different receptors can determine the signaling pathway,[27] thereby suggesting the presence and recruitment of different coactivators or corepressors in different cells. Mutant phenotypes of *alk3* include defects in epiblast proliferation, an absence of mesoderm formation in null mutants, and impaired cardiac and limb development in conditional mutants. Double mutants of *alk3* and *alk6* develop a severe, generalized chondrodysplasia.[39] *Alk3* mutations have been identified in some patients with juvenile polyposis.[46] It has also been reported recently that bone mass is increased in aged, osteoblast-specific, *alk3*-disrupted mice due to reduced bone resorption, as evidenced by reduced bone turnover.[47] Homozygous bmprII mutant embryos are arrested at the egg cylinder without mesoderm.[48] *ActrIIa* disruption causes mild defects in skeletal development and a deficiency in reproduction, while *actrIIb* mutations cause axial patterning defects and left–right asymmetry.[49] Recently, it has been found that ALK3 and ALK6 are the substrate of PP1 and Endofin, which modulates Smad1 phosphorylation, nuclear translocation, and BMP signaling level during osteoblast differentiation by recruiting both Smad1 and PP1c. Endofin mutant bearing PP1c-binding domain enhances ALP activity in C2C12 cells and process of mineralization in human MSCs.[50]

Clinically, approximately 25% of sporadic cases of primary pulmonary hypertension (PPH) have been found to be associated with mutations in BMPRII.[51] Loss of expression of BMPRII is also implicated in tumorigenesis of prostate and breast cancer.[52,53] Due to the clinical linkage and implication of unique Smad-independent pathways, interest in BMPRII has recently increased. The gene, *bmprII*, is located on chromosome 2q33[54–56] and encodes a polypeptide of 1038 amino acids. There are both long and short isoforms of BMPRII. The short isoform, a splice variant of only 530 amino acids that lacks exon 12, is still capable of phosphorylating both the Smads and p38MAPK.[27,57] Thus, like all similar structures among the type II receptor family, long form BMPRII has a longer C-terminal tail following the serine/threonine kinase domain.[56] At least one-third of pathogenic mutations in BMPRII are located within the cytoplasmic domain.[58] Mutations of BMPRII in the C-terminal tail have been found in some familial PPH patients, and may play an important role in the pathogenesis of PPH.[59,60] Physical interaction of the C-terminal tail of BMPRII with LIM kinase 1 (LIMK1) and Tctex1 further supports this notion. LIMK1 regulates actin dynamics through phosphorylation and inactivation of cofilin. BMPRII inhibits the ability of LMK1 to phosphorylate cofilin by promoting dissociation of LIMK1 from the receptors through interaction with its C-terminal tail.[61] Possibly as a result of the difference in cellular context, this hypothesis is somewhat inconsistent with a recently published study, which demonstrated that the interaction between BMPRII and LIMK1 synergizes with the Rho GTPase, Cdc42, to activate LIMK1 catalytic activity.[62] BMPRII also induces phosphorylation of Tctex, a light chain of the motor complex dynein, resulting in movement of Tctex along the microtubules and efficient activation of downstream signal mediators.[63] A recent proteomic analysis of BMPR-II interacting proteins found that several proteins, including PKCβ, tubulin β5, and MAPKKK8, interact with the C-terminal domain of BMPRII.[64] All of these proteins are related to the processes of skeletal development, including migration, differentiation, and apoptosis.[65,66] Thus, the C-terminal tail of BMPRII has conferred unique signaling properties to the receptor. It seems likely that BMPRII's R-Smad-independent functions contribute to the diversity of BMP signaling, and conceivably, the nonclassical pathway in skeletal development.

BMP Signaling and Skeletal Development

BMP Signaling and Condensation

Condensation is the pivotal stage in skeletal development and takes place when a previously dispersed population of mesenchymal cells forms an aggregation of cells, thereby facilitating the selective regulation of genes specific for either chondrogenesis or osteogenesis.[67,68] Condensation is a stepwise set of processes involving initiation, establishment of boundary conditions, cell adhesion, proliferation, growth, and cessation of growth. Condensations must attain a critical size and cells must interact within a condensation for the condensation phase to cease and for differentiation to be initiated.

Condensation size is regulated through signaling pathways involving BMP2/4.[69] Overexpression of either of these growth factors in chick embryos causes dramatic increases in both the size and shape of skeletal elements, as well as enhanced recruitment of mesenchymal precursors to cartilage condensations. Regulation of condensation size by BMP is also partly attributed to skeletal phenotypes in Brachypod (a frameshift mutation in GDF5)[13] and Short Ear (a mutation in BMP5) mice.[12,70]

BMP Signaling and Chondrogenesis

Progressing from condensation to overt differentiation of cells requires downregulation of genes controlling proliferation and upregulation of genes associated with differentiation, via pathways, such as BMP2, BMP4, BMP5. A subsequent transition from proliferating chondroblasts to chondrocytes and hypertrophic chondrocytes is regulated by BMP via Indian hedgehog (Ihh). *Ihh* and *sonic hedge hog(shh)* are highly homologous vertebrate hedgehog genes, and each enhances chondrogenesis and endochondral ossification.[71–73] The anabolic effects of Ihh/Shh and BMP2/4 in metatarsal cultures are analogous, and can be blocked by BMP antagonists. This observation indicates that, in this culture model, local BMPs mediate Ihh/Shh actions on endochondral ossification.[73] BMP signaling induces Ihh and promotes Ihh-mediated osteo/chondrogenic differentiation of a human chondrocytic cell line both *in vivo* and *in vitro*.[74] Constitutively active (CA) BMPRIA, CA BMPRIB, and CA ALK2 overexpression similarly induces Ihh and PTHrP.[38] While BMPs act directly on chondrocytes to induce maturation, this effect is counterbalanced *in vivo* by induction of the Ihh/PTHrP signaling loop.[38] Shh- and BMP4-mediated increases in endochondral ossification can be reversed by treatment with antisense oligonucleotides targeted against Cbfa1,[73] implying that Ihh/Shh likely mediates chondrogenesis through Cbfa1. BMP2 treatment enhances Cbfa1 mRNA expression in chondrocytic cell lines.[75] Cbfa1 binds to an osteoblast-specific *cis*-acting element (OSE2) in the osteocalcin promoter[76] and is expressed in prechondrogenic and preosteogenic condensations. Later, Cbfa1 is strictly confined to the osteogenic cell lineage.[76–78] Runx2-null mice display impaired chondrogenesis,[78] as manifested by interaction between BMP-dependent signaling Smads and Runx2 being necessary for chondrogenesis.[79] The effect of BMPs in chondrogenesis also appears to be mediated by Sox 9, a gene central to chondrogenesis. BMP2 and BMP4 induce Ihh, following which Ihh induces Nkx3.2, a target of left–right signaling. Nkx3.2 in turn induces the expression of Sox 9 that, in the presence of BMP, induces chondrogenesis.[80,81] Sox 9 antisense oligonucleotides blunt the induction of type II and X collagen by BMP in mesenchymal cells.[82] Conversely, Nkx3.2 was recently found to act as a potent sequence-specific repressor of the Runx2 promoter via a regulatory element 0.1 kb upstream from the site of transcriptional initiation, thereby promoting chondrogenic differentiation.[83] Thus, in chondrogenesis, the hierarchy of BMP involvement in the signaling cascade seems to have been established since Runx2 promotes osteogenesis and Sox9 enhances chondrogenesis. Nkx3.2 alters the commitment of mesenchymal progenitor cells by transcriptionally inhibiting the expression of Runx-2 and inducing expression of Sox9. Moreover, in the presence of BMP, Sox9 and Nkx3.2 induce each other's expression, which may be a prerequisite for chondrogenesis initiation.[81]

BMP Signaling and Osteogenesis

Smads have long been known as classical effectors of BMP signaling. New members have been added to the pool of effectors involved in BMP-mediated skeletal development. Targeted disruption of the Runx2 gene, *cbfa1,* leads to disorganized chondrocyte maturation and a complete lack of bone formation due to an arrest in osteoblast development.[78,84,85] During cranial bone development, disruption of *cbfa1* completely eliminates the expression of BMP2 and its downstream genes, Dlx5 and Msx2, in the developing primordium of bone. Runx2 also plays a role in mature osteoblastic function, and transgenic animals overexpressing a dominant negative form of Runx2, under the control of the osteoblastic-specific osteocalcin promoter, display decreased bone formation as a result of impaired osteoblastic function.[86] BMP2 induces Runx2 expression in osteoblast and chondrocyte cultures and, in association with BMP-specific signaling factors, Runx2 mediates BMP2 actions on gene transcription in cells of the osteoblastic lineage.[75,79,87] Interestingly, BMP2/4 promoters contain Runx2 binding sequences, and BMP2 can also be regulated by other BMPs in osteoblasts, implying a positive feedback loop for regulating BMP expression.[88–90] Another important transcription factor involved in osteogenesis is Osterix.[91] *Osterix*-null mice have normal cartilage development but fail to develop a mineralized skeleton. In contrast to *cbfa1*-null mice that do not form osteoblasts, osterix-null mice form cells of the osteoblastic lineage that express Runx2, but the cells do not mature. Loss of Osterix even results in a transformation of periosteal cells into chondrocytes, which is likely caused by tilting the balance between Runx2 and Sox9 to favor chondrogenesis. This would indicate that Osterix has effects on osteoblast maturation that are independent of Runx2, and that osterix acts downstream of Runx2.[91] It has been reported that BMP directs these alternative cell fates by regulating transcriptional coactivator with PDZ-binding motif (TAZ). TAZ coactivates Runx2-dependent gene transcription to drive MSCs to differentiate into osteoblasts, while repressing peroxisome proliferator-activated receptor γ (PPARγ)-dependent gene transcription drives MSCs to differentiate into adipocytes.[92]

A role for homeobox genes has been demonstrated in BMP-mediated osteoblast differentiation and skeletal development. Our previous study demonstrated that Smad3 and Smad4 exhibit different functions in activation of osteopontin (OPN) transcription. Smad3 binds directly to the OPN promoter as a sequence-specific activator, and Smad4 displaces Hoxa9, a homeobox protein and transcription repressor, by formation of a Smad4/Hox complex.[93] Similarly, osteoprotegerin, a secreted receptor of the TNF receptor family, acts as a decoy receptor that binds RANK-L, precluding RANK-L binding to RANK and its resulting effects on osteoclastogenesis and bone resorption. BMPs stimulate osteoprotegerin transcription through two Hoxc8-binding sites. The BMP signaling Smad 1 interacts with Hoxc8 and dislodges it from its binding element, resulting in induction of gene expression.[94] Bapx1 is a non-Hox homeobox gene that plays a central role in axial development of the skeleton. Bapx 1-null mice exhibit defective chondrogenesis and osteogenesis in the axial skeleton, causing a shortening of the vertebral column.[95–97] Bapx1 maintains Ihh expression and regulates the level and pattern of expression of BMP4 in the skeleton.[97] Among members of the Msh family of homeobox genes, Msx 1 and 2 are expressed in skeletal tissue and modulate osteogenesis. Msx 1-null mice display cleft palate, craniofacial, and dental developmental abnormalities. These mice also have defects in skull ossification, which are enhanced in double Msx1/Msx2 mutants.[98–100] Msx 2-null mice have defective chondrogenesis and osteogenesis due to a decreased number of osteoprogenitor cells. The skeletal abnormalities are associated with decreased expression of Runx2, indicating that Msx2 is necessary for osteogenesis and acts upstream of Runx2. In C2C12 cells, BMP2 induces osteoblastic differentiation, and Msx2 mediates this effect. BMP-induced apoptosis in embryonic cells can also be duplicated by the BMP-dependent homeobox gene Msx2, which is expressed at sites of cell replication and programmed cell death.[101,102] Furthermore, the effects of BMPs on osteoblastic differentiation and apoptosis in embryonic cells can be blocked by Msx2 antisense oligonucleotides, suggesting that Msx2 mediates these BMP effects.[102–104] The mammalian homologs of *Drosophila* distalless (Dlx) 5 and 6 are homeobox genes essential for craniofacial and skeletal development.[105] BMP induces Dlx5 expression in osteoblasts, and murine osteoblastic MC3T3 cells overexpressing Dlx5 display increased alkaline phosphatase activity, osteocalcin, and mineralization of the extracellular matrix.[106] Dlx5 mRNA is expressed in osteoblasts after differentiation, concomitant with a decline in Msx 2 mRNA and with the appearance of osteocalcin transcripts.[107] In addition, Dlx5 interferes with the ability of Msx2 to interact with Runx2 in regulating the expression of osteocalcin,[108] while osterix expression induced by BMP2 is mediated by Dlx5 and is independent of Runx2.[109] These results indicate that BMPs differentially mediate expression of skeletal development factors by recruiting different sets of transcription factors and cofactors.

Id genes are BMP-dependent, dominant negative regulators of helix-loop-helix transcription factors and have a direct impact on cell growth and differentiation.[110] Id genes are negative regulators of differentiation and positive regulators of cell proliferation. Downregulation of Id genes is necessary for terminal differentiation of a variety of cellular processes, including bone morphogenesis. Id proteins are potently upregulated by BMPs in mesenchymal cells,[110–113] thereby inhibiting differentiation of mesenchymal cells into myoblasts and adipocytes.[114,115] Moreover, their induction by BMPs may serve as a mechanism to reduce BMP action in cells of osteoblastic lineage.[110,111]

Autoregulation of BMP expression in osteoblasts is apparent, and BMP4 mRNA levels are BMP-dependent. BMPs cause an early, short-lived induction of BMP4 mRNA in osteoblasts followed by an inhibitory effect, suggesting autocrine regulation.[116] BMP2 can also be both upregulated and downregulated by other BMPs in osteoblasts. Runx2-binding sequences in BMP2/4 promoters further suggest the possibility for a positive feedback loop involving Runx2-regulated BMP2/4 expression,[88–90] since BMPs induce Runx2 expression.[87]

Programmed cell death, or apoptosis, is an expected result of cell maturation. The blocking of BMP activity not only arrests osteoblast differentiation but also prevents apoptosis. The proapoptotic effect of BMP2 is PKC-dependent, since BMP2 increases PKC activity. The selective PKC inhibitor calphostin C blocks BMP2-induced increases in Bax/Bcl-2 expression, caspase activity, and apoptosis.[65] The effect of BMPs on apoptosis is not limited to mature osteoblasts. BMPs also induce apoptosis in developing limbs, an event necessary for normal skeletal and joint development.[117,118] Blocking BMP signaling in the developing limb results in a reduction in interdigital apoptosis and, as a consequence, soft tissue webbing.

BMPS AND OTHER GROWTH REGULATORS

Integration of diverse signaling pathways is crucial for cells to interpret cellular context-dependent cues in development and homeostasis. Although BMPs are members of the TGF-β superfamily of polypeptides, TGF-βs and BMPs do not have the same biological activities in cells of the osteoblastic lineage. It appears that the two proteins counterbalance each other in their effects on osteoblast differentiation and maturation. Whereas BMP2 induces the differentiation of stromal cells toward the osteoblastic lineage, TGF-β opposes the effect of BMPs on osteoblastic cell maturation.[119] In some cell cultures, TGF-β induces the differentiation of cells toward the chondrocytic lineage, while in other cells, such as the C2C12 myogenic cell line, TGF-β simply arrests cell maturation.[120]

BMP2 also appears to be a downstream target of other growth factors. In skull bone development, FGF2 treatment of developing bone fronts stimulates Bmp2 gene expression.[121] In syndromic FGF receptor (FGFR)-mediated craniosynostoses (gain-of-function mutations in fibroblast growth factor receptors), noggin expression is suppressed by FGF2. Since noggin misexpression prevents cranial suture fusion, syndromic FGFR-mediated craniosynostoses may be the result of inappropriate downregulation of noggin expression and, possibly, concomitant upregulation of BMP signaling.[122]

BMPs also act in conjunction with other growth factors. BMPs induce the differentiation of cells of the osteoblastic lineage, increasing the pool of mature osteoblasts that can be targeted by IGF-1. BMPs increase IGF-I and IGF-II mRNA levels in osteoblast cultures. IGF-I and IGF-II subsequently increase osteoblastic function, resulting in a coordinated increase in osteoblastic differentiation and function.[123] BMPs also regulate the levels of IGFBPs in skeletal cells. Although the changes vary with the cell line studied, IGFBPs appear to play a role in modulating the anabolic activities of BMPs and IGF-I in bone.[124,125] The insulin- or IGF-1-stimulated PI3-K/Akt pathway is required for BMP-induced osteogenesis in some cell cultures.[110,111,126,127] Thus far, however, no detailed mechanism underlying these findings has been described. Since a Smad anchor for receptor activation (SARA) for BMP signaling, termed SARAb, has been characterized (our unpublished data), its role in BMP signaling explains, in part, the requirement of PI3-K/Akt in BMP signaling. SARAb, like its counterpart in TGF-β signaling, is anchored to PI3P on the inner cell membrane, where it is phosphorylated by PI3K. Inhibition of PI3K phosphorylation activity by Wortmannin results in mislocalization of this anchor protein[128] and a consequent decline in BMP signaling (our unpublished data). However, in hMSC cells treated with PDGF, inhibition of PI3K by Wortmannin in PDGF-treated cells results in enhanced osteoblast differentiation.[129]

Wnts, like BMPs, can induce cell differentiation and act by preventing β-catenin degradation by the ubiquitin–proteasome pathway inhibition of glycogen-synthase kinase-3. This results in β-catenin accumulation, nuclear translocation and association with members of the lymphoid enhancer-binding factor/T cell-specific factor (LEF/TCF) transcription factor family, and the targeting of specific genes. Ectopic canonical Wnt signaling leads to enhanced ossification and suppression of chondrocyte formation. Genetic inactivation of β-catenin results in ectopic formation of chondrocytes at the expense of osteoblast differentiation during both intramembranous and endochondral ossification.[130] β-catenin and LEF/TCF can form a complex with Smad 4 and, as such, have the potential to regulate BMP and TGF-β signaling.[131] The coreceptor of Wnt signaling, LRP 5, is expressed by osteoblasts and stromal cells. Its expression is induced by BMP2 as stromal cells undergo differentiation.[132] LRP 5 mutations that affect Wnt signaling result in decreased bone mass, whereas mutations that create an LRP 5 resistant to Dickkopf inactivation result in sustained Wnt signaling and increased bone mass.[132,133] Chromatin immunoprecipitation studies have shown that β-catenin and TCF1 are recruited to the Runx2 endogenous gene product. Mutational analysis has also demonstrated that a functional TCF regulatory element responsive to canonical Wnt signaling resides in the promoter of *Cbfa1* (–97 to –93). It seems likely that BMP and Wnt signaling converge on Runx2 gene to synergistically regulate osteogenesis and chondrogenesis.[134] It has also been reported that constitutive transcriptional active β-catenin and BMP2 synergistically stimulate new bone formation after subperiosteal injection in mouse calvaria *in vivo*. This synergism with BMP2 in regulating gene transcription occurs without altering the expression of Runx2, suggesting actions independent or downstream of this osteoblast-specific transcription factor. β-catenin, via Tcf/Lef-dependent mechanisms, directs osteogenic lineage allocation by enhancing mesenchymal cell responsiveness to BMP2.[135] In contrast, BMP mediates Wnt signaling by inducing the PTEN-PI3K/Akt pathway. BMP induces PTEN, and PTEN inhibits Akt activity and its downstream signals, including nuclear accumulation of β-catenin in intestinal stem cells. However, no similar phenomenon has been reported in osteochondrogenitor cells.

While convergence of the BMP and MAP kinase signaling pathways has been a primary topic of research in both fields, no clear picture has emerged.[136] Inhibition of PI3K/p70 S6K and p38 MAPK cascades increases osteoblastic differentiation induced by BMP2 in C2C12 cells.[66] In epithelial cells, ERK directly regulates BMP signaling by phosphorylation at serine residues in the Smad linker region, and inhibits both nuclear accumulation and transcriptional activity of Smad1, an observation that has been confirmed *in vivo*.[136] Conversely, BMP2 can stimulate Ras activity and, as a consequence, two MAPKs, ERK and P38, are expressed.[137,138] As a result of Ras/MAPK activation, most members of the Fos/Jun family and activating transcription factor 2 (ATF2) are upregulated and interact with activating protein 1 (AP1) sequences in various genes. The MAP kinase, p38, is essential in BMP2 upregulation of type I collagen, fibronectin, osteopontin, osteocalcin, and alkaline phosphatase activity, whereas ERK mediates BMP2 stimulation of fibronectin and osteopontin. Thus, ERK and p38 differentially mediate BMP2 function in osteoblasts.[139] BMP and MAPK signaling converge on Smads, resulting in differential phosphorylation. EGF signals through a receptor tyrosine kinase (RTK) and strongly activates the Erk family of MAP kinases, which catalyze the phosphorylation of Smad1 in its linker region. Phosphorylation in its linker region prevents Smad1 from translocating into the nucleus.[140,141] Apart from opposing BMP signaling, phosphorylation in the linker region appears to be important for some cell activities. Smad1$^{C/C}$ mutants (mutation in C-terminal motif) recapitulate many Smad1$^{-/-}$ phenotypes, including defective allantois formation and the lack of primordial germ cells (PGC). They also exhibit phenotypes that are both more severe (head and branchial arches) and less severe (allantois growth) than the null mutation. Smad1$^{L/L}$ mutants (mutation in linker region) survive embryogenesis but exhibit defects in gastric epithelial homeostasis, which are correlated with changes in cell contacts, actin cytoskeleton remodeling, and nuclear β-catenin accumulation. MAPKs can regulate independent pathways as well as act interdependently with the Smad pathway to phosphorylate Smads.[137]

Focal adhesion kinase (FAK), a cytosolic nonreceptor tyrosine kinase, can be associated with and activated by β1-integrin. The Ras-ERK pathway downstream of integrin-FAK is involved in Smad1 signals activated by BMP, and possibly provides a mechanism for cooperation between intracellular signals activated by integrin and BMP receptors in osteoblastic cells. Disruption of either collagen synthesis or collagen-α2β1-integrin binding inhibits the stimulatory effect of BMP2 on osteoblastic MC3T3-E1 cells. Ras-ERK signals enhance the transcriptional activity of Smad1 in response to BMP receptor activation in COS-1 cells.[142]

PERSPECTIVES

The availability of natural mutations and genetic loss-of-function and gain-of-function assays, along with other genomics and proteomics technologies, provides powerful tools for investigating the role of BMP in skeletogenesis. Recent studies on the nonclassical pathway, such as regulatory mechanisms at the receptor level and the sharing of effectors among different signaling pathways, have enriched our understanding of the role of BMPs in skeletal development and homeostasis. However, our exploration of the unknown is far from complete, especially the nonclassical, alternative pathways. Challenges for the future include a full understanding of how the cascades of genes that regulate each stage of skeletal development are coordinated. The information gained from future studies will provide the means to manipulate each stage of development to correct skeletal defects that have their origin during development. It is also envisioned that a better understanding of the genes and proteins that regulate skeletal development will provide new modalities for treatment of metabolic bone diseases and other osteopathies.

REFERENCES

1. URIST, M.R. 1965. Bone: formation by autoinduction. Science **150**: 893–899.
2. MIYAZONO, K., S. MAEDA & T. IMAMURA. 2005. BMP receptor signaling: transcriptional targets, regulation of signals, and signaling cross-talk. Cytokine Growth Factor Rev. **16**: 251–263.
3. TEN DIJKE, P., O. KORCHYNSKYI, G. VALDIMARSDOTTIR & M.J. GOUMANS. 2003. Controlling cell fate by bone morphogenetic protein receptors. Mol. Cell Endocrinol. **211**: 105–113.
4. MASSAGUE, J. 1990. The transforming growth factor-beta family. Annu. Rev. Cell. Biol. **6**: 597–641.
5. KAWABATA, M., T. IMAMURA & K. MIYAZONO. 1998. Signal transduction by bone morphogenetic proteins. Cytokine Growth Factor Rev. **9**: 49–61.
6. SCHEUFLER, C., W. SEBALD & M. HULSMEYER. 1999. Crystal structure of human bone morphogenetic protein-2 at 2.7 A resolution. J. Mol. Biol. **287**: 103–115.
7. UZEL, M.I., I.C. SCOTT, H. BABAKHANLOU-CHASE, *et al.* 2001. Multiple bone morphogenetic protein 1-related mammalian metalloproteinases process prolysyl oxidase at the correct physiological site and control lysyl oxidase activation in mouse embryo fibroblast cultures. J. Biol. Chem. **276**: 22537–22543.

8. GIMBLE, J.M., C. MORGAN, K. KELLY, *et al.* 1995. Bone morphogenetic proteins inhibit adipocyte differentiation by bone marrow stromal cells. J. Cell. Biochem. **58:** 393–402.
9. BAHAMONDE, M.E. & K.M. LYONS. 2001. BMP3: to be or not to be a BMP. J. Bone Joint Surg. Am. 83-A Suppl 1: S56–S62.
10. ZHAO, R., A.M. LAWLER & S.J. LEE. 1999. Characterization of GDF-10 expression patterns and null mice. Dev. Biol. **212:** 68–79.
11. DALUISKI, A., T. ENGSTR & M.E. BAHAMONDE, *et al.* 2001. Bone morphogenetic protein-3 is a negative regulator of bone density. Nat. Genet. **27:** 84–88.
12. KINGSLEY, D.M., A.E. BLAND, J.M. GRUBBER, *et al.* 1992. The mouse short ear skeletal morphogenesis locus is associated with defects in a bone morphogenetic member of the TGF beta superfamily. Cell **71:** 399–410.
13. STORM, E.E. & D.M. KINGSLEY. 1999. GDF5 coordinates bone and joint formation during digit development. Dev. Biol. **209:** 11–27.
14. SOLLOWAY, M.J., A.T. DUDLEY, E.K. BIKOFF, *et al.* 1998. Mice lacking Bmp6 function. Dev. Genet. **22:** 321–339.
15. DUDLEY, A.T., K.M. LYONS & E.J. ROBERTSON. 1995. A requirement for bone morphogenetic protein-7 during development of the mammalian kidney and eye. Genes Dev. **9:** 2795–2807.
16. JENA, N., C. MARTIN-SEISDEDOS, P. MCCUE & C.M. CROCE. 1997. BMP7 null mutation in mice: developmental defects in skeleton, kidney, and eye. Exp. Cell Res. **230:** 28–37.
17. LUO, G., C. HOFMANN, A.L. BRONCKERS, *et al.* 1995. BMP-7 is an inducer of nephrogenesis, and is also required for eye development and skeletal patterning. Genes Dev. **9:** 2808–2820.
18. ZHAO, G.Q., K. DENG, P.A. LABOSKY, *et al.* 1996. The gene encoding bone morphogenetic protein 8B is required for the initiation and maintenance of spermatogenesis in the mouse. Genes Dev. **10:** 1657–1669.
19. CHEN, H., S. SHI, L. ACOSTA, *et al.* 2004. BMP10 is essential for maintaining cardiac growth during murine cardiogenesis. Development **131:** 2219–2231.
20. MCPHERRON, A.C., A.M. LAWLER & S.J. LEE. 1999. Regulation of anterior/posterior patterning of the axial skeleton by growth/differentiation factor 11. Nat. Genet. **22:** 260–264.
21. CANALIS, E., A.N. ECONOMIDES & E. GAZZERRO. 2003. Bone morphogenetic proteins, their antagonists, and the skeleton. Endocr. Rev. **24:** 218–235.
22. DERYNCK, R., Y. ZHANG & X.H. FENG. 1998. Smads: transcriptional activators of TGF-beta responses. Cell **95:** 737–740.
23. KRETZSCHMAR, M. & J. MASSAGUE. 1998. SMADs: mediators and regulators of TGF-beta signaling. Curr. Opin. Genet. Dev. **8:** 103–111.
24. LIU, F., F. VENTURA, J. DOODY & J. MASSAGUE. 1995. Human type II receptor for bone morphogenic proteins (BMPs): extension of the two-kinase receptor model to the BMPs. Mol. Cell Biol. **15:** 3479–3486.
25. YAMASHITA, H., P. TEN DIJKE, C.H. HELDIN & K. MIYAZONO. 1996. Bone morphogenetic protein receptors. Bone **19:** 569–574.
26. JAMIN, S.P., N.A. ARANGO, Y. MISHINA, *et al.* 2002. Requirement of Bmpr1a for Mullerian duct regression during male sexual development. Nat. Genet. **32:** 408–410.
27. NOHE, A., S. HASSEL, M. EHRLICH, *et al.* 2002. The mode of bone morphogenetic protein (BMP) receptor oligomerization determines different BMP-2 signaling pathways. J. Biol. Chem. **277:** 5330–5338.

28. ROSENZWEIG, B.L., T. IMAMURA, T. OKADOME, *et al.* 1995. Cloning and characterization of a human type II receptor for bone morphogenetic proteins. Proc. Natl. Acad. Sci. USA **92:** 7632–7636.

29. TEN DIJKE, P., H. YAMASHITA, T.K. SAMPATH, *et al.* 1994. Identification of type I receptors for osteogenic protein-1 and bone morphogenetic protein-4. J. Biol. Chem. **269:** 16985–16988.

30. GREENWALD, J., J. GROPPE, P. GRAY, *et al.* 2003. The BMP7/ActRII extracellular domain complex provides new insights into the cooperative nature of receptor assembly. Mol. Cell **11:** 605–617.

31. NOHE, A., E. KEATING, T.M. UNDERHILL, *et al.* 2003. Effect of the distribution and clustering of the type I A BMP receptor (ALK3) with the type II BMP receptor on the activation of signalling pathways. J. Cell Sci. **116:** 3277–3284.

32. YU, P.B., H. BEPPU, N. KAWAI, *et al.* 2005. Bone morphogenetic protein (BMP) type II receptor deletion reveals BMP ligand-specific gain of signaling in pulmonary artery smooth muscle cells. J. Biol. Chem. **280:** 24443–24450.

33. ONICHTCHOUK, D., Y.G. CHEN, R. DOSCH, *et al.* 1999. Silencing of TGF-beta signalling by the pseudoreceptor BAMBI. Nature **401:** 480–485.

34. BENNETT, D. & L. ALPHEY. 2002. PP1 binds Sara and negatively regulates Dpp signaling in Drosophila melanogaster. Nat. Genet. **31:** 419–423.

35. SHI, W., C. SUN, B. HE, *et al.* 2004. GADD34-PP1c recruited by Smad7 dephosphorylates TGFbeta type I receptor. J. Cell Biol. **164:** 291–300.

36. SHI, W., C. CHANG, S. NIE, *et al.* 2007. Endofin acts as a Smad anchor for receptor activation in BMP signaling. J. Cell Sci. **120:** 1216–1224.

37. GU, Z., E.M. REYNOLDS, J. SONG, *et al.* 1999. The type I serine/threonine kinase receptor ActRIA (ALK2) is required for gastrulation of the mouse embryo. Development **126:** 2551–2561.

38. ZHANG, D., E.M. SCHWARZ, R.N. ROSIER, *et al.* 2003. ALK2 functions as a BMP type I receptor and induces Indian hedgehog in chondrocytes during skeletal development. J. Bone Miner. Res. **18:** 1593–1604.

39. YOON, B.S., D.A. OVCHINNIKOV, I. YOSHII, *et al.* 2005. Bmpr1a and Bmpr1b have overlapping functions and are essential for chondrogenesis *in vivo*. Proc. Natl. Acad. Sci. U SA **102:** 5062–5067.

40. CHEN, D., X. JI, M.A. HARRIS, *et al.* 1998. Differential roles for bone morphogenetic protein (BMP) receptor type IB and IA in differentiation and specification of mesenchymal precursor cells to osteoblast and adipocyte lineages. J. Cell Biol. **142:** 295–305.

41. ENOMOTO-IWAMOTO, M., M. IWAMOTO, Y. MUKUDAI, *et al.* 1998. Bone morphogenetic protein signaling is required for maintenance of differentiated phenotype, control of proliferation, and hypertrophy in chondrocytes. J. Cell Biol. **140:** 409–418.

42. KAWAKAMI, Y., T. ISHIKAWA, M. SHIMABARA, *et al.* 1996. BMP signaling during bone pattern determination in the developing limb. Development **122:** 3557–3566.

43. ZOU, H., K.M. CHOE, Y. LU, *et al.* 1997. BMP signaling and vertebrate limb development Cold Spring. Harb. Symp. Quant. Biol. **62:** 269–272.

44. ZOU, H., R. WIESER, J. MASSAGUE & L. NISWANDER. 1997. Distinct roles of type I bone morphogenetic protein receptors in the formation and differentiation of cartilage. Genes Dev. **11:** 2191–2203.

45. NAMIKI, M., S. AKIYAMA, T. KATAGIRI, et al. 1997. A kinase domain-truncated type I receptor blocks bone morphogenetic protein-2-induced signal transduction in C2C12 myoblasts. J. Biol. Chem. **272:** 22046–22052.
46. HOWE, J.R., J.L. BAIR, M.G. SAYED, et al. 2001. Germline mutations of the gene encoding bone morphogenetic protein receptor 1A in juvenile polyposis. Nat. Genet. **28:** 184–187.
47. MISHINA, Y., M.W. STARBUCK, M.A. GENTILE, et al. 2004. Bone morphogenetic protein type IA receptor signaling regulates postnatal osteoblast function and bone remodeling. J. Biol. Chem. **279:** 27560–27566.
48. BEPPU, H., M. KAWABATA, T. HAMAMOTO, et al. 2000. BMP type II receptor is required for gastrulation and early development of mouse embryos. Dev. Biol. **221:** 249–258.
49. OH, S.P., C.Y. YEO, Y. LEE, et al. 2002. Activin type IIA and IIB receptors mediate Gdf11 signaling in axial vertebral patterning. Genes Dev. **16:** 2749–2754.
50. SHI, W., C. CHANG, S. NIE, et al. 2007. Endofin acts as a Smad anchor for receptor activation in BMP signaling. J. Cell Sci. **120:** 1216–1224.
51. THOMSON, J.R., R.D. MACHADO, M.W. PAUCIULO, et al. 2000. Sporadic primary pulmonary hypertension is associated with germline mutations of the gene encoding BMPR-II, a receptor member of the TGF-beta family. J. Med. Genet. **37:** 741–745.
52. KIM, I.Y., D.H. LEE, D.K. LEE, et al. 2004. Loss of expression of bone morphogenetic protein receptor type II in human prostate cancer cells. Oncogene **23:** 7651–7659.
53. POULIOT, F., A. BLAIS & C. LABRIE. 2003. Overexpression of a dominant negative type II bone morphogenetic protein receptor inhibits the growth of human breast cancer cells. Cancer Res. **63:** 277–281.
54. ASTROM, A.K., D. JIN, T. IMAMURA, et al. 1999. Chromosomal localization of three human genes encoding bone morphogenetic protein receptors Mamm. Genome **10:** 299–302.
55. DENG, Z., F. HAGHIGHI, L. HELLEBY, et al. 2000. Fine mapping of PPH1, a gene for familial primary pulmonary hypertension, to a 3-cM region on chromosome 2q33. Am. J. Respir. Crit. Care Med. **161:** 1055–1059.
56. MACHADO, R.D., M.W. PAUCIULO, N. FRETWELL, et al. 2000. A physical and transcript map based upon refinement of the critical interval for PPH1, a gene for familial primary pulmonary hypertension. The International PPH Consortium Genomics **68:** 220–228.
57. NOHNO, T., T. ISHIKAWA, T. SAITO, et al. 1995. Identification of a human type II receptor for bone morphogenetic protein-4 that forms differential heteromeric complexes with bone morphogenetic protein type I receptors. J. Biol. Chem. **270:** 22522–22526.
58. MACHADO, R.D., M.W. PAUCIULO, J.R. THOMSON, et al. 2001. BMPR2 haploinsufficiency as the inherited molecular mechanism for primary pulmonary hypertension. Am. J. Hum. Genet. **68:** 92–102.
59. DENG, Z., J.H. MORSE, S.L. SLAGER, et al. 2000. Familial primary pulmonary hypertension (gene PPH1) is caused by mutations in the bone morphogenetic protein receptor-II gene. Am. J. Hum. Genet. **67:** 737–744.
60. LANE, K.B., R.D. MACHADO, M.W. PAUCIULO, et al. 2000. Heterozygous germline mutations in BMPR2, encoding a TGF-beta receptor, cause familial primary pulmonary hypertension. The International PPH Consortium. Nat. Genet. **26:** 81–84.

61. FOLETTA, V.C., M.A. LIM, J. SOOSAIRAJAH, *et al.* 2003. Direct signaling by the BMP type II receptor via the cytoskeletal regulator LIMK1. J. Cell Biol. **162:** 1089–1098.
62. LEE-HOEFLICH, S.T., C.G. CAUSING, M. PODKOWA, *et al.* 2004. Activation of LIMK1 by binding to the BMP receptor, BMPRII, regulates BMP-dependent dendritogenesis. EMBO J. **23:** 4792–4801.
63. MACHADO, R.D., N. RUDARAKANCHANA, C. ATKINSON, *et al.* 2003. Functional interaction between BMPR-II and Tctex-1, a light chain of Dynein, is isoform-specific and disrupted by mutations underlying primary pulmonary hypertension. Hum. Mol. Genet. **12:** 3277–3286.
64. HASSEL, S., A. EICHNER, M. YAKYMOVYCH, *et al.* 2004. Proteins associated with type II bone morphogenetic protein receptor (BMPR-II) and identified by two-dimensional gel electrophoresis and mass spectrometry. Proteomics **4:** 1346–1358.
65. HAY, E., J. LEMONNIER, O. FROMIGUE & P.J. MARIE. 2001. Bone morphogenetic protein-2 promotes osteoblast apoptosis through a Smad-independent, protein kinase C-dependent signaling pathway. J. Biol. Chem. **276:** 29028–29036.
66. VINALS, F., T. LOPEZ-ROVIRA, J.L. ROSA & F. VENTURA. 2002. Inhibition of PI3K/p70 S6K and p38 MAPK cascades increases osteoblastic differentiation induced by BMP-2. FEBS Lett. **510:** 99–104.
67. HALL, B.K. & T. MIYAKE. 2000. All for one and one for all: condensations and the initiation of skeletal development. Bioessays **22:** 138–147.
68. SHUM, L., C.M. COLEMAN, Y. HATAKEYAMA & R.S. TUAN. 2003. Morphogenesis and dysmorphogenesis of the appendicular skeleton. Birth Defects Res. C. Embryo Today **69:** 102–122.
69. HALL, B.K. & T. MIYAKE. 1995. Divide, accumulate, differentiate: cell condensation in skeletal development revisited. Int. J. Dev. Biol. **39:** 881–893.
70. KINGSLEY, D.M. 1994. What do BMPs do in mammals? Clues from the mouse short-ear mutation. Trends Genet. **10:** 16–21.
71. CHUNG, U.I., E. SCHIPANI, A.P. MCMAHON & H.M. KRONENBERG. 2001. Indian hedgehog couples chondrogenesis to osteogenesis in endochondral bone development. J. Clin. Invest **107:** 295–304.
72. GRIMSRUD, C.D., P.R. ROMANO, M. D'SOUZA, *et al.* 2001. BMP signaling stimulates chondrocyte maturation and the expression of Indian hedgehog. J. Orthop. Res. **19:** 18–25.
73. KRISHNAN, V., Y. MA, J. MOSELEY, *et al.* 2001. Bone anabolic effects of sonic/Indian hedgehog are mediated by bmp-2/4-dependent pathways in the neonatal rat metatarsal model. Endocrinology **142:** 940–947.
74. UYAMA, Y., K. YAGAMI, M. HATORI, *et al.* 2004. Recombinant human bone morphogenetic protein-2 promotes Indian hedgehog-mediated osteo-chondrogenic differentiation of a human chondrocytic cell line *in vivo* and *in vitro*. Differentiation **72:** 32–40.
75. TAKAZAWA, Y., K. TSUJI, A. NIFUJI, *et al.* 2000. An osteogenesis-related transcription factor, core-binding factor A1, is constitutively expressed in the chondrocytic cell line TC6, and its expression is upregulated by bone morphogenetic protein-2. J. Endocrinol. **165:** 579–586.
76. DUCY, P., R. ZHANG, V. GEOFFROY, *et al.* 1997. Osf2/Cbfa1: a transcriptional activator of osteoblast differentiation. Cell **89:** 747–754.
77. DUCY, P. & G. KARSENTY. 1998. Genetic control of cell differentiation in the skeleton. Curr. Opin. Cell Biol. **10:** 614–619.

78. KIM, I.S., F. OTTO, B. ZABEL & S. MUNDLOS. 1999. Regulation of chondrocyte differentiation by Cbfa1. Mech. Dev. **80:** 159–170.
79. LEE, K.S., H.J. KIM, Q.L. LI, *et al.* 2000. Runx2 is a common target of transforming growth factor beta1 and bone morphogenetic protein 2, and cooperation between Runx2 and Smad5 induces osteoblast-specific gene expression in the pluripotent mesenchymal precursor cell line C2C12. Mol. Cell Biol. **20:** 8783–8792.
80. MURTAUGH, L.C., L. ZENG, J.H. CHYUNG & A.B. LASSAR. 2001. The chick transcriptional repressor Nkx3.2 acts downstream of Shh to promote BMP-dependent axial chondrogenesis. Dev. Cell **1:** 411–422.
81. ZENG, L., H. KEMPF, L.C. MURTAUGH, *et al.* 2002. Shh establishes an Nkx3.2/Sox9 autoregulatory loop that is maintained by BMP signals to induce somitic chondrogenesis. Genes Dev. **16:** 1990–2005.
82. ZEHENTNER, B.K., C. DONY & H. BURTSCHER. 1999. The transcription factor Sox9 is involved in BMP-2 signaling. J. Bone Miner. Res. **14:** 1734–1741.
83. LENGNER, C.J., M.Q. HASSAN, R.W. SERRA, *et al.* 2005. Nkx3.2-mediated repression of Runx2 promotes chondrogenic differentiation. J. Biol. Chem. **280:** 15872–15879.
84. KARSENTY, G., P. DUCY, M. STARBUCK, *et al.* 1999. Cbfa1 as a regulator of osteoblast differentiation and function. Bone **25:** 107–108.
85. KOMORI, T., H. YAGI, S. NOMURA, *et al.* 1997. Targeted disruption of Cbfa1 results in a complete lack of bone formation owing to maturational arrest of osteoblasts. Cell **89:** 755–764.
86. DUCY, P., M. STARBUCK, M. PRIEMEL, *et al.* 1999. A Cbfa1-dependent genetic pathway controls bone formation beyond embryonic development. Genes Dev. **13:** 1025–1036.
87. BANERJEE, C., A. JAVED, J.Y. CHOI, *et al.* 2001. Differential regulation of the two principal Runx2/Cbfa1 n-terminal isoforms in response to bone morphogenetic protein-2 during development of the osteoblast phenotype. Endocrinology **142:** 4026–4039.
88. FENG, J.Q., D. CHEN, A.J. COONEY, *et al.* 1995. The mouse bone morphogenetic protein-4 gene. Analysis of promoter utilization in fetal rat calvarial osteoblasts and regulation by COUP-TFI orphan receptor. J. Biol. Chem. **270:** 28364–28373.
89. GHOSH-CHOUDHURY, N., G.G. CHOUDHURY, M.A. HARRIS, *et al.* 2001. Autoregulation of mouse BMP-2 gene transcription is directed by the proximal promoter element. Biochem. Biophys. Res. Commun. **286:** 101–108.
90. HELVERING, L.M., R.L. SHARP, X. OU & A.G. GEISER. 2000. Regulation of the promoters for the human bone morphogenetic protein 2 and 4 genes. Gene **256:** 123–138.
91. NAKASHIMA, K., X. ZHOU, G. KUNKEL, *et al.* 2002. The novel zinc finger-containing transcription factor osterix is required for osteoblast differentiation and bone formation. Cell **108:** 17–29.
92. HONG, J.H., E.S. HWANG, M.T. MCMANUS, *et al.* 2005. TAZ, a transcriptional modulator of mesenchymal stem cell differentiation. Science **309:** 1074–1078.
93. SHI, X., S. BAI, L. LI & X. CAO. 2001. Hoxa-9 represses transforming growth factor-beta-induced osteopontin gene transcription. J. Biol. Chem. **276:** 850–855.
94. WAN, M., X. SHI, X. FENG & X. CAO. 2001. Transcriptional mechanisms of bone morphogenetic protein-induced osteoprotegrin gene expression. J. Biol. Chem. **276:** 10119–10125.

95. TRIBIOLI, C. & T. LUFKIN. 1997. Molecular cloning, chromosomal mapping and developmental expression of BAPX1, a novel human homeobox-containing gene homologous to Drosophila bagpipe. Gene **203:** 225–233.

96. TRIBIOLI, C., M. FRASCH & T. LUFKIN. 1997. Bapx1: an evolutionary conserved homologue of the Drosophila bagpipe homeobox gene is expressed in splanchnic mesoderm and the embryonic skeleton. Mech. Dev. **65:** 145–162.

97. TRIBIOLI, C. & T. LUFKIN. 1999. The murine Bapx1 homeobox gene plays a critical role in embryonic development of the axial skeleton and spleen. Development **126:** 5699–5711.

98. MAAS, R., Y.P. CHEN, M. BEI, *et al.* 1996. The role of Msx genes in mammalian development. Ann. N. Y. Acad. Sci. **785:** 171–181.

99. SATOKATA, I. & R. MAAS. 1994. Msx1 deficient mice exhibit cleft palate and abnormalities of craniofacial and tooth development. Nat. Genet. **6:** 348–356.

100. SATOKATA, I., L. MA, H. OHSHIMA, *et al.* 2000. Msx2 deficiency in mice causes pleiotropic defects in bone growth and ectodermal organ formation. Nat. Genet. **24:** 391–395.

101. FERRARI, D., A.C. LICHTLER, Z.Z. PAN, *et al.* 1998. Ectopic expression of Msx-2 in posterior limb bud mesoderm impairs limb morphogenesis while inducing BMP-4 expression, inhibiting cell proliferation, and promoting apoptosis. Dev. Biol. **197:** 12–24.

102. MARAZZI, G., Y. WANG & D. SASSOON. 1997. Msx2 is a transcriptional regulator in the BMP4-mediated programmed cell death pathway. Dev. Biol. **186:** 127–138.

103. GOMES, W.A. & J.A. KESSLER. 2001. Msx-2 and p21 mediate the pro-apoptotic but not the anti-proliferative effects of BMP4 on cultured sympathetic neuroblasts. Dev. Biol. **237:** 212–221.

104. JERNVALL, J., T. ABERG, P. KETTUNEN, *et al.* 1998. The life history of an embryonic signaling center: BMP-4 induces p21 and is associated with apoptosis in the mouse tooth enamel knot. Development **125:** 161–169.

105. ROBLEDO, R.F., L. RAJAN, X. LI & T. LUFKIN. 2002. The Dlx5 and Dlx6 homeobox genes are essential for craniofacial, axial, and appendicular skeletal development. Genes Dev. **16:** 1089–1101.

106. MIYAMA, K., G. YAMADA, T.S. YAMAMOTO, *et al.* 1999. A BMP-inducible gene, dlx5, regulates osteoblast differentiation and mesoderm induction. Dev. Biol. **208:** 123–133.

107. RYOO, H.M., H.M. HOFFMANN, T. BEUMER, *et al.* 1997. Stage-specific expression of Dlx-5 during osteoblast differentiation: involvement in regulation of osteocalcin gene expression. Mol. Endocrinol. **11:** 1681–1694.

108. SHIRAKABE, K., K. TERASAWA, K. MIYAMA, *et al.* 2001. Regulation of the activity of the transcription factor Runx2 by two homeobox proteins, Msx2 and Dlx5 Genes. Cells **6:** 851–856.

109. LEE, M.H., T.G. KWON, H.S. PARK, *et al.* 2003. BMP-2-induced Osterix expression is mediated by Dlx5 but is independent of Runx2. Biochem. Biophys. Res. Commun. **309:** 689–694.

110. OGATA, T., J.M. WOZNEY, R. BENEZRA & M. NODA. 1993. Bone morphogenetic protein 2 transiently enhances expression of a gene, Id (inhibitor of differentiation), encoding a helix-loop-helix molecule in osteoblast-like cells. Proc. Natl. Acad. Sci. USA **90:** 9219–9222.

111. HOLLNAGEL, A., V. OEHLMANN, J. HEYMER, *et al.* 1999. Id genes are direct targets of bone morphogenetic protein induction in embryonic stem cells. J. Biol. Chem. **274:** 19838–19845.

112. KORCHYNSKYI, O. & P. TEN DIJKE. 2002. Identification and functional characterization of distinct critically important bone morphogenetic protein-specific response elements in the Id1 promoter. J. Biol. Chem. **277:** 4883–4891.

113. LOPEZ-ROVIRA, T., E. CHALAUX, J. MASSAGUE, et al. 2002. Direct binding of Smad1 and Smad4 to two distinct motifs mediates bone morphogenetic protein-specific transcriptional activation of Id1 gene. J. Biol. Chem. **277:** 3176–3185.

114. JEN, Y., H. WEINTRAUB & R. BENEZRA. 1992. Overexpression of Id protein inhibits the muscle differentiation program: *in vivo* association of Id with E2A proteins. Genes Dev. **6:** 1466–1479.

115. MOLDES, M., F. LASNIER, B. FEVE, et al. 1997. Id3 prevents differentiation of preadipose cells. Mol. Cell Biol. **17:** 1796–1804.

116. PEREIRA, R.C., S. RYDZIEL & E. CANALIS. 2000. Bone morphogenetic protein-4 regulates its own expression in cultured osteoblasts. J. Cell. Physiol. **182:** 239–246.

117. YOKOUCHI, Y., J. SAKIYAMA, T. KAMEDA, et al. 1996. BMP-2/-4 mediate programmed cell death in chicken limb buds. Development **122:** 3725–3734.

118. ZOU, H. & L. NISWANDER. 1996. Requirement for BMP signaling in interdigital apoptosis and scale formation. Science **272:** 738–741.

119. SPINELLA-JAEGLE, S., S. ROMAN-ROMAN, C. FAUCHEU, et al. 2001. Opposite effects of bone morphogenetic protein-2 and transforming growth factor-beta1 on osteoblast differentiation. Bone **29:** 323–330.

120. KATAGIRI, T., A. YAMAGUCHI, M. KOMAKI, et al. 1994. Bone morphogenetic protein-2 converts the differentiation pathway of C2C12 myoblasts into the osteoblast lineage. J. Cell Biol. **127:** 1755–1766.

121. CHOI, K.Y., H.J. KIM, M.H. LEE, et al. 2005. Runx2 regulates FGF2-induced Bmp2 expression during cranial bone development. Dev. Dyn. **233:** 115–121.

122. WARREN, S.M., L.J. BRUNET, R.M. HARLAND, et al. 2003. The BMP antagonist noggin regulates cranial suture fusion. Nature **422:** 625–629.

123. CANALIS, E. & B. GABBITAS. 1994. Bone morphogenetic protein 2 increases insulin-like growth factor I and II transcripts and polypeptide levels in bone cell cultures. J. Bone Miner. Res. **9:** 1999–2005.

124. GABBITAS, B. & E. CANALIS. 1995. Bone morphogenetic protein-2 inhibits the synthesis of insulin-like growth factor-binding protein-5 in bone cell cultures. Endocrinology **136:** 2397–2403.

125. KNUTSEN, R., Y. HONDA, D.D. STRONG, et al. 1995. Regulation of insulin-like growth factor system components by osteogenic protein-1 in human bone cells. Endocrinology **136:** 857–865.

126. GHOSH-CHOUDHURY, N., S.L. ABBOUD, R. NISHIMURA, et al. 2002. Requirement of BMP-2-induced phosphatidylinositol 3-kinase and Akt serine/threonine kinase in osteoblast differentiation and Smad-dependent BMP-2 gene transcription. J. Biol. Chem. **277:** 33361–33368.

127. OSYCZKA, A.M. & P.S. LEBOY. 2005. Bone morphogenetic protein regulation of early osteoblast genes in human marrow stromal cells is mediated by extracellular signal-regulated kinase and phosphatidylinositol 3-kinase signaling. Endocrinology **146:** 3428–3437.

128. SEET, L.F. & W. HONG. 2001. Endofin, an endosomal FYVE domain protein. J. Biol. Chem. **276:** 42445–42454.

129. KRATCHMAROVA, I., B. BLAGOEV, M. HAACK-SORENSEN, et al. 2005. Mechanism of divergent growth factor effects in mesenchymal stem cell differentiation. Science **308:** 1472–1477.

130. DAY, T.F., X. GUO, L. GARRETT-BEAL & Y. YANG. 2005. Wnt/beta-catenin signaling in mesenchymal progenitors controls osteoblast and chondrocyte differentiation during vertebrate skeletogenesis. Dev. Cell **8:** 739–750.

131. NISHITA, M., M.K. HASHIMOTO, S. OGATA, *et al.* 2000. Interaction between Wnt and TGF-beta signalling pathways during formation of Spemann's organizer. Nature **403:** 781–785.

132. GONG, Y., R.B. SLEE, N. FUKAI, *et al.* 2001. LDL receptor-related protein 5 (LRP5) affects bone accrual and eye development. Cell **107:** 513–523.

133. BOYDEN, L.M., J. MAO, J. BELSKY, *et al.* 2002. High bone density due to a mutation in LDL-receptor-related protein 5. N. Engl. J. Med. **346:** 1513–1521.

134. GAUR, T., C.J. LENGNER, H. HOVHANNISYAN, *et al.* 2005. Canonical WNT signaling promotes osteogenesis by directly stimulating RUNX2 gene expression. J. Biol. Chem. **280:** 33132–33140.

135. MBALAVIELE, G., S. SHEIKH, J.P. STAINS, *et al.* 2005. Beta-catenin and BMP-2 synergize to promote osteoblast differentiation and new bone formation. J. Cell. Biochem. **94:** 403–418.

136. AUBIN, J., A. DAVY & P. SORIANO. 2004. *In vivo* convergence of BMP and MAPK signaling pathways: impact of differential Smad1 phosphorylation on development and homeostasis. Genes Dev. **18:** 1482–1494.

137. ATTISANO, L. & J.L. WRANA. 2002. Signal transduction by the TGF-beta superfamily. Science **296:** 1646–1647.

138. WESTON, C.R., D.G. LAMBRIGHT & R.J. DAVIS. 2002. Signal transduction. MAP kinase signaling specificity. Science **296:** 2345–2347.

139. LAI, C.F. & S.L. CHENG. 2002. Signal transductions induced by bone morphogenetic protein-2 and transforming growth factor-beta in normal human osteoblastic cells. J. Biol. Chem. **277:** 15514–15522.

140. KRETZSCHMAR, M., J. DOODY & J. MASSAGUE. 1997. Opposing BMP and EGF signalling pathways converge on the TGF-beta family mediator Smad1. Nature **389:** 618–622.

141. MASSAGUE, J. 2003. Integration of Smad and MAPK pathways: a link and a linker revisited. Genes Dev. **17:** 2993–2997.

142. SUZAWA, M., Y. TAMURA, S. FUKUMOTO, *et al.* 2002. Stimulation of Smad1 transcriptional activity by Ras-extracellular signal-regulated kinase pathway: a possible mechanism for collagen-dependent osteoblastic differentiation. J. Bone Miner. Res. **17:** 240–248.

Nephroblastoma Overexpressed (Nov) Is a Novel Bone Morphogenetic Protein Antagonist

ERNESTO CANALIS[a,b]

[a]Department of Research, Saint Francis Hospital and Medical Center, Hartford, Connecticut, USA

[b]The University of Connecticut School of Medicine, Farmington, Connecticut, USA

ABSTRACT: Nephroblastoma overexpressed (Nov), a member of the CCN family of proteins, is expressed by osteoblasts and its transcription is regulated by transforming growth factor (TGF)-β and bone morphogenetic proteins (BMP). CCN proteins can interact with TGF-β, BMPs, and Wnt. We explored the function of Nov in skeletal cells, *in vitro* and *in vivo*. Constitutive overexpression of Nov in cells of the osteoblastic lineage impaired osteoblastic differentiation, opposed the biological effects of BMP-2 and Wnt 3 and impaired BMP-2 and Wnt signaling, indicating that Nov has BMP-2 antagonistic activity. Transgenic mice overexpressing Nov under the control of the osteocalcin promoter exhibited osteopenia secondary to decreased bone formation, confirming the effects *in vitro*. GST pulldown experiments demonstrated direct Nov–BMP interactions. In conclusion, Nov has BMP antagonistic properties and inhibits osteoblastogenesis and osteoblastic function.

KEYWORDS: bone morphogenetic proteins (BMP); BMP antagonists; CCN proteins; nephroblastoma overexpressed; notch; osteoblastogenesis; Wnt

The number and function of cells present in the bone microenvironment determine skeletal homeostasis and are regulated by systemic hormones, growth factors, and local signals.[1,2] Bone morphogenetic proteins (BMP) are unique since they induce the differentiation of mesenchymal cells toward cells of the osteoblastic lineage, and promote osteoblastic activity.[3,4] BMPs play a fundamental role in endochondral bone formation and chondrogenesis, and stimulate chondrocyte maturation and function.[5–7] The genesis and differentiation of osteoblasts and of osteoclasts are coordinated events, and BMPs also play direct

Address for correspondence: Ernesto Canalis, M.D., Department of Research, Saint Francis Hospital and Medical Center, 114 Woodland Street, Hartford, CT 06105-1299. Voice: 860-714-4068; fax: 860-714-8053.

ecanalis@stfranciscare.org

Ann. N.Y. Acad. Sci. 1116: 50–58 (2007). © 2007 New York Academy of Sciences.
doi: 10.1196/annals.1402.055

and indirect roles in osteoclastogenesis. BMPs interact with type IA or activin receptor-like kinase (ALK)-3 and type IB or ALK-6, and with BMP type II receptor.[8] Upon ligand binding and activation of the type I receptor, dimers of the type I and type II receptor initiate a signal transduction cascade activating the mothers against decapentaplegic (Smad) or the mitogen-activated protein (MAP) kinase signaling pathways.[9,10]

Whereas BMPs play a central role in the regulation of osteoblastogenesis and endochondral bone formation, BMPs in excess can be detrimental.[11,12] Consequently, BMP activity needs to be controlled. The effects of BMPs are regulated by a large group of secreted polypeptides that bind and limit BMP action, as well as by intracellular proteins that regulate BMP signaling. Extracellular BMP antagonists prevent BMP signaling by binding BMPs, precluding their binding to cell surface receptors (TABLE 1).[2] Extracellular BMP antagonists are synthesized by osteoblasts, and they temper BMP activity in the bone microenvironment. The cellular and developmental expression of an antagonist and its interactions with a specific BMP determine the specific functions of the antagonist. Often BMP antagonists oppose the activity of Wnt, another critical signal that regulates osteoblastogenesis. Wnt activity, like that of BMPs, is controlled by intracellular and extracellular antagonists. It is of interest that the synthesis of various BMP antagonists by the osteoblast is induced by BMPs, suggesting the existence of a protective mechanism to prevent skeletal cells from excessive exposure to BMPs.[13,14] The importance of BMP and Wnt antagonists and the need for a tight regulation of the levels and activity of BMPs and Wnt in the bone microenvironment are confirmed by a variety of experimental and clinical observations. Homozygous null mutations of the BMP antagonist *noggin* result in serious developmental and skeletal abnormalities and mutations of *sost*, a gene that encodes the Wnt antagonist sclerostin, and cause sclerosteosis and Van Buchen disease, both characterized by increased bone mass/osteopetrosis.[15–17]

Members of the CCN family of cysteine-rich (CR) secreted proteins include cysteine-rich 61 (Cyr 61), connective tissue growth factor (CTGF),

---

TABLE 1. Extracellular BMP antagonists

Noggin, Follistatin, and Follistatin related
Chordin family (related to CCN)
- Chordin, Chordin-like/neurin
- CR-rich motor neuron (CRIM1), BMPER
- Kielin (Xenopus), Crossveinless (Drosophila)
- Amnionless, Nel

Twisted gastrulation (related to CCN)
Dan/Cerberus family
- Dan, Cerberus, Cer1, WISE, Coco, Gremlin/drm
- Protein related to dan/cerb, Caronte, Dante

Sclerostin

nephroblastoma overexpressed (Nov), and Wnt-inducible secreted proteins (WISP) 1, 2, and 3 (TABLE 2).[18–20] CCN proteins are highly conserved and share four distinct structural modules, an insulin-like growth factor (IGF)-binding domain, a von Willebrand type C domain containing the (CR) domain, a thrombospondin-1 domain, and a C-terminal domain, the latter absent in WISP 2 (FIG. 1).[18–20] The function of each module is not clearly established, although the CR and C-terminal domains are responsible for protein–protein interactions, and the CR domain also for protein folding.[21] CCN proteins are related to certain BMP antagonists, such as twisted gastrulation (Tsg) and chordin, and can have important interactions with regulators of osteoblast cell growth and differentiation (TABLE 1).[22] CTGF binds to BMP-2 and -4 through its CR domain and binds to Wnt coreceptors through its C-terminal domain and interacts with transforming growth factor (TGF)-β.[23,24] Cyr 61 regulates Wnt signaling.[25] The structural similarities between CCN proteins and BMP antagonists of the Tsg and chordin families indicate a functional relationship between CCN peptides and more classic extracellular BMP antagonists. CCN proteins have important functions in cell proliferation and differentiation. Cyr 61 and CTGF play a role in cell adhesion, angiogenesis, and chondrogenesis, and in the development of the embryonic skeleton.[26,27] In addition, CTGF plays a pathogenetic role in osteolytic metastasis of breast cancer.[28] Although CCN proteins share sequence homology, their gene deletions result in distinct phenotypes arguing for nonredundant functions of CCN proteins *in vivo*. WISP 1 enhances the effect of BMP on osteogenesis, and WISP 3 mutations in humans lead to pseudorheumatoid dysplasia.[29,30] These observations substantiate a role of CCN proteins in skeletal cell function with relevance to clinical syndromes.

Nov was identified as aberrantly expressed gene in avian nephroblastoma induced by myeloblastosis-associated virus.[31] It shares 50% sequence homology with Cyr 61 and CTGF, and it is structurally related to the BMP antagonist Tsg. Nov is expressed in a variety of tissues, including bone and cartilage. Nov mRNA and protein levels are downregulated by TGF-β and BMP-2 in osteoblasts, both acting by transcriptional mechanisms.[32] Nov acts as a ligand of integrin receptors, and as a consequence it mediates fibroblast and endothelial cell adhesion and chemotaxis.[33,34] Nov is expressed in hypertrophic cartilage, has angiogenic properties, and enhances TGF-β2 signaling in chondrocytes, but not in osteoblasts.[35] Nov also interacts with the extracellular domain of Notch, leading to the activation of this transmembrane receptor

TABLE 2. CCN proteins

Cyr 61
CTGF
Nov
WISP 1, 2, and 3

FIGURE 1. Structural domains of CCN proteins: IGFB = IGF binding; VWC = von Willebrand or CR = cysteine rich; TSP-1 = thrombospondin; C-T = C-terminal.

and inhibition of myoblast cell differentiation.[36] Further observations support a role of Nov in osteoblastic function, since Nov co-localizes and interacts with Connexin 43, a factor important in cell–cell communications, skeletal development, and osteoblast function.[37–39] The effects of Nov on osteoblastic differentiation and function were recently studied by this laboratory. For this purpose, we examined the effect of Nov overexpression and inactivation on skeletal cells *in vivo* and *in vitro*.

To determine the actions of Nov in osteoblasts, we transduced stromal ST-2 and osteoblastic MC3T3 cells with a pLPCX retroviral vector, which uses the cytomegalovirus promoter to direct *nov* expression. Nov overexpression in ST-2 cells inhibited osteoblastogenesis, impaired mineralized nodule formation, and reduced the effect of BMP on alkaline phosphatase activity, and osteocalcin mRNA expression.[40] To understand possible mechanisms involved in the suppression of osteoblastogenesis by Nov overexpression, we examined whether it modified BMP or Wnt signaling. Nov overexpression impaired the effect of BMP-2 on Smad 1/5/8 phosphorylation and on the transactivation of the BMP/Smad-dependent 12xSBE-Oc-pGL3 reporter construct, containing 12 repeats of Smad-binding sequences directing the expression of luciferase in the context of the osteocalcin minimal promoter. These observations confirm that Nov decreases BMP signaling and activity. Nov increased extracellular receptor kinase (ERK) phosphorylation in the absence and presence of BMP-2. The activation of ERK1/2 could phosphorylate Smad, at nonactivating sites, and reduce the nuclear accumulation of phospho Smads and as a consequence decrease the activity of BMP.[41,42] To test whether activation of ERK by Nov played a role in its BMP inhibitory activity, Nov was tested in the presence and absence of ERK inhibitors. ERK inhibitors did not reverse the effect of Nov on BMP/Smad signaling, suggesting that the inhibition of BMP/Smad signaling by Nov was independent of ERK activation.

In addition to the effects on BMP signaling, Nov overexpression decreased the effect of Wnt 3 on alkaline phosphatase activity, the levels of cytoplasmic β-catenin, and the transactivation of the Wnt/β-catenin-dependent 16xTCF-Luc construct, where 16 repeats of T cell-specific factor 4/lymphoid enhancer-binding factor 1 binding sites direct luciferase expression. These observations indicate that Nov has Wnt/β-catenin antagonistic activity, in addition to its effects on BMP signaling. This was confirmed by demonstrating that Nov decreased the effect of Wnt 3a on the mRNA levels of the Wnt-dependent gene

WISP 1 by 35 to 55%. Nov overexpression also inhibited the terminal differentiation of the more mature MC3T3 cells. In these cells, Nov opposed the effect of BMP-2 on alkaline phosphatase activity, mineralized nodule formation, and the transactivation of the 12xSBE-Oc-pGL3 reporter.

To determine whether Nov had direct interactions with BMP-2 or Wnt, constructs expressing glutathione-s-transferase (GST)-Nov fusion protein were created, and GST pulldown experiments were performed. These demonstrated direct Nov-BMP-2/4 interactions, but not Nov-Wnt 3a or –low-density lipoprotein receptor-related protein (LRP)-6 interactions. The inhibitory effect of Nov on Wnt signaling may be secondary to the inhibition of BMP or to interactions between Nov and the canonical Wnt/β-catenin pathway that are beyond receptor ligand activation. Other BMP antagonists, such as noggin and gremlin, were shown to inhibit BMP as well as Wnt signaling.[43]

Although Nov is a secreted protein, recent observations have demonstrated that Nov accumulates intracellularly as well as in the extracellular matrix.[44] These findings suggest potential mechanisms of Nov action that may be beyond direct Nov/BMP interactions. Addition of recombinant human Nov protein to ST-2 cells cultures mimicked the inhibitory effects on BMP-2 signaling and activity observed in cells transduced with retroviral vectors overexpressing Nov. However, the intracellular accumulation of Nov may lead to additional interactions between Nov and intracellular proteins.

Since Nov was shown to interact with Notch in myogenic cells, Nov/Notch interactions that could explain the effects observed on osteoblastogenesis, BMP, and Wnt signaling were tested.[36] Notch 1 to 4 are transmembrane receptors that regulate cell fate.[45] In osteoblasts, Notch acts as a Wnt antagonist and the effects of Nov on Wnt signaling mimic those we found in ST-2 cells overexpressing the active Notch intracellular domain.[45] Consequently, the contribution of Notch to the inhibitory effects of Nov on Wnt signaling and activity was explored. We determined that Notch impairs osteoblastogenesis, and activates the canonical CBF1/RBP-Jκ/Suppressor of Hairless/Lag 1 signaling pathway, resulting in the transcription of *hes-1*.[45] Contrary to the results reported in myoblasts, Nov did not enhance the transactivation of a 12xCSL-Luc construct, where 12 repeats of CSL binding sites direct luciferase expression, or of a Hes-1 promoter transfected into ST-2 cell lines, and did not increase Hes-1 mRNA levels. Moreover, Nov inhibited the effect of Notch on the transactivation of the 12xCSL-Luc reporter and Hes-1 promoter constructs, indicating that Nov does not enhance, but instead decreases Notch signaling in ST-2 cells. Furthermore, Notch inactivation, using a γ-secretase II inhibitor that precludes release of the intracellular domain of Notch, did not rescue the inhibitory effect of Nov on Wnt 3 signaling. These observations exclude the involvement of Notch activation by Nov in cells of the osteoblastic lineage. Since BMP and Wnt signaling are coordinated events, it is possible that the effects of Nov on Wnt signaling are a simple consequence of its binding to BMP-2/4 tempering its activity.

To define the function of Nov *in vivo*, we created transgenic mice over-expressing Nov under the control of the human osteocalcin promoter.[40] Overexpression of Nov mRNA was documented in calvarial extracts by real time reverse transcription polymerase chain reaction. Calvarial extracts confirmed Nov mRNA levels that were 25- to 250-fold higher in the transgenic line when compared to controls. Four-week-old heterozygous Nov transgenic mice were compared to wild-type controls of identical age, sex, and genetic background. Nov transgenics had normal weight and exhibited a 35% decrease in femoral trabecular bone volume and trabecular number with a similar decrease in osteoblast number per area, but not per surface, indicating no actual change in osteoblast number. Dynamic histomorphometry confirmed a decrease in mineral apposition and bone formation rates, indicating that Nov inhibits osteoblastic function *in vivo*, as it does *in vitro*. The phenotype was not sex-dependent and female and male transgenics exhibited an identical osteopenic phenotype secondary to decreased bone formation. The lack of an effect of Nov on osteoblast number per surface may be a reflection of the osteocalcin promoter used, which is active in mature, nondividing, differentiated osteoblasts. Osteoclast number and eroded surface were not increased, establishing that enhanced bone resorption did not contribute to the phenotype of this particular line. A second line expressing Nov mRNA levels in calvariae that were 1,000- to 9,000-fold higher than in wild types exhibited osteopenia secondary to enhanced bone resorption. However, caution needs to be exerted when interpreting data from transgenics with extreme degrees of gene expression. Stromal cells from transgenics had a decreased response to BMP-2 on phospho Smad 1/5/8 and to Wnt 3a on β-catenin levels, confirming the BMP/Wnt antagonistic activity of Nov.

In conclusion, Nov is a novel BMP antagonist and inhibits the differentiation and function of cells of the osteoblastic lineage. Abnormalities in the BMP/Wnt antagonist axis are important in the pathogenesis of selected bone disorders, and modification of the synthesis or activity of these antagonists may offer new therapeutic alternatives for the management of osteoporosis.

ACKNOWLEDGMENTS

This work was supported by Grant AR21707 from the National Institute of Arthritis and Musculoskeletal and Skin Diseases, and Grant DK45227 from the National Institute of Diabetes and Digestive and Kidney Diseases.

REFERENCES

1. BIANCO, P. & R.P. GEHRON. 2000. Marrow stromal stem cells. J. Clin. Invest. **105:** 1663–1668.
2. CANALIS, E., A.N. ECONOMIDES & E. GAZZERRO. 2003. Bone morphogenetic proteins, their antagonists, and the skeleton. Endocr. Rev. **24:** 218–235.

3. GITELMAN, S.E., M. KIRK, J.Q. YE, *et al.* 1995. Vgr-1/BMP-6 induces osteoblastic differentiation of pluripotential mesenchymal cells. Cell Growth Differ. **6:** 827–836.
4. YAMAGUCHI, A., T. ISHIZUYA, N. KINTOU, *et al.* 1996. Effects of BMP-2, BMP-4, and BMP-6 on osteoblastic differentiation of bone marrow-derived stromal cell lines, ST2 and MC3T3-G2/PA6. Biochem. Biophys. Res. Commun. **220:** 366–371.
5. DE LUCA, F., K.M. BARNES, J.A. UYEDA, *et al.* 2001. Regulation of growth plate chondrogenesis by bone morphogenetic protein-2. Endocrinology **142:** 430–436.
6. GRIMSRUD, C.D., P.R. ROMANO, M. D'SOUZA, *et al.* 2001. BMP signaling stimulates chondrocyte maturation and the expression of Indian hedgehog. J. Orthop. Res. **19:** 18–25.
7. LEBOY, P., G. GRASSO-KNIGHT, M. D'ANGELO, *et al.* 2001. Smad-Runx interactions during chondrocyte maturation. J. Bone Joint Surg. Am. **83-A**(Suppl 1): S15–S22.
8. YAMASHITA, H., P. TEN DIJKE, C.H. HELDIN, *et al.* 1996. Bone morphogenetic protein receptors. Bone **19:** 569–574.
9. DERYNCK, R., Y. ZHANG & X.H. FENG. 1998. Smads: transcriptional activators of TGF-beta responses. Cell **95:** 737–740.
10. NOHE, A., S. HASSEL, M. EHRLICH, *et al.* 2002. The mode of bone morphogenetic protein (BMP) receptor oligomerization determines different BMP-2 signaling pathways. J. Biol. Chem. **277:** 5330–5338.
11. BRUNET, L.J., J.A. MCMAHON, A.P. MCMAHON, *et al.* 1998. Noggin, cartilage morphogenesis, and joint formation in the mammalian skeleton. Science **280:** 1455–1457.
12. SHAFRITZ, A.B., E.M. SHORE, F.H. GANNON, *et al.* 1996. Overexpression of an osteogenic morphogen in fibrodysplasia ossificans progressiva. N. Engl. J. Med. **335:** 555–561.
13. GAZZERRO, E., V. GANGJI & E. CANALIS. 1998. Bone morphogenetic proteins induce the expression of noggin, which limits their activity in cultured rat osteoblasts. J. Clin. Invest. **102:** 2106–2114.
14. PEREIRA, R.C., A.N. ECONOMIDES & E. CANALIS. 2000. Bone morphogenetic proteins induce gremlin, a protein that limits their activity in osteoblasts. Endocrinology **141:** 4558–4563.
15. BRUNKOW, M.E., J.C. GARDNER, N.J. VAN, *et al.* 2001. Bone dysplasia sclerosteosis results from loss of the SOST gene product, a novel cystine knot-containing protein. Am. J. Hum. Genet. **68:** 577–589.
16. LOOTS, G.G., M. KNEISSEL, H. KELLER, *et al.* 2005. Genomic deletion of a long-range bone enhancer misregulates sclerostin in Van Buchem disease. Genome Res. **15:** 928–935.
17. STAEHLING-HAMPTON, K., S. PROLL, B.W. PAEPER, *et al.* 2002. A 52-kb deletion in the SOST-MEOX1 intergenic region on 17q12-q21 is associated with van Buchem disease in the Dutch population. Am. J. Med. Genet. **110:** 144–152.
18. BRIGSTOCK, D.R.. 2003. The CCN family: a new stimulus package. J. Endocrinol. **178:** 169–175.
19. BRIGSTOCK, D.R., R. GOLDSCHMEDING, K.I. KATSUBE, *et al.* 2003. Proposal for a unified CCN nomenclature. Mol. Pathol. **56:** 127–128.
20. GROTENDORST, G.R., L.F. LAU & B. PERBAL. 2000. CCN proteins are distinct from and should not be considered members of the insulin-like growth factor-binding protein superfamily. Endocrinology **141:** 2254–2256.

21. ISAACS, N.W. 1995. Cystine knots. Curr. Opin. Struct. Biol. **5**: 391–395.
22. GARCIA, A.J., C. COFFINIER, J. LARRAIN, *et al.* 2002. Chordin-like CR domains and the regulation of evolutionarily conserved extracellular signaling systems. Gene **287**: 39–47.
23. ABREU, J.G., N.I. KETPURA, B. REVERSADE, *et al.* 2002. Connective-tissue growth factor (CTGF) modulates cell signalling by BMP and TGF-beta. Nat. Cell Biol. **4**: 599–604.
24. MERCURIO, S., B. LATINKIC, N. ITASAKI, *et al.* 2004. Connective-tissue growth factor modulates WNT signalling and interacts with the WNT receptor complex. Development **131**: 2137–2147.
25. LATINKIC, B.V., S. MERCURIO, B. BENNETT, *et al.* 2003. Xenopus Cyr61 regulates gastrulation movements and modulates Wnt signalling. Development **130**: 2429–2441.
26. IVKOVIC, S., B.S. YOON, S.N. POPOFF, *et al.* 2003. Connective tissue growth factor coordinates chondrogenesis and angiogenesis during skeletal development. Development **130**: 2779–2791.
27. MO, F.E., A.G. MUNTEAN, C.C. CHEN, *et al.* 2002. CYR61 (CCN1) is essential for placental development and vascular integrity. Mol. Cell Biol. **22**: 8709–8720.
28. SHIMO, T., S. KUBOTA, N. YOSHIOKA, *et al.* 2006. Pathogenic role of connective tissue growth factor (CTGF/CCN2) in osteolytic metastasis of breast cancer. J. Bone Miner. Res. **21**: 1045–1059.
29. FRENCH, D.M., R.J. KAUL, A.L. D'SOUZA, *et al.* 2004. WISP-1 is an osteoblastic regulator expressed during skeletal development and fracture repair. Am. J. Pathol. **165**: 855–867.
30. KUTZ, W.E., Y. GONG & M.L. WARMAN. 2005. WISP3, the gene responsible for the human skeletal disease progressive pseudorheumatoid dysplasia, is not essential for skeletal function in mice. Mol. Cell Biol. **25**: 414–421.
31. JOLIOT, V., C. MARTINERIE, G. DAMBRINE, *et al.* 1992. Proviral rearrangements and overexpression of a new cellular gene (nov) in myeloblastosis-associated virus type 1-induced nephroblastomas. Mol. Cell Biol. **12**: 10–21.
32. PARISI, M.S., E. GAZZERRO, S. RYDZIEL, *et al.* 2006. Expression and regulation of CCN genes in murine osteoblasts. Bone **38**: 671–677.
33. LIN, C.G., S.J. LEU, N. CHEN, *et al.* 2003. CCN3 (NOV) is a novel angiogenic regulator of the CCN protein family. J. Biol. Chem. **278**: 24200–24208.
34. LIN, C.G., C.C. CHEN, S.J. LEU, *et al.* 2005. Integrin-dependent functions of the angiogenic inducer NOV (CCN3): implication in wound healing. J. Biol. Chem. **280**: 8229–8237.
35. LAFONT, J., C. JACQUES, G.L. DREAU, *et al.* 2005. New target genes for NOV/CCN3 in chondrocytes: TGF-beta2 and type X collagen. J. Bone Miner. Res. **20**: 2213–2223.
36. SAKAMOTO, K., S. YAMAGUCHI, R. ANDO, *et al.* 2002. The nephroblastoma overexpressed gene (NOV/ccn3) protein associates with Notch1 extracellular domain and inhibits myoblast differentiation via Notch signaling pathway. J. Biol. Chem. **277**: 29399–29405.
37. CIVITELLI, R., E.C. BEYER, P.M. WARLOW, *et al.* 1993. Connexin43 mediates direct intercellular communication in human osteoblastic cell networks. J. Clin. Invest. **91**: 1888–1896.
38. FU, C.T., J.F. BECHBERGER, M.A. OZOG, *et al.* 2004. CCN3 (NOV) interacts with connexin43 in C6 glioma cells: possible mechanism of connexin-mediated growth suppression. J. Biol. Chem. **279**: 36943–36950.

39. GELLHAUS, A., X. DONG, S. PROPSON, *et al.* 2004. Connexin43 interacts with NOV: a possible mechanism for negative regulation of cell growth in choriocarcinoma cells. J. Biol. Chem. **279:** 36931–36942.
40. RYDZIEL, S., L. STADMEYER, S. ZANOTTI, *et al.* 2007. Nephroblastoma overexpressed (NOV) inhibits osteoblastogenesis and causes osteopenia. J. Biol. Chem. **282:** 19762–19772.
41. KRETZSCHMAR, M., J. DOODY & J. MASSAGUE. 1997. Opposing BMP and EGF signalling pathways converge on the TGF-beta family mediator Smad1. Nature **389:** 618–622.
42. OSYCZKA, A.M. & P.S. LEBOY. 2005. Bone morphogenetic protein regulation of early osteoblast genes in human marrow stromal cells is mediated by extracellular signal-regulated kinase and phosphatidylinositol 3-kinase signaling. Endocrinology **146:** 3428–3437.
43. GAZZERRO, E., R.C. PEREIRA, V. JORGETTI, *et al.* 2005. Skeletal overexpression of gremlin impairs bone formation and causes osteopenia. Endocrinology **146:** 655–665.
44. GUPTA, R., D. HONG, F. IBORRA, *et al.* 2007. NOV (CCN3) functions as a regulator of human hematopoietic stem or progenitor cells. Science **316:** 590–593.
45. DEREGOWSKI, V., E. GAZZERRO, L. PRIEST, *et al.* 2006. Notch 1 overexpression inhibits osteoblastogenesis by suppressing Wnt/beta-catenin but not bone morphogenetic protein signaling. J. Biol. Chem. **281:** 6203–6210.

The Role of the Perichondrium in Fetal Bone Development

HENRY M. KRONENBERG

Endocrine Unit, Massachusetts General Hospital and Harvard Medical School, Boston, Massachusetts, USA

ABSTRACT: **Most cells in mesenchymal condensations that form endochondral bone become chondrocytes; cells at the outer edges of the condensations, however, become perichondrial cells with distinct properties and functions. Some perichondrial cells become osteoblasts that populate both the future cortical and trabecular bone; others probably become chondrocytes. Perichondrial cells both send signals to the underlying growth cartilage and receive signals from the cartilage. Here I illustrate briefly examples of the complicated interactions between the perichondrium and the underlying growth cartilage.**

KEYWORDS: **perichondrium; growth plate; cartilage; parathyroid hormone-related protein; Indian hedgehog**

INTRODUCTION

Almost all bones form by the process of endochondral bone formation.[1] Mesenchymal cells first form condensations in which these cells move close together, excluding blood vessels. Cells in the center of the condensation differentiate into chondrocytes and proliferate. Cells surrounding the chondrocytes become perichondrial cells. These perichondrial cells have two broad functions: first, they signal to underlying chondrocytes and receive signals from these cells in turn; and second, they provide cells that become osteoblasts and chondrocytes in a carefully orchestrated fashion. In this article, I will review current understanding of these functions of the perichondrium, using illustrative examples.

ENDOCHONDRAL BONE FORMATION

Chondrocytes in the center of the bone anlagen proliferate and secrete a matrix rich in collagen, type II, and aggrecan. When the mass of chondrocytes

Address for correspondence: Henry M. Kronenberg, Endocrine Unit, Thier 1101, Massachusetts General Hospital, 50 Blossom St., Boston, MA 02114. Voice: 617-726-3966; fax: 617-726-7543.
hkronenberg@partners.org

Ann. N.Y. Acad. Sci. 1116: 59–64 (2007). © 2007 New York Academy of Sciences.
doi: 10.1196/annals.1402.059

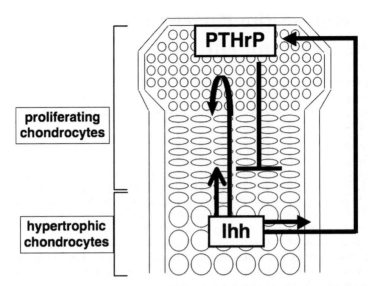

FIGURE 1. Ihh and PTHrP coordinate growth cartilage regulation. PTHrP made at the end of the bone by perichondrial cells and chondrocytes acts on chondrocytes to keep them proliferating. Ihh acts on chondrocytes to increase their rate of proliferation, to hasten the conversion of round proliferating chondrocytes to flat proliferating chondrocytes, and to increase the production of PTHrP. Ihh also acts on perichondrial cells to direct them along the osteoblast lineage.

reaches a characteristic size, different for each bone, chondrocytes in the center of the anlagen stop proliferating and then hypertrophy. These postmitotic, hypertrophic chondrocytes secrete a matrix rich in collagen, type X, and mineralize the matrix that surrounds them. They also signal to the perichondrial region to trigger the invasion of blood vessels and early cells of the osteoblast lineage. The still proliferating chondrocytes nearer the ends of the bones expand the size of the anlagen. Chondrocytes nearest to the hypertrophic zone change their shape and become flat and form orderly columns. The asymmetry of the columns (more cylindrical than spherical) helps sculpt the unique shapes of the individual bones. Perichondrial cells immediately adjacent to the hypertrophic chondrocytes become osteoblasts that both stay on the periphery of the bone and establish the beginning of the cortical shell of the bone and also travel with blood vessels into the center of the anlagen to form the beginning of the primary spongiosa, the precursor of trabecular bone.

PTHRP AND INDIAN HEDGEHOG (IHH): EXAMPLE OF PERICHONDRIAL SIGNALING

Perichondrial cells secrete a series of molecules that control the process just described. Cells in different locations secrete differing regulatory factors. One

of the best characterized is parathyroid hormone-related protein (PTHrP) (see FIG. 1). In early stages of bone development, PTHrP is secreted from perichondrial cells and early chondrocytes near the ends of the forming bones.[2] PTHrP acts on a G protein–coupled receptor, the PTH/PTHrP receptor, synthesized at low levels on flat, proliferating chondrocytes and at high levels in very late proliferating chondrocytes and prehypertrophic chondrocytes. PTHrP acts to keep the chondrocytes in the cell cycle. Chondrocytes only leave the cell cycle when they are sufficiently distant from the source of PTHrP that the PTHrP signal is weak. Then, the prehypertrophic chondrocytes synthesize Ihh, a secreted protein with multiple important functions in bone development. Ihh acts on chondrocytes to increase their proliferation.[3] Ihh also increases the rate at which round proliferating chondrocytes convert to flat, proliferating chondrocytes.[4] Ihh also provides important signals to the perichondrium. Ihh is absolutely required for the synthesis of PTHrP by the perichondrium and chondrocytes at the ends of bone.[5] Ihh also signals to perichondrial cells immediately adjacent to the prehypertrophic and hypertrophic cells, thereby directing them away from the chondrocyte pathway and onto the osteoblast differentiation pathway.[6] Thus, the perichondrial response to Ihh differs, depending on the location of the perichondrial cells; the molecular differences underlying these differing responses is unknown.

The interplay between PTHrP and Ihh regulates the height of the proliferative columns: Ihh increases the movement of round chondrocytes into the flat chondrocyte pool at the top of the columns, and PTHrP keeps flat chondrocytes proliferating and thus delays the exit of flat chondrocytes from the proliferative pool at the bottom of the columns.[7,8] Increased Ihh synthesis,[4] or synthesis of Ihh closer to the sites of PTHrP production[9] increase the amount of PTHrP produced. Though all chondrocytes in early bones apparently have the capacity to respond to Ihh, for example, with an increase in synthesis of patched, the Ihh receptor, only perichondrial cells and chondrocytes at the ends of bones can respond with the synthesis of PTHrP. The spatial separation of the sites of Ihh production and PTHrP production ensures that the conversion of round to flat chondrocytes and then prehypertrophic–hypertrophic chondrocytes occurs at a proper and well-coordinated pace. When PTHrP expression is ablated, not only is the flat layer of chondrocytes dramatically smaller, but the sites across the growth region at which flat cells become hypertrophic are more disorderly than normal, leading to an admixture of proliferating and hypertrophic cells, something that never occurs in normal bones.

The mechanism whereby higher concentrations of Ihh stimulate increased synthesis of PTHrP is not understood in any detail, but recent studies have provided some new insight. As noted above, in the absence of Ihh, no PTHrP is synthesized at all. Strikingly, when the gene encoding the transcription factor, Gli 3, is ablated along with the Ihh gene, then the resulting mice synthesize PTHrP in perichondrial cells but not in chondrocytes.[10] Gli 3 is an important mediator of hedgehog signaling in target cells. The finding that ablation of Gli 3

reverses some of the suppression of PTHrP synthesis in Ihh (-/-) mice suggests that Ihh normally works, at least in part, by suppressing the expression of Gli 3. The data also show that the detailed regulation of PTHrP expression differs in chondrocytes and in perichondral cells. Presumably, other Gli transcription factors complement the role of Gli 3 in Ihh action.

Smoothened is an important trigger of hedgehog action in hedgehog target cells. The removal of smoothened from perichondral cells, using the cre-lox approach, stopped Ihh signaling in perichondral cells. Strikingly, this manipulation stopped the conversion of perichondral cells to osteoblasts.[6] Instead, some of these perichondral cells are transformed into chondrocytes that proliferate and become hypertrophic in abnormal locations. These observations suggest that Ihh signaling influences fate decisions by perichondral cells that have the potential to become either chondrocytes or osteoblasts. When Ihh-producing chondrocytes are ectopically located higher in the growth plate than normal in chimeric mice, then the adjacent perichondral cells become osteoblasts.[11] Thus, the site of Ihh production determines the site of bone collar formation in the perichondrium. Consistent with this idea, when hedgehog protein is widely produced in the growth plate using a recombinant retrovirus to widely express hedgehog protein in the early chick limb, then the entire perichondrium converts to osteoblastic cells, even though, because of high levels of PTHrP, the underlying chondrocytes remain immature proliferating cells.[12]

MORE PERICHONDRIAL SIGNALING

The perichondrium synthesizes and secretes a variety of other paracrine factors that profoundly influence the adjacent cartilage. For example, a series of FGF family members, with the best-studied being FGF18 and FGF9 are synthesized by perichondral cells and activate the FGF receptor 3 found on chondrocytes.[13] This activation serves to both decrease the proliferation of chondrocytes late in fetal development and to delay the differentiation of proliferative chondrocytes into hypertrophic chondrocytes. Activation of the FGFR3 also inhibits the production of Ihh and antagonizes the actions of bone morphogenetic proteins (BMPs).[14] Multiple BMPs and wnts are made in the perichondrium and act on receptors on chondrocytes. Each of these signaling pathways, and certainly others, thus regulate each other to coordinate the development of endochondral bone growth.

CELL FATES IN THE PERICHONDRIUM

Recent lineage studies have shown that marked perichondral cells, when grown with chondrocytes under the renal capsule, contribute both to bone formation in the bone collar and to the invading osteoblasts of the primary

spongiosa.[15] These findings have been confirmed and extended by Maes *et al*., who have reported preliminary lineage studies showing that early cells of the osteoblast lineage can become both bone collar osteoblasts and osteoblasts in the primary spongiosa.[16] The osteoblasts in the primary spongiosa show evidence of responding to Ihh signals, with the expression of the patched gene. These osteoblasts may well be receiving direct signals from Ihh-producing chondrocytes in the growth cartilage.[17] Alternatively, it is possible that this Ihh signaling represents signals that remain in osteoblasts that first were stimulated by Ihh when they were still in the perichondrium. Further studies will be needed to establish whether Ihh stimulates directly both bone collar osteoblasts and primary spongiosa osteoblasts.

CONCLUSION

The perichondrium, thus, has a series of distinct roles in endochondral bone development. A number of paracrine factors are secreted from the perichondrium. The precise nature of those factors varies, depending on the locations of the perichondrial cells, and these sites of synthesis crucially direct the polarity of growth of the underlying cartilage. The perichondrium also responds to signals coming from the growth plate. These signals not only determine the synthetic properties of the perichondrial cells but also direct the fates of these cells into the osteoblast and chondrocyte lineage.

REFERENCES

1. KRONENBERG, H.M. 2003. Developmental regulation of the growth plate. Nature **423:** 332–336.
2. LEE, K., J.D. DEEDS & G.V. SEGRE. 1995. Expression of parathyroid hormone-related peptide and its receptor messenger ribonucleic acids during fetal development of rats. Endocrinology **136:** 453–463.
3. LONG, F., X.M. ZHANG, S. KARP, *et al*. 2001. Genetic manipulation of hedgehog signaling in the endochondral skeleton reveals a direct role in the regulation of chondrocyte proliferation. Development **128:** 5099–5108.
4. KOBAYASHI, T., D.W. SOEGIARTO, Y. YANG, *et al*. 2005. Indian hedgehog stimulates periarticular chondrocyte differentiation to regulate growth plate length independently of PTHrP. J. Clin. Invest. **115:** 1734–1742.
5. ST-JACQUES, B., M. HAMMERSCHMIDT & A.P. MCMAHON. 1999. Indian hedgehog signaling regulates proliferation and differentiation of chondrocytes and is essential for bone formation. [erratum appears in Genes Dev 1999 Oct 1: 13(19).:2617]. Genes Dev. **13:** 2072–2086.
6. LONG, F., U.I. CHUNG, S. OHBA, *et al*. 2004. Ihh signaling is directly required for the osteoblast lineage in the endochondral skeleton. Development **131:** 1309–1318.
7. KARAPLIS, A.C., A. LUZ, J. GLOWACKI, *et al*. 1994. Lethal skeletal dysplasia from targeted disruption of the parathyroid hormone-related peptide gene. Genes Dev. **8:** 277–289.

8. WEIR, E.C., W.M. PHILBRICK, M. AMLING, *et al.* 1996. Targeted overexpression of parathyroid hormone-related peptide in chondrocytes causes chondrodysplasia and delayed endochondral bone formation. Proc. Natl. Acad. Sci. USA **93:** 10240–10245.

9. CHUNG, U.I., E. SCHIPANI, A.P. MCMAHON & H.M. KRONENBERG. 2001. Indian hedgehog couples chondrogenesis to osteogenesis in endochondral bone development [see comments]. J. Clin. Investig. **107:** 295–304.

10. KOZIEL, L., M. WUELLING, S. SCHNEIDER & A. VORTKAMP. 2005. Gli3 acts as a repressor downstream of Ihh in regulating two distinct steps of chondrocyte differentiation. Development **132:** 5249–5260.

11. CHUNG, U.I., B. LANSKE, K. LEE, *et al.* 1998. The parathyroid hormone/parathyroid hormone-related peptide receptor coordinates endochondral bone development by directly controlling chondrocyte differentiation. Proc. Natl. Acad. Sci. USA **95:** 13030–13035.

12. VORTKAMP, A., K. LEE, B. LANSKE, *et al.* 1996. Regulation of rate of cartilage differentiation by Indian hedgehog and PTH-related protein. Science **273:** 613–622.

13. LIU, Z., J. XU, J.S. COLVIN & D.M. ORNITZ. 2002. Coordination of chondrogenesis and osteogenesis by fibroblast growth factor 18. Genes Dev. **16:** 859–869.

14. ORNITZ, D.M. & P.J. MARIE. 2002. FGF signaling pathways in endochondral and intramembranous bone development and human genetic disease. Genes Dev. **16:** 1446–1465.

15. COLNOT, C., C. LU, D. HU & J.A. HELMS. 2004. Distinguishing the contributions of the perichondrium, cartilage, and vascular endothelium to skeletal development. Dev. Biol. **269:** 55–69.

16. MAES, C., T. KOBAYASHI, J. PARUCH & H. KRONENBERG. 2006. Characterization of the osteblast lineage in vivo during early bone development (Abst 1005). J. Bone Mineral Res. **21**(Suppl): S3.

17. MAEDA, Y., E. NAKAMURA, M.T. NGUYEN, *et al.* 2007. Indian Hedgehog produced by postnatal chondrocytes is essential for maintaining a growth plate and trabecular bone. Proc. Natl. Acad. Sci. USA **104:** 6382–6387.

The PTHrP Functional Domain Is at the Gates of Endochondral Bones

ARTHUR E. BROADUS, CAROLYN MACICA, AND XUESONG CHEN

Endocrine Section, Department of Internal Medicine, Yale University School of Medicine, New Haven, Connecticut, USA

ABSTRACT: PTHrP gene-expression products are generally of very low abundance. The PTHrP-lacZ knockin mouse is a useful tool in this regard, identifying PTHrP expression in previously unrecognized sites and serving to score this expression in gene-regulation experiments. These sites include the periosteum and ligament/tendon insertion sites at the surface of endochondral bones, in which PTHrP appears to regulate subjacent bone cell populations. As mesenchymal condensations chondrify, PTHrP/lacZ is also expressed in epiphyseal cartilage (the chondroepiphysis), and this structure contributes PTHrP-expressing chondrocyte populations to both articular cartilage and growth-plate cartilage when these structures take shape postnatally. The Indian hedgehog-PTHrP axis is fully deployed in both of these locations and in articular cartilage appears to protect the joint space from invasion by mineralizing cells. In most of these sites PTHrP is mechanically regulated.

KEYWORDS: PTHrP; articular cartilage; chondroepiphysis; enthesis; growth plate; mechanotransduction; periosteum; synovial joint

INTRODUCTION

The low-abundance problem has been a thorn in the side of those working on PTHrP from the outset. Even in tumors associated with the syndrome of humoral hypercalcemia of malignancy (HHM), abundance is low; the original purification carried out by the Stewart laboratory was 62,000-fold, and using a quantitative RNase protection assay our group measured PTHrP mRNA to be present in an HHM-associated renal carcinoma line at one part in 100,000.[1,2] Further, both of these tumors had been chosen as relatively enriched for PTHrP by bioassays. The low abundance of PTHrP products is now well understood; PTHrP gene transcription is very tightly regulated, including feedback inhibition by some unknown product associated with PTHrP mRNA translation,

Address for correspondence: Arthur E. Broadus, M.D., Ph.D., Yale University, Endocrine Section, Department of Internal Medicine, 333 Cedar Street, New Haven, CT 06520-8020. Voice: 203-785-3966; fax: 203-737-4360.
arthur.broadus@yale.edu

Ann. N.Y. Acad. Sci. 1116: 65–81 (2007). © 2007 New York Academy of Sciences.
doi: 10.1196/annals.1402.061

65

and all PTHrP mRNAs bear multiply AU-rich instability elements in their 3′ UTRs that confer upon them a half-life measured in minutes.[3,4] Since PTHrP is principally a constitutive secretory product, secretion is a direct function of the level of ambient PTHrP mRNA in a cell, and its secretion rate is correspondingly very low.

For the past 15 years, the focus of many in the field has been on normal sites of PTHrP expression and function(s), and here too the low-abundance problem has been pervasive, more so than most in the field seem to recognize. While there exist high-abundance sites of PTHrP expression to be sure (mammary bud, hair and vibrissae follicles, and, to a lesser extent, growth cartilage), in general the normal steady-state levels of PTHrP mRNA and protein are at or beneath the limits of detection of immunohistochemistry (IH) and *in situ* hybridization histochemistry (ISHH).

In preparing the initial PTHrP-lacZ mouse manuscript for publication,[5] we reviewed carefully the IH and ISHH PTHrP literature, which was an eye-opening experience. The basic conclusions of this review were two: (1) in nonabundant sites, these techniques are basically not up to the task of localizing a product in the one part in 10,000–100,000 range, and (2) the findings reported in much of this literature are sufficiently variable and poorly substantiated as to challenge one's confidence in the literature as a whole.[5] Several examples may be useful in illustrating this point. Normal human foreskin keratinocytes in primary culture were the first normal cells shown to produce PTHrP,[6] and squamous cell carcinomas are the most frequent of all HHM-associated tumors seen clinically. These data, coupled with a number of IH reports (reviewed in Ref. 5), have lead to the widespread belief that the interfollicular epidermis is a common and relatively high-abundance site of PTHrP expression.[5,7,8] In fact, the interfollicular epidermis does not express PTHrP.[5,8] Another example involved our own IH experience in the mid 1990s, previously unpublished. Our immunological point-person was Bill Burtis, who published the initial PTHrP IRMA and region-specific assays,[9,10] and who also developed a series of region-specific, affinity-purified antibodies for IH. Many of us in the extended group used these to examine sites, such as skin,[7] mammary tissues, and the brain, with seemingly highly discrete anatomical signals that were reproducible with different antisera. In 1996, we brought on line the so-called rescued PTHrP-null mouse, which was PTHrP-sufficient in chondrocytes and PTHrP-null in all other locations.[11] We reasoned that tissues from the rescued mouse should provide a very powerful negative control for our IH studies in normal mouse tissues, and this proved to be the case to our considerable consternation, as the highly discrete signals we had seen in the normal sections were also seen in the PTHrP-null sections. No one in our extended group has used IH for PTHrP localization in the past 10 years. Lest the reader think that the authors are casting stones from a house of glass, a reference from our own group is included in the IH findings we now question.[7]

THE PTHrP-lacZ MOUSE

We had this mouse on the drawing board from 1997 onward but did not bring it on line until 2004. This is unfortunate, since the PTHrP and bone fields had long since passed the tissue-survey phases, and PTHrP is no longer seen as a fresh face. Yet the findings in this mouse have introduced a number of new areas as well as questions regarding PTHrP functions, to be described in the sections that follow.

The approach was straightforward and used a substitution strategy, in which stop codons followed by an IRES and the lacZ coding sequence were introduced into exon 3 of the murine PTHrP gene by homologous recombination, rendering the allele null as regards PTHrP expression. Per our routine, the allele was outbred onto a CD-1 background, and it is also been fully inbred into a C57 Bl/6 background. The homozygous PTHrP-lacZ mouse is functionally PTHrP-null,[12] and β gal activity is increased in this mouse because of the double-dose of lacZ gene; further, in sites subject to Indian hedgehog (Ihh) feedback, β gal activity is remarkably increased because of the additional features of Ihh-driven PTHrP overexpression as well as Ihh-induced PTHrP-independent proliferation of the target cells in question.[5] We have also bred a so-called single-allele PTHrP-lacZ knockout mouse by introducing a conventional PTHrP-null allele into the mix, creating a PTHrP[lacZ/null] genotype in which the double-dose lacZ aspect has been eliminated; this mouse can serve as a "reporter" for Ihh overexpression in those sites where this occurs.

We used a heat step to inactivate endogenous galactosidase activity.[13] Dr. Chen in our group also developed a means for partially clearing X-gal-stained anatomical specimens so that staining on bone and articular surfaces could be visualized.[5] The attractiveness and sensitivity of the lacZ system are based on the capacity of the β gal enzyme to amplify a signal by generating many molecules of the indole reaction product per molecule of enzyme, providing a combination of sensitivity and simplicity that explains, for example, why R26R reporter mice are so widely used in the field today.[14] Our own estimate of the increase in PTHrP detectability via this system as compared to IH/ISHH is 5- to 10-fold, this being about the same as that reported by another group that used the same strategy to score a low-abundance gene product.[15] A potential downside of this strategy is that the longer half-time conveyed to the system by the β gal readout renders it incapable of registering rapid changes in PTHrP gene expression in either direction. While this is clearly an inherent feature of the system, in practice we have found that β gal activity changes sufficiently rapidly as to be perfectly compatible with the kinds of regulatory experiments we have thus far carried out (see below). Based on these experiments, it is clear that the functional half-time β gal *in vivo* is well less than 24 h; we would estimate this figure at some 8–12 h. We have thus far not explored β gal activity in internal organs and tissues in any detail in the PTHrP-lacZ mouse.[5]

PTHrP/lacZ EXPRESSION IN THE PERIOSTEUM
AND ENTHESES

The scleraxis gene serves as a marker for the connective tissues that join the elements of the muscular and skeletal systems together and to each other.[16,17] These tissues include the ligaments, tendons, capsules, and connective tissues that surround individual bones and muscles. In the axial skeleton, these structures are derived from a fourth somitic compartment known as the syndetome,[16] and in the peripheral skeleton they are derived from precursors that are apparent in the early limb bud.[17] Whether the synovial joints are so derived is unknown.

The periosteum comprises two layers, an outer fibrous layer and a subjacent cambial layer; the latter contains and/or gives rise to the bone cells on the cortical surface. Collectively, the insertion sites of ligaments and tendons are referred to as entheses, and these come in several flavors. The simplest is referred to as a periosteal-muscle insertion site, which corresponds to the attachment of muscles over a broad cortical surface and in which the enthesis itself consists of little more than the two-layered periosteum.[18] So-called fibrous sites are typically associated with bulkier muscles and consist of clear-cut tendons that often insert into cortical bone at sharp angles; these anchor through multiple connective tissue layers or components, and they extend fibers (known as Sharpey's fibers) deep into the mineralized matrix of the bone cortex.[18] Fibrocartilagenous sites are the most complex of the entheses and typify major insertions into an epiphysis (e.g., the Achilles tendon); these comprise both fibrochondrocytes and connective tissue elements and function to buffer somewhat the large mechanical loads they transmit.[18]

PTHrP/lacZ is expressed to one or another extent in all of these insertion sites. In the periosteum, β gal activity is confined to the fibrous layer (FIG. 1), as it is also in the equivalent cell layer in periosteal-muscle insertions. In fibrous sites, both ligamentous and tendinous, PTHrP-lacZ is expressed in pleiomorphic connective tissue cells that lie at the immediate interface of the ligament/tendon with the bony surface or the connective tissue that comprises the so-called cortical depression if one is present (FIG. 2A–C).[5,19] The complexity of such sites is a function of the level of mechanical input a given enthesis experiences, this input being far greater for muscles than for ligaments (Chen et al. manuscript submitted for publication). Thus, the fibrous insertion of a hamstring into the proximal tibial metaphysis extends more deeply into cortical bone and has a greater number of PTHrP/lacZ-expressing cells as well as osteoblasts and osteoclasts than does the nearby tibial insertion of the medial collateral ligament (MCL).[18]

The PTHrP gene has been known to be mechanically induced since the early 1990s, when it was found to be stretch-induced in smooth muscle structures, such as the bladder, uterus, and vessels, in which it serves to slowly relax the smooth muscle structure in question, allowing it to accommodate pressure

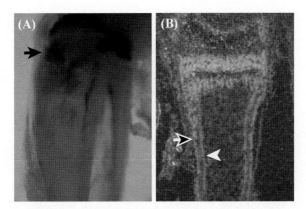

FIGURE 1. PTHrP and PTH1R expression in periosteal sites. (**A**) PTHrP gene-driven β gal activity in the periosteum in the region of the proximal tibial metaphysis in a 3-week-old PTHrP-lacZ mouse. (**B**) PTH1R mRNA expression in periosteal, endosteal, and cartilagenous regions of a metacarpal bone at day 5 (D5). Note that the endosteal signals in the primary spongiosa and the endosteum are continuous, as are the signals in the bone collar and the cambial layer of the periosteum. Modified from Reference[21,22] 19.

and/or gradual filling/distension.[20] Mechanotransduction of the PTHrP gene has also been described in cultured chondrocytes and bone-derived mesenchymal cells, so that this is a cell-autonomous response.[21,22] It was immediately apparent that PTHrP expression in entheses was likely to be the result of mechanical loading, and this was shown to be the case in a variety of such sites by unloading experiments. Unloading was chosen as the experimental approach of choice since these sites were presumed to be loaded in their native state, and we found the two most useful such methods to be tail suspension and/or transection of the entheses near their insertion sites.[19] The results were the most immediate and complete following transection (FIG. 2A,D).[19] In sites such as the tibial insertion of the MCL and the semimembranosus (a hamstring), lacZ-expressing cells are associated with subjacent osteoblasts and alkaline phosphatase activity, and the latter disappears coincident with the rapid unloading-induced decline in PTHrP (FIG. 2E,F). These findings clearly implicate PTHrP in the formation/activity of neighboring osteoblasts.

These various data have led to several working hypotheses as regards putative PTHrP functions in the periosteum and/or entheses, all quite early works-in-progress. First, PTHrP is not expressed uniformly in the periosteum but is found in particular plenty at the metaphyseal-diaphyseal junction (MDJ) during peak growth (FIG. 1A). Here, we propose that PTHrP may mediate the modeling that shapes this region into the adjacent diaphysis during linear growth, a process that is driven by periosteal osteoclastic resorption in the metaphysis and periosteal bone formation in the diaphysis that is distal to the MDJ. It may be also that PTHrP is deployed in the periosteum by mechanical force to remodel

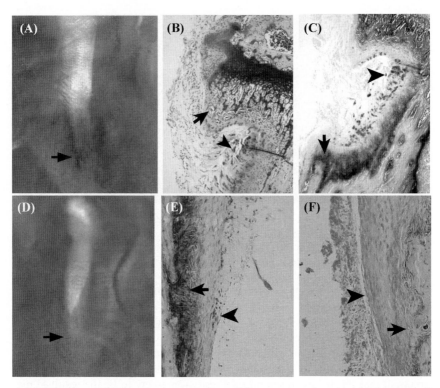

FIGURE 2.   PTHrP expression at tendon and ligament insertion sites. (**A**) Insertion of the (MCL) in the tibial metaphysis in a 4-week-old PTHrP-lacZ mouse. The *arrow* identifies β gal activity at the insertion site. (**B**) Leading edge of the semimembranosus (SM) insertion in the tibial metaphysis in a 6-week-old PTHrP-lacZ mouse. The *arrow* identifies TRAP-positive osteoclasts that are carving away cortical bone in the direction of migration. (**C**) SM insertion site as in panel B. The *arrow* identifies alkaline phosphatase-stained osteoblasts concentrated in the trailing edge of the insertion and the *arrowheads* β gal-positive cells at the junction of the tendon and connective tissue in the cortical depression in which the tendon is anchored. (**D**) Transected MCL in a 4-week-old PTHrP-lacZ mouse; note that the β gal signal at the insertion is gone (*arrow*). (**E&F**) β gal and alkaline phosphatase activities in the control (**E**) and transected (**F**) MCL insertions in 5-week-old PTHrP-lacZ mice. Note that both the β gal (*arrowhead*) and alkaline phosphatase (*arrow*) activities in the transected site have disappeared 7 days after surgery. Modified from Reference 5.

periosteal surfaces and/or that it is a key player in the sexual dimorphism of bone, which is basically a function of androgen-driven periosteal bone formation in the male. In the entheses in general, it would seem reasonable to suppose that PTHrP might serve as a local regulator of force-induced bone cell behavior(s), although it is not yet clear whether this is osteoblastic, osteoclastic, or both. An interesting variation on this theme is the likely involvement of PTHrP in the migration of the insertion sites such as those of the

MCL/semimembranosus in the proximal tibia. Ligamentous and tendinous insertions into a metaphysis must migrate as the epiphysis grows, or they would wind up as functionless strands attached to the mid-diaphysis in the adult.[23,25] This migration takes the form of mechanically driven osteoclastic bone resorption at the leading edge and osteoblastic bone formation at the trailing edge, exactly what we see subjacent to the PTHrP-expressing fibroblast-like cells at these sites (Fig. 2C, E).[19] The acid test of this idea will be enabled by the conditional deletion of PTHrP in such sites, which is a work-in-progress.

CHONDROEPIPHYSIS

Endochondral bones form via a two-step process in which a cartilagenous mold is subsequently replaced by bone. The biological rationale for this two-step process is the capacity of such a structure for linear growth. In endochondral long bones, this growth is mediated at each end by a cartilagenous growth zone, a simple structure that is referred to in the evolutionary literature as a chondroepiphysis, literally an epiphysis composed of cartilage.[26,27] The chondroepiphysis is the only such structure formed in lower species, such as teleosts and primitive tetrapods, and it ossifies along a linear axis via direct expansion of the primary ossification center; this process is referred to as "direct ossification."[26,27] This ancestral structure also lacks a zone of Ranvier, so that it does not form a metaphysis per se. Terrestrial tetrapods and bipeds require a bony epiphysis to withstand the mechanical demands of land mobility, particularly as regards joints, such as the hip, at which loading is not distributed along the long axis of the bone. Thus, in terrestrials, the secondary ossification center and bony epiphysis and metaphysis evolved and brought with them the defined growth plate and articular cartilage structures; these five structures define the modern weight-bearing joint "organ." In the mouse, these structures do not take shape until the first week or so of postnatal life. It follows that the embryonic structure that drives linear growth in modern terrestrials is a chondroepiphysis, which differs from its primitive forerunner principally by having a zone of Ranvier, at least in most locations (FIG. 3B). It is this embryonic structure that has been extensively studied in chicks and mice and in which the details of the Ihh-PTHrP regulatory axis have been so well worked out.[28,29] This is the best-studied and -understood of all PTHrP regulatory functions.[28]

The short bones of the hands and feet provide a window on the evolution of endochondral bone formation, in that they bear a simple chondroepiphysis at one end and a growth-plate at the other (FIG. 3B,C).[26,27] In both sites, PTHrP expression is confined to round proliferative chondrocytes in the subarticular region until the secondary ossification center forms. The secondary center essentially divides this subarticular population of PTHrP-expressing chondrocytes into two subpopulations of round proliferative cells, one that

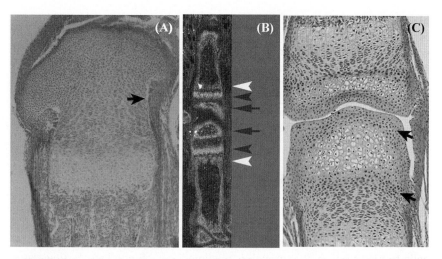

FIGURE 3. The chondroepiphysis. (**A**) Cartilagenous growth zone in the proximal tibia at birth (D0). The embryonic growth zone is a chondroepiphysis, in which β gal activity is concentrated in the round proliferative chondrocytes in the proximal one-third of the epiphyseal cartilage, subjacent to the joint space. The *arrow* identifies the groove of Ranvier. (**B**) PTH1R expression at D5 in a digit, as determined by ISHH. The growth plates flank the metacarpophalangeal (MP) joint, while the proximal region of the metacarpal and the distal portion of the phalanx bear a chondroepiphysis. The secondary ossification center (*arrows*), prehypertrophic chondrocytes of the growth plate (*arrowheads*), and osteoblasts (*white arrowheads*) are indicated. (**C**) MP joint at D5, after the secondary ossification center has begun to form on both sides of the joint. The β gal-positive round proliferative chondrocyte populations are indicated (arrows). Modified from Reference 19 & Chen *et al.*, manuscript submitted for publication.

remains in its initial subarticular location and a second that comes to lie at the top of the columns of flat chondrocytes of the forming growth plate (FIG. 3C). Both subpopulations remain for the life of the mouse, the subarticular subpopulation in the form of articular chondrocytes and the growth-plate subpopulation in a much reduced growth plate structure after skeletal maturity is reached at 12 weeks. It is certainly curious that the secondary ossification center, which is initially defined by a group of differentiating chondrocytes, is formed in the absolute center of the richest source of PTHrP in the chondroepiphysis, and these cells do express the type 1 PTH/PTHrP receptor PTH1R (FIG. 3B).

One is continuously reminded of what is often referred to as the economy of evolution.[30] This is apparent in endochondral bones by the multiple ways in which developmental regulatory molecules are deployed at different stages of development and/or in different sites, and the Ihh-PTHrP regulatory system is an excellent case in point. As noted above, the prototypical example of this system in action in endochondral bone formation is its regulation of the chondrocyte differentiation program in the chondroepiphysis, either in its primitive form or as it serves as the developmental forerunner of the growth plate in higher species.[28,29] Another is to protect the joint space that flanks the chondroepiphysis from invasion by mineralizing chondrocytes and bone cells that are advancing from the primary ossification center. As summarized below, these same two functions are apparent in articular cartilage, which we presume to be a modern-day iteration of the chondroepiphysis. What has evolved more recently is the deployment of this system in the growth plate itself. Here, PTHrP serves to regulate the flow of round chondrocytes into and through the columns and also serves to protect the growth plate from the invasion of mineralizing chondrocytes from the adjacent secondary ossification center; both of these functions are "me to" as regards those in the primitive chondroepiphysis. At this stage, the most rapidly proliferating chondrocyte population is that within the columns, which is presumably being driven in a PTHrP-independent fashion by Ihh via relatively short-range signaling.[29] In the homozygous PTHrP-lacZ mouse, which is functionally PTHrP-null, both ends of endochondral bones ossify by direct and premature invasion of bone cells from the primary ossification center; that is, absent PTHrP, the default pathway is the direct ossification pathway reminiscent of the primitive chondroepiphysis. This must mean that PTHrP-driven delay in the chondrocyte differentiation program is required to provide a window of opportunity for the secondary ossification center to form.

ARTICULAR CARTILAGE

Most synovial joints form by a segmentation process that subdivides the initial mesenchymal condensation into the individual endochondral bones that are specified within it (e.g., the inverted Y condensation of the upper extremity will segment to form the humerus, ulna, and radius, which will articulate via the segmented humeroulnar and humeroradial joints). The histological hallmark of early joint formation is the interzone, a linear disposition of cells that forms perpendicular to the long axis of the bones to be articulated. The interzone is a signaling center that expresses a number of regulatory molecules (the prototype being Gdf-5) and passes through several stages before it cavitates to form the joint space itself.[31–33] There have long been two schools of thought as to the primary source of the articular chondrocytes themselves, one favoring the interzone and another the epiphyseal growth zone/chondroepiphysis, both schools seeing their favorite site as the principal if not sole source of articular chondrocytes. Recent lineage-tracing data suggest that both schools are to some extent correct and that the surface spindle-shaped cells that line the joint space are derived from the interzone and that the subjacent round chondrocytes in the mid-zone and the prehypertrophic chondrocytes in the deep zone of articular cartilage are derived from the chondroepiphysis.[32–34] The data described below indicate that chondroepiphysis is the source of the PTHrP-expressing articular chondrocytes.

As we began to explore the presence of PTHrP in articular cartilage we came to recognize that PTHrP is not exclusively expressed in the chondroepiphyseal ends of chondrifying condensations, as previously assumed, but rather is deployed at the sites of future joint formation, be these at the ends of endochondral bones or elsewhere (Chen *et al.*, manuscript submitted for publication).[5,19] For example, in the forming elbow joint at E14.5, PTHrP/β gal activity is seen at the chondroepiphyseal end of the humerus and in the underlying semilunar fossa of the ulna but not in the olecranon, which constitutes the proximal end of the ulna (Fig. 4A). β gal activity is apparent in this location as early as E12.5, well before the interzone has formed. Another example is provided by the carpals, which form up to six articulations with neighboring carpal and metacarpal bones, each of these nascent articulations being strongly β gal-positive (Fig. 4B). Similarly, the contiguous chondroepiphyseal chondrocyte populations across the forming joints of the digits express β gal activity from the earliest stage of interzone formation. In all of these sites, the PTH1R is expressed in prehypertrophic chondrocytes immediately subjacent to the PTHrP-expressing chondrocytes of the developing articular cartilage in such a fashion that fully hypertrophic chondrocytes are corralled on the bone side of the joint (Fig. 4C). That is, this deployment of ligand and receptor serves to exclude mineralizing chondrocytes from the articular cartilage and the joint space itself. Further, this deployment is an early developmentally programmed event, being apparent at nascent joint sites as soon as they are specified. Taken together, these findings indicate that the chondroepiphysis is but one of several developmental iterations in which Ihh-PTHrP-regulated round chondrocytes are deployed at endochondral bone surfaces, all of which involve nascent synovial joints. The chondroepiphysis itself would appear to be the most ancient of these iterations.

The preceding summary assumes that PTHrP serves to regulate the articular chondrocyte differentiation program much as it does the program in growth chondrocytes and also that it may function in this regard in a regulatory loop with Ihh. These assumptions have been tested in several ways. The first evidence was genetic. As summarized in an earlier section, we created two versions of a lacZ mouse that is functionally PTHrP-null, one the homozygous PTHrP-lacZ mouse and the other the so-called single-allele PTHrP-lacZ PTHrP-null mouse, which bears only one copy of the lacZ allele and can therefore be used as a read-out of Ihh overexpression. That is, in the single-allele mouse, an increase in β gal activity as compared to that in the heterozygous PTHrP-lacZ mouse must reflect either Ihh-driven proliferation and/or overexpression of PTHrP in round chondrocytes. In both the homozygous PTHrP-lacZ and the single-allele PTHrP-lacZ knockout mouse, we observed a marked increase in the numbers of hypertrophic chondrocytes in articular cartilage and also that these cells approached the joint surface more closely than in the heterozygous PTHrP-lacZ mouse (Fig. 5A,B, (Chen *et al.*, manuscript submitted for publication). Nevertheless, the joint space was not invaded by these

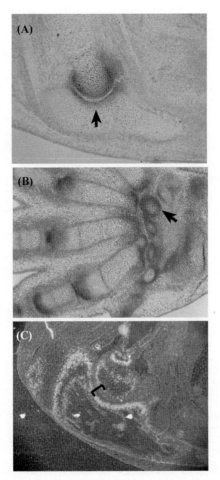

FIGURE 4. β gal activity and PTH1R expression at joint surfaces. (**A**) β gal activity at the elbow joint at embryonic day 14.5 (E14.5). Note that in the ulna β gal activity is concentrated at the semilunar fossa (*arrow*) rather than at the proximal end of the bone (the olecranon). (**B**) β gal activity in the carpals at E14.5. Carpal 4/5 (*arrow*) will form six articulations, each of which expresses PTHrP/lacZ. (**C**) PTH1R mRNA expression by ISHH in the elbow joint at D5. The prehypertrophic chondrocytes that express the receptor lie on the bone side of the PTHrP-expressing chondrocytes of the joint in both the humerus and ulna (*bracket*), so that this deployment of ligand and receptor excludes mineralizing chondrocytes from the articular cartilage and joint space. Modified from Chen *et al.*, manuscript submitted for publication.

cells, and on close inspection in both PTHrP-null versions of the mice we observed several layers of β gal-positive round chondrocytes that were not evident in the heterozygous PTHrP-lacZ joints. We suspected that these layers of round chondrocytes might reflect Ihh-driven PTHrP-independent articular chondrocyte proliferation, just as is seen in growth chondrocytes.[28,29] This

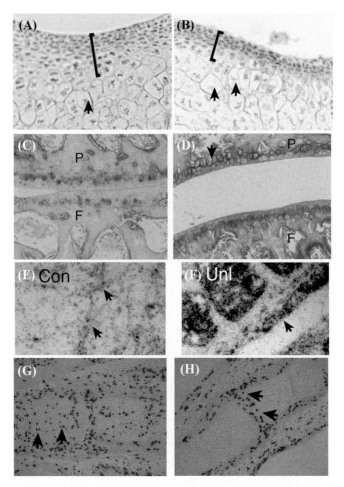

FIGURE 5.   Regulation of the articular chondrocyte differentiation program by PTHrP and Ihh. (**A&B**) Semilunar fossa in a heterozygous PTHrP-lacZ (**A**) and homozygous PTHrP-lacZ (PTHrP-null) (**B**) mouse at E18.5. The *arrows* identify hypertrophic chondrocytes and the *brackets* the cell layers that lie between these cells and the joint space . (**C&D**) Control and unloaded femoropatellar joint in a 7-week-old PTHrP-lacZ mouse sacrificed 7 days after shamming (**C**) or transecting (**D**) the suprapatellar ligament. Unloading was associated with a mean reduction in β gal-positive cells from 360 to 71 cells/mm^2 and an increase in alkaline phosphatase-positive cells from 153 to 281 cells/mm^2 in the control and unloaded sites, respectively. Note also the clustering of hypertrophic chondrocytes (*arrow*) and that these cells approximate but do not enter the joint space on the unloaded side (**D**). The femur (F) and patella (P) are indicated. (**E&F**) Ihh mRNA by ISHH in control (**E**) and unloaded (**F**) femoropatellar joints 4 weeks after a sham or unloading procedure. Note a marked increase in Ihh mRNA expression on unloading. (**G & H**) BrdU analysis of proliferation in wild-type (**G**) and PTHrPnull (**H**) metacarpophalangeal joint at E15.5. Note that proliferation is generalized in the growth cartilage in **G** and concentrated at the forming articular surface in **H**. Modified from Chen *et al.*, manuscript submitted for publication.

appears to be the case based on PCNA as well as BrdU staining, which reveal proliferating round chondrocyte populations throughout the chondroepiphyseal region, including specifically the surface cells that line the nascent joint spaces (the statistical significance of the findings in the lacZ and knockout mice are still under study). Thus, the Ihh-PTHrP seems to be fully deployed in articular chondrocytes and would appear to provide, in combination, a particularly powerful system to protect the joint, driving the proliferation of these cells (Ihh) and also preventing their exit from the cell cycle (PTHrP).

The second line of evidence was functional. In both cleared and histological specimens, we noted that the β gal activity was deployed at the load-bearing sites of joint contact, which in the knee included the femoral condyles, tibial plateaus, and the meniscal surfaces (FIG. 3D). These findings suggest that loading might represent a physiological regulator of PTHrP expression in articular cartilage. Further, since Ihh has also been reported to be mechanically induced, we could imagine several quite different combinations of how the components of the Ihh-PTHrP system might respond to loading/unloading.[35,36] Given that articular cartilage is already loaded in a steady state, we approached this question too by unloading strategies. These included surgical approaches for unloading the patella (and the femoropatellar surface) or the knee itself (and the femorotibial articulation). In both cases, the β gal activity at the articular surface fell dramatically with unloading, and this was accompanied by a marked increase in the numbers of alkaline phosphatase-positive hypertrophic chondrocytes, which approached the articular surface (FIG. 5C,D), much as seen in the PTHrP-null mouse joints (FIG. 5A,B). These changes were rapid (occurring within 3–4 days of unloading), not progressive (they were more-or-less the same at 4 weeks as at 1 week of unloading), and the joint space was spared from actual invasion by mineralizing cells (as in the PTHrP-null system). These findings are in line with previous evidence that prolonged unloading/immobilization does not typically induce irreversible changes in articular cartilage.[37] Clearly, the response of Ihh to unloading was a particularly key issue, and by ISHH Ihh was found to increase in unloaded cartilage in both the patellar and knee systems (FIG. 5E,F). This was accompanied by at least a high-normal level of chondrocyte proliferation in articular chondrocytes, as one would anticipate (this question requires further quantitative study). Thus, the absence of PTHrP feedback clearly trumps any potential independent mechanical effect that loading/unloading might have on Ihh expression in articular cartilage *in vivo*. Equally clearly, the primary physiological response to loading in articular cartilage involves PTHrP signaling, so that in a physiologically loaded, weight-bearing joint at steady state PTHrP would act to maintain articular chondrocytes in abeyance as regards differentiation, and Ihh would be held largely in abeyance as well. It follows that in articular cartilage there is something of a role-reversal of these two signaling partners, in which loading induces PTHrP as the primary signal, and Ihh lies downstream.

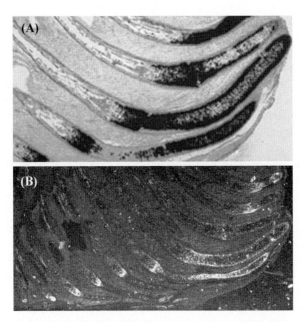

FIGURE 6.   Costal cartilage in a PTHrP-null mouse at E18.5. PTHrP is normally expressed in the perichondrium of costal cartilage and serves to prevent the terminal differentiation and mineralization of the subjacent hyaline chondrocytes of the cartilage. PTHrP-null mice have a pathologically mineralized costal cartilage, and the mouse dies at birth of a shield chest. Panel **A** shows collagen X expression (a marker for hypertrophic chondrocytes) in PTHrP-null costal cartilage at E18.5. Panel **B** shows Ihh mRNA expression in a section contiguous to that in A.

OTHER PERMANENT CARTILAGE STRUCTURES

Articular cartilage is sometimes referred to as "permanent" cartilage, in that hyaline cartilage structures must remain cartilagenous throughout life in order to function, whereas growth cartilage is seen as "transient" as it is normally consumed by the process of ossification.[33,34] Other examples of permanent cartilage structures include the costal cartilage and nasal cartilage. In both of these structures, PTHrP/lacZ is expressed in the perichondrium that surrounds the subjacent highland cartilage.[5] In both sites, one sees promiscuous chondrocyte differentiation as well as mineralization that totally engulfs the cartilagenous structures by E18.5 in the homozygous PTHrP-lacZ (null) mouse (FIG. 6A). There is also a marked increase in Ihh expression in the PTHrP-null costal cartilage (FIG. 6B). The mineralizing costal cartilage induces a shield chest in the PTHrP-null mouse, and this is the cause of its neonatal demise.[5,12,28,38,39]

Thus, in these simple structures, PTHrP-driven prevention of terminal chondrocyte differentiation is the principal if not sole defense against pathological mineralization, in much the same fashion that this regulation occurs in transient growth cartilage. Further, although overexpressed in PTHrP-null costal cartilage, Ihh does not appear to serve as a second line of defense in the structure, as it does in articular cartilage.

SUMMARY AND CONCLUSIONS

PTHrP is a very low-abundance expression product, and the PTHrP-lacZ knockin mouse has proven to be a very useful reporter system in this regard. This utility involves both the identification of previously unrecognized sites of PTHrP gene expression and the capacity to appreciate regulation of gene expression in these sites. It also includes our inability to identify PTHrP in internal bone cell populations, although certainly PTHrP is expressed in other mesenchymal bone cell populations. PTHrP-expressing sites are confined to the structures at which skeletal elements interact with each other and with the muscles that surround and control these elements, that is, "at the gates" of endochondral bones. PTHrP gene expression is regulated in many of these sites by mechanical force. There are a number of functional implications of these findings, as well as the promise of insight into pathophysiological mechanisms.

REFERENCES

1. BURTIS, W.J., T. WU, C. BUNCH, *et al.* 1987. Identification of a novel 17000 dalton PTH-like adenylate cyclase stimulating protein from a tumor associated with humoral hypercalcemia of malignancy. J. Biol. Chem. **262:** 7151–7156.
2. IKEDA, K., M. MANGIN, B.E. EYER, *et al.* 1988. Identification of transcripts encoding a parathyroid hormone-like peptide in messenger RNAs from a variety of human and animal tumors associated with humoral hypercalcemia of malignancy. J. Clin. Invest. **81:** 2010–2014.
3. BROADUS, A.E. & A.F. STEWART. 1994. Parathyroid hormone-related protein structure, processing and physiological actions. *In* The Parathyroids. J.P. Bilezikian, Ed.: 259–294. Raven Press. New York, NY.
4. HOLT, E.H., C. LU, B.E. DREYER, *et al.* 1994. Regulation of parathyroid hormone-related peptide gene expression by estrogen in GH4C1 rat pituitary cells has the pattern of a primary response gene. J. Neurochem. **62:** 1239–1246.
5. CHEN, X., C.M. MACICA, B.E. DREYER, *et al.* 2006. Initial characterization of PTH-related protein gene-driven lacZ in the mouse. J. Bone Min. Res. **21:** 113–123.
6. MERENDINO, J.J., K.L. INSOGNA, L.M. MILSTONE & A.F. STEWART. 1986. Cultured human keratinocytes produce parathyroid hormone-like protein. Science **231:** 388–390.
7. ATILLASOY, E.J., W.J. BURTIS & L.M. MILSTONE. 1991. Immunochemical localization of parathyroid hormone-related protein in normal human skin. J. Invest. Dermatol. **96:** 277–280.

8. CHO, Y-M, D.A. LEWIS, P.F. KOLTZ, *et al.* 2004. Regulation of parathyroid hormone-related protein gene expression by epidermal growth factor-family ligands in primary human keratinocytes. J. Endocrinol. **181:** 179–190.

9. BURTIS, W.J., T.G. BRADY, J.J. ORLOFF, *et al.* 1990. Immunochemical characterization of circulating parathyroid hormone-related protein in patients with humoral hypercalcemia of cancer. N. Engl. J. Med. **322:** 1106–1112.

10. SOIFER, N.E., K.E. DEE, K.L. INSOGNA, *et al.* 1992. Parathyroid hormone-related protein. Evidence for secretion of a novel mid-region fragment by three different cell types. J. Biol. Chem. **267:** 18236–18243.

11. WYSOLMERSKI, J.J., W.M. PHILBRICK, M.E. DUNBAR, *et al.* 1998. Rescue of the parathyroid hormone-related protein knockout mouse demonstrates that parathyroid hormone-related protein is essential for mammary gland development. Development **125:** 1285–1294.

12. KARAPLIS, A.C., A. LUZ, J. GLOWACKI, *et al.* 1994. Lethal skeletal dysplasis from targeted disruption of the parathyroid hormone-related peptide gene. Genes Dev. **8:** 277–289.

13. YOUNG, D.C., S.D. KINSLEY, K.A. RYAN & F.J. DUTKO. 1993. Selective inactivation of eukaryotic b-galactosidase in assays for inhibitors of HIV-1 TAT using bacterial b-galactosidase as a reporter enzyme. Anal. Biochem. **215:** 24–39.

14. MAO, X., F. YUKO & S.H. ORKIN. 1999. Improved resport strain for monitoring Cre recombinase-mediated DNA excisions in mice. Proc. Natl. Acad. Sci. USA **96:** 5037–5042.

15. MOUNTFORD, P., B. ZEVNIK, A.N. DÜWEL, *et al.* 1994. Dicistronic targeting constructs: reporters and modifiers of mammalian gene expression. Proc. Natl. Acad. Sci. USA **91:** 4303–4307.

16. BRENT, A.E., R. SCHWEITZER & C.J. TABIN. 2003. A somitic compartment of tendon progenitors. Cell **113:** 235–248.

17. BENJAMIN, M., T. KUMAI, S. MILZ, *et al.* 2002. The skeletal attachment of tendons-tendon 'entheses'. Comp. Biochem. Physiol. Part A **133:** 931–1045.

18. CHEN, X., C.M. MACICA, A. NASIRI, *et al.* 2007. Mechanical regulation of PTHrP expression in entheses. Bone: In press.

19. BURR, D.B. 1997. Muscle strength, bone mass, and age-related bone loss. J. Bone Min. Res. **12:** 1540–1547.

20. CLEMENS, T.L. & A.E. BROADUS. 2001. Physiologic actions of PTH and PTHrP: IV. Vascular, cardiovascular, and neurologic actions. *In* The Parathyroids, 2nd ed. J.P. Bilezikian, Ed.: 261–274. Academic Press. New York, NY.

21. TANAKA, N., S. OHNO, K. HONDA, *et al.* 2005. Cyclic mechanical strain regulates the PTHrP expression in cultured chondrocytes via activation of the Ca2+ channel. J. Dent. Res. **84:** 64–68.

22. CHEN, X., C.M. MACICA, K.W. NG & A.E. BROADUS. 2005. Stretch-induced PTH-related protein gene expression in osteoblasts. J. Bone Min. Res. **20:** 1454–1461.

23. WEI, X. & K. MESSNER. 1996. The postnatal development of the insertion of the medial collateral ligament in the rat knee. Anat. Embryol. **193:** 53–59.

24. DÖRFL, J. 1980. Migration of tendinous insertions. I. Cause and mechanism. J. Anat. **131:** 179–195.

25. DÖRFL, J. 1980. Migration of tendinous insertions. II. Experimental modifications. J. Anat. **131:** 229–237.

26. RENO, P.L., D.L. MCBURNEY, C.O. LOVEJOY & W.E. HORTON, JR. 2006. Ossification of the mouse metatarsal: differentiation and proliferation in the presence/absence of a defined growth plate. Anat. Rec. Part A. **288A:** 104–118.

27. HAINES, E.W. 1942. The evolution of epiphyses and the endochondral bone. Biol. Rev. **17:** 267–291.
28. KRONENBERG, H.M. 2006. PTHrP and skeletal development. Ann. N. Y. Acad. Sci. **1068:** 1–13.
29. LONG, F., X.M. ZHANG, S. KARP, *et al.* 2001. Genetic manipulation of hedgehog signaling in the endochondral skeleton reveals a direct role in the regulation of chondrocyte proliferation. Development **128:** 5099–5108.
30. NISWANDER, L. 2002. Interplay between the molecular signals that control vertebrate limb development. Int. J. Dev. Biol. **46:** 877–881.
31. STORM, E.E. & D.M. KINGSLEY. 1996. Joint patterning defects caused by a single and double mutations in members of the bone morphogenetic protein (BMP) family. Development **122:** 3969–3979.
32. HYDE, G., S. DOVER, A. ASZODI, *et al.* 2007. Lineage tracing using matrilin-1 gene expression reveals that articular chondrocytes exist as the joint interzone forms. Dev. Biol. **304:** 825–833.
33. PACIFICI, M., E. KOYAMA, Y. SHIBUKAWA, *et al.* 2006. Cellular and molecular mechanisms of synovial joint and articular cartilage formation. Ann. N. Y. Acad. Sci. **1068:** 74–86.
34. KOYAMA E., M. NAGAYAMA, B. YOUNG, *et al.* 2007. Cellular and molecular mechanisms regulating synovial joint formation during mouse limb skeletogenesis. Session IV: mechanisms of skeletal development. Abstract. Ann. N. Y. Acad. Sci. In press.
35. WU, Q-Q, Y. ZHANG & Q. CHEN. 2001. Indian hedgehog is an essential component of mechanotransduction complex to stimulate chondrocyte proliferation. J. Biol. Chem. **276:** 35290–35296.
36. TANG, G.H., A.B. RABIE & U. HAGG. 2004. Indian hedgehog: a mechanotransduction mediator in condylar cartilage. J. Dent. Res. **83:** 434–438.
37. BRANDT, K.D. 2003. Response of joint structures to inactivity and to reloading after immobilization. Arthritis Rheum. **49:** 267–271.
38. KARAPLIS, A.C., J. LUZ, R.J. GLOWACKI, *et al.* 1994. Lethal skeletal dysplasis from targeted disruption of the parathyroid hormone-related peptide gene. Genes Dev. **8:** 277–289.
39. LANSKE, B., A.C. KARAPLIS, K. LEE, *et al.* 1996. PTH/PTHrP receptor in early development and Indian hedgehog-regulated bone growth. Science **273:** 613–622.

Axin1 and Axin2 Are Regulated by TGF-β and Mediate Cross-talk between TGF-β and Wnt Signaling Pathways

DEBBIE Y. DAO, XUE YANG, DI CHEN, MICHAEL ZUSCIK, AND REGIS J. O'KEEFE

Department of Orthopaedics and Department of Pathology, Center for Musculoskeletal Research University of Rochester School of Medicine, Rochester, New York, USA

ABSTRACT: Chondrocyte maturation during endochondral bone formation is regulated by a number of signals that either promote or inhibit maturation. Among these, two well-studied signaling pathways play crucial roles in modulating chondrocyte maturation: transforming growth factor-beta (TGF-β)/Smad3 signaling slows the rate of chondrocyte maturation, while Wingless/INT-1-related (Wnt)/β-catenin signaling enhances the rate of chondrocyte maturation. Axin1 and Axin2 are functionally equivalent and have been shown to inhibit Wnt/β-catenin signaling and stimulate TGF-β signaling. Here we show that while Wnt3a stimulates Axin2 in a negative feedback loop, TGF-β suppresses the expression of both Axin1 and Axin2 and stimulates β-catenin signaling. In Axin2 -/- chondrocytes, TGF-β treatment results in a sustained increase in β-catenin levels compared to wild-type chondrocytes. In contrast, overexpression of Axin enhanced TGF-β signaling while overexpression of β-catenin inhibited the ability of TGF-β to induce Smad3-sensitive reporters. Finally, the suppression of the Axins is Smad3-dependent since the effect is absent in Smad3 -/- chondrocytes. Altogether these findings show that the Axins act to integrate signals between the Wnt/β-catenin and TGF-β/Smad pathways. Since the suppression Axin1 and Axin2 expression by TGF-β reduces TGF-β signaling and enhances Wnt/β-catenin signaling, the overall effect is a shift from TGF-β toward Wnt/β-catenin signaling and an acceleration of chondrocyte maturation.

KEYWORDS: axin; axin1; axin2; TGF-β; Wnt; β-catenin; chondrogenesis

Address for correspondence: Regis J. O'Keefe, M.D., Ph.D., Box 665, Department of Orthopaedics, University of Rochester Medical Center, 601 Elmwood Avenue, Rochester, NY 14642. Voice: 585-273-1261; fax: 585-756-4727.
Regis_okeefe@urmc.rochester.edu

Ann. N.Y. Acad. Sci. 1116: 82–99 (2007). © 2007 New York Academy of Sciences.
doi: 10.1196/annals.1402.082

INTRODUCTION

Chondrocyte development is regulated by a number of signaling pathways that either promote or inhibit maturation. While transforming growth factor-beta (TGF-β) and parathyroid hormone-related peptide (PTHrP) slow the rate of chondrocyte maturation, bone morphogenic proteins (BMPs), thyroid hormone, retinoic acid, and Wingless/INT-1-related (Wnt) proteins enhance the rate of chondrocyte maturation.[1–6] Although these individual pathways have been extensively studied, less work has been completed to define signaling interactions that may be important for cartilage maturation. Understanding these interactions and their molecular targets is essential since chondrocytes in the growth plate simultaneously integrate these various complex signals. Recently, the TGF-β/BMP-related Smad pathways and the Wnt β-catenin pathways have emerged as important signals related to growth and development. Evidence suggests that the TGF-β/BMP Smad pathways and the Wnt/β-catenin pathways control both the proliferation and differentiation of chondrocytes in the growth plate and thus it is likely that these signals are highly interrelated.[2,7–9]

While Wnt/β-catenin signaling inhibits chondrogenesis in the developing skeleton,[10,11] β-catenin signaling is reestablished in the growth plate where numerous Wnts are expressed once cartilage has formed and the skeletal elements have developed.[9,12–14] Cell culture and chick limb bud models suggest that Wnt/β-catenin signaling stimulates chondrocyte maturation. Overexpression of Wnts 4, 8, and 9, β-catenin, and LEF-1 induce *colX*, *alkaline phosphatase*, and other genes associated with chondrocyte hypertrophy.[9,14,15] Inhibition of β-catenin signaling by overexpression of Frzb-1, a secreted decoy receptor, dominant negative Wnt receptors, or dominant negative β-catenin results in delayed maturation.[9,13,14] Thus, cell culture and chick limb bud models show that Wnt/β-catenin signaling is active during endochondral bone formation and suggest that β-catenin stimulates chondrocyte maturation.

Wnt/β-catenin has also been shown to regulate proliferation in numerous cell types and recent findings from our laboratory showed that the induction of proliferation by TGF-β was dependent upon β-catenin.[7] In the absence of Wnt signaling molecules, β-catenin is part of a cytosolic protein complex that includes Axin, adenomatous polyposis coli (APC), and glycogen synthase kinase–3β (GSK).[16,17] In this complex, β-catenin is phosphorylated by GSK,[18] which results in ubiquitination by E3 ubiquitin ligases with subsequent degradation in the 26S proteasome.[17,18] TGF-β signaling results in the phosphorylation of Smad3 and our prior work showed that Smad3 interacts with the E3 ligase, β-TCRP and prevents the ubiquitination of β-catenin.[7] This indirect activation of β-catenin is necessary for the stimulation of proliferation by TGF-β in isolated murine sternal chondroctyes.[7]

Axin is another potential target for TGF-β and β-catenin interactions. Axin is a scaffold for the protein complex, described above, involved in β-catenin catabolism.[17,19] It has interaction sites for numerous proteins involved in Wnt signal transduction, including β-catenin, GSK, casein kinase I, APC, Dishevelled (Dvl), and the LRP receptors.[17,19] During Wnt signaling Axin is recruited to the LRP receptor complex where an interaction with Dvl results in Axin degradation and release of active β-catenin.[17] β-catenin subsequently translocates to the nucleus where it associates with TCF/LEF (T Cell Factor/Lymphoid Enhancer Factor) transcription factors, binds to DNA, and regulates gene transcription.[17]

In addition to its role in β-catenin signaling, Axin also has an important function in TGF-β/Smad signaling. The classic TGF-β−mediated signaling pathway involves Smad activation.[20] Smads are a family of intracellular proteins that comprise three classes of signaling molecules: receptor-associated Smads (2 and 3 for TGF-β; 1, 5, and 8 for BMP signaling), the cofactor Smad4, and the inhibitory Smads (6 and 7).[20,21] Receptor-associated Smads bind to type I receptors and upon ligand binding and activation, are phosphorylated and released into the cytoplasm. Recent data demonstrate that Axin1/Axin2 binds to Smad3 at the type I TGF receptor and is essential for its phosphorylation.[22] The activated receptor-associated Smads then form a heterodimer with Smad4, translocate to the nucleus, and influence gene transcription.[20,21] An additional mechanism through which Axin facilitates TGF-β signaling is by acting as a scaffold for the interaction between the E3 ubiquitin ligase, Arkadia and inhibitory Smad7.[23] Smad7 degradation is accelerated in the presence of Axin, with a subsequent increase in TGF-β/Smad signaling.[23]

Axin has two functionally identical isoforms, Axin1 and Axin2; knockin of *Axin2* into the deleted *Axin1* gene rescues *Axin1*[−/−] mice.[24] *In vitro* studies reveal that both *Axin1* and *Axin2* overexpression reduce β-catenin levels and signaling.[16,25] However, *Axin1* and *Axin2* are differentially expressed and their gene deletions have different phenotypes. *Axin1*[−/−] mice die at embryonic day 9.5 (E9.5) with forebrain truncation, neural tube defects, and axis duplications.[24] In contrast, *Axin2*[−/−] mice complete embryonic development but have craniofacial defects and premature closure of the cranial sutures due to increased β-catenin signaling.[26] The focus of this article is to determine a potential role for the Axins in the coordination of TGF-β and Wnt/β-catenin signaling in cartilage.

METHODS

Chondrocyte Isolation

Primary chondrocytes were isolated from the sterna and ribs of 3-day-old mice as previously described.[2] Tissues were placed in DMEM containing 2 mg/mL pronase for 45 min at 37°C on a 150 rpm shaking platform. Following a wash in HBSS, sterna and ribs were further digested in DMEM with 3mg/mL collagenase D (Sigma, St. Louis, MO) at 37°C and 5% CO_2 for 1.5 h.

Following a wash in HBSS, tissues were transferred into a Petri dish containing DMEM with 3 mg/mL collagenase D and incubated at 37°C and 5% CO_2 in a cell incubator for 6 h. Isolated cells were then resuspended in 5 mL DMEM containing 10% NuSerum IV (Collaborative Biomedical Products, Bedford, MA) and supplemented with 50 U penicillin and streptomycin and 2 mM of L-glutamine. Cells were passed through a cell strainer, centrifuged at 2000 rpm for 3 min at 4°C, and after several washes, cells were plated, in 6-well plates at 5×10^5 cells per well (for protein isolation), or in 12-well plates at 5×10^4 cells per well (for RNA isolation). After 24 h in culture, the culture media were supplemented with 50 ng/mL of ascorbic acid, and, for treatment experiments, human TGF-β1 (Calbiochem, Darmstadt, Germany), Wnt3a (R&D Systems, Inc., Minneapolis, MN), or BMP-2 (Peprotech, Inc., Rockyhill, NJ) was added.

RNA Extraction and Quantitative Reverse-Transcriptase PCR

Total RNA was extracted from primary chondrocyte cultures using the Trizol protocol (Invitrogen, Carlsbad, CA) following the manufacturer's recommendations. One microgram of total RNA was reverse-transcribed using the iscript cDNA synthesis kit (Bio-Rad, Hercules, CA) following the manufacturer's recommended protocol. Two microliters of reverse-transcribed cDNA were used for quantitative PCR. cDNA levels were measured in real-time using the fluorescent dye SYBR Green I (SYBR Green PCR Master Mix, Applied Biosystems, Foster City, CA) using specific primers designed for mouse Axin1 (Forward: 5'- ACG GTA CAA CGA AG CAG AGA GCT -3'; Reverse: 5'- CGG ATC TCC TTT GGC ATT CGG TAA -3'), Axin2 (Forward: 5'- GAG TAG CGC CGT GTT AGT GAC T -3'; Reverse: 5'- CCA GGA AAG TCC GGA AGA GGT ATG -3'), and β-actin (Forward: 5'- TGT TAC CAA CTG GGA CGA CA -3'; Reverse: 5'- CTG GGT CAT CTT TTC ACG GT -3'). The PCR reaction used the RotorGene real-time DNA amplification system (Corbett Research, Sydney, Australia) and the following protocol was used: a 95°C denaturation step for 10 min followed by 45 cycles with denaturation for 30 sec at 95°C, annealing for 30 sec at 55°C, and extension for 30 sec at 72°C. Detection of the fluorescent product occurred after each extension period. PCR products were subjected to a melting curve analysis and the data were analyzed and quantified with the RotorGene analysis software. Gene expression was normalized to β-actin.

Western Blotting

Chondrocytes were treated with TGF-β, Wnt3a, or BMP-2, then at different time points, were washed with cold phosphate-buffered saline and lysed on ice using Golden Lysis Buffer supplemented with protease inhibitor cocktail tablets. The protein concentration of the soluble material was measured using Coomassie Plus Protein Assay kit (Pierce, Rockford, IL). Twenty micrograms of extracts were assayed by SDS-PAGE. After transfer to a nitrocellulose membrane, blots were probed with the following antibodies: mouse anti-total-β-catenin (Santa Cruz Biotechnology, Inc., Santa Cruz, CA) at a concentration of 1:1000, rabbit anti-Axin (Zymed Laboratories, South San Francisco, CA) at a concentration of 1:250, and mouse anti-β-actin (Sigma) at a concentration of 1:5000. Horseradish peroxidase-conjugated goat anti-rabbit or anti-mouse polyclonal antibodies (Bio-Rad Laboratories) were used as secondary antibodies. The immune complexes were detected using Pico or Femto chemiluminescent agents (Pierce).

Transfection and Luciferase Assay

For transfection, primary sternal chondrocytes or chondrocyte cell lines (TMC-23 or C5.18) were plated at a density of 10^5 cells/well in 6-well plates. TMC-23 and C5.18 are well-described murine chondrocyte and rat chondrocyte cell lines, respectively.[27,28] Cultures were transfected 12 h after plating with the TGF-β responsive reporters, p3TP-Luc (500 ng; a gift from Dr. Joan Massague) and 4×SBE or the Topflash reporter of Wnt/β-catenin signaling (500 ng; a gift from Dr. Jennifer Westendorf). In some experiments, cells were cotransfected with S33Y β-catenin (a gift from Dr. Kenneth Kinzler), Axin1 (a gift from Dr. Wei Hsu), or Smad3 (a gift from Dr. Rik Derynck) plasmids. Two hours after transfection, cultures were placed in normal culture media for 24 h, followed by treatment with BMP-2 (50 ng/mL), Wnt3a (100 ng/mL), or TGF-β (5 ng/mL) for the next 24 h. Cells were lysed using passive lysis buffer (Promega, Madison, WI) and luciferase activity in the cell lysate measured using the dual Luciferase assay system (Promega) with an Optocomp luminometer (MGM Instruments, Hamden, CT). Transfection efficiency was determined by cotransfection with pRL vector (Promega) and determining the *Renilla uniformis* luciferase activity. In all experiments, control vectors were used to keep the total amount of transfected DNA identical.

Statistical Analysis

Differences between various treatments or time points were compared using a two-way analysis of variance with Tukey–HSD *post hoc* analysis. Statistical significance was considered present when $P < 0.05$.

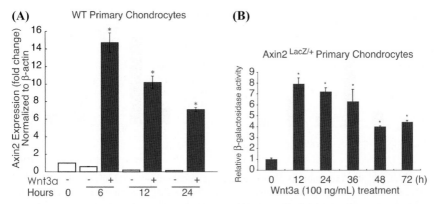

FIGURE 1. Wnt3a induces *Axin2* gene and protein expression. Primary murine sternal chondrocytes were isolated and placed in monolayer culture as described in "Methods." After overnight culture they were treated with Wnt3a (100 ng/mL) or with a vehicle control for 0, 6, 12, and 24 h before total RNA was isolated. Quantitative Reverse-Transcriptase PCR was performed to assess *Axin2* gene expression (normalized to β-actin) (**A**). Primary chondrocytes from Axin2$^{LacZ/+}$ mice were placed in monolayer culture overnight and then treated with 100 ng/mL of Wnt3a. Cultures were harvested after 0, 12, 24, 36, 48, and 72 h and and β-galactosidase activity was measured as a reporter of Axin2 protein expression (**B**). The symbol * represents significance at $P < 0.05$ when compared to 0 h.

RESULTS

Wnt3a Induces Axin2 Gene and Protein Expression in Chondrocytes

Previous findings indicate that Wnt-signaling induces *Axin2* expression.[29,30] To determine if this regulation occurs in chondrocytes, we studied *Axin2* mRNA and protein expression in Wnt3a-treated primary murine chondrocyte cultures. *Axin2* gene expression was rapidly induced by Wnt3a treatment, with a 15-fold induction occurring at 6 h following treatment. *Axin2* levels gradually decreased, but remained elevated at 24 h where a sevenfold induction was observed (**FIG. 1A**). To determine if the induction in gene expression was associated with increased protein levels, primary chondrocytes from Axin2$^{LacZ/+}$ mice were cultured and treated with 100 ng/mL of Wnt3a and β-galactosidase activity was measured as a reporter of Axin2 protein expression at 0, 12, 24, 36, 48, and 72 h following treatment induction. Relative to baseline activity at time zero, β-galactosidase activity following Wnt3a treatment increased nearly eightfold after 12 h (**FIG. 1B**). This induction was not as pronounced at later time points, but remained elevated 72 h later. In contrast, Wnt3a did not induce the expression of Axin1 in chondrocytes (data not shown). Thus, regulation of Axin1 and Axin2 by Wnt3a in chondrocytes is similar to the pattern that is observed with other cell types.[29,30]

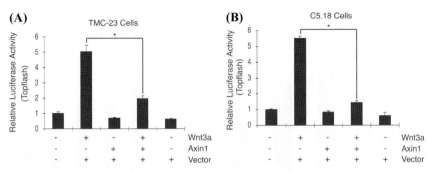

FIGURE 2. Axin inhibits Wnt3a signaling in chondrocytes. TMC23 (**A**) and C5.18 (**B**) chondrocyte cell lines were transfected with the Topflash reporter construct and treated with Wnt3a (100 ng/mL) in the presence or absence of cotransfection with Axin1. Cell extracts were harvested after 48 h and luciferase activity measured. Transfection efficiency was determined by measurement of renilla luciferase activity. The symbol * represents significance at $P < 0.05$.

Axin Regulates β-Catenin Signaling in Chondrocytes

To study the reciprocal interaction of Axin expression on Wnt/β-catenin signaling in chondrocytes, TMC23 and C5.18 chondrocyte cell lines were transfected with the Topflash reporter construct and treated with Wnt3a (100 ng/mL) in the presence or absence of cotransfection with *Axin1*. Luciferase activity was measured 2 days after treatment as an indication of Wnt3a-induced β-catenin signaling activation. As expected, Wnt3a-treated cells had increased luciferase reporter activity compared to controls cells. Overexpression of *Axin1* significantly inhibited Wnt3a-induced Topflash reporter activity in both cell lines (FIG. 2). Since Axin1 and Axin2 have been shown to be functionally equivalent in the regulation of Wnt/β-catenin signaling, the findings suggest that regulation of the Axin genes modulate β-catenin signaling.[24]

Axin Modulates Signaling Cross-talk between the β-Catenin and TGF-β Signaling Pathways

Wnt/β-catenin and TGF-β are two key signaling pathways involved in regulating chondrocyte differentiation.[1–4] Prior work has suggested important interactions between these pathways and previously we showed that TGF-β/Smad3 induces β-catenin signaling in chondrocytes.[7] However, a reciprocal effect of β-catenin on TGF-β/Smad3 signaling has not previously been examined. Initial experiments determined whether β-catenin alters the effect of TGF-β on activation of the Smad3-responsive reporter p3TP-Lux. C5.18 chondrocytes transfected with p3TP-Lux were treated with TGF-β in the presence or absence of cotransfection with Smad3 and/or β-catenin plasmids. Treatment

with TGF-β for 48 h resulted in a 12-fold induction in luciferase activity, and this was increased to 20-fold in cells cotransfected with Smad3. Transfection of the β-catenin plasmid significantly inhibits TGF-β signaling in both the presence ($P < 0.05$) and absence ($P < 0.01$) of Smad3 (FIG. 3A). Combined with prior work, the experiment shows that TGF-β/Smad induces β-catenin, and β-catenin in turn downregulates TGF-β/Smad3 signaling.

To examine if Axin2 participates in this cross-talk, primary chondrocytes isolated from wild-type and Axin2 -/- mice were treated with TGF-β and total β-catenin levels determined by Western blot (FIG. 3B). While the magnitude of TGF-β stimulation of β-catenin in both the wild-type and Axin2 -/- cultures was similar, the pattern of accumulation varied. In wild-type cultures, maximal levels of β-catenin were observed at 2 h and then levels declined progressively with a return to basal levels by 36–48 h following treatment. In contrast, peak β-catenin levels were sustained for 12 h in Axin2 -/- cultures and remained elevated after 48 h after treatment. The findings confirm that TGF-β induces β-catenin accumulation and establish a role for Axin2 in mediating the degradation of β-catenin in chondrocytes.

Axin Enhances TGF-β Signaling

Previous studies have established that Axin binds to type I TGF-β receptor and facilitates the phosphorylation of Smad3.[22] To examine if this interaction is functionally important in chondrocytes, C5.18 chondrocyte cultures were transfected with 4×SBE, a reporter for Smad3 activity, and were treated with 0.02, 0.2, or 2.0 ng/mL of TGF-β in the presence of absence of cotransfection with an Axin1 plasmid. For each concentration of TGF-β, Axin1 cotransfection consistently enhanced reporter activity by approximately 20% (FIG. 4). The findings establish that Axin regulates TGF-β signaling in chondrocytes consistent with results in other cell types.[22]

TGF-β Inhibits Axin1 and Axin2 Expression in Chondrocytes

Since Axin expression modulates TGF-β effects on both β-catenin and Smad3 signals, we examined if TGF-β in turn regulates expression of the *Axins*. Primary murine chondrocytes were treated with either 1 or 5 ng/mL of TGF-β and total RNA isolated from the cultures after 48 h. TGF-β resulted in a marked dose-dependent suppression of *Axin1*, with approximately 70% and 90% inhibitions observed in cultures treated with 1 and 5 ng/mL of TGF-β, respectively(FIG. 5A).

Using a concentration of TGF-β of 5 ng/mL, *Axin1* and *Axin2* mRNA expression was measured following 0, 12, 24, 36, 48, and 72 h of exposure to TGF-β. TGF-β resulted in a time-dependent inhibition of both *Axin1*

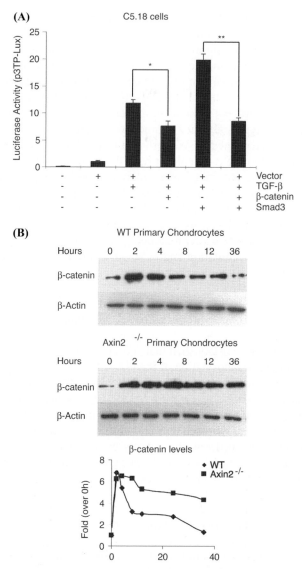

FIGURE 3. β-catenin inhibits TGF-β signaling in chondrocytes while the induction of β-catenin by TGF-β is enhanced in Axin2-deficient chondrocytes. (**A**) Chondrocyte C5.18 cells were transfected with TGF-β responsive reporter, p3TP-Lux in the presence or absence of cotransfections with Smad3 and/or β-catenin and treated with or without TGF-β (5 ng/mL) for 48 h. Transfection of β-catenin significantly inhibited TGF-β or TGF-β+Smad3-induced p3TP-Lux reporter activity (*$P < 0.05$, **$P < 0.01$). (**B**) Protein lysates from TGF-β-treated primary chondrocytes from WT and Axin2-/- mice were assayed for β-catenin by Western blot. TGF-β treatment enhances β-catenin accumulation in both WT and Axin2-/- cells. However, the elevated levels of β-catenin persist in Axin2 -/- cells.

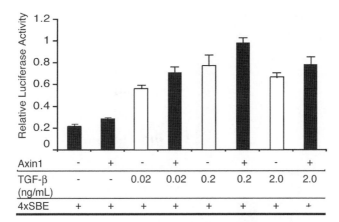

FIGURE 4. Axin enhances TGF-β signaling in chondrocytes. C5.18 chondrocyte cultures were transfected with 4×SBE and treated with 0.02, 0.2, or 2.0 ng/mL of TGF-β in the presence of absence of cotransfection with Axin plasmid. Cultures were harvested after 24 h and show that Axin cotransfection consistently enhanced reporter activity approximately 20% at each TGF-β concentration.

and *Axin2* levels, with increasing inhibition observed over time. Initial *Axin1* inhibition was observed within 12 h (30% inhibition) and was maximal by 36 h (80% inhibition). In contrast, Axin2 inhibition was delayed until 24 h (50% inhibition) and was maximal at 72 h (90% inhibition) (FIG. 5B, C).

To determine if TGF-β also regulates Axin2 protein expression, β-galactosidase activity was measured in Axin2$^{LacZ/+}$ knockin mice following treatment with TGF-β. Similar to the observed effect on mRNA expression, TGF-β resulted in a progressive and sustained decrease in β-galactosidase activity that paralleled the decrease in gene expression. Maximal suppression of protein synthesis also occurred at 72 h (80% inhibition) (FIG. 5D).

BMP Does Not Alter Axin Expression or Regulate β-Catenin Signaling

The delayed inhibition of Axin2 by TGF-β suggests that the effect may be related to alterations in the maturational state of the cells, rather than to a direct signaling effect. For this reason, we examined whether BMP-2, which has effects opposite TGF-β and promotes chondrocyte maturation, regulates the expression of *Axin2*.[2,3] In contrast to the observations made with TGF-β, primary chondrocyte cells from Axin2$^{LacZ/+}$ knockin mice treated with BMP-2 (100 ng/mL) had no elevation of β-galactosidase reporter activity (FIG. 6A).

To further examine BMP/β-catenin interactions, C5.18 cells transfected with the Wnt signaling reporter, Topflash, were treated with BMP-2 in the presence or absence of Wnt3a. Unlike effects observed with TGF-β, BMP-2-treatment did not induce luciferase activity, indicating an absence of stimulation of

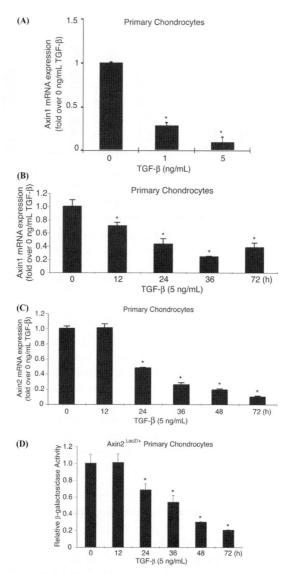

FIGURE 5. TGF-β inhibits *Axin1* and *Axin2* expression in chondrocytes. Primary murine chondrocytes were isolated and placed in overnight culture. The cultures were then treated with TGF-β (1 or 5 ng/mL) and total RNA was harvested at 48 h. Real-time RT PCR was used to determine the effect on *Axin1* gene expression (**A**). Chondrocyte cultures with treated with TGF-β (5 ng/mL) and total RNA harvested at various times between 0 and 72 h and real-time RT-PCR used to determine the effect on expressions of *Axin1* (**B**) and *Axin2* (**C**) gene expression. β-galactosidase activity was measured in primary chondrocytes from Axin2^LacZ/+ knockin mice and revealed that Axin2 protein expression is also negatively regulated by TGF-β progressively over time (**D**). The symbol *represents significance at $P < 0.05$ when compared to 0 ng/mL TGF-β or 0 h.

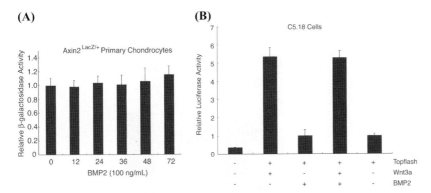

FIGURE 6. BMP does not alter Axin expression or Wnt/β-catenin signaling. Primary chondrocyte cells from Axin2$^{LacZ/+}$ knockin mice were treated with BMP-2 (100 ng/mL) and cell extracts harvested between 0 and 72 h. β-galactosidase reporter activity was measured (**A**). C5.18 cells were transfected with the Wnt Topflash reporter and treated with BMP-2 (100 ng/mL) in the presence or absence of Wnt3a. Cell extracts were harvested after 24 h and luciferase activity was measured. The luciferase activity was normalized to renilla luciferase (**B**).

β-catenin signaling.[7] Furthermore, BMP did not alter the stimulatory effects of β-catenin (FIG. 6B).

Altogether, the findings demonstrate unique crosstalk between the TGF-β/Smad3 and the Wnt/β-catenin pathways and suggest that the regulation of *Axin2* gene and protein levels by TGF-β is due to a pathway-specific signaling event.

TGF-β Inhibits Axin1 and Axin2 through Smad3 Signaling

Since Smad3 has been shown to be a key downstream effector of TGF-β signaling in chondrocytes, we examined whether or not Smad3 plays a role in TGF-β-mediated Axin inhibition.[2,7,31] For these experiments, primary chondrocytes were isolated from Smad3 -/- mice. In contrast to the marked inhibitions of *Axin1* and *Axin2* observed in wild-type cultures treated with TGF-β (FIG. 2), no decrease in the expression of either *Axin1* or *Axin2* occurred in Smad3 -/- chondrocyte cultures at either 48 or 72 h following exposure to TGF-β (FIG. 7A).

To confirm these findings, a Western blot for Axin1 was performed on protein extracts obtained from wild-type and Smad3 -/- chondrocyte cultures treated with and without TGF-β. As previously observed TGF-β stimulated a marked decrease in Axin1 protein expression following 60 h of treatment. In contrast, Axin1 protein levels in Smad3$^{-/-}$ cells remained unchanged when compared to untreated controls (FIG. 7B). These data indicate that the regulation of Axin1 and Axin2 by TGF-β is dependent upon Smad3 signaling.

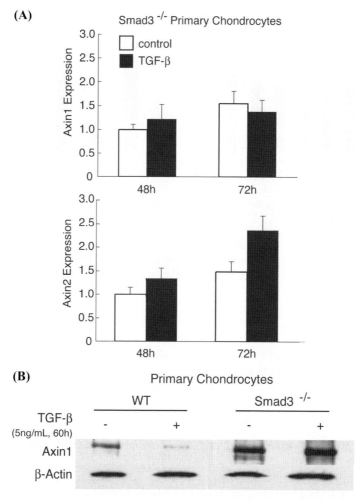

FIGURE 7. Smad3 -/- cells do not exhibit TGF-β-mediated inhibition of Axin expression. Primary chondrocytes from Smad 3-/- mice were isolated and cultured with 5 ng/mL TGF-β treatment for 24 and 48 h. Total RNA was harvested and RT-PCR performed to measure the relative suppression of *Axin1* or *Axin2* gene expression in TGF-β-treated cultures compared to control cultures after 48 h of treatment (**A**). Total cell protein extracts were obtained from wild-type and Smad3 -/- chondrocyte cultures treated with TGF-β or control medium and Axin1 protein levels measured by Western blot. β-actin was used as a loading control (**B**).

DISCUSSION

This study further supports the presence of crosstalk between the TGF-β/Smad3 and Wnt/β-catenin signaling pathways. While we previously demonstrated that TGF-β/Smad3 signaling induced β-catenin signaling, the current findings show that β-catenin, in turn, inhibits TGF-β signaling in chondrocytes.[7] Furthermore, we show that Axin1 and Axin2, two functionally equivalent proteins that modulate both signaling pathways, are involved in the interaction.[24]

TGF-β was found to be a potent inhibitor of both *Axin1* and *Axin2*. In contrast, BMP treatment had no effect on expression of the *Axins*, suggesting that the effect was not related to alterations in the maturational state of the cells, but was related to a specific signaling event downstream of TGF-β. Prior work in our laboratory established that Smad3 induced β-catenin signaling by interfering with the ability of the E3 ubiquitin ligase, β-TCRP, to stimulate degradation of β-catenin.[7] A genetic model using Smad3-deficient chondrocytes was used to definitively demonstrate a role for Smad3 signaling in the suppression of expression of both *Axin1* and *Axin2*. This is the first model that we are aware of establishing TGF-β/Smad3 in the regulation of *Axin* expression.

Prior work suggests a complex role for β-catenin signaling as a modulator of chondrogenesis and chondrocyte maturation.[32] β-catenin signaling is a strong inhibitor of chondrogenesis, but appears to be necessary for the normal function of the growth plate.[10,11] Thus, although this signaling pathway prevents the formation of cartilage, its reestablishment in chondrocytes is necessary.[2–4] In the growth plate β-catenin appears to have a role in the regulation of both proliferation and differentiation.[7,8] *In vivo* models of β-catenin loss of function show reduced chondrocyte proliferation compared to controls.[8] Our recent findings show that the induction of β-catenin signaling is necessary for the stimulation of proliferation by TGF-β.[2] In this model, β-catenin was necessary for the induction of cyclin D1 gene expression in chondrocytes downstream of TGF-β.[7] A role for β-catenin in cell proliferation through induction of cyclin D1 has been established in numerous cell types and has particular importance in cancer models.[8,33–35]

In addition to its effects as a stimulator of proliferation, in chondrocytes a role for β-catenin as a signal involved in acceleration of maturation has also been established.[9,14,36,37] While development models have suggested a role in maturation, *in vitro* studies have clearly demonstrated that gain of β-catenin function results in accelerated chondrocyte maturation while loss of function has resulted in delayed maturation.[9,14,36,37] This has been demonstrated in various models including isolated chick sternal chondrocytes, developing chick limbs *in vivo*, and ectopic ossification models.[9,14,36,37]

Since maturation typically occurs in cells that exit the cell cycle, the dual stimulation of both proliferation and maturation are events that tend to oppose and are not typically dually mediated by a single signaling pathway. However, in the growth plate cell fate decisions are determined by the various growth factors and signaling molecules present in the local environment. Prior work in our laboratory and by others demonstrates that the closely related TGF-β and BMP signaling pathways have opposite effects on chondrocyte maturation.[2,3]

Thus, while the TGF-β–Smad pathways inhibit maturation, the BMP/Smad pathways enhance chondrocyte maturation.[1,3,38] Moreover, our work shows that the absence of Smad3 results in an increase in basal BMP signaling and that the BMP signal is necessary for maturation.[2] Thus, it is clear that the overall effect of the various signals is dependent on the presence or absence of complementary or competing signaling pathways.

β-catenin is a signaling pathway that potentially links and integrates both proliferation and differentiation and may have disparate effects on these two events dependent upon the presence or absence of associated pathways.[32] In this context, TGF-β and the Axins appear to have a particularly important role. In association with prior work, the current studies show that TGF-β stimulates β-catenin signaling while in turn β-catenin signaling inhibits TGF-β signaling.[7] Moreover, TGF-β reduces the expression of the Axins. The cumulative effect of decreased Axin expression makes chondrocytes more sensitive to Wnt/β-catenin signaling and less sensitive to TGF-β Smad signaling. The overall impact of these signaling events is a negative feedback loop on TGF-β signaling and a positive feedback loop on Wnt/β-catenin signaling.

FIGURE 8 summarizes the Axin-mediated interaction between TGF-β/Smad3 and Wnt/β-catenin signaling suggested by our findings. TGF-β signaling activates Smad3. Smad3 was previously shown to be essential for the stimulation of β-catenin expression through an inhibitory role on the E3 ubiquitin ligase, β-TRCP.[7] While Smad3 activation results in dual stimulation of both TGF-β and β-catenin signaling, β-catenin signaling acts to downregulate TGF-β signaling. The Smad3-mediated suppression of Axin expression further inhibits TGF-β/Smad3 signaling, but makes cells more sensitive to Wnt/β-catenin. Thus, TGF-β/Smad3 regulates elements of the β-catenin signaling pathway that act to stimulate β-catenin and result in enhanced responsiveness to Wnt signals. At the same time, these β-catenin pathway molecules suppress TGF-β signaling. The release of cells from TGF-β signaling may be one mechanism leading to maturation and provide a mechanism through which β-catenin shifts toward stimulating the maturation process.

The ability of β-catenin to stimulate maturation probably is dependent upon the presence of other positive regulators of maturation.[32] Prior work suggests that the induction of maturational markers by β-catenin occurs in association with other signals that stimulate maturation, including BMP signals and Runx2.[14,37] Runx2 has been shown to be essential for chondrocyte maturation.[39] Recent work in a chick chondrocyte model has shown a consensus TCF/Lef-binding site in the proximal promoter of Runx2 that is required for induction of Runx2 by β-catenin signaling.[37] Furthermore, it was shown that the activation of the *colX* promoter by β-catenin was dependent upon the Runx2-binding site and that induction of both *colX* and *Runx2* by BMP2 is enhanced by simultaneous β-catenin signaling.[37]

The current findings support the concept that the process of chondrocyte maturation during endochondral ossification is controlled by complex

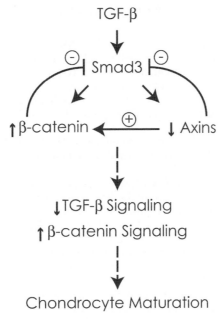

FIGURE 8. Summary of TGF-β/Smad and β-catenin signaling interactions in chondrocytes. TGF-β signaling inhibits chondrocyte maturation and induces Wnt/β-catenin signaling by directly activating β-catenin and reducing expression of *Axin1* and *Axin2*. While β-catenin is necessary for the induction of cyclin D1 and chondrocyte proliferation downstream of TGF-β/Smad3, the induction of β-catenin and suppression of *axins* result in a decrease in TGF-β signaling. The overall effect of enhanced β-catenin signaling and inhibited Smad3 signaling is a release of chondrocytes from the inhibitory effects of TGF-β signaling on maturation. β-catenin in the presence of other stimulators of maturation, including BMP signaling molecules and Runx2, enhances chondrocyte maturation.

interactions between key signaling pathways. The experiments confirm an important role for the Wnt/β-catenin signaling pathway in the integration of TGF-β and BMP signals. The TGF-β/Smad3-mediated induction of β-catenin and inhibition of the Axins result in a feedback inhibition of TGF-β signaling and a stimulation of β-catenin signaling. Further study will define the potential importance of this event in permitting chondrocytes to transition to hypertrophy and progress toward terminal maturation.

ACKNOWLEDGMENTS

The work was supported by National Health Service Awards AR048681 (R.J.O.) and AR053717 (R.J.O.).

REFERENCES

1. FERGUSON, C.M. *et al*. 2000. Smad 2 and 3 mediate TGF-ß1-induced inhibition of chondrocyte maturation. Endocrinology **141:** 4728–4735.
2. LI, T.F. *et al*. 2006. Smad3-deficient chondrocytes have enhanced BMP signaling and accelerated differentiation. J. Bone Miner. Res. **21:** 4–16.
3. KOBAYASHI, T. *et al*. 2005. BMP signaling stimulates cellular differentiation at multiple steps during cartilage development. Proc. Natl. Acad. Sci. USA **102:** 18023–18027.
4. IWAMOTO, M. *et al*. 2003. Runx2 expression and action in chondrocytes are regulated by retinoid signaling and parathyroid hormone-related peptide (PTHrP). Osteoarthr. Cart. **11:** 6–15.
5. IONESCU, A.M. *et al*. 2001. PTHrP modulates chondrocyte differentiation through AP-1 and CREB signaling. J. Biol. Chem. **276:** 11639–11647.
6. BALLOCK, R.T. *et al*. 2000. Thyroid hormone regulates terminal differentiation of growth plate chondrocytes through local induction of bone morphogenetic proteins. Trans. Orthop. Res. Soc. **25:** 160.
7. LI, T.F. *et al*. 2006. Transforming growth factor-beta stimulates cyclin D1 expression through activation of beta-catenin signaling in chondrocytes. J. Biol. Chem. **281:** 21296–21304.
8. AKIYAMA, H. *et al*. 2004. Interactions between Sox9 and beta-catenin control chondrocyte differentiation. Genes Dev. **18:** 1072–1087.
9. ENOMOTO-IWAMOTO, M. *et al*. 2002. The Wnt antagonist Frzb-1 regulates chondrocyte maturation and long bone development during limb skeletogenesis. Dev. Biol. **251:** 142–156.
10. DAY, T.F. *et al*. 2005. Wnt/beta-catenin signaling in mesenchymal progenitors controls osteoblast and chondrocyte differentiation during vertebrate skeletogenesis. Dev. Cell. **8:** 739–750.
11. HILL, T.P. *et al*. 2005. Canonical Wnt/beta-catenin signaling prevents osteoblasts from differentiating into chondrocytes. Dev. Cell. **8:** 727–738.
12. CHURCH, V. *et al*. 2002. Wnt regulation of chondrocyte differentiation. J. Cell Sci. **115:** 4809–4818.
13. HARTMANN, C. & C.J. TABIN. 2000. Dual roles of Wnt signaling during chondrogenesis in the chicken limb. Development **127:** 3141–3159.
14. DONG, Y. *et al*. 2005. Wnt-mediated regulation of chondrocyte maturation: modulation by TGF-beta. J. Cell Biochem. **95:** 1057–1068.
15. TOPOL, L. *et al*. 2003. Wnt-5a inhibits the canonical Wnt pathway by promoting GSK-3-independent beta-catenin degradation. J. Cell Biol. **162:** 899–908.
16. KIKUCHI, A. 1999. Roles of Axin in the Wnt signalling pathway. Cell Signal. **11:** 777–788.
17. CADIGAN, K.M. & Y.I. LIU. 2006. Wnt signaling: complexity at the surface. J. Cell Sci. **119:** 395–402.
18. HA, N.C. *et al*. 2004. Mechanism of phosphorylation-dependent binding of APC to beta-catenin and its role in beta-catenin degradation. Mol. Cell. **15:** 511–521.
19. BHASIN, N. *et al*. 2004. Differential regulation of chondrogenic differentiation by the serotonin2B receptor and retinoic acid in the embryonic mouse hindlimb. Dev. Dyn. **230:** 201–209.
20. ATTISANO, L. & J.L. WRANA. 2002. Signal transduction by the TGF-beta superfamily. Science **296:** 1646–1647.

21. Massague, J. & Y. Chen. 2000. Controlling TGF-beta signaling. Genes Dev. **14:** 627–640.

22. Furuhashi, M. *et al.* 2001. Axin facilitates Smad3 activation in the transforming growth factor beta signaling pathway. Mol. Cell Biol. **21:** 5132–5141.

23. Liu, W. *et al.* 2006. Axin is a scaffold protein in TGF-beta signaling that promotes degradation of Smad7 by Arkadia. EMBO J. **25:** 1646–1658.

24. Chia, I.V. & F. Costantini. 2005. Mouse axin and axin2/conductin proteins are functionally equivalent in vivo. Mol. Cell Biol. **25:** 4371–4376.

25. Kishida, M. *et al.* 1999. Axin prevents Wnt-3a-induced accumulation of beta-catenin. Oncogene **18:** 979–985.

26. Yu, H.M. *et al.* 2005. The role of Axin2 in calvarial morphogenesis and craniosynostosis. Development **132:** 1995–2005.

27. Feng, J.Q. *et al.* 2003. NF-kappaB specifically activates BMP-2 gene expression in growth plate chondrocytes in vivo and in a chondrocyte cell line in vitro. J. Biol. Chem. **278:** 29130–29135.

28. Harada, S. *et al.* 1997. Osteogenic protein-1 up-regulation of the collagen X promoter activity is mediated by a MEF-2-like sequence and requires an adjacent AP-1 sequence. Mol. Endocrinol. **11:** 1832–1845.

29. Lustig, B. *et al.* 2002. Negative feedback loop of Wnt signaling through upregulation of conductin/axin2 in colorectal and liver tumors. Mol. Cell Biol. **22:** 1184–1193.

30. Jho, E.H. *et al.* 2002. Wnt/beta-catenin/Tcf signaling induces the transcription of Axin2, a negative regulator of the signaling pathway. Mol. Cell Biol. **22:** 1172–1183.

31. Yang, X. *et al.* 2001. TGF-beta/Smad3 signals repress chondrocyte hypertrophic differentiation and are required for maintaining articular cartilage. J. Cell Biol. **153:** 35–46.

32. Tamamura, Y. *et al.* 2005. Developmental regulation of Wnt/beta-catenin signals is required for growth plate assembly, cartilage integrity, and endochondral ossification. J. Biol. Chem. **280:** 19185–19195.

33. Tetsu, O. & F. McCormick. 1999. Beta-catenin regulates expression of cyclin D1 in colon carcinoma cells. Nature **398:** 422–426.

34. Saegusa, M. *et al.* 2004. Beta-catenin simultaneously induces activation of the p53-p21WAF1 pathway and overexpression of cyclin D1 during squamous differentiation of endometrial carcinoma cells. Am. J. Pathol. **164:** 1739–1749.

35. Rowlands, T.M. *et al.* 2003. Dissecting the roles of beta-catenin and cyclin D1 during mammary development and neoplasia. Proc. Natl. Acad. Sci. USA **100:** 11400–11405.

36. Kitagaki, J. *et al.* 2003. Activation of beta-catenin-LEF/TCF signal pathway in chondrocytes stimulates ectopic endochondral ossification. Osteoarthr. Cart. **11:** 36–43.

37. Dong, Y.F. *et al.* 2006. Wnt induction of chondrocyte hypertrophy through the Runx2 transcription factor. J. Cell Physiol. **208:** 77–86.

38. Chen, P. *et al.* 1995. Osteogenic protein-1 promotes growth and maturation of chick sternal chondrocytes in serum-free cultures. J. Cell Sci. **108:** 105–114.

39. Yoshida, C.A. *et al.* 2004. Runx2 and Runx3 are essential for chondrocyte maturation, and Runx2 regulates limb growth through induction of Indian hedgehog. Genes Dev. **18:** 952–963.

Synovial Joint Formation during Mouse Limb Skeletogenesis

Roles of Indian Hedgehog Signaling

EIKI KOYAMA,[a] TAKANAGA OCHIAI,[a] RYAN B. ROUNTREE,[b]
DAVID M. KINGSLEY,[b] MOTOMI ENOMOTO-IWAMOTO,[a]
MASAHIRO IWAMOTO,[a] AND MAURIZIO PACIFICI[a]

[a]Department of Orthopaedic Surgery, College of Medicine, Thomas Jefferson
University, Philadelphia, Pennsylvania, USA

[b]Department of Development Biology and HHMI, Stanford University School of
Medicine, Stanford California, USA

ABSTRACT: Indian hedgehog (Ihh) has been previously found to regulate synovial joint formation. To analyze mechanisms, we carried out
morphological, molecular, and cell fate map analyses of interzone and
joint development in wild-type and $Ihh^{-/-}$ mouse embryo long bones.
We found that $Ihh^{-/-}$ cartilaginous digit anlagen remained fused and
lacked interzones or mature joints, whereas wrist skeletal elements were
not fused but their joints were morphologically abnormal. E14.5 and
E17.5 wild-type digit and ankle prospective joints expressed hedgehog
target genes including *Gli1* and *Gli2* and interzone-associated genes including *Gdf5*, *Erg*, and *tenascin-C*, but expression of all these genes was
barely detectable in mutant joints. For cell fate map analysis of joint
progenitor cells, we mated $Gdf5\text{-}Cre^{+/-}/Rosa\ R26R^{+/-}$ double transgenic mice with heterozygous $Ihh^{+/-}$ mice and monitored reporter β-
galactosidase activity and gene expression in triple-transgenic progeny.
In control $Gdf5\text{-}Cre^{+/-}/R26R^{+/-}/Ihh^{+/-}$ limbs, reporter-positive cells
were present in developing interzones, articulating layers, and synovial lining tissue and absent from underlying growth plates. In mutant
$Gdf5\text{-}Cre^{+/-}/R26R^{+/-}/Ihh^{-/-}$ specimens, reporter-positive cells were
present also. However, the cells were mostly located around the prospective and uninterrupted digit joint sites and, interestingly, still expressed
Erg, *tenascin-C*, and *Gdf5*. Topographical analysis revealed that interzone and associated cells were not uniformly distributed, but were much
more numerous ventrally. A similar topographical bias was seen for cavitation process and capsule primordia formation. In sum, Ihh is a critical
and possibly direct regulator of joint development. In its absence, distribution and function of *Gdf5*-expressing interzone-associated cells are
abnormal, but their patterning at prospective joint sites still occurs. The

Address for correspondence: Eiki Koyama of Maurizio Pacifici, Department of Orthopaedic Surgery,
Thomas Jefferson University College of Medicine, 501 Curtis Bldg., 1015 Walnut Street, Philadelphia,
PA 19107. Voice: 215-955-7352; fax: 215-955-9159.
maurizio.pacifici@jefferson.edu

Ann. N.Y. Acad. Sci. 1116: 100–112 (2007). © 2007 New York Academy of Sciences.
doi: 10.1196/annals.1402.063

joint-forming functions of the cells appear to normally involve a previously unsuspected asymmetric distribution along the ventral-to-dorsal plane of the developing joint.

KEYWORDS: synovial joint formation; limb skeletogenesis; Indian hedgehog; interzone; joint progenitor cells; hedgehog signaling; Gli proteins; transcription factor Erg

INTRODUCTION

The biology and developmental biology of synovial joints continue to be the focus of intense research activity owing to the fact that joints are essential for skeletal function and quality of life and are susceptible to malfunction during natural aging and in congenital or acquired conditions including osteoarthritis. Classic embryological studies showed several years ago that the onset of limb skeletogenesis involves formation of an uninterrupted Y-shaped mesenchymal condensation made of a seemingly uniform population of cells.[1] The proximal arm of the Y-shaped condensation corresponds to the future femur or humerus and the two arms of the Y will give rise to tibia/fibula or radius/ulna, but the condensation initially displays no overt morphological sign of knee or elbow joint. Additional uninterrupted mesenchymal condensations appear in the autopod, are termed *digital rays*[2] and represent the primordia of tarsal/carpal and phalangeal elements. The first overt evidence of joint formation is the emergence of the interzone at each future joint location.[3–5] The interzone consists of closely associated and flat-shaped mesenchymal cells and provides a clear demarcation between adjacent cartilaginous elements. The interzone is widely believed to be essential for joint formation,[3] but its specific roles are not fully clear.[6] To address this and related issues, genetic cell fate mapping studies were recently carried out in which *Rosa R26R* reporter mice[7] were mated with mice expressing Cre-recombinase in incipient joints under the control of growth and differentiation factor-5 (*Gdf5*) regulatory sequences.[8] The data indicated that *Gdf5*-expressing and interzone-associated progenitor cells are likely to give rise to several joint tissues, including articular cartilage and synovial lining.

Indian hedgehog (Ihh) is widely recognized as a critical regulator of long bone development and growth. *Ihh* is expressed in the prehypertrophic zone of growth plate[9–11] and regulates a number of processes. We found that Ihh promotes osteogenic cell differentiation, leading us to first propose that one key role for Ihh is to induce intramembranous bone collar formation around the developing diaphysis.[10,12] Other studies indicated that Ihh regulates the rates of chondrocyte proliferation and maturation in concert with periarticular-derived parathyroid hormone-related protein (PTHrP).[11,13] Studies on *Ihh*[−/−] mouse embryos provided further and more conclusive support for such multiple Ihh roles in long bone development.[14,15] Strikingly, these studies led to the realization that Ihh also regulates synovial joint formation.[14] The autopod region of $Ihh^{-/-}$ mouse embryos was found to contain uninterrupted cartilaginous digital rays lacking obvious joints, when in fact joints were forming in the digits of wild-type littermates. More moderate defects involving partial joint fusion or other defects were observed in more proximal mutant joints, such as elbow and knee. Despite the remarkable nature of these observations and their potentially fundamental implications, it remains largely obscure how Ihh regulates joint formation and what specific roles it actually has. This study was carried out to tackle these interesting and fundamental issues, using a combination of gene expression analyses and genetic cell fate map approaches in wild-type and $Ihh^{-/-}$ mouse embryos.

RESULTS

Anatomical and Gene Expression Studies

Whole mount inspection of mouse embryo limbs was used to further analyze whether development of interphalangeal joints is more affected by lack of hedgehog signaling than that of more proximal limb joints. Staining with alcian blue did show that the cartilaginous anlagen of E18.5 $Ihh^{-/-}$ mouse embryo toes were uninterrupted, lacked overt interzones, and stained uniformly with alcian blue (FIG. 1B). In contrast, skeletal elements in neighboring mutant ankle were distinct and separated from each other, as revealed by the presence of alcian blue-negative mesenchymal tissue (FIG. 1B, *arrows*). Standard joint formation processes were well under way in wild-type littermate toes and ankle that displayed obvious alcian blue-negative joint-associated tissues (FIG. 1A).

To examine hedgehog signaling, we determined the gene expression patterns of hedgehog target and effector genes by *in situ* hybridization. In early E14.5 wild-type joints, interzones were apparent between adjacent cartilaginous anlagen and displayed clear levels of *Gli1* and *Gli3* transcripts (FIG. 2A, C, E, *arrows*). *Gli2* expression was not as prominent (FIG. 2D), expression of hedgehog receptor *Patched-1* demarcated the cartilaginous anlagen (FIG. 2B) and expression of hedgehog signaling receptor *Smoothened* was widespread (FIG. 2F). At E17.5 when the cartilaginous anlagen become physically separated, expression of *Gli1*, *Gli3*, and *Patched-1* was still appreciable and seemed to be more prominent on the concave side of the developing joints (FIG. 2G, H, I, K, *arrows*). In E17.5 $Ihh^{-/-}$ limbs, however, *Patched-1*, *Gli1*, and *Gli2* transcripts were essentially undetectable (FIG. 2M–P); interestingly, *Gli3* and *Smoothened* transcripts were present, but largely occupied the mesenchymal tissues flanking the prospective and uninterrupted joint site (FIG. 2Q, R, *arrowheads*).

To corroborate these latter findings, we examined expression of *Gdf5* and

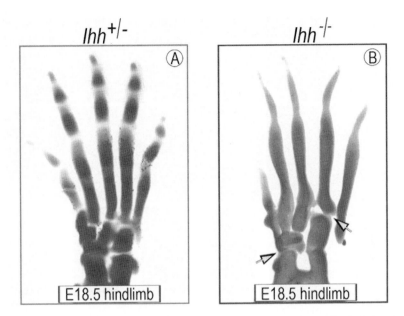

FIGURE 1. Anatomical examination of E18.5 wild-type and *Ihh*$^{-/-}$ mouse embryo hindlimbs. Specimens were stained with alcian blue to distinguish cartilaginous from non-cartilaginous structures. **(A)** Wild-type specimen in which the alcian blue-positive phalangeal and ankle skeletal elements are clearly interrupted at prescribed locations by alcian blue-negative joints. **(B)** *Ihh*$^{-/-}$ specimen in which the digit primordia are uniformly stained and uninterrupted and lack obvious joints. *Arrows* point to alcian blue-negative tissue surrounding the mutant ankle elements. (In color in Annals online).

also *Erg* and *tenascin-C*. Erg is an *ets* transcription factor family member that we found to be associated with joint and articular cartilage formation and function,[16–20] and tenascin-C is a pericellular matrix protein abundant in developing an adult articular cartilage.[16,17,21] In both E15.5 and E17.5 wild-type digits, *Gdf5, Erg,* and *tenascin-C* were all expressed in the interzones and associated articular layers (FIG. 3A–F). However, in *Ihh*$^{-/-}$ digits, expression of the three genes was essentially restricted to the mesenchymal tissues flanking the prospective and uninterrupted joint sites (FIG. 3G–M, *arrowheads*), thus resembling the gene expression pattern of *Gli3* and *Smoothened*.

As pointed out in FIG. 1B, mesenchymal tissue separates the various skeletal elements in *Ihh*$^{-/-}$ ankles and wrists. To clarify its nature, we processed control and mutant specimens for *in situ* hybridization. The tissue present in control E15.5 wrists was conspicuous and well organized (FIG. 4F) and strongly expressed *Gdf5* and *Patched-1* (FIG. 4G–H). In contrast, the tissue present in the *Ihh*$^{-/-}$ wrists was thin and inconspicuous (FIG. 4M) and expressed barely detectable levels of *Patched-1* and *Gdf5* (FIG. 4N–O).

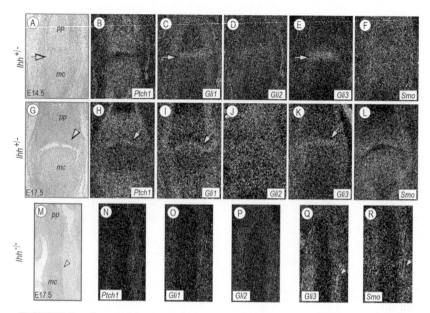

FIGURE 2. Gene expression patterns of hedgehog target and effector genes. (**A–F**) E14.5 wild-type forelimb digit early joints in which nascent interzone cells clearly display *Gli1* and *Gli3* transcripts (*arrow*). *Patched-1* expression characterizes contiguous cartilaginous tissue (**B**), *Gli2* expression is low (**D**), and *Smoothened* (*Smo*) expression is widespread (**F**). (**G–L**) E17.5 wild-type forelimb digit joints in which the cartilaginous elements are now well separated from each other. Note that transcripts for *Patched-1, Gli1*, and *Gli3* seem to be slightly more abundant on the convex side of the developing joint (H, I, K, *arrow*). (**M–R**) E17.5 *Ihh*$^{-/-}$ forelimb specimen from littermates in which joints are absent and prospective joint sites are uninterrupted. Note that *Patched-1, Gli1* and *Gli2* transcripts are essentially undetectable (**M–P**), while *Gli3* and *Smoothened* transcripts are present but flank the prospective joint site (**Q–R**, *arrowhead*). *pp*, proximal phalange; *mc*, metacarpal element. (In color in Annals online.)

Genetic Cell Fate Map Analysis

Next, we carried out genetic cell fate map analysis of joint progenitor cells in wild-type and *Ihh*$^{-/-}$ littermates by mating *Gdf5-Cre*$^{+/-}$/*Rosa R26R*$^{+/-}$ double transgenic mice with heterozygous *Ihh*$^{+/-}$ mice and analyzing their triple-transgenic progenies. Resulting embryonic limbs were processed for detection of reporter β-galactosidase activity by whole mount staining and histochemistry.[22] As shown in FIGURE 4A, reporter activity was clearly prominent and restricted to the developing interphalangeal and wrist joints in control *Gdf5-Cre*$^{+/-}$/*R26R*$^{+/-}$/*Ihh*$^{+/-}$ heterozygous embryos. Identical patterns were seen in *Ihh* wild-type (*Gdf5-Cre*$^{+/-}$/*R26R*$^{+/-}$/*Ihh*$^{+/+}$) mice (not shown). Histochemical analysis showed that reporter-positive cells were present in interzone-associated articulating layers and developing lining tissue but were absent

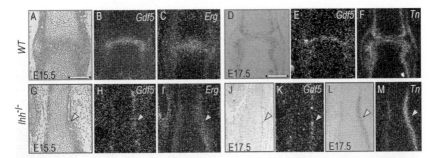

FIGURE 3. Gene expression patterns of joint marker genes. **(A–F)** E15.5 and E17.5 wild-type forelimb digit joints in which interzone and associated cells strongly express *Gdf5, Erg,* and *tenascin-C* (*Tn*). **(G–M)** E15.5 and E17.5 *Ihh⁻/⁻* forelimb digits in which transcripts for *Gdf5, Erg,* and *tenascin-C* (*Tn*) are present in tissue flanking the prospective joint sites (*arrowheads*). (In color in Annals online.)

in underlying shaft and growth plate cartilage (FIG. 4B) in agreement with previous findings.[8] A similar distribution of reporter-positive cells was seen in the wrists (FIG. 4F). When we examined mutant *Gdf5-Cre*⁺/⁻ /*R26R*⁺/⁻/*Ihh*⁻/⁻ specimens by whole mount, we were surprised to observe a band of reporter-positive cells located at the prospective location of metacarpal–phalangeal joints (FIG. 4I, *arrow*), a finding seemingly to indicate that digit joints were actually forming in the *Ihh*⁻/⁻ embryos. However, histochemical analysis revealed that the reporter-positive cells were mostly located outside and around the prospective joint site and were largely absent from the uninterrupted cartilaginous tissue (FIG. 4J, *arrow*). A moderate number of reporter-positive cells characterized the mutant wrist joints (FIG. 4M). It is interesting to note that no reporter-positive cells were detected at the location of the interphalangeal joints (FIG. 4I, *arrowhead*). This suggests that while the more proximal metacarpal–phalangeal joints could at least be patterned in the absence of Ihh, the more distal interphalangeal joints could not, in line with a previously proposed model of sequential joint formation in autopods.[23]

To more clearly understand the topography of distribution of reporter-positive cells, E15.5 control and mutant triple-transgenic autopods were cross-sectioned so that the dorsoventral distribution of the cells could be appreciated and analyzed. Interestingly, we found that the reporter-positive cells were not distributed uniformly along such dorsoventral plane, but appeared to be more numerous and concentrated in the ventral half in control joints (FIG. 4C–E, *arrowheads*). The ventral side displayed also conspicuous bilateral masses of mesenchymal cells (FIG. 4E, *double arrowhead*) and, interestingly, the cavitation process was well under way ventrally but not dorsally (FIG. 4E, *arrow*). The primordia of extensor (*et*) and flexor (*ft*) tendons were clearly visible at their characteristic dorsal and ventral locations (FIG. 4C–E). When we examined *Ihh*⁻/⁻ prospective digit joint sites, we found that the reporter-positive

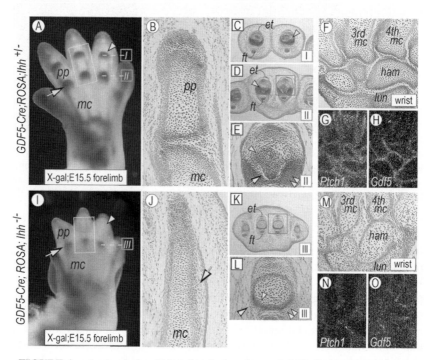

FIGURE 4. Anatomical and histochemical analyses of *Gdf5-Cre/ROSA R26R* control and *Ihh*$^{-/-}$ autopods. (**A–E**) Control E15.5 autopods showing obvious presence of reporter activity (blue color) at prescribed interphalangeal joint locations (**A**). Histochemical analysis (**B**) shows that reporter-positive cells represent interzone cells, adjacent most-epiphyseal chondrocytes and joint lining cells, but the cells are absent in underlying growth plate and shaft. Joint cross sections (**C–E**) corresponding to location I and II in panel **A** reveal that the ventral half of the joints contain more numerous reporter-positive cells (*arrowhead*), large ventrolateral mesenchymal cell masses (*double arrowhead*), and ongoing cavitation process (*arrow*). (**F–H**) Sections of control E15.5 wrists reveal pericartilaginous distribution of reporter-positive cells and well-contoured and reciprocally shaped developing joints (**F**) that express *Patched-1* and *Gdf5* (**G–H**). (**I–L**) E15.5 *Ihh*$^{-/-}$ autopods showing surprising presence of reporter activity at prospective joint site (*arrow* in I), but histochemical analysis reveals that the reporter-positive cells are located around the uninterrupted site (*arrow* in J). Cross-sections corresponding to location III in panel **I** show some crowding of reporter-positive cells in the ventral aspect of the uninterrupted joint site (*arrowhead* in **L**) and absence of mesenchymal masses (*double arrowhead* in **L**) and cavitation (*arrow* in **L**). (**M–O**) Sections of E15.5 *Ihh*$^{-/-}$ wrists showing that there are few reporter-positive cells and that the prospective joints are completely misshaped and lack reciprocal concave–convex contours (**M**) and express barely detectable levels of *Patched-1* and *Gdf5* (**N–O**). *pp*, proximal phalange; *mc*, metacarpal element; *et*, extensor tendon; *ft*, flexor tendon; *ham*, hamate; *lun*, lunate. (In color in Annals online.)

cells also appeared to be slightly more numerous on the ventral aspect of the uninterrupted site (FIG. 4K–L). However, there was no overt sign of cavitation or presence of ventral–lateral mesenchymal masses (FIG. 4K–L, *arrow* and *double arrowhead*, respectively).

DISCUSSION

Since it was discovered about 10 years ago and found to be expressed by prehypertrophic chondrocytes in the growth plate, Ihh has received a great deal of well-deserved attention. It can safely be said that Ihh is a true master regulator of limb skeletogenesis. Its expression in the prehypertrophic zone places it in a developmentally strategic position. We theorized long ago that the prehypertrophic zone is critical for the overall orchestration, coordination, and temporospatial unfolding of the chondrocyte maturation program and long bone anlaga development.[24] Ihh likely represents a key element in such prehypertrophic chondrocytes' function, and it is now well established and recognized that Ihh regulates central processes and functions in developing long bones that include intramembranous bone collar formation, chondrocyte proliferation and maturation rates, expression of PTHrP, and endochondral ossification.[25,26]

The data in this study now provide further insights into the fact that Ihh is also important for synovial joint formation and that joint initiation and morphogenesis are disrupted in $Ihh^{-/-}$ limbs. We find that wild-type interzones and early joints express *Gli1* and *Gli3*, and *Patched-1* expression characterizes the contiguous cartilaginous tissue in line with previous studies.[27–29] Given that these genes are mediators of hedgehog signaling, the findings indicate that hedgehog signaling is locally involved in joint formation. Indeed, *Gli1* and *Patched-1* are not expressed at appreciable levels in mutant $Ihh^{-/-}$ autopods, and this is associated with, and possibly causally linked to, absence or defects in joint formation. It remains to be clarified whether Ihh action in joint formation is direct or indirect, the former implying that Ihh itself diffuses and reaches the joint-forming sites. Ihh, and hedgehog proteins in general, are known to be able to travel from their site of synthesis to far-away targets.[30,31] This task is rather demanding and difficult since hedgehog proteins contain C-terminal- and N-terminal-bound cholesterol and palmitoylate tails and tend to stay close to their synthetic site.[32] One diffusion and traveling strategy characterizing hedgehog proteins is the possible formation of micelles by which the hydrophobic tails would be buried on the inside and the more hydrophilic protein moiety would face the extracellular fluids.[33] An additional strategy is likely provided by the fact that hedgehog proteins contain a heparin-binding domain, and interaction with heparan sulfate proteoglycans modulates their diffusion.[34] Previous work from our group and others has shown that growth plate chondrocytes express a number of heparan sulfate proteoglycans. Syndecan-3 and -4 are particularly evident in the proliferative zone[35,36] and syndecan-2 characterizes the perichondrium.[37] Given their binding characteristics and location, these macromolecules could attract Ihh and regulate its diffusion along specific spatial directions. Indeed, we and others have shown that Ihh diffuses away from its site of synthesis and is present in the proliferative zone and surrounding perichondrial tissue[30,31] and that defects in heparan sulfate proteoglycan synthesis or expression are accompanied by aberrant distribution of Ihh.[38,39] It is thus possible that mechanisms such as these could allow Ihh to influence joint development.

Our cell fate tracking data using *Gdf5-Cre/Rosa R26R* mice now reveal that reporter-positive cells are present at prospective joint sites in mutant *Ihh*$^{-/-}$ digits and are distributed outside of the uninterrupted cartilaginous tissue. Gene expression analysis shows that the cells express joint marker genes including *Erg, tenascin-C*, and *Gdf5*. Thus, these cells exhibit interzone-like characteristics and may represent interzone precursors or full-fledged interzone cells, implying that the cells become developmentally determined and patterned even in the absence of overt joint formation. Why then would the cells be largely located around the periphery of the prospective mutant joint site? One possibility is that in the absence of Ihh signaling, the cells may not be able to participate in interzone formation because chondrocytes occupying the site do not undergo dedifferentiation and maintain the skeletal anlaga uninterrupted. If so, this would suggest that interzone formation may normally involve participation of perijoint precursor cells migrating into the joint site under the auspices of hedgehog signaling, a possibility also raised in previous studies.[40,41]

Whatever the explanation, it is noteworthy that our data provide novel and unexpected insights into interzone and joint organization and functioning. We show that reporter-positive cells are not uniformly distributed across the dorsoventral axis of developing control digit joints and display a gradient-like distribution, with more numerous cells present in the ventral half. Our data also show that the cavitation process appears to follow a similar ventral-to-dorsal pattern and is evident in the ventral side already at E15.5 but unappreciable on the dorsal side. Interestingly, there appears to be some ventral crowding of reporter-positive cells around prospective (uninterrupted) *Ihh*$^{-/-}$ digit joint sites, suggesting that such distribution is independent of hedgehog signaling and overt joint formation. Several previous studies have focused on the important question of how the dorsal–ventral axis of limb development and symmetry is regulated, and factors, such as *Wnt7 a* and *Lmx1* proteins, are critical players.[42–44] Thus, it is possible that interzone precursors and associated cells are patterned by similar mechanisms and attain the differential distribution revealed by our data. Why should the cells be differentially distributed along the ventral-to-dorsal axis? One obvious possibility is that this asymmetric distribution is linked to, and needed for, digit joint formation. The digits can only bend in one direction (toward the ventral side) and thus their joints must develop accordingly to permit and direct such unidirectional movement.

Presence of more numerous reporter-positive cells in the ventral half, presence of conspicuous ventrolateral cell masses, and the ventral-to-dorsal direction of the cavitation process may all be needed for appropriate formation, organization, and functioning of digit joints.

As previously realized and reiterated here, developing limb joints are differentially sensitive to the absence of Ihh. Interphalangeal joints fail to form in $Ihh^{-/-}$ embryos and the intervening cartilaginous tissue remains uninterrupted. Instead, the mutant wrist and ankle sites contain cartilaginous elements separated by mesenchymal tissue (FIG. 1) and, as shown previously, elbow and knee sites contain separated or partially separated cartilaginous elements.[14] It is been known for a while that though limb synovial joints are comparable biomechanical instruments, their development and organization are responsive to, and directed by, somewhat distinct mechanisms. For instance, major defects are seen in interphalangeal joints in mice lacking *Gdf5*, whereas the more proximal joints are minimally affected.[45] Proximal joints are significantly affected only in double *Gdf5/Gdf6*-null mice,[46] strongly indicating that *Gdf5* and *Gdf6* have redundant function in development of proximal joints, whereas development of interphalangeal joints largely relies on *Gdf5*. Regulatory sequences specific for expression of *GdF* genes in proximal but not distal joints have also been identified, suggesting that different mechanisms exist for generating specific interzone expression patterns at different anatomical locations in the limb.[47] It is possible then that the different limb joints also have distinct compensatory mechanisms or redundant pathways that would render them differentially sensitive to Ihh deficiency. Clearly, much more work is needed to sort out this important issue. In any case, it is interesting to note that even though they are not fully ablated by Ihh absence, the developing mutant wrist (and ankle) joints are profoundly misshaped and disorganized. The reciprocal concave–convex contours that characterize normal wrist elements and sculpt the two sides into functional and interlocking instruments are absent in the mutants and both sides display a convex shape; in addition, the intervening mesenchymal tissue is poorly organized and thin (see FIG. 4M). Thus, Ihh signaling appears to be essential for joint morphogenesis, and this suggestion is reiterated by our observation that the opposing sides of wild-type digit joints display small but appreciable differences in gene expression of hedgehog target and receptor genes (see FIG. 2G–L).

A major future task is to clarify how articular cartilage, ligaments, synovial lining, and other differentiated joint tissues form during joint development and to what extent they do or do not derive from interzone cells. Based on matrillin-1 gene expression patterns and genetic tracking, it has been proposed recently that articular chondrocytes do not derive from interzone cells but are descendants of the original mesenchymal cell condensations.[48] Interzone cells would instead give rise to the other joint tissues, such as synovial lining. Our *Gdf5-Cre/Rosa R26R*-based cell fate studies do not allow us to confirm or refute such possibility, given that both interzone cells and most-epiphyseal early chondrocytes express *Gdf5* and would become reporter-positive. In other studies, we have shown that Erg is likely to have an important role in establishing the stable permanent status of articular chondrocytes.[20] Important roles are also played by TGF-β signaling mechanisms.[49,50] These and related studies emphasize the fact that joint formation is very complex and requires a multitude of mechanisms regulating: joint site patterning and determination; interzone formation and joint morphogenesis; and formation of differentiated joint tissues. This multitude of mechanisms and processes needs to be coordinated and orchestrated spatiotemporally such that appropriate joints form at each specific location along the longitudinal axis of the limb. The challenge now is to figure out exactly how this feat is accomplished and whether disturbances in these fundamental mechanisms underlie congenital joint defects.

ACKNOWLEDGMENTS

This study was supported by NIH grants AR046000, AG025868, and AR042236. Dr. Roundtree's present address is: BN-Immuno Therapeutics, Garcia Avenue, Mountain View, California, USA.

REFERENCES

1. HINCHLIFFE, J.R. & D.R. JOHNSON. 1980. The Development of the Vertebrate Limb: 72–83. Oxford University Press. New York, NY.
2. HAMRICK, M.W. 2001. Primate origins: evolutionary change in digital ray patterning and segmentation. J. Hum. Evol. **40:** 339–351.
3. HOLDER, N. 1977. An experimental investigation into the early development of the chick elbow joint. J. Embryol. Exp. Morphol. **39:** 115–127.
4. MITROVIC, D.R. 1977. Development of the metatarsalphalangeal joint in the chick embryo: morphological, ultrastructural and histochemical studies. Am. J. Anat. **150:** 333–348.
5. MITROVIC, D. 1978. Development of the diathrodial joints in the rat embryo. Am. J. Anat. **151:** 475–485.
6. PACIFICI, M., E. KOYAMA & M. IWAMOTO. 2005. Mechanisms of synovial joint and articular cartilage formation: recent advances, but many lingering mysteries. Birth Defects Res., Pt. C. **75:** 237–248.
7. SORIANO, P. 1999. Generalized lacZ expression with the ROSA26 Cre reporter strain. Nat. Genet. **21:** 70–71.
8. ROUNTREE, R.B. et al. 2004. BMP receptor signaling is required for postnatal maintenance of articular cartilage. PLoS Biol. **2:** 1815–1827.
9. BITGOOD, M.J. & A.P. MCMAHON. 1995. Hedgehog and Bmp genes are coexpressed at many diverse sites of cell-cell interaction in the mouse embryo. Dev. Biol. **172:** 126–138.
10. KOYAMA, E. et al. 1996. Early chick limb cartilaginous elements possess polarizing activity and express Hedgehog-related morphogenetic factors. Dev. Dynam. **207:** 344–354.
11. VORTKAMP, A. et al. 1996. Regulation of rate of cartilage differentiation by Indian hedgehog and PTH-related protein. Science **273:** 613–622.

12. NAKAMURA, T. *et al.* 1997. Induction of osteogenic differentiation by hedgehog proteins. Biochem. Biophys. Res. Commun. **237:** 465–469.
13. LANSKE, B. *et al.* 1996. PTH/PTHrp receptor in early development and Indian hedgehog-regulated bone growth. Science **273:** 663–666.
14. ST-JACQUES, B. M. HAMMERSCHMIDT & A.P. MCMAHON. 1999. Indian hedgehog signaling regulates proliferation and differentiation of chondrocytes and is essential for bone formation. Genes Dev. **13:** 2076–2086.
15. CHUNG, U.-I. *et al.* 2001. Indian hedgehog couples chondrogenesis to osteogenesis in endochondral bone development. J. Clin. Invest. **107:** 295–304.
16. KOYAMA, E. *et al.* 1996. Expression of syndecan-3 and tenascin-C: possible involvement in periosteum development. J. Orthop. Res. **14:** 403–412.
17. PACIFICI, M. 1995. Tenascin-C and the development of articular cartilage. Matrix Biol. **14:** 689–698.
18. IWAMOTO, M. *et al.* 2000. Transcription factor ERG variants and functional diversification of chondrocytes during long bone development. J. Cell Biol. **150:** 27–39.
19. IWAMOTO, M. *et al.* 2005. The balancing act of transcription factors C-1-1 and Runx2 in articular cartilage development. Biochem. Biophys. Res. Commun. **328:** 777–782.
20. IWAMOTO, M. *et al.* 2007. Transcription factor ERG and joint and articular cartilage formation during mouse limb and spine skeletogenesis. Dev. Biol. **305:** 40–51.
21. SAVARESE, J.J., H. ERICKSON & S.P. SCULLY. 1996. Articular chondrocyte tenascin-C production and assembly into *de novo* extracellular matrix. J. Orthop. Res. **14:** 273–281.
22. LOBE, C.G. *et al.* 1999. Z/AP, a double receptor for Cre-mediated recombination. Dev. Biol. **208:** 281–292.
23. SPITZ, F. & D. DUBOULE. 2001. The art of making a joint. *Science* **291:** 1713–1714.
24. PACIFICI, M. *et al.* 1990. Hypertrophic chondrocytes. The terminal stage of differentiation in the chondrogenic cell lineage? *In* Cell Lineages in Development, Vol. 599. F.A. Pepe, *et al.*, Eds.: 45–57. New York Academy of Sciences. New York, NY.
25. IWAMOTO, M., M. ENOMOTO-IWAMOTO & K. KURISU. 1999. Actions of hedgehog proteins on skeletal cells. Crit. Rev. Oral Biol. Med. **10:** 477–486.
26. KRONENBERG, H.M. 2003. Developmental regulation of the growth plate. Science **423:** 332–336.
27. CHUANG, P.-T. & A.P. MCMAHON. 1999. Vertebrate hedgehog signaling modulated by induction of a hedgehog-binding protein. Nature **397:** 617–621.
28. IWASAKI, M., A.X. LE & J.A. HELMS. 1997. Expression of Indian hedgehog, bone morphogenetic protein 6 and gli during skeletal morphogenesis. Mech. Dev. **69:** 197–202.
29. SPATER, D. *et al.* 2006. Role of canonical Wnt-signaling in joint formation. Eur. Cells Materials **12:** 71–80.
30. GRITLI-LINDE, A. *et al.* 2001. The whereabouts of a morphogen: direct evidence for short- and graded long-range activity of hedgehog signaling peptides. Dev. Biol. **236:** 364–386.
31. YIN, M. *et al.* 2002. Antiangiogenic treatment delays chondrocyte maturation and bone formation during limb skeletogenesis. J. Bone Min. Res. **17:** 56–65.
32. PETERS, C. *et al.* 2004. The cholesterol membrane anchor of the Hedgehog protein confers stable membrane association to lipid-modified proteins. Proc. Natl. Acad. Sci. USA **101:** 8531–8536.

33. PEPINSKY, R.B. *et al.* 1998. Identification of a palmitic acid-modified form of human Sonic hedgehog. J. Biol. Chem. **273:** 14037–14045.
34. BELLAICHE, Y., I. THE & N. PERRIMON. 1998. Tout-velu is a Drosophila homologue of the putative tumor suppressor EXT-1 and is needed for Hh diffusion. Nature **394:** 85–88.
35. SHIMAZU, A. *et al.* 1996. Syndecan-3 and the control of chondrocyte proliferation during endochondral ossification. Exp. Cell Res. **229:** 126–136.
36. SHIMO, T. *et al.* 2004. Indian hedgehog and syndecan-3 coregulate chondrocyte proliferation and function during chick limb skeletogenesis. Dev. Dyn. **229:** 607–617.
37. DAVID, G. *et al.* 1993. Saptial and temporal changes in the expression of fibroglycan (syndecan-2) during mouse embryonic development. Development **119:** 841–854.
38. KOZIEL, L. *et al.* 2004. Ext1-dependent heparan sulfate regulates the range of Ihh signaling during endochondral ossification. Dev. Cell. **6:** 801–813.
39. KOYAMA, E. *et al.* 2007. Conditional Kif3 a ablation causes abnormal hedgehog signaling topography, growth plate dysfunction and ectopic cartilage formation in mouse cranial base synchondroses. Development **134:** 2159–2169.
40. KOYAMA, E. *et al.* 1995. Syndecan-3, tenascin-C and the development of cartilaginous skeletal elements and joints in chick limbs. Dev. Dynam. **203:** 152–162.
41. NIEDERMAIER, M. *et al.* 2005. An inversion involving the mouse Shh locus results in brachydactyly through dysregulation of Shh expression. J. Clin. Invest. **115:** 900–909.
42. DEALY, C.N. *et al.* 1993. Wnt-5 a and Wnt-7 a are expressed in the developing chick limb bud in a manner that suggests roles in pattern formation along the proximodistal and dorsoventral axes. Mech. Dev. **43:** 175–186.
43. PARR, B.A. & A.P. MCMAHON. 1995. Dorsalizing signal wnt-7 a required for normal polarity of D-V and A-P axes of the mouse limb. Nature **374:** 350–353.
44. RIDDLE, R.D. *et al.* 1995. Induction of the LIM homeobox gene Lmx1 by WNT7 a establishes the dorsoventral pattern in the vertebrate limb. Cell **83:** 631–640.
45. STORM, E.E. *et al.* 1994. Limb alterations in brachypodism mice due to mutations in a new member of the TGFb-superfamily. Nature **368:** 639–643.
46. STORM, E.E. & D.M. KINGSLEY. 1996. Joint patterning defects caused by single and double mutations in members of the bone morphogenetic protein (BMP) family. Development **122:** 3969–3979.
47. MORTLOCK, D.P., C. GUENTHER & D. M. KINGSLEY. 2003. A general approach for identifying distant regulatory elements applied to the Gdf6 gene. Genome Res. **13:** 2069–2081.
48. HYDE, G. *et al.* 2007. Lineage tracing using matrillin-1 gene expression reveals that articular chondrocytes exist as the joint interzone forms. Dev. Biol. **304:** 825–833.
49. SERRA, R. *et al.* 1997. Expression of a truncated, kinase-defective TGF-b type II receptor in mouse skeletal tissue promotes terminal chondrocyte differentiation and osteoarthritis. J. Cell Biol. **139:** 541–552.
50. SPAGNOLI, A. *et al.* 2007. TGF-b signaling is essential for joint morphogenesis. J. Cell Biol. **177:** 1105–1117.

Morphogen Receptor Genes and Metamorphogenes

Skeleton Keys to Metamorphosis

FREDERICK S. KAPLAN,[a] JAY GROPPE,[b] ROBERT J. PIGNOLO,[c] AND EILEEN M. SHORE[d]

[a]Departments of Orthopaedic Surgery and Medicine, The University of Pennsylvania School of Medicine, Philadelphia, Pennsylvania, USA

[b]Department of Biochemistry, University of Texas Health Science Center at San Antonio, San Antonio, Texas, USA

[c]Department of Medicine, The University of Pennsylvania School of Medicine, Philadelphia, Pennsylvania, USA

[d]Departments of Orthopaedic Surgery and Genetics, The University of Pennsylvania School of Medicine, Philadelphia, Pennsylvania, USA

ABSTRACT: Morphogen receptors are nodal points in signal transduction pathways that regulate morphogenesis during embryonic development. A recent discovery identified a recurrent missense mutation in a gene encoding a morphogen receptor responsible for the elusive process of skeletal metamorphosis in humans. Metamorphosis, the postnatal transformation of one normal tissue or organ system into another, is a biological process rarely seen in higher vertebrates or mammals, but exemplified pathologically by the disabling autosomal dominant disorder, fibrodysplasia ossificans progressiva (FOP). Individuals with FOP experience episodes of spontaneous or trauma-induced metamorphosis that convert normal functioning aponeuroses, fascia, ligaments, tendons, and skeletal muscles into a highly ramified and disabling second skeleton of heterotopic bone. The recurrent single nucleotide missense mutation in the gene encoding activin receptor IA/activin-like kinase 2 (ACVR1/ALK2), a bone morphogenetic protein (BMP) type I receptor that causes FOP in all classically affected individuals worldwide, is one of the most specific disease-causing mutations in the human genome and the first identified human metamorphogene. These findings provide deep insight into a signaling pathway that regulates tissue and organ stability following morphogenesis, and that when dysregulated in a specific manner, orchestrates the metamorphosis of one normal tissue or organ system into

Address for correspondence: Frederick S. Kaplan, M.D., Isaac and Rose Nassau Professor of Orthopaedic Molecular Medicine, The University of Pennsylvania School of Medicine, The Department of Orthopaedic Surgery, Hospital of the University of Pennsylvania, Silverstein 2, 3400 Spruce Street, Philadelphia, PA 19104. Voice: 215-349-8726; fax: 215-349-5298.

frederick.kaplan@uphs.upenn.edu

Ann. N.Y. Acad. Sci. 1116: 113–133 (2007). © 2007 New York Academy of Sciences.
doi: 10.1196/annals.1402.039

another. The study of skeletal metamorphosis in FOP provides profound insight into the molecular mechanisms that ensure phenotypic stability following morphogenesis and that ordinarily lay deeply hidden in the highly conserved signaling pathways that regulate cell fate. Such insight is applicable to a broad range of human afflictions.

KEYWORDS: metamorphosis; morphogen; morphogen receptor; meta-morphogene; bone morphogenetic protein; BMP; BMP signaling pathways; BMP receptor; heterotopic ossification; ACVR1; fibrodysplasia ossificans progressiva

My soul would sing of metamorphoses;
But since, o gods,
You were the source of these bodies
Becoming other bodies,
Breathe your breath into my book of changes.
— Ovid
The Metamorphoses

Only when he was already in the doorway
did he turn his head–not completely,
for he felt his neck stiffening;
He soon made the discovery
That he could no longer move at all.
— Franz Kafka
The Metamorphosis

Something deeply hidden
had to be behind things
— Albert Einstein

INTRODUCTION: METAPHOR AND METAMORPHOSIS

The term *metamorphosis* refers to any striking developmental change of an animal's form or structure. The concept of metamorphosis has captivated mankind since antiquity. Ovid's masterpiece, *The Metamorphoses*, was the unifying theme of a monumental mythology that depicted morphologic transformation in classical legend more than 2,000 years ago.[1] In more modern accounts of *The Metamorphosis*, Kafka reveals a truth we know instinctively but are loathe to confront—that life can be transformed rapidly and tragically into a nightmare that redefines reality.[2]

While the only defined biological examples of metamorphoses are those in insects and amphibia, the medical concept of metamorphosis implies a

pathological process that transforms one normal tissue or organ system into another.[3] Anyone who has ever met a child with the catastrophically disabling condition fibrodysplasia ossificans progressiva (FOP) knows that Kafka's story was no fiction. The condition has been known for centuries, and its elusive mysteries have taunted physicians and scientists for as long.[4]

The childhood victims of this musculoskeletal sabotage seem normal at birth except for a telltale malformation of the great toes. Soon, the children succumb to progressive waves of ectopic skeletogenesis that transform the body's soft connective tissues (including aponeuroses, fascia, ligaments, tendons, and skeletal muscles) into an armament-like encasement of bone. Ribbons, sheets, and plates of heterotopic bone progressively seize the body's soft connective tissues, cross the joints, lock them in place, and render movement impossible. Attempts to remove this heterotopic bone lead to catastrophic exacerbations of the disease. Presently, there is no effective treatment or prevention. Clinical management is symptomatic and supportive.[4-9]

This article will focus on the biological process of skeletal metamorphosis in humans as exemplified by FOP. For those seeking a detailed exposition of FOP, many recent and comprehensive reviews are available.[4-10] Here, we will explore emerging knowledge of the genetic, molecular, and cellular basis of this rare and illustrative process and examine important questions that are being investigated.

Jules Rosenstirn wrote of FOP in 1918: "One does not wonder that a disease, so baffling in its course from the first causes to its ultimate state, should invite the speculative as well as the patiently investigating observer to lift the obscuring veil and solve this embarrassing puzzle."[11] As the patiently investigating observer will see, the obscuring veil has just been lifted, but the puzzle has not yet been solved.

MORPHOGENESIS AND METAMORPHOSIS

Much scientific attention has been devoted to the developmental process of morphogenesis, the establishment of pattern and form during embryogenesis.[3] However, little attention has been devoted to the medical aspects of metamorphosis—the dramatic and pathologic transformation of one normal tissue or organ system into another following morphogenesis. For insight, we turn our attention to the study of rare diseases.[12]

Rare diseases, such as progeria, proteus syndrome, and promyelocytic leukemia, have provided robust insight into complex biological processes, such as aging, regeneration, and oncogenesis, respectively. Similarly, FOP provides profound insight into the complex biological process of skeletal metamorphosis. The specificity of a rare disease, such as FOP, often permits a causative genetic factor to be isolated in a complex regulatory network, thus identifying and defining the network itself. Such insight is often the catalyst for dissecting

the interdependent signaling pathways that are influenced by common genetic variations and that lead to some of the most common diseases of mankind.[12]

Genetic diseases do not usually create new functional tissue or organs, but in FOP, several normal connective tissues undergo a profound metamorphosis into mature heterotopic bone through a postnatal transformation. Metamorphosis thus joins the ranks of wound healing, regeneration, oncogenesis, and aging as one of the key biological processes in which to study and therapeutically manipulate tissue behavior following morphogenesis.

As Thomas Maeder wrote in an article in *The Atlantic Monthly*:

> "FOP and its problems lie at the crossroads of several seemingly unrelated disciplines. Answers to questions that FOP poses will also address grander issues of how the body first creates its shape and then knows where to stop, how tissues decide to become what they are, and why they don't turn into something else."[13]

THE PATHOLOGY OF SKELETAL METAMORPHOSIS

The process of skeletal metamorphosis, as exemplified by FOP, does not involve the transdifferentiation of one cell into another, but rather a pathological process in which the normal structure and function of one tissue or organ is destroyed and replaced by that of another functioning tissue or organ.[14]

The pathologic stages of skeletal metamorphosis have been well described in FOP, and correspond to the bone morphogenetic protein (BMP)-induced lesions described by Urist,[15] Wozney *et al.*[16] Kaplan *et al.*[17] Gannon *et al.*[18–19] Glaser *et al.*[20] and Kan *et al.*[21] Skeletal metamorphosis in FOP begins with an inflammatory infiltrate, proceeds to muscle cell death, replacement with a highly angiogenic fibroproliferative lesion, and maturation through a cartilage anlage that ends with the formation of a new skeletal element.[14–21]

Despite these detailed descriptions, there has been a persistent misconception in the medical and scientific community about the process. For example, it is commonly believed that the process of skeletal metamorphosis involves a transdifferentiation of muscle cells into bone cells. That, in fact, does not appear to occur *in vivo*. While studies by Katagiri and others document that C2C12 premyoblasts can transiently express an osteogenic phenotype in the presence of high concentrations of recombinant BMP or constitutively active BMP type I receptors *in vitro*, that is not the process by which skeletal metamorphosis occurs in FOP or in BMP-induced heterotopic ossification *in vivo*.[22–23] Rather, normal functioning connective tissues, such as aponeuroses, fascia, ligaments, tendons, and skeletal muscles, are replaced by heterotopic bone through a complex process that involves inflammation, muscle cell degeneration and death, fibroproliferation from an as yet unidentified connective tissue progenitor cell population, intense angiogenesis, condensation of fibroproliferative tissue to form cartilage, calcification of cartilage, and eventual replacement of

calcified cartilage with mature heterotopic bone containing marrow elements.[14-21] Thus, normal muscles cells do not become normal bone cells. Rather, normal muscle tissue is replaced by normal bone tissue in a complex pathological process of skeletal metamorphosis, as outlined above.

REPAIR, REGENERATION, AND METAMORPHOSIS

Metamorphosis is at the far-end of a spectrum of cellular and tissue response to injury that includes repair and regeneration.[24] Repair and regeneration are adaptive biological processes that involve inflammatory stimuli (FIG. 1). In mammals, however, the regenerative response is blunted, and most tissues respond to injury with a repair process that involves the formation of scar tissue.[24] Bone, however, is one of the only tissues in mammals that responds to injury with complete regeneration.[25] Hair, is the other.[26,27] Even liver, which may exhibit a regenerative response following injury, forms a scar at the sight of injury.

For the past century, histologic evidence has supported that the embryogenesis of the skeleton and its regeneration following injury are closely related processes.[25] Two independent studies strongly suggest that while inflammatory signals are superfluous for skeletal formation, they are required for the initiation of skeletal regeneration and heterotopic ossification.[28,29] In addition, a recent study showed that BMP2 activity was required for the initiation of skeletal regeneration following fractures.[30] Previous studies have indicated that BMP stimulation of the prostaglandin pathway through the action of cyclooxygenase 2 was essential for fracture healing, thus linking the BMP signaling pathway with the skeletal regeneration-inflammatory pathway postnatally, but not during morphogenesis.[28] Thus, the same signaling pathway that exhibits robustness and refractoriness to stochastic changes during morphogenesis has evolved to be exquisitely sensitive to environmental signals postnatally.[28-31]

Cunningham *et al.* showed that BMP4 was chemotactic to monocytes, further implicating an association between the BMP pathway, inflammation, and skeletal regeneration postnatally.[32] Studies by Wozney and by Glaser showed that recombinant BMPs implanted subcutaneously stimulate heterotopic ossification with an early inflammatory response, nearly identical to that seen in FOP lesions.[16,20] Endogenous prostanoids are elevated in the serum of FOP patients during disease flare-ups (data not shown) and are derived from local tissue beds.[33] While BMP signaling is required for skeletal formation as well as regeneration, evidence from histologic, biochemical, and BMP pathway studies suggests that an inflammatory response is required for skeletal regeneration and at least permissive for skeletal metamorphosis.[28-34] It is tempting to speculate that specific inflammatory microenvironments are also required for skeletal metamorphosis, and such studies are presently under way.

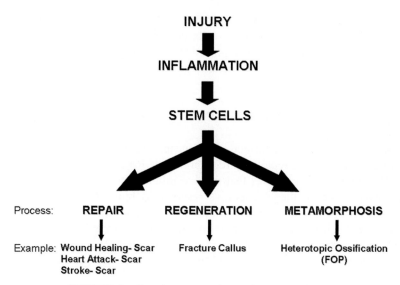

FIGURE 1.   Repair, regeneration, and metamorphosis.

METAMORPHOSIS AND ONCOGENESIS

The clinical and pathological features of skeletal metamorphosis are often mistaken for oncogenesis, especially in its early stages.[7,17,35] A recent survey of FOP patients worldwide indicated that nearly 90 percent of FOP patients were misdiagnosed, most commonly with soft tissue sarcoma or aggressive juvenile fibromatosis, and that 50 percent of those patients experienced permanent and lifelong harm as a result of diagnostic and therapeutic errors that resulted from the misdiagnosis. While skeletal metamorphosis and primary skeletal oncogenesis share some clinical and pathological features, their molecular basis and natural histories are distinctly different (TABLE 1).

MORPHOGEN RECEPTOR GENES
AND METAMORPHOGENES

The BMP signaling pathway is one of the most highly conserved signaling pathways in nature and regulates a myriad of developmental and postdevelopmental processes beginning in early embryogenesis and continuing through adult life.[16,36–43] A large body of evidence supports that BMPs act as morphogens in vertebrate development.[16,31,40–43] BMPs signal by binding to and activating heterotetrameric transmembrane complexes of type I and type II BMP receptors. Both type I and type II BMP receptors are serine/threonine kinases that have similar functional domains. Ligand binding occurs preferentially at the N-terminal extracellular domain of the BMP type I receptor, which

TABLE 1. Clinical and pathological comparison of skeletal metamorphosis and skeletal oncogenesis

	Metamorphosis	Oncogenesis
Type of lesion	Reactive	Clonal
Lesional evolution	Rapid	Slow
Replacement of one normal tissue with another normal tissue	Yes	No
Inflammatory trigger	Yes	No
Early inflammatory infiltrate	Yes	No
Causative mutation(s)	ACVR1/ALK2 (R206H)	Many
Angiogenesis	Yes	Yes
Growth in the absence of growth signals	Yes	Yes
Growth despite stop commands	Yes	Yes
Evasion of autodestruct mechanisms	No	Yes
Immorality of lesional cells	No	Yes
Invasive ability of lesional cells	Yes	Yes
Metastatic ability of lesional cells	No	Yes

is connected by a single transmembrane region to the C-terminal cytoplasmic kinase domain.[40–43] A unique feature of type I receptors is a cytoplasmic juxtamembrane region rich in glycine and serine residues (GS domain). Following ligand binding, serines and threonines in the GS domain are phosphorylated by the constitutively active type II BMP receptor. The BMP type I receptor is activated by these phosphorylation events and transmits downstream BMP signals through BMP pathway-specific Smads (Smads1,5,8) and p38 MAPK signaling pathways to regulate transcription of BMP-responsive target genes. BMP signaling can be mediated by four known type I receptors: TSR (ALK1); ACVR1 (ALK2); BMPR1A (ALK3); and BMPR1B (ALK6). Numerous comprehensive reviews are available on this seminal signaling pathway that will orient the reader to the concepts that follow.[40–43]

A large body of work has supported dysregulated BMP signaling in the pathogenesis of FOP.[44–51] We recently mapped FOP to chromosome 2q23–24 by linkage analysis and identified an identical heterozygous missense activating mutation (c.617G>A; R206H) in the glycine-serine (GS) activation domain of activin receptor 1A/activin-like kinase 2 (ACVR1/ALK2), a BMP type I receptor in all classically affected individuals worldwide.[52]

This single nucleotide missense mutation transforms a morphogen receptor gene into a metamorphogene (TABLE 2). The resultant mutant protein alters the basal set point and ligand-dependent sensitivity for BMP signaling in a cell-autonomous manner in putative connective tissue progenitor cells.[51] Identification of the mutant transmembrane receptor (and remarkably a single substituted amino acid residue in that receptor) provides a basis for elucidating the molecular pathophysiology of dysregulated BMP signaling and resultant skeletal metamorphosis in this illustrative and disabling condition.[51,52]

TABLE 2. Oncogenes and metamorphogenes

Normal gene type	Regulated process	Mutated gene type	Type of mutation	Example	Phenotype
Proto-oncogene	Cell cycle	Oncogene	Dominant activating	Fos	Osteosarcoma
Tumor-suppressor gene	Cell cycle	Tumor-suppressor gene	Recessive inactivating (2-hit)	Rb	Osteosarcoma
Morphogen receptor gene	Cell fate	Metamorphogene	Dominant activating	ACVR1/ ALK2	FOP
Secreted morphogen antagonist	Cell fate	Metamorphosis-suppressor gene	Recessive inactivating (2-hit?)	Noggin?	?

Although ACVR1/ALK2 has been recognized as a BMP receptor, investigations of its functions in embryonic development and in regulating cell differentiation have been limited (Reviewed in 52). ACVR1/ALK2 is expressed in many tissues including skeletal muscle and chondrocytes. Constitutive activation of ACVR1/ALK2 induces alkaline phosphatase activity in C2C12 cells, upregulates BMP4, downregulates BMP antagonists, expands cartilage elements, induces ectopic chondrogenesis, and stimulates joint fusions.[53] Constitutive ACVR1/ALK2 expression (similar to that seen in FOP), in embryonic chick limbs induces expansion of chondrogenic anlage, suggesting that promiscuous ACVR1/ALK2 signaling alters cell fate and induces undifferentiated mesenchyme to form cartilage.[53] Enhanced ACVR1/ALK2 activation in FOP is supported by recent data showing increased pathway-specific Smad phosphorylation and expression of BMP transcriptional targets in FOP cells as well as partial rescue of an ACVR1/ALK2-homologue loss of function phenotype in mutant zebrafish embryos (data not shown).

ACVR1/ALK2 is one of seven activin-like kinases (ALKs) in the human genome that all encode TGF-β/BMP type I serine/threonine transmembrane receptors involved in the determination of cell fate and differentiation in a wide variety of cells and tissues during embryonic development and postnatal life.[54] Alternate names may be more familiar to most (TABLE 3), and mutations are known in several.

METAMORPHOGENES AND MOLECULAR MECHANISMS OF METAMORPHOSIS

Protein homology modeling of the mutated ACVR1/ALK2 receptor in FOP predicts changes in both ligand-independent BMP signaling and in ligand-stimulated BMP signaling in FOP cells.[55] The canonical FOP mutation (ACVR1c.617G>A; R206H), an arginine residue replaced by a histidine residue at amino acid 206 in the GS domain of ACVR1/ALK2, is

TABLE 3. Activin-like kinases

ALKs	OMIM no.	Alternate names	Signaling through R-Smads	Available inhibitors	Disease-associated phenotypes
ALK1	601284	TSR1	1,5,8	–	Hereditary hemorrhagic telangiectasia, Type II
ALK2	102576	ACVR1	1,5,8	–	FOP
ALK3	601299	BMPRIA	1,5,8	–	Juvenile polyposis syndrome
ALK4	601300	ACVR1B	2,3	+	Pancreatic carcinoma
ALK5	190181	TGF-βR1	2,3	+	Loeys-Dietz syndrome
ALK6	603248	BMPRIB	1,5,8	–	Symphalangism syndromes
ALK7	608981	ACVR1C	2,3	+	None known

hypothesized to affect the binding and downstream function of coregulatory proteins.

The GS domain of all TGF-β/BMP type I receptors is a critical site for activation of pathway-specific Smad signaling proteins by constitutively active TFG-β/BMP type II receptors.[56–59] FKBP12 stabilizes the inactive confirmation of all type I TGF-β/BMP receptors including ACVR1/ALK2. When bound to the GS domain, FKBP12 prevents leaky activation of type I receptors in the absence of ligand.[56–59] Importantly, FKBP12 also serves as a docking protein for the Smad–Smurf complexes that mediate ubiquitination, internalization, and degradation of ACVR1/ALK2, and is predicted to regulate the concentration of ACVR1/ALK2 and its BMP type I receptor oligomerization partners at the cell membrane.[60,61]

The FOP mutation is predicted to impair FKBP12 binding and/or activity with resultant increased basal activity of BMP signaling in the absence of ligand as well as hyperresponsiveness of BMP signaling following ligand stimulation, two features of BMP signal dysregulation that have recently been demonstrated in FOP cells.[51] Thus one scenario is that FKBP12 interaction with the GS domain may be altered in FOP leading to promiscuous ACVR1/ALK2 activity. Recent preliminary data strongly support this hypothesis (data not shown), and are the subject of intensive investigation.

METAMORPHOGENES AND METAMORPHOSIS-SUPPRESSOR GENES

The existence of a dominant metamorphogene (as in ACVR1c.617G> A;R206H) raises the intriguing question of whether there may also be corresponding recessive metamorphosis-suppressor genes, analogous to recessive tumor-suppressor genes (TABLE 2). To explore this concept, let us first examine the BMP signaling pathway for potential candidates and ask whether any extant animal models may shed light on this important question.[40,47,62,63]

Morphogen gradients are determined not only by morphogens and their receptors, but also by secreted morphogen antagonists in highly regulated negative feedback pathways.[31,40,47,62,63] The BMP signaling pathway harbors a rich repertoire of secreted BMP antagonists including noggin, gremlin, chordin, DAN, cerebrus, and others.[40] Homozygous knockouts of these genes are commonly lethal during embryogenesis or during early postnatal life indicating the critical role that these molecules play in morphogenesis.[62]

Mutations of the *noggin* gene are worth examining in some detail. Mice with homozygous deletions of the noggin gene fail to form joints and die at birth due to failure of expansion of the chest wall.[62] In addition, many of the Noggin ($^-/^-$) animals have expansion of the normal borders of osteogenesis in the limbs and axial regions and have exuberant orthotopic and heterotopic bone, compatible with overactivity of BMP signaling.[62] What is not entirely clear, however, is whether the aponeurosis, fascia, ligaments, tendons, and skeletal muscles in these animals are specified normally during embryogenesis and then undergo a metamorphosis later in fetal life, or whether the soft connective tissues fail to be specified properly following gastrulation. Preliminary data are inconclusive, but strongly suggest the latter rather than the former. However, the lethality of these germline knockouts prohibits using these models to explore their potential role in metamorphosis, a postnatal phenomenon.

To explore the potential role of secreted morphogen antagonists in metamorphosis, a more fruitful approach might be to ask: if recessive metamorphosis-suppressor genes exist, might they function analogously to recessive tumor-suppressor genes in a classic two-hit model?[64-68] In such a scenario, for example, a germline inactivating mutation of one *noggin* allele might be followed by a somatic inactivating mutation of the second *noggin* allele in a connective tissue stem cell or progenitor cell, resulting in homozygous deletion of the *noggin* gene in that particular stem cell and its progeny and a phenotype of focal heterotopic ossification. Genetically engineered mice with haploinsufficiency of the *noggin* gene have no obvious phenotype, but would provide an excellent model for examining this two-hit hypothesis. It is interesting to note that patients with heterozygous mutations in the *noggin* gene have multiple symphalangism syndromes, a condition fully compatible with life, but also highly conducive, it would seem, to developing loss of heterozygosity at the second *noggin* allele in a connective tissue progenitor cell.[69]

It would be interesting to determine whether focal heterotopic ossification ever develops postnatally in mice or in humans with heterozygous inactivating mutations in *noggin*, and if so, whether such heterotopic ossification was due, as predicted, to loss-of-function of Noggin in a clonal population of somatic cells. While no such mice or humans have been reported to date, one would predict that such a phenotype might arise given a sufficient background population.

Pharmacologic and experimental physiologic data suggest that skeletal metamorphosis (as in FOP) may be driven by an imbalance in a morphogen gradient that favors agonist activity over antagonist activity regardless of the mechanism,[20,21] and that recessive metamorphosis-suppressor genes, analogous to recessive tumor-suppressor genes would likely exist (TABLE 2). Plausibly, somatic mutagenesis experiments in mice with heterozygous inactivating mutations in *noggin* might provide an excellent experimental model for examining this two-hit hypothesis *in vivo*.

MAPPING THE METAMORPHOSIS GENOME

ACVR1/ALK2 (c.617G>A;R206H) is the first human metamorphogene, but it is not the only one. Recent studies have identified patients with atypical FOP and FOP variants that harbor a unique array of mutations in the ACVR1/ALK2 gene that also cause skeletal metamorphosis (data not shown). Perhaps other types of tissue or organ metamorphosis will be discovered through widespread mutagenesis screens in various animal models, or may already exist in the natural world, but have not yet been documented.[70] The identification of genes and signaling pathways involved in skeletal metamorphosis (and other types of metamorphosis, should they exist) will facilitate a more comprehensive understanding of the regulatory pathways involved in tissue stability following morphogenesis.

METAMORPHOSIS AND EPIGENETICS

While the underlying genetics of FOP has recently been elucidated, the process of skeletal metamorphosis in FOP raises intriguing epigenetic questions as well. Epigenetic processes guide embryonic cells with limitless potential into restricted repertoires of adult fates.[71–74] The process of skeletal metamorphosis abolishes those fates and replaces them with another. Thus, the process of skeletal metamorphosis raises several critical questions: Are the epigenetic imprints on tissue-specific cells (in aponeuroses, fascia, ligaments, tendons, and skeletal muscles) reprogrammed, as in cloning? Or rather, are the tissue-specific cells that are imprinted with distinct epigenetic fates destroyed following an inflammatory trigger, and the remaining tissue scaffolds repopulated from surviving pluripotent cells that are programmed with a new epigenetic identity? The histopathologic and experimental data overwhelmingly support the latter rather than the former.[14,16–21,29,46] Genetic programming of a new epigenetic identity, triggered in part by inflammatory events and driven by postnatal imbalance in morphogenetic signaling pathways in multipotent tissue progenitor cells, may be a central theme in the pathophysiology of skeletal metamorphosis.

METAMORPHOSIS AND THE IMMUNE SYSTEM

While dysregulation of the BMP signaling pathway can explain many features of skeletal metamorphosis in FOP, other puzzling features of the disease strongly implicate an underlying inflammatory trigger and/or a conducive inflammatory microenvironment (FIG. 2).[34]

A recent study showed that aberrant expression of BMPs in soft tissue causes focal hypoxia and hypoxic stress within the target tissue, a prerequisite for the differentiation of stem cells to chondrocytes and subsequent heterotopic bone formation.[75] Flare-ups of FOP are frequently associated with muscle-fatigue and trauma-associated hypoxia.[5–7] Such physiologic hypoxic stress is predicted to exacerbate tissue damage, the release of local inflammatory mediators including BMPs and free radicals, and the subsequent stimulation of mutant ACVR1/ALK2 in as yet uncharacterized connective tissue progenitor cells that would potentiate skeletal metamorphosis (FIG. 2).[75–82]

A major new area of investigation of FOP research is beginning to focus on the relationship of inflammatory triggers, mutant ACVR1/ALK2 receptors, and local environmental factors, such as free radicals, pO_2, and pH in the episodic flare-ups of the disease.[34, 55, 75–82] Recent protein modeling studies by Groppe et al. predict that the canonical FOP mutation creates a pH-sensitive switch within the cytoplasmic domain of the mutant ACVR1/ALK2 receptor that leads not only to ligand-independent activation, but also to ligand-dependent hyperresponsiveness of mutant ACVR1/ALK2 in the microenvironment of a lowered intracellular pH.[55]

Our working hypothesis on the pathophysiology of heterotopic skeletogenesis in FOP is that soft tissue injury creates an inflammatory and acidic microenvironment in which prostaglandins, free radicals, and hypoxia stimulate resting connective tissue progenitor cells (CTPs) to become activated CTPs. These CTPs express the mutant ACVR1/ALK2 receptor that is putatively hyperactive in a mildly acidic intracellular environment and that results in misregulated BMP signaling through increased basal leakiness and conditional hyperresponsiveness to environmental BMPs (FIG. 2).[29] Once formed, the fibroproliferative cells in the FOP lesion produce robust amounts of BMP4, and overactivity of the BMP signaling is sufficient to drive the process of endochondral ossification to completeness in the absence of a continued inflammatory stimulus (FIG. 2).[29]

The recent discovery of the FOP metamorphogene in all familial and sporadic cases of classic FOP suggests a cell-autonomous basis for skeletal metamorphosis in FOP.[52] Protein modeling of the mutant receptor predicts destabilization of the GS activation domain consistent with basally leaky and conditional hyperresponsive BMP signaling as the cause of the ectopic chondrogenesis, osteogenesis, and joint fusion seen in FOP. Furthermore, these findings allow us to hypothesize that trauma and inflammation recruit osteogenic CTPs in which the activating mutation is expressed. These findings

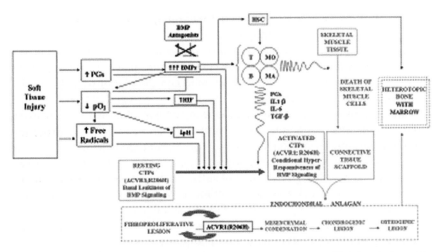

FIGURE 2. Hypothetical model of skeletal metamorphosis in fibrodysplasia ossificans progressiva. PGs = prostaglandins; pO_2 = tissue oxygen tension; HIF = hypoxia-inducing factor; CTPs = connective tissue progenitor cells; HSC = hematopoietic stem cells; T = T cells; B = B cells; Mo = monocytes; MA = mast cells; IL-1β = interleukin-1β; IL-6 = interleukin 6; TGF-β = transforming growth factor-beta; Straight arrows = denote progression; Wavy arrows = denote influence.

are consistent with *in vivo* observations in animal models that showed that once the endochondral anlagen are induced, abrogation of the inflammatory response will not inhibit the formation of heterotopic bone.[29]

Interactions between hematopoietic cells and mesenchymal cells in a microenvironment that is presumptively inductive and conducive to the initiation, progression, and sustenance of bone formation in FOP probably occurs at several steps (FIG. 2). Recent studies have strongly suggested that migration of stem cells to sites of inflammation is a key step in normal and disordered regenerative responses and likely plays a role as well in the pathology of skeletal metamorphosis.[24,29]

STEM CELLS AND METAMORPHOSIS

It is well established in FOP and in BMP-induced skeletal metamorphosis that mature connective tissue cells do not transdifferentiate into bone cells. Rather, normal functioning connective tissue is replaced with normal functioning bone tissue in a complex pathologic process, previously outlined in this article. Stem cells and progenitor cells must therefore lie at the very heart of the process of metamorphosis (FIG. 2).

Hematopoietic cells have been implicated in the skeletal metamorphosis of FOP, and their replacement has been postulated as a possible cure (reviewed in Ref. 29). However, the definitive contribution of hematopoietic cells to the pathogenesis of skeletal metamorphosis has, until recently, remained obscure. A recent study in a patient with FOP who coincidentally developed intercurrent aplastic anemia demonstrated that bone marrow transplantation does not cure FOP, most likely because the hematopoietic cell population is not the site, or at least not the dominant site, of the intrinsic dysregulation of the BMP signaling pathway in FOP. However, following transplantation of bone marrow from a normal donor, immunosuppression of the immune system appeared to ameliorate activation of skeletal metamorphosis in a genetically susceptible host. Thus, cells of hematopoietic origin may contribute to the formation of an ectopic skeleton, although they are not sufficient to initiate the process alone. Moreover, even a normal functioning immune system is apparently sufficient to trigger an FOP flare-up in a genetically susceptible host.[29]

Which cells contribute to the fibroproliferative and chondrogenic mesenchymal anlagen in skeletal metamorphosis? The question is fascinating, important, and unresolved. Recent studies performed with two independent routes of investigation support the contention that such cells are not of hematopoietic origin, but arise from a different pool of connective tissue progenitor cells residing in skeletal muscle and associated connective tissues with possible lineage origins in endothelial cells, smooth muscle cells, satellite cells, neural cells, or other connective tissue progenitor cells. We cannot rule out the possibility that some hematopoietically derived circulating osteoprogenitor cells contribute to the heterotopic bone, but they are not likely to be the predominant contributing cell population. Therefore, multiple sources of pluripotent stem cells or progenitors may contribute to the formation of an ectopic skeleton in the process of skeletal metamorphosis.[29] Detailed lineage-tracing experiments in transgenic mice with stable lineage markers will be necessary to definitively determine the origin of these cells. The development of a knockin mouse that replicates the identical mutation of classic FOP is presently under way and will facilitate the identification of the autonomous connective tissue progenitor cell(s) that are the sites of dysregulated BMP signaling in FOP.

Taken together, skeletal metamorphosis appears to be a stem cell or progenitor cell disorder. One normal tissue is replaced with another, but first, nearly all vestiges of the original tissue, except its soft tissue scaffolding, neurovascular infrastructure, and progenitor cell repository are destroyed and then replaced with a different tissue and a different epigenetic identity. Recent studies in FOP suggest that connective tissue progenitor cells resident in the local tissues are responsible for this metamorphosis although the process may be triggered by local inflammatory signals.[29,34] Essentially, skeletal metamorphosis is a disorder of dysregulated tissue repair. Metamorphosis thus unmasks the deep developmental restraints that must exist in normal tissue repair, and that are liberated, to the detriment of the host, by the specific recurrent mutation in the ACVR1/ALK2 (R206H) metamorphogene.[29,52]

PREVENTION AND TREATMENT OF SKELETAL METAMORPHOSIS

The successful prevention of skeletal metamorphosis will ultimately be based on at least one of four principles: blocking the renegade receptors and downstream signaling pathways in responsive cells, suppressing the immunological triggers that activate the responsive cells and pathways, inhibiting the responsive osteoprogenitor cells in the target tissue(s), and/or modifying the tissue environment so that it is less conducive to skeletal metamorphosis.

The identification of the recurrent heterozygous missense mutation that causes skeletal metamorphosis in all classically affected individuals with FOP provides a specific druggable target and a rational point of intervention in a critical signaling pathway in nature.[83] Plausible therapeutic approaches to inhibiting the promiscuous BMP signaling that leads to skeletal metamorphosis in FOP include inhibitory RNA technology, monoclonal antibodies, and most plausibly orally available small molecule selective signal transduction inhibitors (STIs) of ACVR1/ALK2.[83]

Small molecule STIs have proven incisive for investigating signal transduction pathways. Clinically, there is a great need for small molecule inhibitors that could control renegade BMP signaling. Such inhibitors will be important not only for treating FOP and perhaps other related disorders of skeletal metamorphosis, but also in probing the dysregulated BMP signaling pathway that may contribute to more common forms of acquired skeletal metamorphosis.[84–87]

METAMORPHOSIS AND TISSUE ENGINEERING

ACVR1/ALK2 (R206H) is the first identified human metamorphogene and provides a genetic basis for understanding the general biological principles that orchestrate the pathological transformation of one normal organ system into another. The immediate goal of FOP research is to understand the molecular and cellular pathophysiology of this process and to use that knowledge to develop pharmacologic methods to block it. Conversely, it may be possible to harness the gene mutation that causes skeletal metamorphosis to create new skeletal elements in a controlled way—for patients who have osteoporotic fractures, for those with severe bone loss from trauma or neoplasms, for those with fractures that fail to heal, for those with spinal fusions that are slow to heal, or for those with congenital malformations of the spine and limbs. The discovery of the FOP gene, the first human metamorphogene, is a monumental milestone in an epic journey to understand the biological principles that regulate tissue stability and skeletal metamorphosis following embryogenesis.

ARE THERE OTHER EXAMPLES OF SKELETAL METAMORPHOSIS IN HUMANS?

Classic FOP and its phenotypic and genotypic variants are the most dramatic, but not the only examples of skeletal metamorphosis in humans. There are, for example, many acquired forms of skeletal metamorphosis that are spatially and temporally limited in their extent. Some are fairly common, such as the heterotopic ossification that occurs following closed head injury, spinal cord injury, total hip replacement, athletic injury, and blast injuries from war.[84–88] Additionally, approximately 13 percent of individuals with end-stage valvular heart disease form mature heterotopic bone in the aortic valve.[84] To date, there are few studies on the molecular pathogenesis in these more common disorders of skeletal metamorphosis. Where it has been examined, the histology is similar in all, and it would not be surprising if the ACVR1/ALK2 signaling pathway was involved in the more common as in the rarer forms of skeletal metamorphosis.

Nature does not use different genes, molecules, and pathways for common conditions or for rare ones. Rather, it is often the rare disease that reveals which gene, molecule, or pathway nature hijacks in its common infirmities.[12] William Harvey, the discoverer of the circulatory system wrote in 1657,

> "Nature is nowhere accustomed more openly to display her secret mysteries than in cases where she shows traces of her workings apart from the beaten path; nor is there any better way to advance the proper practice of medicine than to give our minds to the discovery of the usual law of nature by the careful investigation of cases of rarer forms of disease."[89]

SUMMARY

Metamorphosis, the postnatal transformation of one normal tissue or organ system into another, is a biological process rarely studied in higher vertebrates or mammals, but exemplified pathologically by the autosomal dominant disorder FOP and its variants. The recurrent single nucleotide missense mutation in the gene encoding ACVR1/ALK2, a BMP type I receptor that causes skeletal metamorphosis in all classically affected individuals worldwide, is the first identified human metamorphogene. The identification of this metamorphogene is beginning to provide deep insight into a highly conserved signaling pathway that regulates tissue stability following morphogenesis, and that when damaged at a highly specific locus, permits the renegade metamorphosis of one normal tissue or organ system into another. A deeper understanding of the process of skeletal metamorphosis, as revealed by the rare condition FOP, will likely lead to the development of more effective treatments for FOP and possibly for more common disorders of skeletal metamorphosis. The discovery of this first human metamorphogene also provides dramatic new insight into the mechanisms of tissue stability and developmental plasticity following morphogenesis that will provide a greater understanding of the repertoire of normal and pathological mechanisms of tissue remodeling and repair.

ACKNOWLEDGMENTS

This work was supported in part by the Center for Research In FOP and Related Disorders, the International FOP Association, The Ian Cali Endowment, The Weldon Family Endowment, The Isaac and Rose Nassau Professorship of Orthopaedic Molecular Medicine, and by a grant from The National Institutes of Health (RO1-AR40196).

REFERENCES

1. MANDELBAUM, A. 1993. The Metamorphoses of Ovid. Harcourt, New York.
2. KAFKA, F. 1948. The Metamorphosis. Schocken Books. New York.
3. GILBERT, S. 2000. Developmental Biology. Sinauer Associates Sunderland, MA.
4. KAPLAN, F.S. *et al.* 1998. Editorial comment. Clin. Orthop. Rel. Res. **346:** 2–4.
5. CONNOR, J.M. & D.A. EVANS. 1982. Fibrodysplasia ossificans progressiva: the clinical features and natural history of 34 patients. J. Bone Joint Surg. Br. **64:** 76–83.
6. COHEN, R.B. *et al.* 1993. The natural history of heterotopic ossification in patients who have fibrodysplasia ossificans progressiva (FOP) J. Bone Joint Surg. Am. **75:** 215–219.
7. KAPLAN, F.S. *et al.* 2005. The phenotype of fibrodysplasia ossificans progressiva (FOP). Clin. Rev. Bone Miner. Metab. **3:** 183–188.
8. GLASER, D.L. & F.S. KAPLAN. 2005. Treatment considerations for the management of fibrodysplasia ossificans progressiva. Clin. Rev. Bone Miner. Metab. **3:** 243–250.
9. KAPLAN, F.S. *et al.* 2005. Introduction. Clin. Rev. Bone Miner. Metab. **3:** 175–177.
10. KAPLAN, F.S. & D.L. GLASER. 2005. Thoracic insufficiency syndrome in patients with fibrodysplasia ossificans progressiva. Clin. Rev. Bone Miner. Metab. **3:** 213–216.
11. ROSENSTIRN, J.A. 1918. A contribution to the study of myositis ossificans progressiva. Ann. Surg. **68:** 485–520; 591–637.
12. KAPLAN, F.S. 2006. The key to the closet is the key to the kingdom. Orphan Disease Update **24:** 1–9.
13. MAEDER, T. 1998. A few hundred people turned to bone. Atlantic Monthly **281:** 81–89.
14. PIGNOLO, R.J. *et al.* 2005. The fibrodysplasia ossificans progressiva lesion. Clin. Rev. Bone Miner. Metab. **3:** 195–200.
15. URIST, M.R. 1965. Bone formation by autoinduction. Science **150:** 893–899.
16. WOZNEY, J.M. *et al.* 1988. Novel regulators of bone formation: molecular clones and activities. Science **242:** 1528–1534.
17. KAPLAN, F.S. *et al.* 1993. The histopathology of fibrodysplasia ossificans progressiva: an endochondral process. J. Bone Joint Surg. Am. **75:** 220–230.

18. GANNON, F.H. *et al.* 1998. Acute lymphocytic infiltration in an extremely early lesion of fibrodysplasia ossificans progressiva. Clin. Orthop. Rel. Res. **346:** 19–25.

19. GANNON, F.H. *et al.* 2001. Mast cell involvement in fibrodysplasia ossificans progressiva. Hum. Pathol. **32:** 842–848.

20. GLASER, D.L. *et al.* 2003. *In vivo* somatic call gene transfer of an engineered noggin mutein prevents BMP4-induced heterotopic ossification. J. Bone Joint Surg. Am. **85:** 2332–2342.

21. KAN, L. *et al.* 2004. Transgenic mice overexpressing BMP4 develop a fibrodysplasia ossificans progressiva (FOP)–like phenotype. Am. J. Pathol. **165:** 1107–1115.

22. KATAGIRI, T. *et al.* 1994. Bone morphogenetic protein-2 converts the differentiation pathway of C2C12 myoblasts into the osteoblast lineage. J. Cell Biol. **127:** 1755–1766.

23. AKIYAMA, S. *et al.* 1997. Constitutively active BMP type I receptors transduce BMP2 signals without the ligand in C2C12 myoblasts. Exp. Cell Res. **235:** 362–369.

24. STOICK-COOPER, C.L. 2007. Advances in signaling in vertebrate regeneration as a prelude to regenerative medicine. Genes Dev. **21:** 1292–1315.

25. FERGUSON, C. *et al.* 1999. Does adult fracture repair recapitulate embryonic skeletal formation? Mech. Dev. **87:** 57–66.

26. ITO, M. *et al.* 2007. Wnt-dependent *de novo* hair follicle regeneration in adult mouse skin after wounding. Nature **47:** 316–320.

27. CHUONG, C-M. 2007. Regenerative biology: new hair from healing wounds. Nature **447:** 265–266.

28. ZHANG, X. *et al.* 2002. Cyclooxyenase-2 regulates mesenchymal cell differentiation into the osteoblast lineage and is critically involved in bone repair. J. Clin. Invest. **109:** 1405–1415.

29. KAPLAN, F.S. *et al.* 2007. Hematopoietic stem-cell contribution to ectopic skeletogenesis. J. Bone Joint Surg. Am. **89:** 347–357.

30. TSUJI, K. *et al.* 2006. BMP2 activity, although dispensable for bone formation is required for the initiation of fracture healing. Nat. Genet. **38:** 1424–1429.

31. LANDER, A.D. 2007. Morpheus unbound: reimagining the morphogen gradient. Cell **128:** 245–256.

32. CUNNINGHAM, N.S. *et al.* 1992. Osteogenin and recombinant bone morphogenetic protein 2B are chemotactic for human monocytes and stimulate transforming growth factor beta1 mRNA expression. Proc. Natl. Acad. Sci. USA **89:** 11740–11744.

33. MAXWELL, W.A. *et al.* 1978. Elevated prostaglandin production in cultured cells from a patient with fibrodysplasia ossificans progressiva. Prostaglandins **15:** 123–129.

34. KAPLAN, F.S. *et al.* 2005. Immunological features of fibrodysplasia ossificans progressiva (FOP) and the dysregulated BMP4 pathway. Clin. Rev. Bone Miner. Metab. **3:** 189–193.

35. KITTERMAN, J.A. *et al.* 2005. Iatrogenic harm caused by diagnostic errors in fibrodysplasia ossificans progressiva (FOP). Pediatrics **116:** 654–661.

36. MASSAGUÉ, J. 2000. How cells read TGF-β signals. Nat. Rev. Mol. Cell Biol. **1:** 169–178.

37. HOFFMAN, A. & G. GROSS. 2001. BMP signaling pathways in cartilage and bone formation. Critical Rev. Eukaryotic Gene Exp. **11:** 23–45.

38. TEN DIJKE, P. *et al.* 2003. Controlling cell fate by bone morphogenetic protein receptors. Mol. Cell Endocrinol. **211:** 105–113.

39. KLOEN, P. *et al.* 2003. BMP signaling components are expressed in human fracture callus. Bone **33:** 362–371.

40. CANALIS, E. *et al.* 2003. Bone morphogenetic proteins, their antagonists, and the skeleton. Endocrine Rev. **24:** 218–235.

41. SHI, Y. & J. MASSAGUÉ. 2003. Mechanisms of TFG-β signaling from cell membrane to the nucleus. Cell **113:** 685–700.

42. MASSAGUÉ, J. *et al.* 2005. Smad transcription factors. Genes Dev. **19:** 2783–2810.

43. HARTUNG, A. *et al.* 2006. Yin and yang in BMP signaling: impact on the pathology of disease and potential for regeneration. Signal Transduction **6:** 314–328.

44. KAPLAN, F.S. *et al.* 1990. Fibrodysplasia ossificans progressiva (FOP): a clue from the fly? Calcif. Tiss. Int. **47:** 117–125.

45. SHAFRITZ, A.B. *et al.* 1996. Overexpression of an osteogenic morphogen in fibrodysplasia ossificans progressiva. N. Engl. J. Med. **335:** 555–561.

46. GANNON, F.H. *et al.* 1997. Bone morphogenetic protein 2/4 in early fibromatous lesions of fibrodysplasia ossificans progressiva (FOP). Hum. Pathol. **28:** 339–343.

47. AHN, J. *et al.* 2003. Paresis of a bone morphogenetic protein—antagonist response in a genetic disorder of heterotopic skeletogenesis. J. Bone Joint Surg. Am. **85:** 667–674.

48. SERRANO DE LA PEÑA, L. *et al.* 2005. Fibrodysplasia ossificans progressiva (FOP): a disorder of ectopic osteogenesis misregulates cells surface expression and trafficking of BMPRIA. J. Bone Miner. Res. **20:** 1168–1176.

49. FIORI, J.L. *et al.* 2006. Dysregulation of the BMP-p38 MAPK signaling pathway in cells from patients with fibrodysplasia ossificans progressiva. J. Bone Miner. Res. **21:** 902–909.

50. KAPLAN, F.S. *et al.* 2006. Dysregulation of the BMP4 signaling pathway in fibrodysplasia ossificans progressiva. Ann. N.Y. Acad. Sci. **1068:** 54–65.

51. BILLINGS, P.C. *et al.* 2008. Dysregulated BMP signaling and enhanced osteogenic differentiation of connective tissue progenitor cells from patients with fibrodysplasia ossificans progressiva. J. Bone Miner. Res. Submitted.

52. SHORE, E.M. *et al.* 2006. A recurrent mutation in the BMP type I receptor ACVR1 causes inherited and sporadic fibrodysplasia ossificans progressiva. Nat. Genet. **38:** 525–527.

53. ZHANG, D. *et al.* 2003. Alk2 functions as a BMP type I receptor and induces Indian hedgehog in chondrocytes during skeletal development. J. Bone Miner. Res. **18:** 1593–1604.

54. HARRADINE, K.A. & R.J. AKHURST. 2006. Mutations of TFG-β signaling molecules in human disease. Annals Med. **38:** 403–414.

55. GROPPE, J.C. *et al.* 2007. Functional modeling of the ACVR1 (R206H) mutation in FOP. Clin. Orthop. Rel. Res. In press.

56. WANG, T. *et al.* 1996. The immunophilin FKBP12 functions as a common inhibitor of the TFG-β family type I receptors. Cell **86:** 435–444.

57. CHEN, Y.-G. *et al.* 1997. Mechanism of TGF-β receptor inhibition by FKBP12. EMBO J. **16:** 3866–3876.

58. HUSE, M. *et al.* 1999. Crystal structure of the cytoplasmic domain of the type I TGF-β receptor complex with FKBP12. Cell **96:** 425–436.

59. HUSE, M. *et al.* 2001. The TGF-β receptor activation process: an inhibitor-to-substrate binding switch. Mol. Cell **8:** 671–682.

60. EBISAWA, T. *et al.* 2001. Smurf1 interacts with transforming growth factor-β type 1 receptor through Smad 7 and induces receptor degradation. J. Biol. Chem. **276:** 12477–12480.

61. YAMAGUCHI, T. *et al.* 2006. FKBP12 functions as an adaptor of the Smad7-Smurf1 complex on activin type I receptor. J. Mol. Endocrinol. **36:** 569–579.

62. BRUNET, L.J. *et al.* 1998. Noggin, cartilage morphogenesis, and joint formation in the mammalian skeleton. Science **280:** 1455–1457.

63. GROPPE, J. *et al.* 2002. Structural basis of BMP signaling inhibition by the cystine knot protein noggin. Nature **420:** 636–642.

64. KNUDSON, A.G. 1996. Hereditary cancer: two hits revisited. J. Cancer Res. Clin. Oncol. **122:** 135–140.

65. KNUDSON, A.G. 2001. Two genetic hits (more or less) to cancer. Nat. Rev. Cancer **1:** 157–162.

66. JONES, P.A. & S.B. BAYLIN. 2002. The fundamental role of epigenetic events in cancer. Nat. Rev. Cancer **3:** 414–428.

67. FUTREAL, P.A. *et al.* 2004. A census of human cancer genes. Nat. Rev. Cancer **4:** 177–183.

68. VOGELSTEIN, B. & K.W. KINZLER. 2004. Cancer genes and the pathways they control. Nat. Med. **10:** 789–799.

69. GONG, Y. *et al.* 1999. Heterozygous mutations in the gene encoding noggin affect human joint morphogenesis. Nat. Genet. **21:** 302–304.

70. KAPLAN, F.S. *et al.* 2005. Animal models of fibrodysplasia ossificans progressiva. Clin. Rev. Bone Miner. Metab. **3:** 229–234.

71. RAUCH, C. *et al.* 2002. C2C12 myoblast/osteoblast transdifferentiation steps enhanced by epigenetic inhibition of BMP2 endocytosis. Am. J. Physiol. Cell Physiol. **283:** 235–243.

72. FEINBERG, A.P. 2007. Phenotypic plasticity and the epigenetics of human disease. Nature **447:** 433–440.

73. GOSDEN, R.G. & A.P. FEINBERG. 2007. Genetics and epigenetics—nature's pen-and-pencil set. N. Engl. J. Med. **356:** 731–733.

74. REIK, W. 2007. Stability and flexibility of epigenetic gene regulation in mammalian development. Nature **447:** 425–432.

75. OLMSTED-DAVIS, E. *et al.* 2007. Hypoxic adipocytes pattern early heterotopic bone formation. Am. J. Pathol. **170:** 620–632.

76. HAKIM, M. *et al.* 2005. Dexamethasone and recovery of contractile tension after a muscle injury. Clin. Orthop. Rel. Res. **439:** 235–242.

77. JÄRVINEN, T.A.H. *et al.* 2005. Muscle injuries: biology and treatment. Am. J. Sports Med. **33:** 745–764.

78. WYNN, T.A. 2007. Common and unique mechanisms regulate fibrosis in various fibroproliferative diseases. J. Clin. Invest. **117:** 524–529.

79. VANDEN BOSSCHE, L.C. *et al.* 2007. Free radical scavengers are more effective than indomethacin in the prevention of experimentally induced heterotopic ossification. J. Orthop. Res. **25:** 267–272.

80. SCHIPANI, E. *et al.* 2001. Hypoxia in cartilage: HIF-1 alpha is essential for chondrocyte growth arrest and survival. Genes Dev. **15:** 2865–2876.

81. PFANDER, D. *et al.* 2003. HIF-1-alpha controls extracellular matrix synthesis by epiphyseal chondrocytes. J. Cell Sci. **116:** 1819–1826.

82. PROVOT, S. *et al.* 2007. Hif-1-alpha regulates differentiation of limb bud mesenchyme and joint development. J. Cell Biol. **177:** 451–464.

83. KAPLAN, F.S. *et al.* 2007. A new era for fibrodysplasia ossificans progressiva: a druggable target for the second skeleton. Exp. Opin. Biol. Ther. **7:** 705–712.
84. MOHLER, E.R., III *et al.* 2001. Bone formation and inflammation in cardiac valves. Circulation **103:** 1522–1528.
85. KAPLAN, F.S. *et al.* 2004. Heterotopic ossification. J. Am. Acad. Orthop. Surg. **12:** 116–125.
86. PIGNOLO, R.J. & K.L. FOLEY. 2005. Nonhereditary heterotopic ossification: implications for injury, arthropathy, and aging. Clin. Rev. Bone Miner. Metab. **3:** 261–266.
87. WOLF, J.M. 2006. The genetics key to a rare disease and its impact on orthopedics. Orthopedics **29:** 672.
88. POTTER, B.K. *et al.* 2006. Heterotopic ossification in the residual limbs of traumatic and combat-related amputees. J. Am. Acad. Orthop. Surg. **14:** S191–S197.
89. WILLIS, R. 1989. The Works of William Harvey. University of Pennsylvania Press. Philadelphia, PA.

Differential Effects of Vascular Endothelial Growth Factor on Joint Formation during Limb Development

GLORIA EMELÍ CORTINA-RAMÍREZ AND JESÚS CHIMAL-MONROY

Instituto de Investigaciones Biomédicas, Universidad Nacional Autónoma de México, Ciudad Universitaria, México DF, México

ABSTRACT: During skeletal development the proliferating chondrocytes choose between two fates: to differentiate into prejoint cells or undergo hypertrophy. Vascular endothelial growth factor (VEGF) participates in endochondral ossification, but it is unknown whether it has a role in joint development. We evaluated the VEGF effects on joints, implanting VEGF-beads in the presumptive limb joints of chick embryos. The wrist and elbow showed partial and complete fusions. Using the model of ectopic digits induced by transforming growth factor-β (TGF-β), VEGF inhibited joint formation when it was applied after TGF-β. These data suggest differential effects of VEGF on different joints during development.

KEYWORDS: joint formation; cartilage; chondrocyte; digit development; VEGF; TGF-β

INTRODUCTION

During development of the appendicular skeleton, cartilage differentiation begins with aggregation of prechondrogenic mesenchyme in response to chondrogenic factors, giving rise to a blastema that prefigures the future bone skeletal elements.[1] These cartilage elements are only formed by proliferating chondrocytes. When joint formation starts, the proliferating chondrocytes at specific regions of the cartilaginous primordium exit the cell cycle and become prejoint cells, delimiting the boundary of each individual skeletal element. Later, the cells in the center of the new skeletal element, instead of becoming prejoint cells, adopt an alternative fate and become prehypertrophic; undergoing a process of hypertrophy, mineralization, and vascular invasion. This process is called endochondral ossification.[2,3]

Address for correspondence: Jesús Chimal-Monroy, Instituto de Investigaciones Biomédicas, Universidad Nacional Autónoma de México, Apartado Postal 70228, Ciudad Universitaria, México DF 04510, México. Voice: 5255-5622-3819; fax: 5255-5622-3897.

jchimal@servidor.unam.mx

Ann. N.Y. Acad. Sci. 1116: 134–140 (2007). © 2007 New York Academy of Sciences.
doi: 10.1196/annals.1402.024

It is known that vascular endothelial growth factor (VEGF) is a potent pro-moter of endochondral ossification, and since it is able to induce invasion of vascular vessels into cartilage, it is important for chondrocyte survival and matrix degradation.[4-6] Also the condensation of mesenchymal cells and the chondrogenesis require vascular regression in the mesenchyme of developing limbs.[7] Even though VEGF participates in the control of hypertrophic carti-lage differentiation or in the condensation of mesenchymal cells, it is unknown whether VEGF has an effect on developing joints.

To determine the role of VEGF on joint development, its effects were evalu-ated on different developing joints showing partial and complete fusions in the wrist and elbow joints and in an ectopic digit induced by transforming growth factor-β (TGF-β) in the interdigital tissue[8] but no effects were observed in shoulder and hip joints. These data, all together, suggest a differential role of VEGF on joint formation depending on joint location during limb skeletogene-sis and an effect on joint formation at early steps of ectopic digit development.

MATERIALS AND METHODS

Embryo Manipulations and VEGF Treatment

Fertilized White Leghorn chicken eggs (ALPES, Puebla México) were in-cubated at 38°C and staged according to Hamburger and Hamilton.[9] Heparin-acrylic beads (Sigma, St. Louis, MO, USA) incubated for 1 h at room tem-perature in human VEGF 165 at 1 μg/μL (Peprotech, Mexico City, Mexico) were implanted into different locations of right fore- and hind-limb buds of chick embryos at stages 22–24; the left limb bud was used as an internal con-trol. Also PBS beads were implanted as controls of manipulation procedure and no effects were observed. After bead implantation, chick embryos were returned to the incubator and analyzed by skeletal staining, histology, and *in situ* hybridization.

TGF-β Extra Digit Induction and VEGF Treatment

Heparin–acrylic beads were incubated for 1 h at room temperature in hu-man TGF-β1 200 ng/μL implanted into the third interdigit of hind-limbs at stage 29 in chick embryos and 6 h later VEGF-soaked beads (1 μg/μL) were implanted in the interdigital mesenchyme between the TGF-β1 bead and the apical ectodermal ridge (AER).

The morphology of the ectopic digits induced by TGF-β1 alone was used as controls to compare with those digits induced by TGF-β1 and treated with VEGF. After 4 days of TGF-β1 treatment, the hind-limbs were dissected out and analyzed for Alcian Blue/Alizarin Red staining.

Histology

Limbs were dissected out and fixed in 4% paraformaldehyde at 4°C overnight, dehydrated, and embedded in Paraplast by standard techniques. Tissue sections were then stained with Safranin O and Fast Green.

Growth and Differentiation Factor 5 (Gdf5) In Situ Hybridization

The expression of *Gdf5* was studied by whole-mount *in situ* hybridization. Specific chick probe for *Gdf5* was used.[10] Samples were treated with 10 μg/mL proteinase K (preincubated at 37°C) for 25 min at 20°C. Hybridization with digoxigenin-labeled antisense RNA probe was performed at 68°C overnight. Reactions were developed with BM purple AP substrate (Roche Applied Science, Mannheim, Germany).

RESULTS

To determine the role of VEGF on different joints, VEGF beads were implanted into the three different regions of limb: the stylopod, the zeugopod, and the autopod. When VEGF-soaked beads were implanted into the stylopod in the presumptive shoulder (FIG. 1A, B) and hip (data not shown) regions of stages HH 22–24 embryos no effect was observed. However, VEGF beads implanted into the zeugopod of stage HH 24 embryos in the presumptive elbow region caused joint fusion between humerus and radius (FIG. 1C, D). Also when VEGF beads were implanted into the knee joint region no effect was observed (data not shown). Finally, implantation of the VEGF beads in the presumptive wrist area of stage HH 24 embryos produced joint fusion between radius and carpal zone (FIG. 1E, F), while in the presumptive ankle region at same stage no effect was observed (data not shown).

To evaluate the role of VEGF on joint formation during digit development we induced an ectopic digit by TGF-β1 bead implantation in the third interdigital mesenchyme of stage HH 29 chick embryos (FIG. 2A). After 6 h of TGF-β1 implantation a bead of VEGF was implanted above TGF-β1 bead, causing a complete absence of joints in the ectopic digits (FIG. 2B). Apart from this effect, VEGF produced abnormal digit morphology, and a shorter and wider ectopic digit without its characteristic tip was observed (FIG. 2A–D). To establish whether the VEGF treatment inhibits the expression of *Gdf5*, a joint marker, we evaluated the expression of this gene. The results showed that *Gdf5* was expressed, although any histological evidence of joint formation was not observed (FIG. 2C–F).

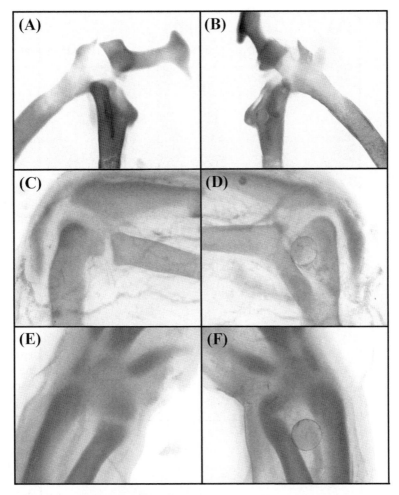

FIGURE 1. Effects of VEGF on different joints. (**A–F**). Limbs stained with Alcian Blue/Alizarin Red to show skeletal morphology of untreated shoulder (**A**), elbow (**C**), and wrist (**E**) joints and treated with VEGF (**B, D,** and **F**).

DISCUSSION

The differential effects of VEGF on different joints showed in this study suggest that although the mechanisms about joint formation are similar in all locations, they may have differences at molecular levels that control joint development at each location, as it has been described for GDF5/6/7 genes that show a differential expression on different joints.[11] Also there are diverse joint defects in single and double mutants in the mouse lacking *Gdf5* and *Gdf6* genes.[11]

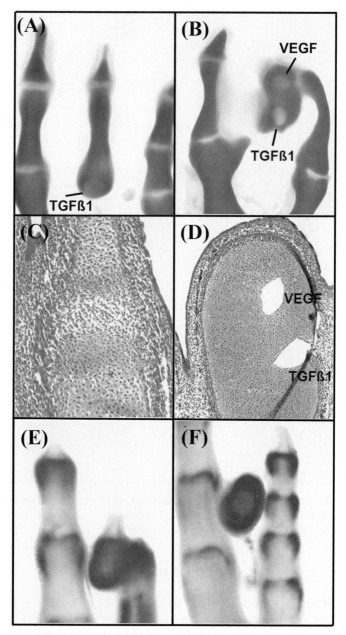

FIGURE 2. Effects of VEGF on joint development in ectopic digits. Limbs stained with Alcian Blue/Alizarin Red to show ectopic digits induced by TGF-β1 alone (**A**) or TGF-β1 and VEGF (**B**). Histological sections from ectopic digits untreated (**C**) and treated with VEGF (**D**) and stained with Safranin and Fast Green. *In situ* hybridization for Gdf5 in ectopic digit untreated (**E**) and treated with VEGF (**F**).

It has been described that vascular regression is a fundamental process for initial condensation of mesenchymal cells during formation of the appendicular skeleton.[7] In this study the results suggest that vascular regression is not a prerequisite for joint formation, since the presence of VEGF inhibited some joints, but not others. Then it may be that VEGF may have other functions different to vascular vessel formation. Whether this factor allows cartilage differentiation of proliferating chondrocytes toward prehypertrophic cartilage instead of prejoint cells is still unknown.

Finally the results of this study suggest that the model of ectopic digit induced by TGF-β1 is an appropriate model to understand more about the different factors that regulate joint formation.

ACKNOWLEDGMENTS

This work was partially supported by grants 42568-Q (J C-M) from Consejo Nacional de Ciencia y Tecnología (CONACYT), México, and IN200205 (J C-M) from DGAPA, Universidad Nacional Autónoma de México (UNAM). G. E. C-R. was supported by a long-term fellowship from CONACYT, Mexico 42568-Q. We thank Dora Copeland for correcting the English version of the manuscript.

REFERENCES

1. KARSENTY, G. & E.F. WAGNER. 2002. Reaching a genetic and molecular understanding of skeletal development. Dev. Cell **2:** 389–406.
2. GARCIADIEGO-CAZARES, D. *et al.* 2004. Coordination of chondrocyte differentiation and joint formation by alpha5beta1 integrin in the developing appendicular skeleton. Development **131:** 4735–4742.
3. CHIMAL-MONROY, J. *et al.* 2005. Coordination of joint formation and cartilage differentiation in the appendicular skeleton. Trends Dev. Biol. **1:** 47–53.
4. ZELZER, E. *et al.* 2004. VEGFA is necessary for chondrocyte survival during bone development. Development **131:** 2161–2171.
5. ZELZER, E. *et al.* 2002. Skeletal defects in VEGF(120/120) mice reveal multiple roles for VEGF in skeletogenesis. Development **129:** 1893–1904.
6. ZELZER, E. *et al.* 2001. Tissue specific regulation of VEGF expression during bone development requires Cbfa1/Runx2. Mech. Dev. **106:** 97–106.
7. YIN, M. & M. PACIFICI. 2001. Vascular regression is required for mesenchymal condensation and chondrogenesis in the developing limb. Dev. Dyn. **222:** 522–533.
8. GANAN, Y. *et al.* 1996. Role of TGF betas and BMPs as signals controlling the position of the digits and the areas of interdigital cell death in the developing chick limb autopod. Development **122:** 2349–2357.
9. HAMBURGER, V. & H.L. HAMILTON. 1992. A series of normal stages in the development of the chick embryo. 1951. Dev. Dyn. **195:** 231–272.

10. MERINO, R. *et al.* 1999. Expression and function of Gdf-5 during digit skeletoge-
    nesis in the embryonic chick leg bud. Dev. Biol. **206:** 33–45.
11. SETTLE, S.H., JR. *et al.* 2003. Multiple joint and skeletal patterning defects caused
    by single and double mutations in the mouse Gdf6 and Gdf5 genes. Dev. Biol.
    254: 116–130.

Negative Regulation of Otic Capsule Chondrogenesis

It Can Make You Smad

WEI LIU,[a] SYDNEY BUTTS,[a] HAROLD KIM,[b] AND DOROTHY A. FRENZ[a]

[a] Department of Otorhinolaryngology, Head, and Neck Surgery, Albert Einstein College of Medicine, Bronx, New York, USA

[b] Wilson Ear Clinic, Portland, OR, USA; Department of Anatomy and Structural Biology, Albert Einstein College of Medicine, Bronx, New York, USA

ABSTRACT: The transforming growth factor-beta (TGF-β) superfamily, including TGF-β1 and bone morphogenetic protein (BmP2, BmP4), participates in the regulation of the developing cartilaginous otic capsule, which prefigures the endochondral bony labyrinth of the inner ear. This study investigates Smad-6 and -7, downstream components of the TGF-β/BMP signaling pathway, in otic capsule chondrogenic control, and supports a function for these inhibitory Smads as negative regulators of capsule chondrogenesis. The importance of otic capsule chondrogenic control and implications of Smad signaling for otosclerosis, a disease affecting the endochondral bony labyrinth, are indicated.

KEYWORDS: Smad-6 and -7; otic capsule chondrogenesis; negative regulation

INTRODUCTION

Development of the cartilaginous capsule of the mouse inner ear prefigures the formation of the endochondral bony labyrinth and is regulated by inductive interactions between otic epithelium and its surrounding periotic mesenchyme. During these inductive events, signaling molecules from the otic epithelium act on the underlying periotic mesenchyme to promote tissue induction and differentiation. These signaling molecules include members of the transforming growth factor-beta (TGF-β) superfamily.[1–3] TGF-β elicits its biological effects by binding to heterodimeric serine–threonine kinase receptor complexes.[4] TGF-β initiates signaling by binding to the type II TGF-β receptor (TGF-βRII) and thereby provoking transphosphorylation and activation of the type I

Address for correspondence: Dorothy A. Frenz, Ph.D., Albert Einstein College of Medicine, Kennedy Center 301, 1410 Pelham Parkway South, Bronx, New York 10461. Voice: 718-430-4082; fax: 718-430-4258.

frenz@aecom.yu.edu

Ann. N.Y. Acad. Sci. 1116: 141–148 (2007). © 2007 New York Academy of Sciences.
doi: 10.1196/annals.1402.005

receptor (TGF-βRI).[5] Once activated, TGF-βRI can signal to Smad proteins, downstream substrates of TGF-βRI and essential components of intracellular signaling.[6] Smad proteins are classified according to their functional role in signaling of TGF-β family members, with different Smad proteins conveying distinct biological signals. Three functional classes of Smad proteins mediate cellular responses to TGF-β: receptor-regulated Smads, common Smads, and inhibitory Smads. Receptor-regulated Smad-2 and Smad-3 act as positive regulators of otic capsule chondrogenesis in the developing mouse inner ear.[7] However, the function of Smad-6 and -7 is currently unknown. In this study, we explore the putative role of these inhibitory Smads in otic capsule formation, and discuss its implications in the pathogenesis of otosclerosis.

MATERIALS AND METHODS

Experimental Animals

C57/BL6 female mice were intercrossed with CBA males (Charles River Laboratories, Wilmington, MA). Gestational age of embryos was estimated by the vaginal plug method, with the day of plug occurrence designated as day 1 (E1). After death of gravid female mice by isoflurane inhalation followed by cervical dislocation, embryos were harvested and immediately placed into Dulbecco's phosphate-buffered saline (Gibco [now Invitrogen], Carlsbad, CA). Embryonic age was determined by a combination of external features and somite count.

Otosclerotic Bone Specimens

Surgical specimens of otosclerotic bone were obtained from patients who received stapedectomies at the Montefiore Medical Center, the University Hospital of the Albert Einstein College of Medicine. These specimens were not collected specifically for the proposed research project and did not contain a code derived from personal information. The use of the otosclerotic bone specimens was reviewed by the Albert Einstein College of Medicine Institutional Review Board, and classified as exempt in accordance with the Office for Human Research Protections Guidance on Research Involving Coded Private Information or Biological Specimens.

Immunohistochemistry

Mouse inner ear specimens or specimens of human otosclerotic bone were fixed in methacarn, dehydrated, cleared in Histoclear, and embedded in paraffin. Deparaffinized sections were processed by using the avidin–biotin complex (ABC) method (Vector).[1] The deparaffinized sections were pretreated for

30 min in 100% methanol + 0.3% peroxide, following the blocking of nonspecific binding sites with normal rabbit serum. Sections were covered with an optimal dilution of primary Smad-6 or Smad-7 antibody (Santa Cruz Biotechnologies, Santa Cruz, CA) and incubated overnight at 4°C. Controls were prepared by omission of primary antibody or by replacing it with bovine serum albumin. After incubation with secondary biotinylated antisera and ABC reagent, sections were subjected to 0.05% 3'3-diaminobenzidine tetrahydrochloride in 0.05 M Tris buffer with 0.01% hydrogen peroxide. Sections were stained for 15 min, then counterstained with Mayers hematoxylin, and mounted in Crystalmount, with final mounting in Permount.

Tissue Culture

Periotic mesenchyme with added otic epithelium (periotic mesenchyme + otic epithelium) was isolated from CBA/C57 BL6 mouse embryos of embryonic age 10.5 days (E10.5) and cultured as described.[8] Periotic mesenchyme was dissociated then resuspended at a density of 2.5×10^7 cells/mL in Ham's F-12 culture medium supplemented with 10% fetal bovine serum. Ten microliter droplets of cell suspension were plated in a 4-well tissue culture plate (Nunc). After a 1-h incubation at 37°C, cultures were maintained in 1 mL of medium containing or not containing Smad-specific oligonucleotides (described below). Medium was exchanged every other day.

Oligonucleotides

Oligonucleotides, as listed below, were prepared as purified products by Invitrogen: Smad-6 antisense oligonucleotide complementary to position 298–312 of the murine Smad-6 gene—5'-TTTGGACCTGAACAT-3' (GenBank accession no. NM008542); Smad-6 sense oligonucleotide -5'-ATGTTCAGGTCCAAA-3'; Smad-7 antisense oligonucleotide complementary to position 1–15 of the murine Smad-7 gene—5'-CCTGGCGGCGGATCC-3' (GenBank accession no. AJ404961); Smad-7 sense oligonucleotide—5'-GGATCCGCCGCCAGG-3'. The specificity of selected oligonucleotides was confirmed by sequence comparisons in the NCBI GenBank database using the BLAST algorithm. The use of oligonucleotides in cultured periotic mesenchyme + otic epithelium, including assessment of uptake by cultured cells, has been described.[3,7,9,10]

Quantitative Alcian Blue Staining of Cultures

Cultures were fixed on day 7 with a solution of 10% formalin + 1% cetyl pyridinium chloride, then stained with Alcian blue 8GX, pH 1.0, a stain that at

TABLE 1. Effect of Smad6, -7 antisense oligonucleotide

Treatment	OD of Alcian blue	Treatment	OD of Alcian blue
Control	0.313 ± 0.01 ($n = 8$)	Control	0.334 ± 0.05 ($n = 4$)
Smad-6 sense	0.306 ± 0.01 ($n = 2$)	Smad-7 sense	0.304 ± 0.01 ($n = 2$)
Smad-6 antisense	0.413 ± 0.02 ($n = 3$)	Smad-7 antisense	0.464 ± 0.05 ($n = 3$)

Cultured E10.5 periotic mesenchyme + otic epithelium was grown in the presence of Smad-6 or Smad-7 antisense oligonucleotide (AS) (60 µg/mL) or corresponding sense oligonucleotides with medium exchanges every other day. Some cultures served as untreated controls. Values represent the mean optical density of matrix-bound Alcian blue stain ± SEM.

this pH binds specifically to sulfated glycosaminoglycans in the matrix of chondrifying cells. After washing cultures to remove unbound stain, matrix-bound stain was extracted and measured spectrophotometrically, as described.[10]

Values for optical density of bound Alcian blue stain in TABLE 1 were normalized to control for differences in the amount of otic epithelium at cell seeding between experimental setups.

RESULTS

Expression of Inhibitory Smads in the Developing Mouse Inner Ear

We examined the pattern of Smad-6 and -7 protein distribution in the developing mouse inner ear at embryonic ages 10.5 days (E10.5), E12, and E14, stages that correspond to otic epithelial–periotic mesenchymal interactions, periotic mesenchymal cell condensation, and capsule chondrification, respectively. Pale immunostain for Smad-7 was evident in the yet uncondensed periotic mesenchyme surrounding the E10.5 otic vesicle (FIG. 1A). This periotic mesenchyme will form the cartilaginous otic capsule that prefigures the endochondral bony labyrinth. At E12, when the now shield-shaped otic vesicle comprises distinct cochlear and vestibular portions, Smad-7 continues to be localized to the periotic mesenchyme, most notably in the condensing mesenchyme just dorsal and lateral to the forming horizontal semicircular duct (FIG. 1B). Dense reaction product for inhibitory Smads (6 and 7) was evident in the chondrifying otic capsule that contours around the E14 inner ear (FIG. 1C), consistent with a putative function in the turning off of chondrogenesis to promote appropriate capsule modeling.

Blockade of Smad-6 and -7 in Culture

We hypothesized that the function of Smad-6 and -7 in inner ear development may be to prevent overgrowth of otic capsule cartilage, and have examined the

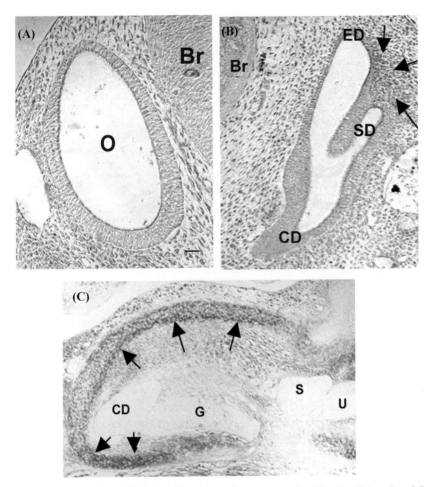

FIGURE 1. The developing mouse inner ear, immunostained for Smad-6 or Smad-7. **(A)** E10.5 otocyst, showing pale immunolabel for Smad-7 in the yet uncondensed periotic mesenchyme surrounding the otocyst (O) epithelium. **(B)** E12.5 inner ear, showing continued localization of Smad-7 to the now condensing capsular mesenchyme, with greatest intensity (*arrows*) just dorsal and lateral to the forming horizontal semicircular duct (SD). **(C)** Dense reaction product for Smad-6 is evident in the chondrifying capsule (*arrows*) of the E14 inner ear. At each developmental age studied (E10.5, E12, E14), similar patterns of Smad-6 and -7 were observed. Br = brain; ED = endolymphatic duct; CD = cochlear duct; G = ganglia; S = saccule; U = utricle. Bar, 100 μm.

effect of blockade of Smad-6 or Smad-7 in a high-density culture model of otic capsule chondrogenesis. Cultured E10.5 periotic mesenchyme containing otic epithelium (periotic mesenchyme + otic epithelium) was maintained in the presence of Smad-6- or Smad-7-specific antisense oligonucleotide for

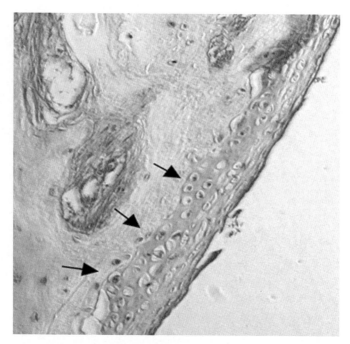

FIGURE 2. Specimen of human otosclerotic bone, showing that Smad-6 protein is present in osteoblasts (*arrows*) in regions of new bone formation.

the duration of the culture period (7 days). Smad-6 and -7 sense oligonu-cleotides served as oligonucleotide treatment controls. A comparable extent of chondrogenesis, assayed by binding of Alcian blue stain, pH 1.0, was noted in sense oligonucleotide-treated control and untreated control cultures (TABLE 1). However, in cultures treated with Smad-6 or Smad-7 antisense oligonucleotide, enhancement of chondrogenesis occurred (TABLE 1) in comparison to their cor-responding sense oligonucleotide-treated cultures.

Inhibitory Smad-6 in Otosclerosis

The signaling molecules that control otic capsule chondrogenesis during embryonic development are contended to play a role in the pathogenesis of otosclerosis, a disease where chondrogenic overgrowth and inappropriate bone remodeling results in lesions to the labyrinthine capsule of the inner ear. We therefore investigated the pattern of expression of Smad-6 in surgical speci-mens of human otosclerotic bone, and show that this inhibitory Smad is present in regions of new bone formation in otosclerotic specimens consisting of the stapedial footplate (FIG. 2).

DISCUSSION

The role of TGF-β signaling molecules in initiation and regulation of otic capsule formation has been well established.[1-3,11-13] However, precisely how chondrogenic control is exerted to prevent cartilage overgrowth and ensure appropriate modeling of the otic capsule is not yet completely understood. Our findings support a function for inhibitory Smads in this context, as suggested by the pattern of Smad-6 or Smad-7 distribution *in vivo*, most notably at a time (E14) when chondrification is ongoing (FIG. 1), and the ability of Smad-6 and -7 antisense oligonucleotide blockade to stimulate chondrogenesis *in vitro* (TABLE 1). These observations allude to the possibility that when signaling molecules that normally participate in regulation of chondrogenesis go awry (e.g., downregulation of inhibitory Smads), inappropriate capsular cartilage can form.

Chondrogenic control is important to prevent the formation of excess cartilage not only during otic embryogenesis, but also in the mature inner ear. The inappropriate bone remodeling of otosclerosis is unique to the endochondral bone layer of the otic capsule,[14] which forms from a cartilaginous substrate precursor. Members of the TGF-β superfamily, including TGF-β1, BMP-2, and downstream Smad-2, which function in regulation of chondrogenesis during otic capsule formation,[1,3,7] are present in otosclerotic specimens of the labyrinthine capsule in regions where new bone has formed.[15] This finding, combined with our current observation of the localization of inhibitory Smad-6 to otosclerotic bone (FIG. 2), may likely reflect a recapitulation of the signaling activity of otic capsule formation. It will be important for future studies to examine levels of expression of inhibitory Smads in otosclerotic bone in comparison to appropriate control bone, to ascertain whether their hypothesized downregulation may contribute to capsular remodeling, which typically occurs at only low levels in capsular endochondral bone.[16] By gaining insight into the molecular basis of otic capsule formation, it is anticipated that we can also gain insight into the molecular mechanisms that underlie otosclerosis and into effective strategies that used experimentally to limit chondrogenic defects, may be translated into therapeutics to effectively ameliorate inappropriate bone turnover and remodeling in patients with otosclerosis.

ACKNOWLEDGMENT

Supported by a grant from the NIDCD (R01DC-04706).

REFERENCES

1. FRENZ, D.A., V. GALINOVIC-SCHWARTZ, W. LIU, *et al.* 1992. Transforming growth factorβ1 is an epithelial-derived signal peptide that influences otic capsule formation. Dev. Biol. **153**: 324–336.

2. FRENZ, D.A., W. LIU, J.D. WILLIAMS, *et al.* 1994. Induction of chondrogenesis: requirement for synergistic interaction of basic fibroblast growth factor and transforming growth factor-beta. Development **120:** 4115–4224.

3. LIU, W., S.H. OH, Y.K. KANG, *et al.* 2003. Bone morphogenetic protein 4 (BMP4): a regulator of capsule chondrogenesis in the developing mouse inner ear. Dev. Dyn. **226:** 427–438.

4. ABDOLLAH, S., M. MACIAS-SILVA, T. TSUKAZAKI, *et al.* 1997. TbetaRI phosphorylation of Smad2 on Ser465 and Ser467 is required for Smad2-Smad4 complex formation and signaling. J. Biol. Chem. **272:** 27678–27685.

5. MASSAGUE, J. 1998. TGF-beta signal transduction. Annu. Rev. Biochem. **67:** 753–791.

6. DENNLER, S., S. ITOH, D. VIVIEN, *et al.* 1998. Direct binding of Smad3 and Smad4 to critical TGF beta-inducible elements in the promoter of human plasminogen activator inhibitor-type 1 gene. EMBO J. **17:** 3091–3100.

7. BUTTS, S.C., W. LIU, G. LI & D.A. FRENZ. 2005. Transforming growth factor-β1 signaling participates in the physiological and pathological regulation of mouse inner ear development by all-trans retinoic acid. Birth Defects Res. (Part A) **73:** 218–228.

8. FRENZ, D.A. & T.R. VAN DE WATER. 1991. Epithelial control of periotic mesenchyme chondrogenesis. Dev. Biol. **144:** 38–46.

9. FRENZ, D.A. & W. LIU. 1998. Role of FGF3 in otic capsule chondrogenesis *in vitro*: an antisense oligonucleotide approach. Growth Factors **15:** 173–182.

10. LIU, W., G. LI, J.S. CHIEN, *et al.* 2002. Sonic hedgehog regulates otic capsule chondrogenesis and inner ear development in the mouse embryo. Dev. Biol. **248:** 240–250.

11. FRENZ, D.A., W. LIU & M. CAPPARELLI. 1996. Role of BMP-2a in otic capsule chondrogenesis. Ann. N.Y. Acad. Sci. **785:** 256–258.

12. PARADIES, N.E., L.P. SANFORD, T. DOETSCHMAN & R.A. FRIEDMAN. 1998. Developmental expression of the TGF betas in the mouse cochlea. Mech. Dev. **79:** 165–168.

13. CHANG, W., P. TEN DIJKE & D. WU. 2002. BMP pathways are involved in otic capsule formation and epithelial-mesenchymal signaling in the developing chicken inner ear. Dev. Biol. **251:** 380–394.

14. VAN DEN BOGAERT, K., P.J. GOVAERTS, E.M. LEENHEER, *et al.* 2002. Otosclerosis: a genetically heterogeneous disease involving at least three different genes. Bone **30:** 624–630.

15. FRENZ, D.A. 2001. Growth factor control of otic capsule chondrogenesis. Einstein Quart. J. Biol. Med. **18:** 7–13.

16. ZEHNDER, A.F., A.G. KRISTIANSEN, J.C. ADAMS, *et al.* 2005. Osteoprotegrin in the inner ear may inhibit bone remodeling in the otic capsule. Laryngoscope **115:** 172–177.

A Novel Transgenic Mouse Model to Study the Osteoblast Lineage *in Vivo*

CHRISTA MAES, TATSUYA KOBAYASHI, AND HENRY M. KRONENBERG

Endocrine Unit, Massachusetts General Hospital, Harvard Medical School, Boston, Massachusetts, USA

ABSTRACT: Over the past few decades, osteoblast differentiation has been studied extensively in a variety of culture systems and findings from these experiments have shaped our understanding of the bone-forming cell lineage. However, *in vitro* assays are bound by intrinsic limitations and are unable to effectively mirror many aspects related to osteoblasts *in vivo*, including their origin, destiny, and life span. Therefore, these fundamental questions strongly advocate the need for novel models to characterize the osteoblast lineage *in vivo*. Here, we developed a transgenic mouse system to study stage-specific subsets of osteoblast lineage cells. We believe that this system will prove to be a helpful tool in deciphering multiple aspects of osteoblast biology *in vivo*.

KEYWORDS: osteoblast; lineage tracing; inducible Cre; tamoxifen; Cre-ERt; perichondrium; endochondral ossification; bone development; osteoprogenitor; osteoblast differentiation

INTRODUCTION

Bone Development . . . Featuring the Osteoblast

During embryonic development, the establishment of the skeleton is achieved by the formation of as many as 206 separate bones at sites distributed all over the body. Bones can develop through two distinct mechanisms, intramembranous or endochondral ossification. Both processes depend on the coordinate growth, differentiation, function, and interaction of various cell types but ultimately rely on osteoblast lineage cells for the synthesis and mineralization of bone matrix. In intramembranous bones, which

Address for correspondence: Christa Maes, Ph.D., Harvard Medical School, Massachusetts General Hospital, Endocrine Unit, 50 Blossom Street, Thier 1101, Boston, MA 02114. Voice: 001-617-726-3966; fax: 001-617-726-7543.

cmaes1@partners.org

Future address (1/08): K.U. Leuven, Laboratory of Experimental Medicine and Endocrinology, Herestraat 49, B-3000 Leuven, Belgium. Voice: 0032-16-345972; fax: 0032-16-345934.

christa.maes@med.kuleuven.

Ann. N.Y. Acad. Sci. 1116: 149–164 (2007). © 2007 New York Academy of Sciences.
doi: 10.1196/annals.1402.060

are generally considered to be restricted to the flat bones of the skull (calvaria bones and mandibles) and parts of the clavicles, the osteoblasts differentiate directly from the mesenchymal precursor cells that aggregate at the site of the future bone formation. In contrast, in endochondral bones—comprising all the long bones of the axial and appendicular skeleton—the initial mesenchymal condensations that prefigure the future bones develop first into cartilage, while osteoblasts appear only later in the process (for a more extensive overview of endochondral ossification see Ref. 1). Briefly, the transitional cartilage "anlagen" enlarges and prefigures the shape of the future bone through chondrocyte proliferation and matrix production. Chondrocytes in the center (diaphysis) of the bone model then stop proliferating, undergo further maturation and ultimately become hypertrophic. Whereas cartilage itself is typified by its avascular nature, the terminal hypertrophic chondrocytes become actually a target for vascular invasion, most probably due to the production of high levels of angiogenic factors, such as VEGF.[2–4] At this time (around embryonic day (E)14 in mice, depending on the particular long bone; see below), cells in the surrounding perichondrium differentiate into osteoblasts that deposit mineralized bone matrix around the cartilage template, the "bone collar". Concomitantly, vascular endothelial cells and osteoclasts/chondroclasts accumulate within the perichondrial region. Through an as yet undefined stimulus, these events converge into the initiation of the actual "endochondral ossification" process, literally being the formation of true bone tissue within the cartilage template: the mid-diaphyseal hypertrophic cartilage undergoes apoptosis and becomes invaded by endothelial cells and osteoclasts/chondroclasts, the cartilage matrix is degraded, and osteoblasts and marrow cells start to populate the newly excavated area, which thereby is transformed into the primary ossification center (POC) (FIG. 1). Thus, the avascular cartilage template is replaced by highly vascularized bone and marrow tissues, a process that continues to take place at the growth plates. Here, continual deposition of mineralized bone matrix by osteoblasts on the remnants of calcified cartilage, followed by osteoclast-mediated bone remodeling, results in the formation of spongiform trabecular bone. In contrast, the process by which the perichondrial/periosteal bone collar surrounding the growing longitudinal bone expands, thickens, and solidifies into dense cortical bone may be considerably different and has been debated to resemble intramembranous ossification. In any case, certain differences seem to exist in the structure and ontology of cortical versus trabecular bone, as well as in the origin, differentiation and/or bone-forming activity of osteoblasts in different anatomical locations (for instance, calvaria versus long bones). Yet, strikingly little is currently known about the underlying regulation and the potentially specific features of the osteoblastic cell populations in these different locations. Moreover, despite the well-documented line of progression of osteoblast differentiation *in vitro* (see below), many unknowns remain regarding their origin, life span, and fate *in vivo*.

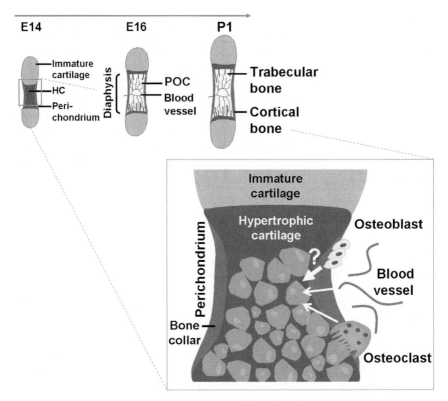

FIGURE 1. Early events during endochondral ossification. In the cartilage template of the future long bone, immature chondrocytes differentiate into hypertrophic chondrocytes (HC), while osteoblasts develop in the surrounding perichondrium. Around E14 in mice, blood vessels and osteoclasts accumulate within the perichondrium, and subsequently invade and erode the hypertrophic cartilage to start the formation of the POC. Although perichondrium-derived osteoblasts are assumed to populate POC, their movement into the POC and the extent to which they contribute to trabecular and cortical bone formation has not been characterized *in vivo*. (In color in Annals online.)

Inherent Caveats Associated with the Study of Osteoblast Development in Vivo

A combination of biological and technical reasons may explain why we still know noticeably little about the path of life and death of osteoblasts *in vivo*. First, the lack of a clear-cut spatial organization of osteoblast lineage development within the bone and marrow environment evidently hampers its characterization. This is particularly opposed to the study of chondrocyte differentiation that is markedly facilitated by the demarcated

differentiation-staged layer structure in the growth plate. Second, histological analysis of bone tissue can be quite challenging, given that the discrepancy between the rigid calcified structures and the soft vascular and marrow components in its surroundings complicate sectioning. The heterogeneity of the tissue also adds technical difficulties and considerable variability to other research approaches, such as analysis of gene expression and cellular composition. Third, once more in contrast to chondrocyte biology, only a few cellular markers can be used at the histological level to identify, spatially localize, characterize, and stage osteoblasts *in vivo*. Altogether, these caveats associated with the study of the osteoblast lineage *in vivo* have favored the use of *in vitro* models to characterize the properties of osteoblastic cells.

Osteoblast Differentiation (in Vitro)

Several osteoblastic cell lines have been extensively characterized in their properties to proliferate and differentiate toward mature, mineralizing osteoblasts. These lines include the nontransformed mouse MC3T3-E1 and rat UMR201 osteoblastic cells, human (MG-63, SaOS-2, and TE85) and rat (ROS17/2.8 and UMR106) osteosarcoma-derived cell lines, and pluripotent murine mesenchymal precursor cell lines, such as C2C12 and C3H10T1/2, that can differentiate to osteoblasts as well as to other lineages (for detailed review, see Refs. 5, 6). In addition, primary cells isolated from different anatomical locations can be put in culture to study osteoblast characteristics. The most extensively used may be the osteoblastic cells isolated from newborn mouse/rat calvaria and the adherent stromal cells isolated from bone marrow. In addition, osteoblasts can be retrieved from trabecular bone fragments, from the periosteum, or from a variety of other sources.[5,6] These models, together with genetic *in vivo* studies performed over the last decade, have shaped our present understanding of the osteoblast differentiation process and provided the basis for a stepwise progression model in which osteoblasts at subsequent stages of maturation are characterized by specific gene expression profiles.[7] The current scheme starts from an initial osteochondroprogenitor cell that has the capacity to differentiate into either chondrocytic or osteoblastic committed progenitor cells (FIG. 2). The key transcription factors involved in driving the cell to a chondrocytic fate are Sox9 together with L-Sox5 and Sox6, while osteoblastic differentiation requires the action of Runx2 and later Osterix (Osx). Other factors involved in the cell fate decision are being identified, as exemplified by the recent implication of β-catenin (a prime transcriptional activator in the canonical Wnt pathway) in the control of chondrocyte and osteoblast differentiation (for recent reviews, see Refs. 8–11).

Runx2, a member of the runt family and previously known as Cbfa1, serves as the earliest transcriptional regulator of osteoblast differentiation and is viewed as the central regulator of ossification.[12–14] The earliest osteoblast progenitors

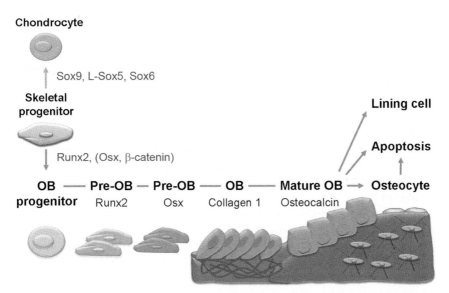

FIGURE 2. Osteoblast differentiation. Osteoblasts (OB) derive from mesenchymal progenitors that can differentiate into chondrocytes or osteoblasts. The transcription factors involved in driving the cell to a chondrocyte fate are Sox9, L-Sox5, and Sox6, while osteoblast differentiation requires Runx2 and later Osx and β-catenin. Progressive osteoblast differentiation from committed osteoblast progenitor cells is characterized by changes in gene expression, providing markers that typify specific stages. Runx2 is the earliest marker of preosteoblasts, followed by Osx. Preosteoblasts further differentiate into osteoblasts that produce large amounts of type I collagen (Collagen 1) and later mineralize the osteoid matrix as mature osteoblasts expressing osteocalcin. Eventually, the cells can become entrapped within the bone matrix as osteocytes, remain quiescently at the surface of the bone as lining cells, or die by apoptosis. (In color in Annals online.)

or preosteoblasts are consequently typified by their expression of Runx2 in the absence of expression of more mature osteoblast markers. Immediately downstream of Runx2 is another crucial transcription factor controlling osteoblastogenesis, Osx.[15] This zinc finger-containing transcription factor is required for osteoblast differentiation, as evidenced by the fact that Osx null mice do not show any bone formation. Interestingly, mesenchymal progenitor cells in Osx null mice do express Runx2, whereas conversely, Osx is not expressed in Runx2 null mice. This finding demonstrated that Osx acts downstream of Runx2.[15] As the preosteoblasts expressing Osx differentiate further along the osteoblast lineage, they are thought to gradually stop proliferating and start producing abundant matrix proteins, most notably the main bone constituent type I collagen (Collagen1, Col1). The differentiation process culminates in the development of mature osteoblasts that typically express osteocalcin. Probably at this stage the cells become endowed with the rather unique capacity to mineralize the osteoid matrix, generating true bone. It is thought that mature osteoblasts can ultimately undergo either of three fates: become embedded in the bone matrix as osteocytes, remain quiescently at the surface of the bone as lining cells, or die by apoptosis (FIG. 2).[7]

Of note, it is often hard to determine the extent to which *in vitro* findings effectively mirror the *in vivo* situation. Similarly, it is not always clear how much of the history and characteristics of a specific osteoblastic cell isolated from its local milieu is actually transmitted into the culture model and whether the original features of isolated primary cells are well-preserved *in vitro*. Alternatively, the cells may represent a rather generalized population that has the potential, tendency, or even predisposition to differentiate along the osteoblast lineage upon exposure to a specific environment provided by the growth medium. In this regard it may be feasible that potentially subtle differences between "subsets" of osteoblasts *in vivo*, for instance, stage-specific populations or cells at different anatomical locations, are too delicate to be analyzed in culture. In recognition of these inherent limitations related to *in vitro* approaches, we here aimed at developing novel tools to characterize the osteoblast lineage in their *in vivo* environment.

METHODOLOGY

Lineage Tracing Using Temporal- and Spatial-Controlled Transgene Expression

Over the last decade, the Cre-loxP system has been extensively used for gene manipulation in mice based on site-specific recombination catalyzed by the Cre integrase enzyme, between two of its consensus 34-bp recognition sites (loxP sites).[16] With the addition of this tool to mouse genetics, the function of genes could be analyzed based on any desired gene modification, including overexpression, alteration, and inactivation. With respect to the latter, an advantage of the Cre-loxP system over the generation of germline null mice was brought by the fact that mutations could be induced in a conditional manner as opposed to being ubiquitous. As such, a myriad of mouse lines have been engineered using cell type- or tissue-specific promoters to drive the expression of the Cre enzyme exclusively in the tissue of interest to study the function of a gene without being hampered by its effects elsewhere in the body (see Ref. 17 and database developed by Dr. A. Nagy at http://nagy.mshri.on.ca/cre/index.php/). Still, in many cases tissue-specific gene targeting results in (embryonic) lethality, thereby precluding the study of the gene's function at a later stage and, for instance, in adult pathology models. To circumvent this problem and add an extra sophisticated level of fine-tuning of the gene targeting, modifications of the Cre-loxP system have been developed to allow inducible activation of Cre activity (for review of the different systems, see Refs. 17, 18). In one nuclear receptor fusion protein approach, the temporal control of the recombinase activity of Cre is achieved by its covalent attachment to the modified estrogen receptor ligand-binding domain (ERt) creating the fusion protein Cre-ERt.[19–23] The ERt domain responds only to the synthetic estrogen receptor ligand tamoxifen or 4-hydroxytamoxifen (4OHT), but not to endogenous estrogens. In line with the classical mode of action of nuclear steroid hormone receptors, proteins containing the ERt domain remain in the cytoplasm in the absence of the ligand, thereby precluding the fused Cre enzyme from recombining loxP-flanked sequences embedded in the genome. Upon administration of tamoxifen, the binding of this ligand to the ERt domain mediates the nuclear translocation of the complex, thus allowing Cre action to take place (FIG. 3).[20–23] The clearance of tamoxifen from the cell after a certain time mediates the return of the Cre-ERt complex to the inactive cytoplasmic state, thus creating a window of time in which temporal transgene modulation can be achieved. The recombination event can be monitored using a Cre reporter transgene, such as ROSA26R (R26R),[24] allowing to visualize cells in which the Cre-ERt transgene has been active. R26R mice carry a construct containing a "floxed" transcriptional stop (neo) cassette immediately upstream of the β-galactosidase (LacZ) sequence at the ubiquitously active Rosa26 genomic locus. Only after Cre-mediated excision of this stop cassette, the LacZ gene is transcribed and it will continue to be expressed as a heritable genomic trait throughout the life of the cell and its progeny. The activity of the LacZ enzyme on the chromogenic substrate 5-bromo-4-chloro-3-indolyl-β-D-galactopyranoside (X-gal) can be used to visualize the recombined cells based on the generation of a blue insoluble product upon cleavage of the colorless X-gal.

Using this inducible system it is thus possible to genetically mark Cre-ERt-expressing cells at a certain point in time, and subsequently follow the track of these cells and their descendants. Such lineage tracing studies first require the Cre-ERt transgene to be tightly controlled by the administration of tamoxifen, with no tamoxifen-independent leaky expression occurring. Ideally, the induced state should be reached rapidly after the ligand injection and be fully reversible. Second, the levels of Cre should be sufficiently high to result in efficient labeling of cells belonging to the lineage under investigation. Partial recombination in less than 100% of the lineage cells, leading to mosaic labeling is, however, not a restraint for the purpose of tracing experiments. Third, the ligand should not interfere with endogenous pathways or have adverse effects on the animal. In this regard, toxicity of tamoxifen has been reported at high doses,[25,26] while termination of pregnancy can still occur upon administration of 2 mg 4OHT per mouse, a dose routinely used for Cre-ERt activation in embryos (personal observations). Last, tracing experiments require the investigator to have quite a good sense of the kinetics of the tamoxifen-induced recombination event (see below). In recent years, the Cre-ERt model has been successfully applied to following the fate of diverse cell types, including hematopoietic stem cells,[27,28] neuronal progenitors,[29–31] pancreatic acinar cells[32] and β-cells,[33,34] follicular keratinocytes,[35] endothelial cells,[36] and cardiac progenitors.[37,38]

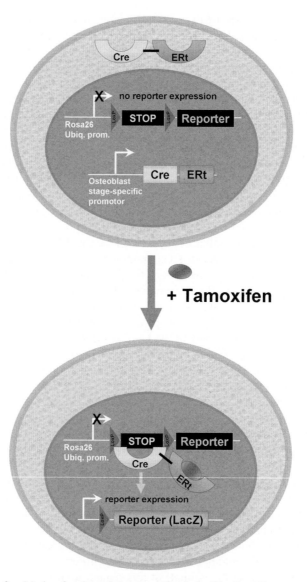

FIGURE 3. Mode of action of the tamoxifen-inducible Cre-ERt transgene system. In the absence of tamoxifen, the Cre-ERt fusion protein remains in the cytoplasm of the cell. Upon administration of tamoxifen, binding of this ligand to the ERt domain mediates the nuclear translocation of the complex. Only then, the Cre enzyme can excise the loxP-flanked transcriptional stop cassette in the reporter transgene and allow its expression. The LacZ transgene of Rosa26R reporter mice is expressed from the ubiquitous Rosa26 locus, while the promoter driving the Cre-ERt expression can confer tissue-specificity to the system. In our study, Cre-ERt is expressed in stage-specific subsets of osteoblast lineage cells. (In color in Annals online.)

Generation and Characterization of Osteoblast-Targeted Cre-ERt Transgenic Mice

As outlined above, markers of subsequent stages of osteoblast differentiation have been identified. The transcription factors Osx and Runx2 are early markers of the osteoblast lineage, Col1 expression is seen in late proliferating osteoblast precursors and mature osteoblasts, and osteocalcin is typically expressed in mature osteoblasts. We have used the mouse Osx- and 3.2 kb Col1-promoters to drive the expression of the inducible Cre-ERt transgene in stage-specific subsets of osteoblast-lineage cells.[39] Osx-Cre-ERt and Col1-Cre-ERt mice (collectively designated osteoblast (OB)-Cre-ERt mice) were intercrossed with R26R reporter mice, thus rendering targeted osteoblastic cells and their descendants detectable by X-gal staining of LacZ activity. First, we confirmed the specificity and ligand dependency of the R26R reporter expression. Whole mount X-gal staining of OB-Cre-ERt(Tg/−);R26R/+ embryos showed intense skeleton-specific LacZ activity (blue staining) exclusively after pulsing the pregnant mother with 4OHT. In our experiments, intraperitoneal injection of 2 mg 4OHT (per $\sim$30 g mouse) was effective in inducing Cre-ERt activity and nontoxic in most animals, although occasional loss of embryos was observed. As expected, no staining was detected in OB-Cre-ERt(−/−);R26R/+ littermates lacking the Cre-ERt transgene. These observations verified the tightly controlled, Cre- and tamoxifen-dependent LacZ activation in the embryonic skeleton.

Next, we designed experiments to assess the kinetics of the induction of LacZ activity in the embryos after 4OHT administration to pregnant female mice. On the one hand, we aimed to define the minimal time required for visual induction of LacZ activity by 4OHT. Therefore, we pulsed a set of females at E13.5 and retrieved the offspring at various time intervals following 4OHT injection, ranging from 4 h to multiple days. These experiments led us to conclude that the induction occurs within hours, as judged by careful analysis of whole mount X-gal staining of the embryos. On the other hand, we defined the end point of the labeling time window following 4OHT injection in the OB-Cre-ERt mice. With respect to cell tracing studies this information is crucial, because legitimate conclusions can only be drawn from pulse-chase studies if the specifics in the timing of the "pulse" versus the "chase" are known. Particularly, the transition period between the phase in which cells become genetically marked by the 4OHT-induced Cre-ERt activity and the subsequent phase in which these cells and their descendants can be followed up on should ideally be strictly constrained, with no *de novo* labeling of cells occurring beyond a given time point after 4OHT administration. The characterization of this parameter is challenging, yet working on the skeleton offers at this point an unprecedented advantage. Indeed, by virtue of the differential developmental timing of the various skeletal elements it is possible to analyze the kinetics of the R26R induction using this site-dependent development progression as an internal clock. For instance, the long bones of the appendicular skeleton develop along a proximodistal time sequence, with a slightly advanced development seen in the forelimbs compared to the corresponding elements (stylopod, zeugopod, and autopod, respectively) in the hind limbs. As such, the initial invasion of the hypertrophic core of the cartilage anlagen—the event starting the formation of the POC—occurs approximately at E14.5 in the humerus and around E15.5 in radius and ulna, while the femur reaches this developmental stage only by E15.0 and the tibia by E16.0. Metatarsals do not undergo initial invasion until E17.5–E18.0 (the exact time points may vary somewhat depending on the mouse strain). Accordingly, the events prior to the POC formation proceed along the same relative temporal sequence. Thus, osteoblast differentiation starts sooner in the humerus than in any other bone of the limbs. As a result, injecting 4OHT in OB-Cre-ERt mice at a given time can result in LacZ labeling exclusively in the humerus, while injecting at a later time point will lead to X-gal staining in additional anatomical locations. Based on this principle, we performed sets of experiments varying the timing of 4OHT injection and of sacrifice. We found that 24-h intervals of injection led to distinct X-gal staining patterns that furthermore were maintained over time. Thus, from these findings we conclude that no additional labeling occurs beyond 24 h after 4OHT administration, thereby rendering the marked cells traceable from this time point onward. These findings are consistent with other recent reports on the time period of tamoxifen activity,[28,40–42] with the reported 6-h half-life of 4OHT in mice,[43] and with pharmacological studies indicating rapid internal tissue accumulation of the compound likely decreasing its concentration in the uterus.[44,45] A more detailed description of the characteristics and induction kinetics of the OB-Cre-ERt mouse lines will be given elsewhere.

Current Approach: Analysis of Perichondrial Osteoblastogenesis and Destiny

The earliest osteoblast differentiation during endochondral bone formation is observed in the perichondrium surrounding the cartilaginous bone model (FIG. 1). Although these osteoblasts are assumed to populate the POC developing later on, their movement into the POC and the extent to which they may contribute to trabecular and cortical bone formation still remain enigmatic. Indeed, due to the tight temporal and spatial coupling of the three key events of POC initiation—chondrocyte hypertrophy, osteoblast differentiation, and initial vascular invasion—it is difficult to elucidate their precise interactions and individual contributions to the bone developmental process. In one very elegant strategy by Colnot et al.,[46] an ex vivo tissue tracing system was designed to address this question. Embryonic limbs were isolated at E14.5 and transplanted under the renal capsule of adult mice. Bone formation proceeded similarly to the *in vivo* events but prolonged in time. Tissue manipulations using Rosa26 host or donor mice allowed the authors to dissect the contributions of the perichondrium and the vasculature to POC formation. This study revealed that cells derived from the perichondrium do populate the POC as osteoblast lineage cells (osteoblasts and osteocytes) in this *ex vivo* system.[46] However, the movement of defined lineage cells from the perichondrium to the primary spongiosa within the growing marrow space *in vivo* and the extent to which these cells contribute to bone formation and maintenance at later stages have not yet been characterized.

To elucidate the differentiation, migration, and bone-forming potential of osteoblasts originating in the perichondrium, we injected pregnant mice carrying Osx-Cre-ERt(Tg/−);R26R/+ or Col1-Cre-ERt(Tg/−);R26R/+ embryos with 4OHT at early time points (12.5–13.5 dpc) to label perichondrial osteoblast cells selectively before vascular invasion of the bone template.[39] The respective transgenes marked osteoblast lineage cells at different stages of differentiation in the perichondrium. Pulse-chase studies, isolating the fetal bones at various time points after the labeling period, then allowed tracking of these cells by histological analysis of X-gal stained samples. As will be described in detail elsewhere, this study identified different stage-specific subsets of osteoblast lineage cells developing in the perichondrium that vary substantially in their capacity to enter the POC and thus are endowed with differential capacities to become cortical or trabecular bone-forming cells. This transgenic mouse model can be further employed for the study of these (pre-)osteoblasts *in vivo* during bone development and beyond.

Potential Applications of the OB-Cre-ERt System

As one can imagine, the cell (stage)-specific and inducible transgenic system developed here has a myriad of potential applications and may prove an extremely helpful tool in the characterization of the osteoblast lineage *in vivo* in a variety of settings. As exemplified by the current approach described above, the model can aid in the study of osteoblast origin, development, migration, and destination during embryogenesis. Studies on the life span and survival of the labeled cells, as well as on their interaction with other cell types can be included as well. Such applications can be directed at both endochondral and intramembranous bones. Furthermore, given the inducible nature of the recombination event, studies on the osteoblast lineage in the postnatal or adult animal should be equally possible. The feasibility of such approaches may depend on the expression levels of the Cre-ERt trangene at these stages and hence on the postnatal activity of the driving promoters. Another caveat may be related to the visualization of the labeled cells, as the use of reporter genes could be quite challenging in adult bone. Indeed, the applicability of the R26R reporter mouse in the adult skeleton is hampered by the endogenous β-galactosidase activity that is seen in postnatal mouse bones and that appears to be particularly associated with osteoclastic cells (Refs. 47, 48, and personal observations). Similarly, the autofluorescent background of bone and marrow tissues may hinder clear-cut analysis of osteoblasts labeled using fluorescent reporters, such as the green fluorescent protein (GFP), although this limitation may be avoided by using appropriate tissue processing (e.g., preparing frozen sections) or suitable fluorescent isomers (e.g., GFP topaz, sapphire, or emerald).[49] Alternatively, different reporter lines can be used, such as the Z/AP mouse,[50] that harbors a heat-resistant alkaline phosphatase transgene (switched on by excision of a floxed cassette containing a LacZ gene) and that has been successfully applied to visualize Cre activity in postnatal bone.[51–53] Thus, tools are available to overcome the technical caveats related to cell-tracing approaches in postnatal bone tissue. As such, the OB-Cre-ERt model can be applied to study the osteoblast lineage *in vivo* during stages of post-natal growth or adult bone homeostasis, as well as in pathological conditions. Assessing the reaction of labeled osteoblasts to catabolic conditions, such as osteoporosis, or the reactivation of these cells in anabolic settings (e.g., fracture repair) are only a few examples of how this system can help to shed light on important bone pathologies.

A second major field of application of the OB-Cre-ERt model is the site- and time-specific manipulation of genes of interest in osteoblastogenesis and bone biology. Upon tamoxifen-mediated activation, the Cre-recombinase can excise floxed sequences designed to inactivate, activate, or modulate a given gene specifically in the cell (stage)-specific populations expressing the Cre-ERt fusion protein. In combination with the use of a reporter gene (i.e., triple transgenic mice) this allows investigators to analyze potential perturbations in (pre-)osteoblast differentiation, migration, survival, etc in the context of a gene mutation. Such tracing strategies include the possibilities to assess the track of (pre-)osteoblasts at any given time either with a null background for a certain gene, or in combination with a conditional mutation specifically in the cell lineage that is the subject of tracing. For example, it can be analyzed in detail how modulation of the (transmission of the) signals known to influence perichondrial osteoblastogenesis, such as Ihh, FGF18, and VEGF (see chapter by Dr. Kronenberg and Refs. 54–61) would affect the fate of the perichondrium-derived osteoblasts.

CONCLUSION

In conclusion, despite tremendous progress in our understanding of os-teoblast differentiation, many aspects related to the origin, destiny, and life span of osteoblasts *in vivo* to date remain enigmatic. The OB-Cre-ERt mice introduced here are powerful tools to study the osteoblast lineage during fetal development, postnatal growth, and bone homeostasis and pathology.

ACKNOWLEDGMENTS

This work was supported by NIH grant DK056246 to H.M.K. and a Postdoctoral Fellowship of the Research Foundation—Flanders (F. W. O.-Vlaanderen) and the K. U. Leuven to C.M.

REFERENCES

1. KRONENBERG, H.M. 2003. Developmental regulation of the growth plate. Nature **423:** 332–336.
2. GERBER, H.P. & N. FERRARA. 2000. Angiogenesis and bone growth. Trends Cardiovasc. Med. **10:** 223–228.
3. KARSENTY, G. & E.F. WAGNER. 2002. Reaching a genetic and molecular understanding of skeletal development. Dev. Cell **2:** 389–406.
4. MAES, C. & G. CARMELIET. 2007. Vascular and nonvascular roles of VEGF in bone development. *In* VEGF and development. C. Ruhrberg, Ed.: 1–12. Landes Bioscience. Austin, TX. Available at http://www.eurekah.com/chapter/3112.
5. KARTSOGIANNIS, V. & K.W. NG. 2004. Cell lines and primary cell cultures in the study of bone cell biology. Mol. Cell Endocrinol. **228:** 79–102.
6. MAJESKA, R.J. & G.A. GRONOWICZ. 2002. Current methodologic issues in cell and tissue culture. *In* Principles of Bone Biology, Vol. 2. J.P. Bilezikian, L.G. Raisz & G.A. Rodan, Eds.: 1529–1541. Academic Press. San Diego, CA.
7. AUBIN, J.E. & J.T. TRIFFITT. 2002. Mesenchymal stem cells and osteoblast differentiation. *In* Principles of Bone Biology, Vol. 1. J.P. Bilezikian, L.G. Raisz & G.A. Rodan, Eds.: 59–81. Academic Press. San Diego, CA.
8. KOBAYASHI, T. & H. KRONENBERG. 2005. Minireview: transcriptional regulation in development of bone. Endocrinology **146:** 1012–1017.
9. KOMORI, T. 2006. Regulation of osteoblast differentiation by transcription factors. J. Cell Biochem. **99:** 1233–1239.
10. HARTMANN, C. 2006. A Wnt canon orchestrating osteoblastogenesis. Trends Cell Biol. **16:** 151–158.
11. LIAN, J.B., G.S. STEIN, A. JAVED, *et al.* 2006. Networks and hubs for the transcriptional control of osteoblastogenesis. Rev. Endocr. Metab. Disord. **7:** 1–16.
12. DUCY, P., M. STARBUCK, M. PRIEMEL, *et al.* 1999. A Cbfa1-dependent genetic pathway controls bone formation beyond embryonic development. Genes Dev. **13:** 1025–1036.
13. KOMORI, T., H. YAGI, S. NOMURA, *et al.* 1997. Targeted disruption of Cbfa1 results in a complete lack of bone formation owing to maturational arrest of osteoblasts. Cell **89:** 755–764.
14. OTTO, F., A.P. THORNELL, T. CROMPTON, *et al.* 1997. Cbfa1, a candidate gene for cleidocranial dysplasia syndrome, is essential for osteoblast differentiation and bone development. Cell **89:** 765–771.
15. NAKASHIMA, K., X. ZHOU, G. KUNKEL, *et al.* 2002. The novel zinc finger-containing transcription factor osterix is required for osteoblast differentiation and bone formation. Cell **108:** 17–29.
16. NAGY, A. 2000. Cre recombinase: the universal reagent for genome tailoring. Genesis **26:** 99–109.

17. ALBANESE, C., J. HULIT, T. SAKAMAKI, *et al.* 2002. Recent advances in inducible expression in transgenic mice. Semin. Cell Dev. Biol. **13:** 129–141.
18. GARCIA, E.L. & A.A. MILLS. 2002. Getting around lethality with inducible Cre-mediated excision. Semin. Cell Dev. Biol. **13:** 151–158.
19. METZGER, D., J. CLIFFORD, H. CHIBA, *et al.* 1995. Conditional site-specific recombination in mammalian cells using a ligand-dependent chimeric Cre recombinase. Proc. Natl. Acad. Sci. USA **92:** 6991–6995.
20. INDRA, A.K., X. WAROT, J. BROCARD, *et al.* 1999. Temporally-controlled site-specific mutagenesis in the basal layer of the epidermis: comparison of the recombinase activity of the tamoxifen-inducible Cre-ER(T) and Cre-ER(T2) recombinases. Nucleic Acids Res. **27:** 4324–4327.
21. DANIELIAN, P.S., D. MUCCINO, D.H. ROWITCH, *et al.* 1998. Modification of gene activity in mouse embryos *in utero* by a tamoxifen-inducible form of Cre recombinase. Curr. Biol. **8:** 1323–1326.
22. FEIL, R., J. BROCARD, B. MASCREZ, *et al.* 1996. Ligand-activated site-specific recombination in mice. Proc. Natl. Acad. Sci. USA **93:** 10887–10890.
23. ZHANG, Y., C. RIESTERER, A.M. AYRALL, *et al.* 1996. Inducible site-directed recombination in mouse embryonic stem cells. Nucleic Acids Res. **24:** 543–548.
24. SORIANO, P. 1999. Generalized lacZ expression with the ROSA26 Cre reporter strain. Nat. Genet. **21:** 70–71.
25. GUO, C., W. YANG & C.G. LOBE. 2002. A Cre recombinase transgene with mosaic, widespread tamoxifen-inducible action. Genesis **32:** 8–18.
26. HONG, S.B., M. FURIHATA, M. BABA, *et al.* 2006. Vascular defects and liver damage by the acute inactivation of the VHL gene during mouse embryogenesis. Lab Invest. **86:** 664–675.
27. GOTHERT, J.R., S.E. GUSTIN, M.A. HALL, *et al.* 2005. In vivo fate-tracing studies using the Scl stem cell enhancer: embryonic hematopoietic stem cells significantly contribute to adult hematopoiesis. Blood **105:** 2724–2732.
28. SAMOKHVALOV, I.M., N.I. SAMOKHVALOVA & S. NISHIKAWA. 2007. Cell tracing shows the contribution of the yolk sac to adult haematopoiesis. Nature **446:** 1056–1061.
29. BATTISTE, J., A.W. HELMS, E.J. KIM, *et al.* 2007. Ascl1 defines sequentially generated lineage-restricted neuronal and oligodendrocyte precursor cells in the spinal cord. Development **134:** 285–293.
30. MASAHIRA, N., H. TAKEBAYASHI, K. ONO, *et al.* 2006. Olig2-positive progenitors in the embryonic spinal cord give rise not only to motoneurons and oligodendrocytes, but also to a subset of astrocytes and ependymal cells. Dev. Biol. **293:** 358–369.
31. BURNS, K.A., A.E. AYOUB, J.J. BREUNIG, *et al.* 2007. Nestin-CreER mice reveal DNA synthesis by nonapoptotic neurons following cerebral ischemia-hypoxia. Cereb. Cortex. [Epub ahead of print]. pmID (17259645).
32. DESAI, B.M., J. OLIVER-KRASINSKI, D.D. DE LEON, *et al.* 2007. Preexisting pancreatic acinar cells contribute to acinar cell, but not islet beta cell, regeneration. J. Clin. Invest. **117:** 971–977.
33. DOR, Y., J. BROWN, O.I. MARTINEZ, *et al.* 2004. Adult pancreatic beta-cells are formed by self-duplication rather than stem-cell differentiation. Nature **429:** 41–46.
34. STROBEL, O., Y. DOR, A. STIRMAN, *et al.* 2007. Beta cell transdifferentiation does not contribute to preneoplastic/metaplastic ductal lesions of the pancreas by genetic lineage tracing *in vivo*. Proc. Natl. Acad. Sci. USA **104:** 4419–4424.

35. LEVY, V., C. LINDON, Y. ZHENG, *et al.* 2007. Epidermal stem cells arise from the hair follicle after wounding. FASEB J. **21:** 1358–1366.

36. GOTHERT, J.R., S.E. GUSTIN, J.A. VAN EEKELEN, *et al.* 2004. Genetically tagging endothelial cells in vivo: bone marrow-derived cells do not contribute to tumor endothelium. Blood **104:** 1769–1777.

37. LAUGWITZ, K.L., A. MORETTI, J. LAM, *et al.* 2005. Postnatal isl1+ cardioblasts enter fully differentiated cardiomyocyte lineages. Nature **433:** 647–653.

38. SUN, Y., X. LIANG, N. NAJAFI, *et al.* 2007. Islet 1 is expressed in distinct cardiovascular lineages, including pacemaker and coronary vascular cells. Dev. Biol. **304:** 286–296.

39. MAES, C., T. KOBAYASHI, J. PARUCH, *et al.* 2006. Characterization of the osteoblast lineage in vivo during early bone development. J. Bone Miner. Res. **21** (Suppl): S3.

40. NAKAMURA, E., M.T. NGUYEN & S. MACKEM. 2006. Kinetics of tamoxifen-regulated Cre activity in mice using a cartilage-specific CreER(T) to assay temporal activity windows along the proximodistal limb skeleton. Dev. Dyn. **235:** 2603–2612.

41. XU, H., F. CERRATO & A. BALDINI. 2005. Timed mutation and cell-fate mapping reveal reiterated roles of Tbx1 during embryogenesis, and a crucial function during segmentation of the pharyngeal system via regulation of endoderm expansion. Development **132:** 4387–4395.

42. ZHANG, Z., T. HUYNH & A. BALDINI. 2006. Mesodermal expression of Tbx1 is necessary and sufficient for pharyngeal arch and cardiac outflow tract development. Development **133:** 3587–3595.

43. ROBINSON, S.P., S.M. LANGAN-FAHEY, D.A. JOHNSON, *et al.* 1991. Metabolites, pharmacodynamics, and pharmacokinetics of tamoxifen in rats and mice compared to the breast cancer patient. Drug Metab. Dispos. **19:** 36–43.

44. LIEN, E.A., E. SOLHEIM & P.M. UELAND. 1991. Distribution of tamoxifen and its metabolites in rat and human tissues during steady-state treatment. Cancer Res. **51:** 4837–4844.

45. KISANGA, E.R., J. GJERDE, J. SCHJOTT, *et al.* 2003. Tamoxifen administration and metabolism in nude mice and nude rats. J. Steroid Biochem. Mol. Biol. **84:** 361–367.

46. COLNOT, C., C. LU, D. HU, *et al.* 2004. Distinguishing the contributions of the perichondrium, cartilage, and vascular endothelium to skeletal development. Dev. Biol. **269:** 55–69.

47. ODGREN, P.R., C.A. MACKAY, A. MASON-SAVAS, *et al.* 2006. False-positive beta-galactosidase staining in osteoclasts by endogenous enzyme: studies in neonatal and month-old wild-type mice. Connect. Tissue Res. **47:** 229–234.

48. KOPP, H.G., A.T. HOOPER, S.V. SHMELKOV, *et al.* 2007. Beta-galactosidase staining on bone marrow. The osteoclast pitfall. Histol. Histopathol. **22:** 971–976.

49. CLARK, S. & D. ROWE. 2002. Application of transgenic mice to problems of skeletal biology. *In* Principles of Bone Biology, Vol. 2. J.P. Bilezikian, L.G. Raisz & G.A. Rodan, Eds.: 1491–1502. Academic Press. San Diego, CA.

50. LOBE, C.G., K.E. KOOP, W. KREPPNER, *et al.* 1999. Z/AP, a double reporter for cre-mediated recombination. Dev. Biol. **208:** 281–292.

51. MIAO, D., B. HE, Y. JIANG, *et al.* 2005. Osteoblast-derived PTHrP is a potent endogenous bone anabolic agent that modifies the therapeutic efficacy of administered PTH 1–34. J. Clin. Invest. **115:** 2402–2411.

52. ZHANG, M., S. XUAN, M.L. BOUXSEIN, et al. 2002. Osteoblast-specific knockout of the insulin-like growth factor (IGF) receptor gene reveals an essential role of IGF signaling in bone matrix mineralization. J. Biol. Chem. **277:** 44005–44012.

53. HE, B., R.A. DECKELBAUM, D. MIAO, et al. 2001. Tissue-specific targeting of the pthrp gene: the generation of mice with floxed alleles. Endocrinology **142:** 2070–2077.

54. MAES, C., P. CARMELIET, K. MOERMANS, et al. 2002. Impaired angiogenesis and endochondral bone formation in mice lacking the vascular endothelial growth factor isoforms VEGF164 and VEGF188. Mech Dev. **111:** 61–73.

55. HINOI, E., P. BIALEK, Y.T. CHEN, et al. 2006. Runx2 inhibits chondrocyte proliferation and hypertrophy through its expression in the perichondrium. Genes Dev. **20:** 2937–2942.

56. COLNOT, C., L. DE LA FUENTE, S. HUANG, et al. 2005. Indian hedgehog synchronizes skeletal angiogenesis and perichondrial maturation with cartilage development. Development **132:** 1057–1067.

57. CHUNG, U.I., E. SCHIPANI, A.P. MCMAHON, et al. 2001. Indian hedgehog couples chondrogenesis to osteogenesis in endochondral bone development. J. Clin. Invest. **107:** 295–304.

58. ST-JACQUES, B., M. HAMMERSCHMIDT & A.P. MCMAHON. 1999. Indian hedgehog signaling regulates proliferation and differentiation of chondrocytes and is essential for bone formation. Genes Dev. **13:** 2072–2086.

59. VORTKAMP, A., K. LEE, B. LANSKE, et al. 1996. Regulation of rate of cartilage differentiation by Indian hedgehog and PTH-related protein. Science **273:** 613–622.

60. LIU, Z., K.J. LAVINE, I.H. HUNG, et al. 2007. FGF18 is required for early chondrocyte proliferation, hypertrophy and vascular invasion of the growth plate. Dev. Biol. **302:** 80–91.

61. ZELZER, E., W. MCLEAN, Y.S. NG, et al. 2002. Skeletal defects in VEGF(120/120) mice reveal multiple roles for VEGF in skeletogenesis. Development **129:** 1893–1904.

Balanced Regulation of Proliferation, Growth, Differentiation, and Degradation in Skeletal Cells

HARRY C. BLAIR,[a] LI SUN,[b] AND RONALD A. KOHANSKI[c]

[a]Department of Pathology, University of Pittsburgh, Pittsburgh, Pennsylvania, USA

[b]Division of Bone Metabolism, Mount Sinai School of Medicine, New York, New York, USA

[c]Biology of Aging Program, National Institute on Aging, National Institutes of Health, Bethesda, Maryland, USA

ABSTRACT: In cartilage and bone-producing cells, proliferation and growth are balanced with terminal differentiation. Maintaining this balance is essential for modeling, growth, and maintenance of the skeleton. Cartilage growth follows a program regulated by hormones and cytokines interacting with a counter-regulatory system in which hedgehog and parathyroid hormone (PTH)-rP signals are key elements. This maintains chondrocyte proliferation and, at specific sites, allows differentiation. Bone is produced by differentiation of mesenchymal stem cells on a scaffold of mineralizing cartilage. However, bone, once formed, is continually resorbed and replaced. Thus, maintenance of bone mass requires retention of stem cells and preosteoblasts in undifferentiated division-competent stages. Maintenance of the undifferentiated states is poorly understood, whereas the rate of osteoblast formation is regulated in part by PTH and insulin-like growth factor. The precursor pool is also subject to depletion by differentiation of mesenchymal stem cells to nonbone cells including adipocytes. In the aging skeleton, disordered balance between bone formation and resorption is in major part due to immune dysregulation that increases formation of bone-degrading osteoclasts; tumor necrosis factor (TNF)-α is a major intermediate in this process.

KEYWORDS: TGF-β; BMP; wnt; frizzled; β-catenin; chondrocyte; parathyroid hormone; osteoclast; T lymphocyte

INTRODUCTION

Some specific mechanisms that regulate skeletal mass are reviewed here. To allow a useful depth of discussion, we will focus on mechanisms relating

Address for correspondence: Harry C. Blair, M.D., Department of Pathology, 705 Scaife Hall, University of Pittsburgh, Pittsburgh, PA 15261. Voice: 412-383-9616; fax: 412-647-8567.
hcblair@imap.pitt.edu

Ann. N.Y. Acad. Sci. 1116: 165–173 (2007). © 2007 New York Academy of Sciences.
doi: 10.1196/annals.1402.029

to the cell fate decision of precursor cells: to mature to form bone or cartilage matrix versus remaining in a precursor pool. This will include discussion of mechanisms that regulate the decision of chondrocytes to progress from growth to terminal differentiation. Following this, we discuss, similarly, the decision of osteoblast precursors to progress from division-competent stem cells or osteoblasts to terminal differentiation and bone formation. Finally, we discuss regulation of the balance between bone formation and degradation after the skeleton is formed, concluding with some recent evidence on the contributions of the immune system. However, to place these topics in context, we will first review briefly some related issues in bone developmental biology and pathology.

The sizes and densities of bone that are formed during development depend on how the bones are modeled and on regulation of remodeling. For most skeletal components, the early stages of bone modeling produce cartilaginous anlage. Later, growth within cartilaginous plates (which are retained after the major part of the skeleton is replaced by bone) is critical to elongation of bones. Cartilaginous modeling and cartilaginous growth are vulnerable to defects in chrondrocyte growth and differentiation, as well as to availability of nutrients, vitamin D, and hormones, particularly growth hormone, which leads to dwarfism when it is deficient (reviewed elsewhere Ref.1,2). We will concentrate on local growth factors and signaling molecules, which recent work has shown are of importance in determining where chondrocytes continue to grow and divide relative to the terminal differentiation pathway related to mineralization and replacement by bone.

Replacement of the primary skeleton with mineralized bone is affected by separate genetic defects, such as mutations in type I collagen, that cause osteogenesis imperfecta. In development of the mineralized skeleton, systemic deficiencies including nutritional and vitamin deficits are also important.[3–5] However, there are fundamental defects in osteoblast differentiation that are affected by recently described transcriptional factors. These may regulate the retention of division-competent stem cells and preosteoblastic cells in periosteal and endosteal compartments. These precursor cells are required to allow continued formation of new bone with aging.

After the skeleton reaches its adult size, loss of cartilage is mainly an issue in arthritis. However, bone mass remains plastic throughout life. It is subject to changes due to the relative rates of bone formation and bone resorption. Bone loss in the adult, especially with aging, is extremely important and is sensitive mainly to interactions of the endocrine and immune systems. Immune signals, such as TNF-α, are increasingly seen as important in regulating the balance between continued osteoblast differentiation and bone resorption. However, it should be noted that immune mechanisms that are clearly important in progressive arthritides have some overlap with immune mechanisms involved in bone loss,[6] albeit with different distributions and consequences of the immune response in localized joint diseases.

CHONDROCYTE PROLIFERATION VERSUS CHONDROCYTE DIFFERENTIATION

Cartilage in the growth plates of bone must, under the control of growth hormone and other systemic influences, continue in proliferation and production of matrix. While this is absolutely necessary, at the same time, chondrocytes in specific locations relative to the shaft of the bone and periosteum must undergo a developmental program leading to organization of bone. This involves rearrangement of chondrocytes into columns oriented along the axis of the bone, calcification of the matrix between columns of chondrocytes with secretion of cytokines initiating bone formation (including VEGF), followed by apoptosis of the chondrocytes. Fundamental regulators of this process are signals—PTHrP and Indian hedgehog (Ihh)—that diffuse from perichondrial undifferentiated pluripotent cells into cartilage, and from cartilage into adjacent perichondrial cells and differentiating chondrocytes, respectively.

The PTH receptor in chondrocytes is a key element in the balance of cartilage growth, cartilage maturation, and bone differentiation. Its ligand, PTHrP, is the ancestral membrane-associated signal for the PTH, later adapted for secondary roles in calcium homeostasis. The PTH receptor is a member of the seven transmembrane-domain receptor family. This large protein family was adapted from early evolution for cell differentiation and cell–cell signaling. This receptor family also includes smoothened, frizzled, and the glycoprotein hormone receptors. Many of the glycoprotein hormone receptors are, in higher vertebrates, concentrated in the pituitary-endocrine axis, although selective expression in ancestral tissues occurs as well, reflecting the origin of these receptors in cell–cell recognition in the earliest organized multicellular organisms, including the slime molds.[7] These receptors are G-protein coupled, although cell response varies widely, and are linked to the many secondary cell regulatory systems.

A specific signal of the hedgehog family of proteins, in cartilage, Ihh is another key to this process. Hedgehog proteins are gradient-dependent signaling proteins (morphogens) that allow development of axial, dorsoventral, and skeletal elements to be regulated by size and location during development.[8] The hedgehog signal, whose diffusion is regulated in part by posttranslational modification, signals through accessory transmembrane receptors via smoothened,[8] a seven-transmembrane domain regulatory protein as discussed above. Smoothened signaling is itself complex; as with most receptors in this family; signaling is G-protein coupled but it is also linked to other signaling systems.[9,10] Ihh and PTHrP are counterregulatory signals; Ihh signals in large part via gli-family transcription factors. PTHrP supports maintenance of proliferation-competent chondrocytes, while Ihh is an inhibitor of its own expression, allowing chondrocyte maturation at proper distances from PTHrP. This involves induction of cartilage-specific proteins, including type II collagen and Sox9 (a crucial transcription factor for cartilage formation).[11]

The PTHrP and Ihh-dependent mechanisms interact with other systems that determine where differentiation occurs, a key system being the bone morphogenic proteins (BMPs) and related transforming growth factor-β (TGF-β) signaling.[12] BMP and TGF proteins are also strongly regulated by hedgehog proteins, although the interections involved are complex. BMPs are expressed in a spatiotemporally specific manner in the axial skeleton under the control of homeotic (Hox) genes and a Wnt3a clock (which response is dependent on frizzled).[13] These interactions are subjects of active investigation (recent developments are discussed by O'Keefe elsewhere in this volume.[14–17]) In brief, the developing story includes signaling downstream from BMP/TGF-β receptors, which is mainly dependent on Smads, interacting with kinase pathways and resulting in activation of β-catenin, which is often seen as primarily downstream of Wnt/frizzled signaling. This new paradigm will help reconcile many difficult-to-understand observations. There are likely many further related stories; for example, β-catenin is complexed with cell membrane proteins including connexins, and is potentially involved in regulation related to cell membrane complexes. These considerations are not only relevant to osteoblasts, well known to have abundant gap junctions, but also to chondrocytes that express connexin43 in a regulated manner.[18]

MAINTAINING THE OSTEOBLAST PRECURSOR POOL

This topic, despite its importance, includes a large number of unknowns. It is not known, for example, if cells capable of differentiating to form all components of the embryo may exist in adult bone marrow.[19] Leaving aside, as unresolvable at present, the question of whether there are truly pluripotent stem cells in the adult marrow, there are a great many interesting issues in what clearly does occur: the continuing differentiation of bone from mesenchymal stem cells or preosteoblasts in the marrow.

A supply of undifferentiated cells capable of forming osteoblasts exists in the peritrabecular space. Whether mesenchymal stem cells that are distributed widely in the marrow can repopulate the osteoblast precursor space is uncertain. Despite the certainty that many mesenchymal stem cells in the delicate connective tissue stroma that separates hematopoietic and adipose marrow compartments, replacement of osteoblasts post fracture appears to be almost exclusively from tissue originating in the periosteal callus rather than from adjacent parts of the bone marrow. Marrow mesenchymal stem cells can form bone *in vitro* without the necessity of significant numbers of cells of other lineages being involved. However, the dramatic periosteal expansion in fractures [20] and the existence of cytokines and fibrocartilage in the healing fracture make it likely that the developmental program for this is far better supported by differentiation from periosteal callus than occurs in the endosteum. This may reflect, however, the necessity for programmed formation of signals including

BMPs[21] and VEGF[22] at the proper locations, rather than limitations on the potential for differentiation in any individual mesenchymal stem cell.

That said, in the normal continuity of bone resorption and new bone formation, it appears likely that most osteoblasts are derived from the endosteal, or medullary, stem cells, and preosteoblasts. Mechanisms allowing some stem cells to remain undifferentiated while permitting site-directed osteoblast formation are not clear. There is evidence that quiescence and self-renewal are promoted by the physical characteristics of the niche, including low oxygen tension,[23] while a number of lines of investigation have pointed to gap-junction communication, including with adjacent endothelial cells, as probably important in the process of the preosteoblast exiting the self-renewing niche.[24,25]

Nevertheless, there must be more to maintenance of the precursor pool versus differentiation than VEGF and gap junctional communication. In the hematopoietic components of marrow, a number of transcription factor and cell cycle regulators,[26–28] as well as a rhoGTPase,[29] are believed to be involved in maintenance and self-renewal of stem cell populations. The self-renewal factors for the mesenchymal stem cell component will undoubtedly be different; interferonα/β is proposed to control undifferentiation to some extent, possibly via Stat1.[30] In more distal differentiation, the TNF family receptor DR3 is proposed as a signal that may maintain committed osteoblasts in an undifferentiated, presumably division competent, state, although the transcriptional mechanism by which this might act is not established, and alternate cell death outcomes may also be mediated by DR3.[31]

In addition, factors regulating osteoblast formation from quiescent or self-renewing mesenchymal stem cells and preosteoblastic populations may include stimuli for differentiation to other terminal fates. Possibilities here include the transcription factors PPARγ[32] and O/E-1,[33] which promote terminal adipogenesis. While the mechanisms are likely to be complex, suppression of PPARγ stimulates osteoblastic differentiation,[34] and effects in aging mice were consistent with this finding.[35] Thus, the idea of a PPARγ-dependent switch in the preosteoblast is certainly engaging. On the other hand, in osteoblasts, PPARγ activation regulates important osteogenic products including VEGF. Thus, at a minimum, PPARγ interacts in differentiation- or time-dependent manners with other regulators of development. Osteoblastic versus alternative (e.g., adipocytic) differentiation, is considered in more detail later in this volume by Horowitz *et al.*, including the potential role of transcription factors that also regulate B lymphocyte development.[36]

BALANCING BONE FORMATION AND DEGRADATION

Relatively little progress has been made in stimulating osteoblast differentiation when the bone formation rate is exceeded by bone degradation, the process ultimately leading to osteoporosis and fracture. The most promising approach is intermittent PTH therapy.[37] The mechanism is unknown, but osteoblasts are PTH receptor-bearing cells and PTH has a role in inducing osteoclast formation through upregulation of RANKL expression in osteoblasts and related cells. PTH has long been known to stimulate both formation of bone and degradation, so it is not surprising that a low and/or intermittent dose of PTH is capable of stimulating differentiation of preosteoblasts without greatly accelerating osteoclast formation.

There are undoubtedly cosignaling pathways, but a "formula" that will specifically activate osteoblast formation while maintaining precursor cell niche occupancy remains an unsolved problem. It is clear, however, that the chondrocyte maturation paradigm may not directly be applicable to the problem in that the frizzled coreceptor Lrp5 is not required for the anabolic effect of PTH.[38] On the other hand, it was recently shown that wnt signaling stimulates osteoblastogenesis in a cell line *in vitro*, arguing that either a noncanonical response or a conditional expression of wnt-responsive receptor proteins may exist and may be important in the osteoblast differentiation decision, in parallel to the role of wnt signaling in cartilage differentiation.[39]

The anabolic PTH effect includes cyclin activation in preosteoblasts,[40] but it is mediated by response of differentiated osteoblasts to PTH. Mice overexpressing Cbfa1, a key osteoblast transcription factor, under the control of the type I collagen promoter, expressed only in osteoblasts at the matrix-secretory stage, do not respond to PTH as an anabolic signal and are osteoporotic.[41] Candidate substances that may mediate proliferation of early osteoblast precursors in response to PTH signaling on osteoblasts at later differentiation stages include insulin-like growth factor-1.[42] It is thus likely that the pathway involved in the anabolic function of PTH involves entirely different signals than those required for stimulation of osteoclast formation by PTH, and there is experimental support for this view.[43]

The other side of the equation of bone mass maintenance is osteoclast formation. Mouse models show that estrogen deficiency leads to an expansion of the peripheral pool of T cells with activation of TNF-α production, while TNF-α acts as an accelerator of osteoclast formation.[44] Although TNF-α is also implicated in osteoporosis related to estrogen deprivation in humans, it is likely that the mechanism will differ somewhat in the human as thymic activity is typically absent when estrogen deficiency occurs. In addition, there is direct evidence for an alternative pathway involving CD40 upregulation of RANKL production in the human.[45] The relative importance in the human of CD40 and TNF-α-related pathways will be interesting issues in comparative physiology. In any case, TNF-α effects include induction of RANKL,[46] and also directly affects osteoclast differentiation.[47,48] Later in this volume, recent findings relating to immune mediators of osteoclast formation are reviewed in detail by Weitzmann and Pacifici.[44]

CONCLUSIONS

Regulation of precursor pool maintenance, both in chondrocyte and osteoblast lineages, is crucial to forming a normal skeleton and to maintaining skeletal density throughout life. While the processes regulating cell division and differentiation are still incompletely understood, the crucial role of balanced hedgehog and PTHrP signaling in maintaining chondrocyte undifferentiation has been demonstrated. Less is known about maintenance of osteoblast precursor pools, although it is clear that PTH and insulin-like growth factor-1 signals are involved in regulating the rate of osteoblast formation, and that alternative differentiation pathways, such as formation of adipocytes, can deplete the early precursor pool. Finally, the balance of bone formation and resorption can be tipped by hyperactive immune function, which is mediated, at least in part, by the production of TNF-α.

ACKNOWLEDGMENTS

This work was supported in part by the National Institutes of Health (USA), National Institutes on Aging and of Arthritis and Musculoskeletal Diseases, and by the Department of Veteran's Affairs.

REFERENCES

1. BAITNER, A.C. *et al.* 2000. The genetic basis of the osteochondrodysplasias. J. Pediatr. Orthop. **20:** 594–605.
2. NILSSON, O. *et al.* 2005. Endocrine regulation of the growth plate. Horm. Res. **64:** 157–165.
3. SPECKER, B. 2004. Nutrition influences bone development from infancy through toddler years. J. Nutr. **134:** 691S–695S.
4. HOLICK, M.F. 2006. Resurrection of vitamin D deficiency and rickets. J. Clin. Invest. **116:** 2062–2072.
5. BISHOP, N., A. SPRIGG & A. DALTON. 2007. Unexplained fractures in infancy: looking for fragile bones. Arch. Dis. Child. **92:** 251–256.
6. WILLIAMS, R.O. 2007. Collagen-induced arthritis in mice: a major role for tumor necrosis factor-α. Methods Mol. Biol. **361:** 265–284.
7. BLAIR, H.C., A. WELLS & C.M. ISALES. 2007. Pituitary glycoprotein hormone receptors in non-endocrine organs. Trends Endocrinol. Metab. **18:** 227–233.
8. RIOBO, N.A. & D.R. MANNING. 2007. Pathways of signal transduction employed by vertebrate Hedgehogs. Biochem. J. **403:** 369–379.
9. VORTKAMP, A. *et al.* 1996. Regulation of rate of cartilage differentiation by Indian hedgehog and PTH-related protein. Science **273:** 613–622.
10. MAU, E. *et al.* 2007. PTHrP regulates growth plate chondrocyte differentiation and proliferation in a Gli3 dependent manner utilizing hedgehog ligand dependent and independent mechanisms. Dev. Biol. **305:** 28–39.

11. LAI, L.P. & J. MITCHELL. 2005. Indian hedgehog: its roles and regulation in endo-chondral bone development. J. Cell Biochem. **96:** 1163–1173.
12. YOON, B.S. & K.M. LYONS. 2004. Multiple functions of BMPs in chondrogenesis. J. Cell Biochem. **93:** 93–103.
13. AULEHLA, A. & B.G. HERRMANN. 2004. Segmentation in vertebrates: clock and gradient finally joined. Genes Dev. **18:** 2060–2067.
14. DONG, Y. *et al.* 2005. Wnt-mediated regulation of chondrocyte maturation: mod-ulation by TGF-β. J. Cell Biochem. **95:** 1057–1068.
15. DONG, Y.F. *et al.* 2006. Wnt induction of chondrocyte hypertrophy through the Runx2 transcription factor. J. Cell Physiol. **208:** 77–86.
16. LI, T.F. *et al.* 2006. Transforming growth factor-β stimulates cyclin D1 expression through activation of β-catenin signaling in chondrocytes. J. Biol. Chem. **281:** 21296–21304.
17. LI, T.F. *et al.* 2006. Smad3 deficient chondrocytes have enhanced BMP signaling and accelerated differentiation. J. Bone Miner. Res. **21:** 4–16.
18. TONON, R. & P. D'ANDREA. 2000. Interleukin-1β increases the functional expres-sion of connexin 43 in articular chondrocytes: evidence for a Ca^{2+}-dependent mechanism. J. Bone Miner. Res. **15:** 1669–1677.
19. KUCIA, M. *et al.* 2007. Morphological and molecular characterization of novel pop-ulation of CXCR4+ SSEA-4+ Oct-4+ very small embryonic-like cells purified from human cord blood: preliminary report. Leukemia **21:** 297–303.
20. ALLEN, M.R., J.M. HOCK & D.B. BURR. 2004. Periosteum: biology, regulation, and response to osteoporosis therapies. Bone **35:** 1003–1012.
21. KUGIMIYA, F. *et al.* 2005. Involvement of endogenous bone morphogenetic protein (BMP) 2 and BMP6 in bone formation. J. Biol. Chem. **280:** 35704–35712.
22. STREET, J. *et al.* 2002. Vascular endothelial growth factor stimulates bone repair by promoting angiogenesis and bone turnover. Proc. Natl. Acad. Sci. USA **99:** 9656–9661.
23. D'IPPOLITO, G. *et al.* 2006. Sustained stromal stem cell self-renewal and osteoblas-tic differentiation during aging. Rejuv. Res. **9:** 10–19.
24. VILLARS, F. *et al.* 2002. Effect of HUVEC on human osteoprogenitor cell differ-entiation needs heterotypic gap junction communication. Am. J. Physiol. Cell Physiol. **282:** C775–C785.
25. GUILLOTIN, B. *et al.* 2004. Human primary endothelial cells stimulate human os-teoprogenitor cell differentiation. Cell Physiol. Biochem. **14:** 325–332.
26. CHENG, T. *et al.* 2000. Hematopoietic stem cell quiescence maintained by p21cip1/waf1. Science **287:** 1804–1808.
27. LACORAZZA, H.D. *et al.* 2005. The transcription factor MEF/ELF4 regulates the quiescence of primitive hematopoietic cells. Cancer Cell **9:** 175–187.
28. JANKOVIC, V. *et al.* 2007. Id1 restrains myeloid commitment, maintaining the self-renewal capacity of hematopoietic stem cells. Proc. Natl. Acad. Sci. USA **104:** 1260–1265.
29. YANG, L. *et al.* 2007. Rho GTPase Cdc42 coordinates hematopoietic stem cell quiescence and niche interaction in the bone marrow. Proc. Natl. Acad. Sci. USA **104:** 5091–5096.
30. HATZFELD, A. *et al.* 2007. A sub-population of high proliferative potential-quiescent human mesenchymal stem cells is under the reversible control of in-terferon α. Leukemia **21:** 714–724.
31. BORYSENKO, C.W. *et al.* 2006. Death receptor-3 mediates apoptosis in human ost-eoblasts under narrowly regulated conditions. J. Cell Physiol. **209:** 1021–1028.

32. SHI, X.M. *et al.* 2000. Tandem repeat of C/EBP binding sites mediates PPARγ2 gene transcription in glucocorticoid-induced adipocyte differentiation. J. Cell Biochem. **76:** 518–527.

33. AKERBLAD, P. *et al.* 2002. Early B-cell factor (O/E-1) is a promoter of adipogenesis and involved in control of genes important for terminal adipocyte differentiation. Mol. Cell Biol. **22:** 8015–8025.

34. YAMASHITA, A. *et al.* 2006. Transient suppression of PPARγ directed ES cells into an osteoblastic lineage. FEBS Lett. **580:** 4121–4125.

35. MOERMAN, E.J. *et al.* 2004. Aging activates adipogenic and suppresses osteogenic programs in mesenchymal marrow stroma/stem cells: the role of PPAR-γ2 transcription factor and TGF-β/BMP signaling pathways. Aging Cell **3:** 379–389.

36. HOROWITZ, M.C. *et al.* 2005. B cells and osteoblast and osteoclast development. Immunol. Rev. **208:** 141–153.

37. COMPSTON, J.E. 2006. Skeletal actions of intermittent parathyroid hormone: effects on bone remodelling and structure. Bone **40:** 1447–1452.

38. IWANIEC, U.T. *et al.* 2007. PTH stimulates bone formation in mice deficient in Lrp5. J. Bone Miner. Res. **22:** 394–402.

39. KANG, S. *et al.* 2007. Wnt signaling stimulates osteoblastogenesis of mesenchymal precursors by suppressing CCAAT/enhancer-binding protein-α and peroxisome proliferator-activated receptor-γ. J. Biol. Chem. **282:** 14515–14524.

40. DATTA, N.S. *et al.* 2007. Cyclin D1 as a target for the proliferative effects of PTH and PTHrP in early osteoblastic cells. J. Bone Miner. Res. **22:** 951–964.

41. MERCIRIS, D. *et al.* 2007. Overexpression of the transcriptional factor Runx2 in osteoblasts abolishes the anabolic effect of parathyroid hormone *in vivo*. Am. J. Pathol. **170:** 1676–1685.

42. LOCKLIN, R.M. *et al.* 2003. Mediators of the biphasic responses of bone to intermittent and continuously administered parathyroid hormone. J. Cell Biochem. **89:** 180–190.

43. LIU, F. *et al.* 2007. CREM deficiency in mice alters the response of bone to intermittent parathyroid hormone treatment. Bone **40:** 1135–1143.

44. WEITZMANN, M.N. & R. PACIFICI. 2005. The role of T lymphocytes in bone metabolism. Immunol. Rev. **208:** 154–168.

45. LOPEZ-GRANADOS, E. *et al.* 2007. Osteopenia in X-linked hyper-IgM syndrome reveals a regulatory role for CD40 ligand in osteoclastogenesis. Proc. Natl. Acad. Sci. USA **104:** 5056–5061.

46. KHOSLA, S. 2001. The OPG/RANKL/RANK system. Endocrinology **142:** 5050–5055.

47. FUJIKAWA, Y. *et al.* 2001. The effect of macrophage-colony stimulating factor and other humoral factors (interleukin-1, -3, -6, and -11, tumor necrosis factor-α, and granulocyte macrophage-colony stimulating factor) on human osteoclast formation from circulating cells. Bone **28:** 261–267.

48. SABOKBAR, A., O. KUDO & N.A. ATHANASOU. 2003. Two distinct cellular mechanisms of osteoclast formation and bone resorption in periprosthetic osteolysis. J. Orthop. Res. **21:** 73–80.

49. YASUDA, E. *et al.* 2005. PPAR-γ ligands up-regulate basic fibroblast growth factor-induced VEGF release through amplifying SAPK/JNK activation in osteoblasts. Biochem. Biophys. Res. Commun. **328:** 137–143.

Control of Postnatal Bone Mass by the Zinc Finger Adapter Protein Schnurri-3

LAURIE H. GLIMCHER,[a,b] DALLAS C. JONES,[a] AND MARC N. WEIN[a]

[a]*Department of Immunology and Infectious Diseases, Harvard School of Public Health, Boston, Massachusetts, USA*

[b]*Department of Medicine, Harvard Medical School, Boston, Massachusetts, USA*

ABSTRACT: The completed skeleton undergoes continuous remodeling for the duration of adult life. Rates of bone formation by osteoblasts and bone resorption by osteoclasts determine adult bone mass. Abnormalities in either the osteoblast or osteoclast compartment affect bone mass and result in skeletal disorders, the most common of which is osteoporosis, a state of low bone mass. Much is known about the molecular control of bone formation and resorption from rare single gene disorders resulting in elevated or reduced bone mass. Such genetic disorders can be attributed either to osteoclast deficiencies, collectively termed "osteopetrosis," or to intrinsically elevated osteoblast activity, termed "osteosclerosis." However, an increasing need for anabolic therapies to prevent age-induced bone loss has stimulated a search for additional genes that act at the level of the osteoblast to regulate matrix synthesis. Recently, we have discovered a zinc finger adaptor protein called Schnurri-3 (Shn3) that potently regulates adult bone mass. Mice that lack Shn3 have normal skeletal morphogenesis but display profoundly elevated bone mass that increases with age. The molecular mechanism was revealed to be the recruitment of WWP1, a Nedd4 family E3 ubiquitin ligase, by Shn3 to the major transcriptional regulator of the osteoblast, Runx2. In the absence of Shn3, Runx2 degradation by WWP1 is inhibited resulting in increased levels of Runx2 protein and enhanced expression of Runx2 target genes leading to increased osteoblast synthetic activity. Small molecules that inhibit Shn3 or WWP1 may be attractive candidates for the treatment of diseases of low bone mass.

KEYWORDS: Schnurri-3; WWP1; osteoblast; skeletal remodeling; E3 ligases

INTRODUCTION

Regulation of osteoblast differentiation and function involves a complex network of developmental cues and environmental factors.[1] The transcription

Address for correspondence: Laurie H. Glimcher, Department of Immunology and Infectious Diseases, Harvard School of Public Health, 651 Huntington Ave, FXB 205, Boston, MA 02115. Voice: 617-432-0622; fax: 617-432-0084.

lglimche@hsph.harvard.edu

Ann. N.Y. Acad. Sci. 1116: 174–181 (2007). © 2007 New York Academy of Sciences.
doi: 10.1196/annals.1402.044

factor Runx2 interacts with a number of nuclear transcription factors, coactivators, and adaptor proteins to interpret these extracellular signals to appropriately regulate osteoblast differentiation.[3,4] Runx2 is an essential component of skeletogenesis as mutations in Runx2 cause the human autosomal dominant disease cleidocranial dysplasia.[5-7] Runx2$^{-/-}$ mice exhibit a complete lack of both intramembranous and endochondral ossification, which results in an unmineralized skeleton.[7,8]

In contrast to the significant progress in understanding the molecular mechanisms responsible for osteoblast differentiation during embryonic development, only a small number of genes are known to regulate postnatal osteoblast function.[9-11] Runx2 plays a central role in osteoblast differentiation during embryonic development and also is a key regulator of osteoblast activity in adult mice.[12] This latter function is mediated in part through its induction of transcription factor ATF4 that regulates collagen biosynthesis in mature osteoblasts.[13] Several of these factors function during embryogenesis to regulate the commitment of mesenchymal stem cells to the osteoblast lineage.[3,10,14,15] These include Runx2, Twist, and Osterix. Others, including the Wnt coreceptor LRP5 and the transcription factor ATF4, function later in mature osteoblasts by controlling their synthetic function during bone remodeling.

Recently, the E3 ubiquitin ligase Smurf1, a member of the Nedd4 family of HECT domain-containing E3 ligases, has been shown to inhibit adult osteoblast activity via its impact on MEKK2/JNK/AP-1 signaling.[16] Although Smurf1 has been reported to interact with Runx2 and promote its ubiquitination,[15,17] osteoblasts derived from Smurf1$^{-/-}$ mice exhibit normal levels of Runx2 protein suggesting that other E3 ligases within the Nedd4 family may regulate Runx2 protein levels.[16]

Schnurri-3 (Shn3) is one of three known mammalian homologues of the Drosophila protein Shn, which acts during embryogenesis as an essential nuclear cofactor for Decapentaplegic (Dpp) signaling, the Drosophila homologue of BMP/TGF-β[18] (FIG. 1). While Shn3, a ZAS family protein, was originally identified by others as a DNA-binding protein of the heptameric recombination signal sequence required for VDJ recombination of immunoglobulin genes,[19] we have not been able to confirm that it directly binds DNA. Instead, we have described that it functions as an adaptor protein in the immune system. Shn3 interacts with TRAF2 to inhibit NF-κB and JNK-mediated responses, including apoptosis and TNF-α gene expression.[20] In T lymphocytes, Shn3 regulates IL-2 gene expression through its ability to partner with c-Jun and potentiate AP-1 transcriptional activity.[21] Although Shn3 has been demonstrated to regulate the activities of these important transcription factors *in vitro*, it remained to be seen what the *in vivo* role(s) of Shn3 were.

We therefore generated a strain of mice with a targeted disruption of the Shn3 gene. While the abnormalities in the immune system proved to be minimal, we uncovered an essential role for Shn3 in regulating postnatal bone formation *in vivo*. Homozygous Shn3 mutant (Shn3$^{-/-}$) mice exhibited a profound high

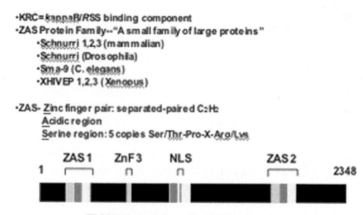

FIGURE 1. Schematic of Shn3 structure.

bone mass phenotype that was secondary to augmented osteoblast synthetic activity and bone formation.[2]

OSTEOGENESIS AND SHN3

Mice lacking Shn3 exhibited evidence of increased bone mass as early as 2 weeks of age and this phenotype progressed over time to virtual obliteration of the long bone marrow (BM) cavity by 7 months. Micro-qCT analysis revealed increased number, thickness, and mineralization of trabeculae with a moderate increase in cortical thickness (FIG. 2). Dual calcein-labeling experiments yielded an approximately fivefold increase in bone formation rate. *In vitro* cultured neonatal Shn3$^{-/-}$ calvarial osteoblasts recapitulated the *in vivo* phenotype and displayed increased expression of bone sialoprotein (BSP) and osteocalcin (OCN) mRNA but similar levels of ALP mRNA compared to WT osteoblasts. The expression levels of the important osteoblast transcription factors Osterix and ATF4 were elevated as well.[13,22] Taken together, these observations demonstrated an important role for Shn3 in regulating the expression of genes that control bone formation and mineralization.

RUNX2 IS A TARGET OF SHN3

An examination of the genes that were elevated in Shn3$^{-/-}$ osteoblasts revealed that they were all direct Runx2 targets,[4,13] suggesting that Runx2 might mediate the effect of Shn3. Indeed, Shn3$^{-/-}$ mice possess a bone formation

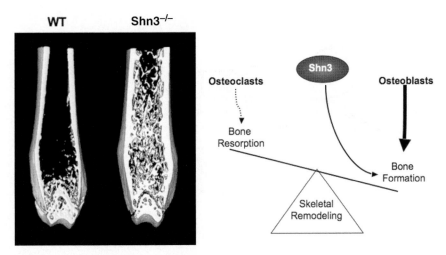

FIGURE 2. Increased bone mass in Shn3$^{-/-}$ mice. Reproduced with permission from Lian *et al.*[3]

phenotype *in vivo* opposite to that observed in mice expressing an osteoblast-targeted Runx2 dominant negative transgene.[12] Runx2 mRNA levels were comparable between Shn3$^{-/-}$ and WT osteoblasts, however, levels of Runx2 protein were increased by approximately twofold in the absence of Shn3 and this led to increased Runx2 DNA binding. Conversely, pulse chase-labeling experiments revealed increased degradation of Runx2 protein when Shn3 was overexpressed. Ectopic expression of Shn3 also resulted in inhibition of Runx2 transactivation function. The inhibitory effect of Shn3 on Runx2 activity was accompanied by a direct physical association of the two proteins as evidenced by their coimmunoprecipitation from primary osteoblasts.

As a master regulator of both osteoblast lineage commitment and activity of committed osteoblasts, Runx2 interacts with a number of nuclear transcription factors, coactivators, and adaptor proteins to interpret extracellular signals to appropriately regulate osteoblast function.[3,4] A number of protein–protein interactions that are inhibitory to Runx2 function have been described,[3,10,14,15] although only the inhibitory roles of HDAC4, Twist1, and Stat1 have been verified by *in vivo* loss of function evidence.

SHN3 RECRUITS THE E3 LIGASE WWP1 TO RUNX2

Hence, Shn3 directly associates with Runx2 and the effect of this association is to reduce Runx2 activity by enhancing its degradation (FIG. 3). The degradation of Runx2 was accompanied by its ubiquitination raising the possibility that Shn3 was an E3 ubiquitin ligase. However, Shn3 itself contained none of the typical motifs characteristic of E3 ligase family members[23,24] and

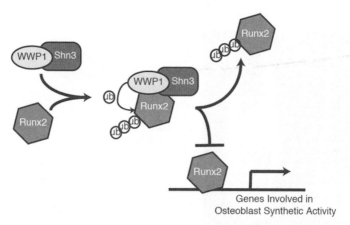

FIGURE 3. Schematic of Shn3/WWP1/Runx2 interaction.

recombinant Shn3 protein fragments did not ubiquitinate Runx2 *in vitro*. An alternative possibility was that Shn3 might promote Runx2 ubiquitination indirectly via recruiting a ubiquitous cellular factor, such as an E3 ligase of the Nedd4 family[25] to a Runx2/Shn3 complex. When we analyzed Shn3's ability to interact with members of the Nedd4 family (FIG. 4), Shn3 interacted only with WWP1. Additionally, coexpression of Shn3 and WWP1 led to high levels of Runx2 ubiquitination. Finally, reducing endogenous levels of WWP1 with shRNA-mediated knockdown caused a posttranscriptional increase in Runx2 protein levels, and significantly enhanced extracellular matrix synthesis in primary osteoblasts. Although a role for Smurf1 as a Runx2 E3 ubiquitin ligase had been suggested from overexpression studies, Smurf1-deficient osteoblasts do not display any alteration in Runx2 levels[16] consistent with our failure to detect a Smurf1/Shn3 physical or functional interaction.

E3 ubiquitin ligases of the SCF and APC family are known to function in large heteromeric complexes;[26] however, we believe that this is the first example of an adaptor protein functioning to augment the activity of a HECT domain-containing ubiquitin ligase. Additionally, in overexpression studies, we find that all the members of the Nedd4 family of HECT E3 ligases can associate with Runx2 (unpublished data), although only WWP1 and Shn3 promote robust Runx2 polyubiquitination. Indeed, it has previously been reported that the association of other HECT E3 ligases with a potential substrate is insufficient to induce ubiquitination.[27] Future studies designed to learn more about the interaction between Nedd4 family ubiquitin ligases and Shn3, and to understand how these interactions influence E3 ligase activity will be necessary.

Recent work has suggested an alternative mechanism in which Nedd4 family E3 ligases are regulated at a posttranslational level. Itch is another member of this family that has recently been positioned in the TGF-β pathway.[28] In

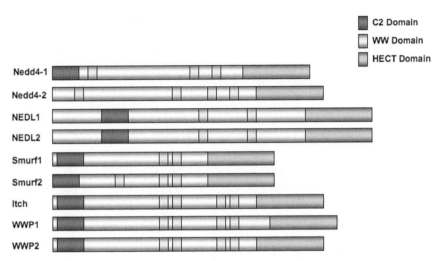

FIGURE 4. NEDD4 family members.

CD4$^+$ T cells, Itch is directly phosphorylated by JNK, and this phosphorylation event serves to augment the ability of Itch to promote ubiquitination of AP-1 components c-Jun and JunB.[29] Since we have previously described a role for Shn3 in regulating TNF-induced JNK activation,[20] it will be interesting to determine if regulation of WWP1 activity by osteoblast stress-induced kinases exists, and if phosphorylation-dependent ubiquitination promoted by WWP1 is regulated at any level by Shn3.

SUMMARY AND CONCLUSIONS

Our experiments revealed that Shn3 functions in a multimeric complex and comprises Shn3, Runx2, and the Nedd4 family E3 ubiquitin ligase WWP1 in osteoblasts. This complex acts to inhibit Runx2 function through the ability of WWP1 to promote Runx2 polyubiquitination and proteasome-dependent degradation. Shn3 is an integral and required component of this multimeric complex, as revealed most unequivocally in mice lacking this protein that display profoundly increased rates of bone formation. *Ex vivo*, the absence of Shn3 in osteoblasts results in elevated Runx2 protein levels, augmented Runx2 transcriptional activity, and elevated transcription of Runx2 target genes.

Our studies establish a role for Shn3 as a potent inhibitor of bone formation via an osteoblast-intrinsic mechanism. Our findings lead us to conclude that Shn3 regulates osteoblast function through an association with WWP1 to regulate the ubiquitination and hence degradation of Runx2 although we have preliminary evidence for the existence of additional Shn3/WWP1 substrates. These observations suggest that compounds designed to block Shn3 function may be possible therapeutic agents for the treatment of osteoporosis. Additional definition of the functional domains of Shn3 will be necessary to make this very large protein amenable for drug discovery. However, the availability of the crystal structure of WWP1[30] has facilitated the identification of WWP1 inhibitory molecules that have anabolic activity *in vitro*.

ACKNOWLEDGMENTS

This work was supported by NIH grants AI29673 (LHG), AR46983 (LHG), postdoctoral fellowships from the Arthritis Foundation (DJ), and the Medical Scientist Training Program at Harvard Medical School (MW).
Competing Interest: None declared

REFERENCES

1. KARSENTY, G. & E.F. WAGNER. 2002. Reaching a genetic and molecular understanding of skeletal development. Dev. Cell. **2:** 389–406.
2. JONES, D.C., M.N. WEIN, M. OUKKA, *et al.* 2006. Regulation of adult bone mass by the zinc finger adapter protein Schnurri-3. Science **312:** 1223–1227.
3. LIAN, J.B., A. JAVED, S.K. ZAIDI, *et al.* 2004. Regulatory controls for osteoblast growth and differentiation: role of Runx/Cbfa/AML factors. Crit. Rev. Eukaryot. Gene Expr. **14:** 1–41.
4. STEIN, G.S., J.B. LIAN, A.J. VAN WIJNEN, *et al.* 2004. Runx2 control of organization, assembly and activity of the regulatory machinery for skeletal gene expression. Oncogene **23:** 4315–4329.
5. LEE, B., K. THIRUNAVUKKARASU, L. ZHOU, *et al.* 1997. Missense mutations abolishing DNA binding of the osteoblast-specific transcription factor OSF2/CBFA1 in cleidocranial dysplasia. Nat. Genet. **16:** 307–310.
6. MUNDLOS, S, F. OTTO, C. MUNDLOS, *et al.* 1997. Mutations involving the transcription factor CBFA1 cause cleidocranial dysplasia. Cell **89:** 773–779.
7. OTTO, F., A.P. THORNELL, T. CROMPTON, *et al.* 1997. Cbfa1, a candidate gene for cleidocranial dysplasia syndrome, is essential for osteoblast differentiation and bone development. Cell **89:** 765–771.
8. KOMORI, T, H. YAGI, S. NOMURA, *et al.* 1997. Targeted disruption of Cbfa1 results in a complete lack of bone formation owing to maturational arrest of osteoblasts. Cell **89:** 755–764.
9. YOSHIDA, Y., S. TANAKA, H. UMEMORI, *et al.* 2000. Negative regulation of BMP/Smad signaling by Tob in osteoblasts. Cell **103:** 1085–1097.
10. KIM, S., T. KOGA, M. ISOBE, *et al.* 2003. Stat1 functions as a cytoplasmic attenuator of Runx2 in the transcriptional program of osteoblast differentiation. Genes Dev. **17:** 1979–1991.
11. JOHNSON, M.L., K. HARNISH, R. NUSSE & W. VAN HUL. 2004. LRP5 and Wnt signaling: a union made for bone. J. Bone Miner. Res. **19:** 1749–1757.

12. Ducy, P., M. Starbuck, M. Priemel, *et al.* 1999. A Cbfa1-dependent genetic pathway controls bone formation beyond embryonic development. Genes Dev. **13:** 1025–1036.

13. Yang, X., K. Matsuda, P. Bialek, *et al.* 2004. ATF4 is a substrate of RSK2 and an essential regulator of osteoblast biology; implication for Coffin-Lowry Syndrome. Cell **117:** 387–398.

14. Alliston, T., L. Choy, P. Ducy, *et al.* 2001. TGF-beta-induced repression of CBFA1 by Smad3 decreases cbfa1 and osteocalcin expression and inhibits osteoblast differentiation. EMBO J. **20:** 2254–2272.

15. Zhao, M., M. Qiao, B.O. Oyajobi, *et al.* 2003. E3 ubiquitin ligase Smurf1 mediates core-binding factor alpha1/Runx2 degradation and plays a specific role in osteoblast differentiation. J. Biol. Chem. **278:** 27939–27944.

16. Yamashita, M., S.X. Ying, G.M. Zhang, *et al.* 2005. Ubiquitin ligase Smurf1 controls osteoblast activity and bone homeostasis by targeting MEKK2 for degradation. Cell **121:** 101–113.

17. Zhao, M., M. Qiao, S.E. Harris, *et al.* 2004. Smurf1 inhibits osteoblast differentiation and bone formation in vitro and *in vivo*. J. Biol. Chem. **279:** 12854–12859.

18. Affolter, M., T. Marty, M.A. Vigano & A. Jazwinska. 2001. Nuclear interpretation of Dpp signaling in Drosophila. EMBO J. **20:** 3298–3305.

19. Wu, L.C., C.H. Mak, N. Dear, *et al.* 1993. Molecular cloning of a zinc finger protein which binds to the heptamer of the signal sequence for V(D)J recombination. Nuc. Acid Res. **21:** 5067–5073.

20. Oukka, M., S.T. Kim, G. Lugo, *et al.* 2002. A mammalian homologue of Drosophila *schnurri*, KRC, regulates TNF receptor-driven responses and interacts with TRAF2. Mol. Cell **9:** 121–131.

21. Oukka, M., M.N. Wein & L.H. Glimcher. 2004. Schnurri-3 (KRC) interacts with c-Jun to regulate the IL-2 gene in T cells. J. Exp. Med. **199:** 15–24.

22. Nakashima, K., X. Zhou, G. Kunkel, *et al.* 2002. The novel zinc finger-containing transcription factor osterix is required for osteoblast differentiation and bone formation. Cell **108:** 17–29.

23. Patterson, C. 2002. A new gun in town: the U box is a ubiquitin ligase domain. Sci STKE **2002:** PE4.

24. Pickart, C.M. 2004. Back to the future with ubiquitin. Cell **116:** 181–190.

25. Ingham, R.J., G. Gish & T. Pawson. 2004. The Nedd4 family of E3 ubiquitin ligases: functional diversity within a common modular architecture. Oncogene **23:** 1972–1984.

26. Jackson, P.K. & A.G. Eldridge. 2002. The SCF ubiquitin ligase: an extended look. Mol. Cell **9:** 923–925.

27. Schwarz, S.E., J.L. Rosa, M. Scheffner. 1998. Characterization of human hect domain family members and their interaction with UbcH5 and UbcH7. J. Biol. Chem. **273:** 12148–12154.

28. Chae, H.J., H.R. Kim, C. Xu, *et al.* 2004. BI-1 regulates an apoptosis pathway linked to endoplasmic reticulum stress. Mol. Cell **15:** 355–366.

29. Gao, M., T. Labuda, Y. Xia, *et al.* 2004. Jun turnover is controlled through JNK-dependent phosphorylation of the E3 ligase Itch. Science **306:** 271–275.

30. Verdecia, M.A., C.A. Joazeiro, N.J. Wells, *et al.* 2003. Conformational flexibility underlies ubiquitin ligation mediated by the WWP1 HECT domain E3 ligase. Mol. Cell **11:** 249–259.

Suppression of PPAR Transactivation Switches Cell Fate of Bone Marrow Stem Cells from Adipocytes into Osteoblasts

ICHIRO TAKADA,[a] MIYUKI SUZAWA,[a] KUNIHIRO MATSUMOTO,[b] AND SHIGEAKI KATO[a,c]

[a]Institute of Molecular and Cellular Biosciences, University of Tokyo, Tokyo, Japan

[b]Department of Molecular Biology, Graduate School of Science, Nagoya University, Nagoya, Japan

[c]ERATO, Japan Science and Technology, Saitama, Japan

ABSTRACT: Osteoblasts and adipocytes differentiate from common pleiotropic mesenchymal stem cells under transcriptional controls by numerous factors and multiple intracellular signalings. However, cellular signaling factors that determine cell fates of mensenchymal stem cells in bone marrow remain to be largely uncovered, though peroxisome proliferator-activated receptor-γ (PPAR-γ) is well established as a prime inducer of adipogenesis. Here, we describe two signaling pathways that induce the cell fate decision into osteoblasts from adipocytes. One signaling is a TAK1/TAB1/NIK cascade activated by TNF-α and IL-1, and the activated NF-κB blocked the DNA binding of PPAR-γ, attenuating the activated PPAR-mediated adipogenesis. The second signaling is the noncanonical Wnt pathway through CaMKII-TAK1/TAB2-NLK. Activated NLK by a noncanonical Wnt ligand (Wnt-5a) transrepresses PPAR transactivation through a histone methyltransferase, SETDB1. Wnt-5a induces phosphorylation of NLK, leading to the formation of a corepressor complex that inactivates PPAR function through histone H3-K9 methylation. Thus, two signaling pathways lead to an osteoblastic cell lineage decision from mesenchymal stem cells through two distinct modes of PPAR transrepression.

KEYWORDS: PPAR-gamma; osteoblastogenesis; bone marrow

INTRODUCTION

Osteoblasts are cytodifferentiated from mesenchymal stem cells in bone marrow through multiple steps of commitments. The bone marrow stem cells

Address for correspondence: Shigeaki Kato, Ph.D., Institute of Molecular and Cellular Biosciences, The University of Tokyo, 1-1-1 Yayoi, Bunkyo-ku, Tokyo 113-0032, Japan. Voice: 81-3-5841-8478; fax: 81-3-5841-8477.

uskato@mail.ecc.u-tokyo.ac.jp

Ann. N.Y. Acad. Sci. 1116: 182–195 (2007). © 2007 New York Academy of Sciences.
doi: 10.1196/annals.1402.034

are pleiotrophic to be differentiated into the other cell types like chondro-cytes and adipocytes (FIG. 1). Thus, it is evident that there are regulators to govern the cell fate switching, and particularly in normal bone continuous osteoblastogenesis is maintained while adipogenesis appears suppressive. Osteoblastogenesis is governed by a number of regulators including cytokines and Wnt peptide ligands.[1–3] Canonical and noncanonical Wnt signaling pathways are activated by multiple Wnt ligands through binding to frizzled (Fzd) plasma membrane receptors (FIG. 2). During activation of the canonical pathway, stabilization and nuclear translocation of the intracellular transducer β-catenin is induced to associate with members of the T cell factor/lymphoid enhancer factor (TCF/LEF) family of transcriptional factors for transcriptional activation.[4] Activated canonical Wnt signaling is shown to stimulate osteoblastic differentiation at several steps of cytodifferentiation. When compared to the canonical Wnt signaling, the downstream cascades from the noncanonical signal[5,6] are barely described and their physiological impact in cell fate decision of mesenchymal stem cell remains totally obscure.

It is known that adipogenesis is abnormally enhanced in the bone marrow in osteoporotic or elder states and under chronic treatments of a PPAR-γ agonist for diabetes. PPAR-γ is a prominent nuclear receptor (NR) in many biological events, such as cell differentiation and cell lineage decision.[7–9] Particularly, activated PPAR-γ is well established as a prime regulator to stimulate adipogenesis from pleiotrophic mesenchymal stem cells.[10] PPAR-γ function appears to be modulated by cross talk through other cellular signaling pathways.[11] PPAR-γ is one of the members of the NR gene superfamily. NRs are ligand-inducible transcription factors to control expression of target genes through choromatin remodeling and histone modifications. A number of coregulators are considered to support the ligand-dependent transcriptional controls, but the enzymes of histone acetylation/deacetylation have been well characterized as transcriptional coregulators.[12–14] In the absence of cognate ligands, NRs are transcriptionally silent, associating with corepressors/corepressor complexes that often contain histone deacetylase (HDAC), which renders histones hypoacetylated. Ligand binding to NRs induces clearance of such HDAC corepressors, leading to recruitment of coactivator/coactivator complexes, in which histone acetyltransferase (HAT) activity is often detectable.[12–14] More recently, histone methylation/demethylation has been reported as a key event to trigger ligand-dependent transcriptional controls by NRs.[15] Thus, ligand-induced transcriptional controls look to be tightly coupled with histone modifications.

Histone modification is one of the epigenetic marks on chromatin governing transcriptional regulations in the programming of chromosomal DNA. Chromatin remodeling appears to occur in response to histone modifications (FIG. 3). A number of posttranslational and covalent modifications of histones have already been illustrated: methylation of lysine (K) and arginine (R) residues: acetylation of K; phosphorylation of serine (S) and threonine (T) residues; sumoylation of K; etc. Accumulating evidence has established that

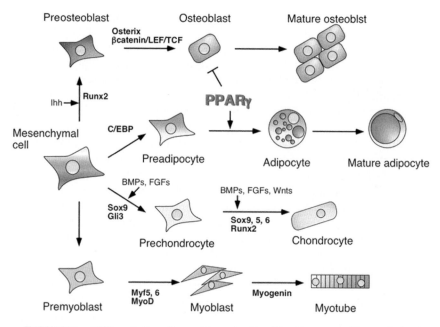

FIGURE 1. Differentiation of mesenchymal cells of four lineages and its regulation by transcription factors. Transcription factors important for each differentiation are indicated by bold block letters and signaling molecules are indicated by italic letters. PPAR-γ acts as an inducer of adipocyte differentiation and inhibits osteoblast differentiation.

hyperacetylation of histone in specific chromosomal regions generally results in activation of a given chromatin area in gene regulation, whereas hypoacetylated histones are indicators of inactive chromatin.[16–18] Other histone modifications are less predictable in how they activate chromatin; however, specific combinations of several histone modifications at certain histone residues are considered to constitute a "histone code" that defines chromatin states for transcriptional control.[16] Among such histone modifications, lysine methylation results in unique transcriptional outcomes depending on the methylation sites, acting as docking signals for recruiting chromatin remodelers/modifiers.[17–19] Among methylated sites on mammalian chromatin, methylated H3-K9, H3-K27, and H4-K20 are considered as hallmarks of a condensed chromatin state.[17] Furthermore, H3-K9 methylation by histone methyltransferase (HKMTs) has been shown to trigger heterochromatin formation and transcriptionally silence euchromatic regions by recruiting heterochromatin proteins (FIG. 4).[18,19] Reflecting the critical roles of methylated lysines at specific sites, multiple HKMTs have been identified that recognize the same lysine residue for mono-, di-, and/or tri-methylations, although the biological role of each HKMT still remains elusive.[19,20] Histone modifications are altered during cell lineage decisions, and rearrangements of histone modifications take

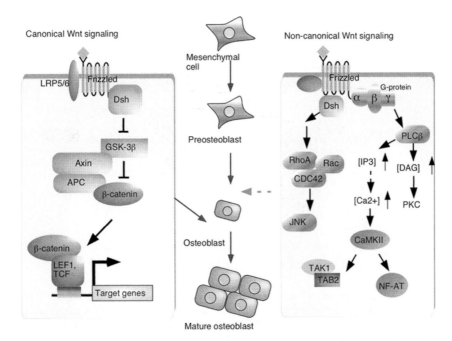

FIGURE 2. Signaling mechanisms of the Wnt/β-catenin (canonical) signaling pathway (left panel) and β-catenin independent (noncanonical) signaling pathways (right panel). LRP = low-density lipoprotein receptor-related protein; GSK-3 = serine/threonine kinase glycogen synthase kinase-3; APC = adenomatous polyposis coli; Dsh = Dishevelled; LEF1 = lymphocyte enhancer factor 1; TCF = T cell factor; JNK = c-jun N-terminal kinase; NF-AT = nuclear factor of activated T cells; PLC = phospholipase C; CaMKII = calcium/calmodulin-dependent protein kinase II; TAK1 = TGFβ−activating kinase 1; TAB2 = TAK1-binding protein 2; NLK = nemo-like kinase.

place in response to changes in the extracellular environment.[21] However, the molecular link of such histone modifications to the cell fate decision of bone marrow mesenchymal stem cells remains to be uncovered.

A TAK1/TAB1/NIK CASCADE ACTIVATED BY TNF-α/IL-1 TRANSREPRESS PPAR-γ FUNCTION WITH ATTENUATING ADIPOGENESIS

TNF-α/IL-1 Block Adipogenesis from Bone Marrow Mesenchymal Stem Cells

Osteoblastogenesis is dominant over adipogenesis in the bone marrow of young animals, and the cell differentiation balance between the two cell types is prone to be reversed in bone marrow of elder and/or osteoporotic animals.[3] Therefore, we reasoned that PPAR-γ function is suppressed in the bone marrow of young animals. To test this idea, we used ST2 cells derived from

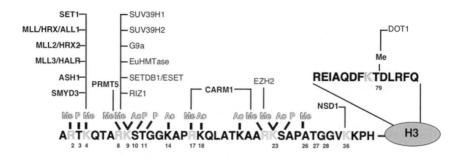

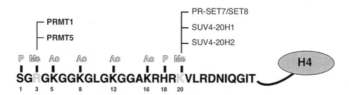

FIGURE 3.   Histone H3 and H4 modification and histone methyltransferases. Modified residues were shown. Ac = acetylation; Me = methylation; P = phosphorylation. Histone methyltransferases activating transcriptional function are indicated in bold black letters and repressing transcriptional functions are indicated in italic letters.

mouse bone marrow stromal cells. This cell line mimics the responses to extracellular stimulations in primary bone marrow mesenchymal stem cells, and is cytodifferentiated into either adipocytes through PPAR-γ activation or osteoblasts if induced by several cytokines.[22] We first screened several cytokines that are physiologically abundant in bone marrow, and found that TNF-α and IL-1 were potent to inhibit adipogenesis from the mensenchymal stem cells upon activation of PPAR-γ by a synthetic agonist, troglitazone (Tro). Consistently, TNF-α/IL-1 treatments were found to transrepress Tro-induced transactivation function of PPAR-γ. Interestingly, the Tro treatment was also stimulatory for osteoblastogenesis in ST2 cells. Thus, activated PPAR-γ appeared to stimulate cytodifferentiation of bone marrow progenitor cells into osetoblasts as well as adipocytes. As these cytokines are known to activate NF-κB in the nuclear and the nuclear NF-κB is indispensable for osteoclastogenesis from hematopoetic stem cells, they look physiologically important for cell fate decision of mesenchymal stem cells. We therefore studied the downstream signaling cascade of TNF-α/IL-1.

TNF-α/IL-1 Activates NF-κB Through TAK1/TAB1/NIK Cascade for Blocking Adipogenesis in ST2 Cells

As several cascades activated by TNF-α/IL-1 were illustrated, we tested which cascade is responsible for the suppressive action of TNF-α/IL-1 in

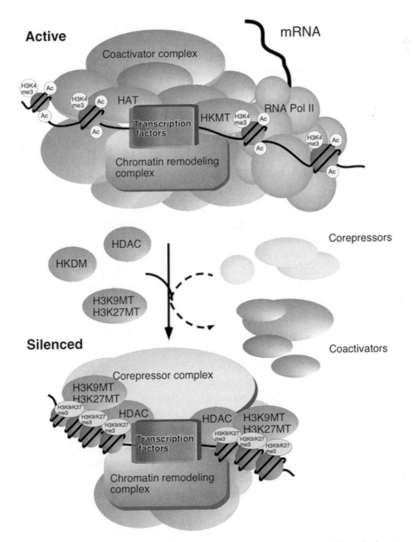

FIGURE 4. Model of histone modification and chromatin remodeling during transcriptional activation. In the active state, coactivator complex was recruited to DNA-bound transcription factor and histone H3 lysine4 was methylated by histone lysine methyltransferase (HKMT; MLL1–5, etc.) and many histone lysine residues were acetylated by histone acetyltransferase (HAT; CBP/p300, p160 famly, PCAF/GCN5, Tip60, etc.). Then nucleosome was opened by ATP-dependent chromatin remodeling complex and transcription was activated. In the silenced state, corepressor complex was recruited and histone H3 lysine9 was methylated by histone H3 lysine9 methyltransferase (H3K9MT; SUV39, G9a, SETDB1, etc) and H3 lysine27 was methylated by histone H3 lysine27 methyltransferase (H3K27MT; EZH2). Histone deacetylases (HDACs) and histone demethylases (HKDMs) were also involved during this process.

adipogenesis of ST2 cells. Similarly, transrepressive function of the downstream factors for PPAR-γ was also monitored. Consequently, we found that the TAK1/TAB1/NIK(NF-κB-inducible kinase) cascade activated by TNF-α/IL-1 is inhibitory for adipogenesis and Tro-induced transactivation of PPAR-γ in ST2 cells. Though it was previously reported that phosphorylation of PPAR-γ by MAP kinase results in repression of the PPAR-γ function,[11] NIK was not a kinase for the PPAR-γ phosphorylation.

Activated NF-κB Through TAK1/TAB1/NIK Blocks DNA Binding of PPAR-γ, Leading to Suppression of Adipogenesis in Mensenchymal Stem Cells

The activated TAK1/TAB1/NIK was found to suppress the Tro-induced expression of endogenous PPAR-γ target genes as seen in a luciferase reporter gene assay. Transrepression of NR was known to link to HDAC corepressor function, but the TNF-α/IL-1 treatment was unlikely to induce recruitment of known HDAC corepressor complex components to PPAR-γ in a ChIP of the PPAR-γ target gene promoters. Instead, we uncovered that DNA binding of PPAR-γ is blocked by treatment of the cytokines as well as the downstream factors. Further downstream of the TAK1/TAB1/NIK, NF-κB was activated. By Electrophoretic Mobility Shift Assay (EMSA), we demonstrated that activated NF-κB blocks DNA binding of PPAR-γ. Together with the previous reports that activated PPAR-γ by agonist binding is inhibitory for DNA binding of NF-κB,[23] it appears that association of activated PPAR-γ with nuclear NF-κB results in the blocking of DNA bindings of both factors (FIG. 5). Thus, we presume that TNF-α/IL-1 triggers activation of NF-κB through TAK1/TAB1/NIK axis, leading blocking PPAR-γ DNA binding through physical association with NF-κB. Since PPAR-γ is a prime regulator of adipogenesis, suppression of PPAR-γ function may inhibit adipogenesis, and consequently cell fate decision is shifted more to osteoblasts than adipocytes in bone marrow.

NONCANONICAL WNT SIGNAL INDUCES OSTEOBLASTOGENESIS THROUGH PPAR-γ TRANSREPRESSION BY A HISTONE METHYLTRANSFERASE

A Noncanonical Wnt Ligand, Wnt-5a, Transrepresses PPAR-γ Function with Attenuating a PPAR-γ Agonist-Induced Adipogenesis

We then tested a negative cross talk of PPAR-γ function with Wnt signalings. First, expression of Wnt ligands and their receptors was tested in ST2 cells. Several frizzled receptors and Wnt ligands genes are expressed at significant levels in ST2 cells as well as in mouse bone marrow cell primary culture,

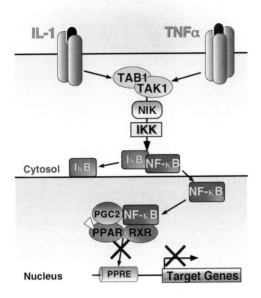

FIGURE 5. A mechanistic model for IL-1– and TNF-α–dependent suppression of PPAR-γ function. Activated NFκB associates with PPAR-γ and inhibits the DNA-binding ability of PPAR-γ.

and interestingly a noncanonical Wnt ligand, Wnt-5a, was found to express at significant levels (FIG. 6). A canonical Wnt ligand (Wnt-3a) did not affect transactivation function of PPAR-γ induced by a synthetic PPAR-γ agonist, troglitazone (Tro), while Wnt-5a was capable of transrepressing PPAR-γ function in the synthetic as well as natural PPAR-γ target gene promoters. We then explored the downstream signaling of Wnt-5a in terms of transrepression of PPAR-γ, leading to identification of CaMKII-TAK1/TAB2-NLK axis.

Since Wnt-5a has recently been identified as a candidate for genetic determination of diet-induced obesity,[24] we therefore tested the effect of Wnt-5a on ST2 cell differentiation. Treatment with Tro expectedly induced adipogenesis, lipid accumulation, and glycerol-3-phosphate dehydrogenase (GPDH) activity as reported previously.[22,25] Wnt-5a was potent to transdifferentiate adipoprogenitors into osteoblastic cells expressing alkaline phosphatase with induction of Runx2, a critical osteoblastogenic transcription factor. Consistently, Wnt-5a was inhibitory for activated PPAR-γ-mediated induction of mature marker genes of adipocytes (GPDH, aP2, and LXR-α) and chondrocytes (type II collagen and aggrecan). Although TAZ is an osteoblastogenic factor of mesenchymal stem cell differentiation by transrepressing PPAR-γ function,[26] the activated noncanonical Wnt signaling by Wnt-5a appeared not

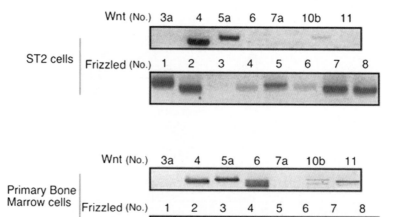

FIGURE 6. Expression levels of Wnts and Frizzled mRNA in ST2 cells and primary mice bone marrow mesenchymal cells. After the extraction of total RNA from these cells, RT-PCR was performed.

to modulate the TAZ expression during osteoblastic differentiation of the ST2. It is likely that the TAZ-mediated cascade does not converge with that of Wnt-5a in the osteoblastic differentiation of mesenchymal stem cells. Thus, activated PPAR-γ appears to potentiate cytodifferentiation of mesenchymal stem cells into adipo-/osteoprogenitors.

Wnt-5a Haploinsufficiency in Mice Resulted in Increase of Bone Marrow Adipocytes with Bone Loss

Reflecting the physiological impact of PPAR-γ on adipogenesis from mesenchymal stem cells in bone marrow,[27] haploinsufficiency of PPAR-γ in mice (PPAR-γ[+/−]) was reported to reduce adipogenesis in bone marrow but stimulate osteoblastogenesis. Disruption of canonical Wnt signaling factors in mice is known to cause bone loss, presumably owing to lowered osteoblastogenesis, however, no increase of adipogenesis was detected. Canonical Wnt signaling appears likely to stimulate osteoblastic differentiation but not affect adipogenesis from adipo-/osteoprogenitor cells.

Supporting *in vitro* observations of Wnt-5a action, Wnt-5a[+/−] mice[28] showed a clear bone loss phenotype, with decreased trabecular bone mass in the femur (FIG. 7). In the bone marrow, a significant increase of adipocyte numbers was seen (FIG. 7), confirming our idea that Wnt-5a induces the cell lineage fate of bone marrow mesenchymal stem cells into osteoblasts from adipocytes.

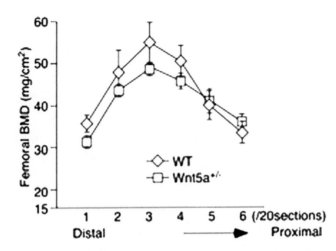

FIGURE 7. The femora were collected from 18-week-old WT and Wnt-5a$^{+/-}$ mice and fixed in 70% ethanol. Femoral bone mineral densities (BMD) of each of 20 equal longitudinal divisions were determined by dual-energy X-ray absorptiometry and distal region (1–6) was shown.

NLK Activated by Wnt-5a Forms a Complex with a HKMT, SETDB1

As NLK was the downstream factor of the Wnt-5a signaling, we then explored the molecular basis of the transrepressive function of NLK for PPAR-γ. An HDAC inhibitor, tricostatine A, was unable to reverse NLK-mediated suppression of PPAR-γ function in ST2 cells, indicating possible involvement of the other inactivating histone-modifying enzymes. NLK-containing complexes were biochemically purified from nuclear extracts of KCl-treated HeLa cells expressing FLAG-tagged NLK using a glycerol gradient centrifugation fractionation.[29,30] An NLK-nuclear protein complex with a molecular weight of around 400–500 kDa was isolated with the other two components. By MALDI-TOF MS analysis of two proteins (220 kDa and 170 kDa), the 220 kDa band was identified as a DEAH-box and CHD domain-containing ATPase protein, CHD7;[31] the 170 kDa band was an HKMT (SETDB1) known to be able to suppress transcription through histone H3 methylation at K9.[32,33] Complex formation of endogenous NLK, SETDB1, and CHD7 with PPAR-γ was seen only when ST2 cells were treated with Wnt-5a.

SETDB1 Activated by Wnt-5a Transrepresses the Ligand-Induced Transactivation of PPAR-γ Through H3-K9 Methylation

In a ChIP analysis of endogenous transcriptional factors and histone modifications at the PPAR-γ –binding site (PPRE) of the aP2 gene promoter,[10] Tro

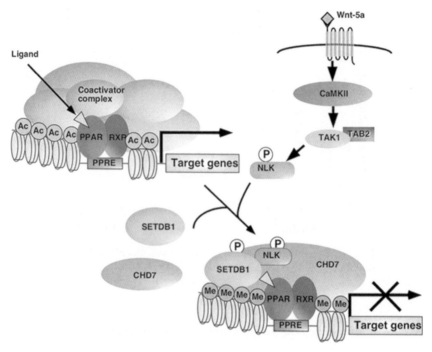

FIGURE 8. A mechanistic model for Wnt-5a-dependent suppression of PPAR-γ function.

treatment induced recruitment of known PPAR-γ coactivator SRC-1 and corepressor N-CoR. Treatment with Wnt-5a for 6 h in the presence of Tro induced recruitment of NLK, CHD7, and SETDB1 to the proximal PPRE region. Consistently, an increase in histone di- and tri-methylation at histone H3-K9 was observed together with hypoacetylation of histone. Such coordinated histone-inactivating modification was more visible under a longer treatment of Wnt-5a for 7 days that were long enough to induce osteoblastogenesis. Overexpression of either NLK or SETDB1 in the presence of Tro was potent to induce osteoblastogenesis rather than adipogenesis. Reversely, knockdown of either NLK, SETDB1 of CHD7 potentiated more Tro-induced adipogenesis rather than osteoblastogenesis even in the presence of Wnt-5a (FIG. 8).

Noncanonical Wnt Signaling Activated by Wnt-5a Induces Cell Fate Decision into Osteoblasts from Adipocytes in Bone Marrow

In this study, we observed that Wnt-5a induced osteoblastogenesis through attenuating PPAR-γ-induced adipogenesis in mesenchymal stem cells of bone marrow. Wnt-5a is a Wnt ligand to activate a noncanonical Wnt signaling cascade mediated through CaMKII-TAK1/TAB2-NLK, which is distinct from the canonical Wnt signaling through the β-catenin/TCF-mediated pathway. Several recent reports have demonstrated that the canonical Wnt cascade through LRP5/β-catenin is indispensable for osteoblastogenesis.[34–36] Wnt signaling through both canonical and noncanonical cascades is thus considered to support osteoblastogenesis from bone marrow mesenchymal cells. However, only noncanonical Wnt cascade appears significant in the prevention of activated PPAR-γ –mediated adipogenesis in bone marrow.

SETDB1 Is a Signal-Dependent Corepressor for PPAR-γ

An HKMT, SETDB1, was biochemically identified as an interacting partner of NLK. SETDB1 appears to form a complex with CHD7 and phosphorylated NLK upon Wnt-5a-induced activation of noncanonical Wnt signaling. The SETDB1 complex associated with PPAR-γ to methylate H3-K9 in the PPAR-γ target gene promoters, leading to chromatin inactivation through consequent histone-inactivating modifications. Thus, this complex is presumed as a new type of HKMT corepressor complexes for nuclear receptors in terms of the signaling dependency. SETDB1 may be a nuclear target activated by signalings of cell membrane receptors to corepress several classes of transcriptional factors.

ACKNOWLEDGMENTS

We thank Dr. A.P. Kouzmenko, M. Kim, and R. Fujiki for helpful discussions and Ms. H. Higuchi for article preparation. And we also thank Dr. S. Ishii (RIKEN Tsukuba Institute) for helpful advices and T Komori (Nagasaki University) for kindly gifted Runx expression vector and reporter vector. This work was supported in part by a Grant-In-Aid for Basic Research Activities for Innovative Biosciences (BRAIN) and Priority Areas from the Ministry of Education, Science, Sports, and Culture of Japan (to S.K.).

REFERENCES

1. CHIEN, K.R. & G. KARSENTY. 2005. Longevity and lineages: toward the integrative biology of degenerative diseases in heart, muscle, and bone. Cell **120:** 533–544.
2. HARADA, S. & G.A. RODAN. 2003. Control of osteoblast function and regulation of bone mass. Nature **423:** 349–355.
3. RALSTON, S.H. & B. DE CROMBRUGGHE. 2006. Genetic regulation of bone mass and susceptibility to osteoporosis. Genes Dev. **20:** 2492–2506.
4. BEHRENS, J. *et al.* 1996. Functional interaction of beta-catenin with the transcription factor LEF-1. Nature **382:** 638–642.

5. Ishitani, T. *et al.* 2003. The TAK1-NLK mitogen-activated protein kinase cascade functions in the Wnt-5a/Ca(2+) pathway to antagonize Wnt/beta-catenin signaling. Mol. Cell Biol. **23:** 131–139.
6. Veeman, M.T., J.D. Axelrod & R.T. Moon. 2003. A second canon. Functions and mechanisms of beta-catenin-independent Wnt signaling. Dev. Cell **5:** 367–377.
7. Evans, R.M., G.D. Barish & Y.X. Wang. 2004. PPARs and the complex journey to obesity. Nat. Med. **10:** 355–361.
8. Feige, J.N. *et al.* 2006. From molecular action to physiological outputs: peroxisome proliferator-activated receptors are nuclear receptors at the crossroads of key cellular functions. Prog. Lipid Res. **45:** 120–159.
9. Lehrke, M. & M.A. Lazar. 2005. The many faces of PPARgamma. Cell **123:** 993–999.
10. Tontonoz, P. *et al.* 1994. mPPAR gamma 2: tissue-specific regulator of an adipocyte enhancer. Genes Dev. **8:** 1224–1234.
11. Hu, E. *et al.* 1996. Inhibition of adipogenesis through MAP kinase-mediated phosphorylation of PPARgamma. Science **274:** 2100–2103.
12. Dilworth, F.J. & P. Chambon. 2001. Nuclear receptors coordinate the activities of chromatin remodeling complexes and coactivators to facilitate initiation of transcription. Oncogene **20:** 3047–3054.
13. McKenna, N.J. & B.W. O'Malley. 2002. Combinatorial control of gene expression by nuclear receptors and coregulators. Cell **108:** 465–474.
14. Rosenfeld, M.G., V.V. Lunyak & C.K. Glass. 2006. Sensors and signals: a coactivator/corepressor/epigenetic code for integrating signal-dependent programs of transcriptional response. Genes Dev. **20:** 1405–1428.
15. Garcia-Bassets, I. *et al.* 2007. Histone methylation-dependent mechanisms impose ligand dependency for gene activation by nuclear receptors. Cell **128:** 505–518.
16. Fischle, W., Y. Wang & C.D. Allis. 2003. Histone and chromatin cross-talk. Curr. Opin. Cell Biol. **15:** 172–183.
17. Margueron, R., P. Trojer & D. Reinberg. 2005. The key to development: interpreting the histone code? Curr. Opin. Genet. Dev. **15:** 163–176.
18. Martin, C. & Y. Zhang. 2005. The diverse functions of histone lysine methylation. Nat. Rev. Mol. Cell. Biol. **6:** 838–849.
19. Bannister, A.J. & T. Kouzarides. 2005. Reversing histone methylation. Nature **436:** 1103–1106.
20. Metzger, E., M. Wissmann & R. Schule. 2006. Histone demethylation and androgen-dependent transcription. Curr. Opin. Genet. Dev. **16:** 513–517.
21. Wu, R.C., C.L. Smith & B.W. O'Malley. 2005. Transcriptional regulation by steroid receptor coactivator phosphorylation. Endocr. Rev. **26:** 393–399.
22. Suzawa, M. *et al.* 2003. Cytokines suppress adipogenesis and PPAR-gamma function through the TAK1/TAB1/NIK cascade. Nat. Cell Biol. **5:** 224–230.
23. Ricote, M. *et al.* 1998. The peroxisome proliferator-activated receptor-gamma is a negative regulator of macrophage activation. Nature **391:** 79–82.
24. Almind, K. & C.R. Kahn. 2004. Genetic determinants of energy expenditure and insulin resistance in diet-induced obesity in mice. Diabetes **53:** 3274–3285.
25. Mueller, E. *et al.* 1998. Terminal differentiation of human breast cancer through PPAR gamma. Mol. Cell **1:** 465–470.
26. Hong, J.H. *et al.* 2005. TAZ, a transcriptional modulator of mesenchymal stem cell differentiation. Science **309:** 1074–1078.

27. SCHWARTZ, A.V. *et al.* 2006. Thiazolidinedione use and bone loss in older diabetic adults. J. Clin. Endocrinol. Metab. **91:** 3349–3354.
28. YAMAGUCHI, T.P. *et al.* 1999. A Wnt5a pathway underlies outgrowth of multiple structures in the vertebrate embryo. Development **126:** 1211–1223.
29. KITAGAWA, H. *et al.* 2003. The chromatin-remodeling complex WINAC targets a nuclear receptor to promoters and is impaired in Williams syndrome. Cell **113:** 905–917.
30. YANAGISAWA, J. *et al.* 2002. Nuclear receptor function requires a TFTC-type histone acetyl transferase complex. Mol. Cell **9:** 553–562.
31. VISSERS, L.E. *et al.* 2004. Mutations in a new member of the chromodomain gene family cause CHARGE syndrome. Nat. Genet. **36:** 955–957.
32. SCHULTZ, D.C. *et al.* 2002. SETDB1: a novel KAP-1-associated histone H3, lysine 9-specific methyltransferase that contributes to HP1-mediated silencing of euchromatic genes by KRAB zinc-finger proteins. Genes Dev. **16:** 919–932.
33. WANG, H. *et al.* 2003. mAM facilitates conversion by ESET of dimethyl to trimethyl lysine 9 of histone H3 to cause transcriptional repression. Mol. Cell **12:** 475–487.
34. GLASS, D.A., II *et al.* 2005. Canonical Wnt signaling in differentiated osteoblasts controls osteoclast differentiation. Dev. Cell **8:** 751–764.
35. GONG, Y. *et al.* 2001. LDL receptor-related protein 5 (LRP5) affects bone accrual and eye development. Cell **107:** 513–523.
36. ROSS, S.E. *et al.* 2000. Inhibition of adipogenesis by Wnt signaling. Science **289:** 950–953.

Transcriptional Regulation of Osteoblasts

RENNY T. FRANCESCHI,[a] CHUNXI GE,[a] GUOZHI XIAO,[b]
HERNAN ROCA,[c] AND DI JIANG[a]

[a]University of Michigan School of Dentistry, Ann Arbor, Michigan, USA

[b]University of Pittsburg School of Medicine, Pittsburg, Pennsylvania, USA

[c]University of Michigan School of Medicine and Cancer Center, Ann Arbor, Michigan, USA

ABSTRACT: The differentiation of osteoblasts from mesenchymal precursors requires a series of cell fate decisions controlled by a hierarchy of transcription factors. Among these are RUNX2, Osterix (OSX), ATF4, and a large number of nuclear coregulators. During bone development, initial RUNX2 expression coincides with the formation of mesenchymal condensations well before the branching of chondrogenic and osteogenic lineages. Given that RUNX2 is expressed so early and participates in several stages of bone formation, it is not surprising that it is subject to a variety of controls. These include regulation by nuclear accessory factors and posttranslational modification, especially phosphorylation. Specific examples of RUNX2 regulation include interactions with DLX proteins and ATF4 and phosphorylation by the ERK/MAP kinase pathway. RUNX2 is regulated via phosphorylation of critical serine residues in the P/S/T domain. MAPK activation of RUNX2 was also found to occur *in vivo*. Transgenic expression of constitutively active MEK1 in osteoblasts accelerated skeletal development while a dominant-negative MEK1 retarded development in a RUNX2-dependent manner. These studies allow us to begin understanding the complex mechanisms necessary to fine-tune bone formation in response to extracellular stimuli including ECM interactions, mechanical loads, and hormonal stimulation.

KEYWORDS: bone; osteoblast; transcription; RUNX2; extracellular matrix

Osteoblast and chondrocyte differentiation is controlled by a hierarchy of transcription factors that are expressed in a defined temporal sequence (FIG. 1). RUNX2, an essential factor for bone and hypertrophic cartilage formation, is expressed very early in skeletal development, first appearing with the

Address for correspondence: Dr. Renny T. Franceschi, Department of Periodontics and Oral Medicine, University of Michigan School of Dentistry, 1011 N. University Ave. Ann Arbor, MI 48109-1078. Voice: 734-763-7381; fax 734-763-5503.
rennyf@umich.edu

Ann. N.Y. Acad. Sci. 1116: 196–207 (2007). © 2007 New York Academy of Sciences.
doi: 10.1196/annals.1402.081

Transcription Factor Control of Skeletal Lineages

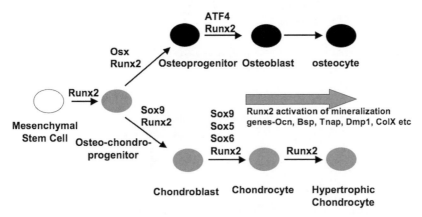

FIGURE 1. Transcription factor control of skeletal lineages. Major transcription factors that, based on genetic studies, are involved in osteoblast and chondrocyte differentiation are included in this chart. Also shown is the sequential nature of transcription factor expression with RUNX2 persisting throughout osteoblast and chondrocyte lineages (see text).

formation of mesenchymal condensations in areas destined to become bone and persisting through subsequent stages of bone formation.[1] Several other transcription factors function together with RUNX2 to move cells down chondrocyte or osteoblast lineages. For osteoblasts, this is accomplished by Osterix (Osx), which commits osteochondroprogenitor cells to the osteoblast lineage.[2] Subsequently, another factor, ATF4, controls the transcriptional activity of mature osteoblasts.[3] RUNX2 also participates in the chondrogenic lineage. However, at early stages, it is likely suppressed by the chondrocyte-specific factors, Sox 8/9.[4] However, Sox factors are downregulated to permit RUNX2 to coordinate chondrocyte hypertrophy. Evidence for this model largely comes from genetic studies showing bone phenotypes of increasing severity as *Runx2, Osx,* or *Atf4* are knocked out. Thus, skeletal development in *Runx2*-deficient embryos fails to progress beyond the cartilage anlage stage.[5,6] Fetuses, which die at birth of respiratory failure, have no detectable mineral in either bones or cartilage. In contrast, Osx -/- embryos have hypertrophic cartilage and normal levels of RUNX2, suggesting that this factor is downstream of RUNX2.[2] ATF4 -/- mice have a milder phenotype consistent with this factor having a regulatory function in bone formation.[3]

An interesting feature of this scheme is that most of the known RUNX2 target genes are only expressed at specific stages of chondrogenic or osteogenic lineages, yet RUNX2 is present throughout both lineages. RUNX2 mRNA is first expressed at E9.5, peaks at E12.5 well before any mineralization has occurred, and continues to be present through the later stages of development.[1] This essentially constitutive expression of RUNX2 once skeletal development

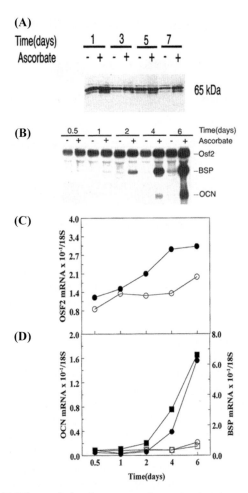

FIGURE 2. RUNX2 protein levels are not well correlated with transcriptional activity. MC3T3-E1 clone 4 preosteoblast cells were grown in control (–) or ascorbate-containing medium (+). At the times indicated, RUNX2 protein levels were determined by Western blotting (**A**) while *Runx2* (Osf2), *Ocn* and *Bsp* mRNA levels (**B–D**) were measured on Northern blots (**B**) and quantified by densitometry. (**C**) Runx2; (**D**) Ocn and Bsp. Legend: open symbols = control; closed symbols = ascorbate. For panel D, Ocn mRNA (m, l), Bsp mRNA (o, n). From Xiao *et al.* 1998.[7]

has commenced implies that other factors or signals must be regulating its activity. FIGURE 2 shows an example of how other factors can regulate RUNX2 activity. In this experiment, MC3T3-E1 preosteoblast cells were induced to differentiate by growth in ascorbic acid, thereby allowing cells to secrete a collagenous ECM. Over time, as ECM accumulates, there is a dramatic induction of osteoblast marker genes, such as bone sialoprotein and osteocalcin,

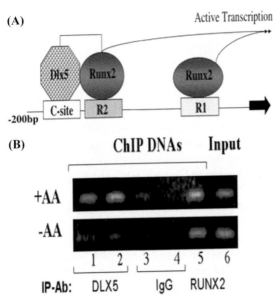

FIGURE 3. *In vivo* binding of RUNX2 and DLX5 to chromatin sites in differentiated and undifferentiated cells. **(A)** Schematic of the proximal *Bsp* promoter. RUNX2 and homeodomain (DLX5) protein-binding sites are indicated. **(B)** Comparison of *Bsp* chromatin occupancy by RUNX2 and DLX5. Chromatin immunoprecipitation (ChIP) assays were used to detect RUNX2 and DLX5 bound to the proximal *Bsp* promoter in control (–AA) and differentiated (+AA) MC3T3-E1 clone 4 cells. Antibodies used for ChIP are indicated. Note that RUNX2 remains chromatin associated regardless of differentiated state while Dl×5 is only present in differentiated cells. From Roca *et al.*, 2005.[8]

but RUNX2 mRNA and protein levels only increase 1.5- to 2-fold. Similarly, the amount of RUNX2 bound to the promoter of target genes as measured by chromatin immunoprecipitation also remains relatively unchanged during differentiation.[7] FIGURE 3 shows RUNX2 bound to the proximal Bsp promoter. We identified two RUNX2 sites and a DLX5 homeodomain protein-binding site in this region and showed that these sites cooperate to control expression in osteoblasts.[8] Surprisingly, when RUNX2 chromatin occupancy was compared in control versus differentiated osteoblasts, no obvious differences were observed. In contrast, DLX5 chromatin occupancy was only observed in differentiated cells. However, total DLX5 protein measured by Western blot was equivalent in control and differentiated cells indicating that the affinity of DLX5 for chromatin (and possibly Runx2) increases with differentiation.

There are at least two possible ways RUNX2 transcriptional activity could be regulated in the absence of changes in RUNX2 protein levels. First, as suggested by the DLX5 result discussed above, levels of cofactors or ability of cofactors to interact with RUNX2 could be regulated. Second, covalent modification could alter RUNX2 transcriptional activity.

The first possibility is probably a very common control mechanism especially when we consider the broad range of factors known to interact with RUNX2. For example, CBFβ forms heterodimers with all members of the Runx family.[9] RUNX2 also interacts with ATF4 and this factor may mediate some of the responses of osteoblasts to PTH.[10,11] SMAD proteins, mediators of BMP/TGFβ actions, can also stimulate RUNX2 activity.[12] In addition, there are a number of inhibitory factors. SOX9, mentioned above, suppresses RUNX2-dependent chondrocyte hypertrophy.[4] TWIST may prevent RUNX2 stimulation of mineralization in developing cranial bones to prevent craniosynostosis.[13] Last, histone deacetylases (HDACs) and HDAC accessory factors, such as mSin3a, are known to bind RUNX2 and keep chromatin in a deacetylated, inactive state.[14]

Posttranslational modification represents the second major mechanism for controlling RUNX2 activity. Over the past decade, we described a pathway involving integrin-mediated activation of the ERK/MAP kinase pathway that results in phosphorylation and stimulation of RUNX2 transcriptional activity. Integrins provide information to cells about the ECM environment and mediate cell attachment and spreading.[15] Integrins are also major mediators of mechanical loads experienced by cells and may transduce information about ECM stiffness to control differentiation.[16,17] In addition to providing continuity between the ECM and the actin-containing cytoskeleton, integrins are important signal transduction molecules that activate Ras-ERK and p38 MAP kinase pathways, calcium channels, and mechanosensors.[15]

Osteoblast differentiation requires elaboration of a collagenous ECM. This explains the well-known requirement for ascorbic acid in osteoblast differentiation since this vitamin cofactor is essential for secretion of the collagenous ECM.[18] An example of ECM-mediated osteoblast differentiation was shown in FIGURE 2. Bone sialoprotein and osteocalcin, two osteoblast markers, are only expressed in cells grown in ascorbate-containing medium. In studies that will not be extensively discussed in this article, we showed that the response of osteoblasts to ECM is mediated by α2β1 integrins and the ERK/MAP kinase pathway.[7,19,20] We specifically showed that osteoblast differentiation could be blocked with specific inhibitors of collagen synthesis and integrin-collagen binding. Furthermore, inhibition of MAPK signaling using either pharmacological inhibitors or dominant-negative pathway intermediates blocked differentiation. In contrast, induction of osteoblast gene expression was induced with a constitutively active MAPK intermediate, MEK1.

Initially, studies with the osteocalcin gene were used to relate MAPK responsiveness to RUNX2. We found that MAPK activation via transfection of cells with constitutively active MEK1 (MEK-SP) could induce the RUNX2-responsive Ocn gene while dominant-negative MEK1 (MEK-DN) was inhibitory.[20] An examination of the Ocn promoter for MAPK-responsive sequence elements identified two RUNX2-binding sites called OSE2a and b.

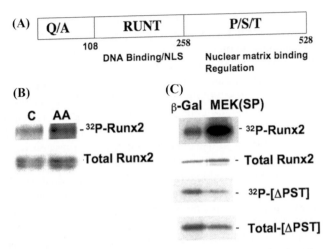

FIGURE 4. RUNX2 phosphorylation. (**A**) Cartoon showing major domains of RUNX2 and their known functions. (**B**) RUNX2 phosphorylation increases with osteoblast differentiation. MC3T3-E1 clone 4 cells were grown in control or differentiation (AA-containing) medium for 8 days before metabolic labeling with $[^{32}P]$ orthophosphate. RUNX2 was immunoprecipitated from nuclear extracts and visualized by autoradiography. Total RUNX2 was measured by Western blotting with RUNX2-specific antibody. (**C**) ERK/MAPK increases phosphorylation in the RUNX2 P/S/T domain. COS7 cells were transfected with RUNX2 and constitutively active MEK1 (MEK(SP)) expression vectors as indicated. RUNX2 phosphorylation was measured as in panel B. Phosphorylation of a truncated RUNX2 lacking the entire PST domain (ΔPST) is also shown (*bottom*). Note lack of MEK(SP)-dependent phosphorylation. From Franceschi *et al.*, 2003.[28]

Mutation of either of these sites reduced MAPK responsiveness with the proximal site, OSE2a, having the strongest effect on Ocn promoter activity.[21]

As might be expected, activation of MAPK, either by ascorbate addition or transfection of cells with MEK-SP results in increased RUNX2 phosphorylation (FIG. 4 and Ref. 20). Examination of different truncations of the RUNX2 protein showed that the C-terminal proline/serine/threonine region (P/S/T domain) of RUNX2 was required for both MAPK responsiveness and phosphorylation (FIG. 4 and Ref. 20). More detailed deletional analysis further localized a minimal region for MAPK responsiveness between amino acids 254 and 320 (result not shown). The specific identification and functional significance of ERK/MAPK phosphorylation sites in RUNX2 will be reported separately.

A number of other stimuli that act through the ERK/MAPK pathway have been shown to stimulate RUNX2 phosphorylation and transcriptional activity. For example, we showed that FGF2 via its receptor, FGFR2, can rapidly stimulate ERK1/2 and RUNX2 phosphorylation and Ocn gene expression.[22] A related study showed that PKCδ, a second kinase activated by the FGFR2, is also required for Ocn stimulation and that this kinase uses a distinct site in RUNX2 (S247) that can be resolved from ERK/MAPK sites.[23] Last, exposure of osteoblasts to mechanical loading was shown to stimulate RUNX2 phosphorylation and osteoblast gene expression.[24-26]

All the studies discussed above were conducted in cell culture. To establish the importance of the ERK/MAPK pathway and RUNX2 phosphorylation to osteoblast function *in vivo*, we took a transgenic approach.[27] Transgenes were constructed using a 0.6 kb mOG2 promoter to specifically drive MEK-SP and MEK-DN expression in osteoblasts. Transgene expression was detected only in bones and had a time course of expression during development that parallels expression of the endogenous Ocn gene. Examination of ERK1/2 phosphorylation in calvarial osteoblasts derived from these animals showed a 50% increase in cells from *Mek-sp* mice and a 50% decrease in *Mek-dn* cells (see Ref. 27). This indicates that transgene expression leads to subtle changes in MAPK activity that resemble fluctuations normally induced by physiological stimuli. Examination of skeletal whole mounts of transgenic mice revealed that *Mek-dn* decreased skeletal size and calvarial mineralization while these parameters were increased in *Mek-sp* mice (FIG. 5A–F). Histology of long bones revealed an additional interesting difference between wild-type and transgenic mice (FIG. 5G). At E15.5, long bones are normally undergoing endochondral ossification in diaphyseal regions. However, in *Mek-dn* mice, this process is drastically delayed with only early bony collar formation being visible. In contrast, in the *Mek-sp* mice, endochondral bone formation is accelerated.

We next examined if transgenic modification of osteoblast MAPK activity could alter RUNX2 phosphorylation and transcriptional activity (FIG. 6). For this experiment, we isolated calvarial osteoblasts from transgenic mice and conducted metabolic labeling with [^{32}P] orthophosphate followed by immunoprecipitation. Cells were also transfected with a *Runx2* reporter gene to measure transcriptional activity. Clearly, RUNX2 phosphorylation was increased in *Mek-sp* cells as was luciferase activity of the *Ocn* reporter gene. Also, as expected, *Mek-sp* stimulated *in vitro* osteoblast differentiation as measured by induction of osteoblast marker mRNAs or mineralization while differentiation was inhibited in Mek-sp cells (see Ref. 27).

To provide evidence that MAPK effects on skeletal development are mediated by RUNX2, we took a genetic approach. *Runx2* +/− mice are known to have a characteristic phenotype (hypoplastic clavicles, patent fontanelles) that is a phenocopy of the human genetic disease, cleidocranial dysplasia or CCD.[5] We reason that if MAPK acts by altering RUNX2 activity, calvaria and clavicles should be selectively sensitive to the *Mek* transgene when RUNX2 is limiting (i.e., in *Runx2* +/− mice). To test this, *Runx2* +/− mice were crossed with *Mek-sp* or *Mek-dn* transgenic lines and the resulting skeletal phenotypes were examined at E19 (FIG. 7). Note the reduced calvarial mineralization and tiny clavicles in the *Runx2* +/− embryos. The presence of the *Mek-sp* trans-

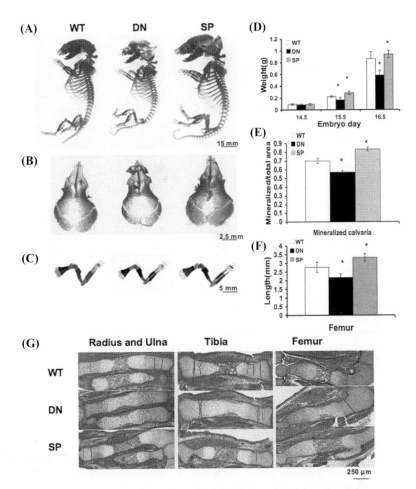

FIGURE 5. Altered skeletal development in *TgMek-dn* and *TgMek-sp* mice. (**A**) Whole mounts of E15.5 skeletons stained with alcian blue and alizarin red. (**D**) Effects of transgene expression on embryo weights. (**B, E**) Cranial bones showing differences in mineralization (**B**) and quantification of mineralized area (expressed as percent of total calvarial area, **E**). (**C, F**) Hindlimbs showing differences in the size of bones with transgene expression (**C**) and quantification of femur lengths (**F**). (**G**) Histology of long bones from wild-type, *TgMek-dn* and *TgMek-sp* mice. Note delay in bony collar and trabecular bone in *TgMek-dn* embryos. Bar = 250 μm. Statistical analysis values are expressed as means ± SD, $n =$ 8/group. * significantly different from wild-type at $P < 0.01$. From Ge *et al.*, 2007.[27]

gene led to a clear rescue of the CCD phenotype with increased clavicle size and calvarial mineralization. In the presence of *Mek-dn*, effects on *Runx2* +/− mice were even more dramatic. In this case, the *Mek-dn* transgene exacerbated effects of *Runx2* haploinsufficiency with a further reduction in calvarial mineralization and near disappearance of clavicles. Notably, *Runx2* +/−; *Mek-*

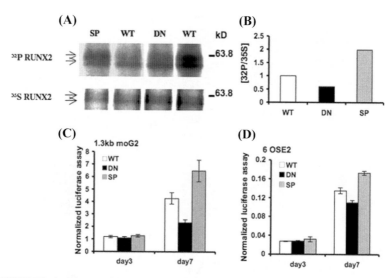

FIGURE 6. Changes in RUNX2 phosphorylation and transcriptional activity in os-teoblasts from *TgMek-dn* and *TgMek-sp* mice. (**A, B**) Regulation of RUNX2 phospho-rylation. Calvarial cells were grown under differentiating conditions for 10 days before metabolic labeling with [^{32}P]-orthophosphate or [^{35}S]-methionine/cysteine (to normalize for total RUNX2) and immunoprecipitation with an anti-RUNX2 antibody. Each IP reaction contained 500 μg total protein. Normalized ^{32}P incorporation into RUNX2 is shown in **B**. (**C, D**) RUNX2-dependent transcriptional activity. Cells were transfected with p1.3mOG2-luc (**C**) or p6OSE2mOG2-luc (**D**) plasmids and grown under differentiating conditions for the times indicated before the measurement of luciferase activity. From Ge *et al.*, 2007.[27]

sp embryos did not survive the birth process due to severe skeletal defects. These experiments provide strong evidence that the ERK/MAPK pathway, via actions on RUNX2 transcriptional activity, is important for normal osteoblast differentiation and bone formation *in vivo*.

SUMMARY

Osteoblast differentiation is controlled by the sequential expression of sev-eral transcription factors. Among these is RUNX2, which is associated with commitment of mesenchymal cells to bone lineages and subsequent differ-entiation of osteoblasts and hypertrophic chondrocytes. RUNX2 is expressed throughout osteoblast and chondrogenic differentiation and appears to control different events at different stages of cell maturation. Rather than being con-trolled at transcriptional or translational levels, RUNX2 activity is controlled by its interaction with accessory factors and by posttranslational modifications, such as phosphorylation. Interaction of preosteoblasts with a type I collagen-containing ECM stimulates differentiation by an integrin-mediated pathway

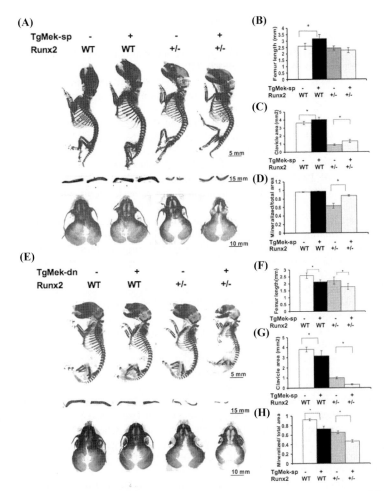

FIGURE 7. Genetic interactions between *Mek-dn* and *Mek-sp transgenes and Runx2.* *TgMek-dn* or *TgMek-sp* mice were crossed with *Runx2+/−* mice to generate the genotypes indicated. (**A–D**) Partial rescue of CCD phenotype in *Runx2+/−* mice with *Mek-sp.* (**A**) Skeletal whole mounts of newborn mice stained with alcian blue and alizarin red (*top*), isolated clavicles (*middle*), and crania (*bottom*). (**B–D**) Measurements of femur length (**B**), clavicle areas, (**C**) and mineralized area of calvaria (expressed as a fraction of total calvarial area). (**E–H**) Increased severity of CCD phenotype with *Mek-dn*. Groups are as in panels A–D. Statistical analysis values are expressed as means ± SD, $n = 8$/group. Comparisons are indicated by bars. * significantly different at $P < 0.01$. From Ge *et al.*, 2007.[27]

that, via activation of ERK/MAP kinase, phosphorylates RUNX2 in the C-terminal P/S/T domain to increase transcriptional activity. This regulation, first identified in osteoblast cell cultures, also apparently occurs *in vivo* in that transgenic modification of ERK/MAPK activity affects skeletal development via a mechanism requiring RUNX2.

REFERENCES

1. DUCY, P. *et al.* 1997. Osf2/Cbfa1: a transcriptional activator of osteoblast differentiation [see comments]. Cell **89:** 747–754.
2. NAKASHIMA, K. *et al.* 2002. The novel zinc finger-containing transcription factor osterix is required for osteoblast differentiation and bone formation. Cell **108:** 17–29.
3. YANG, X. *et al.* 2004. ATF4 is a substrate of RSK2 and an essential regulator of osteoblast biology; implication for Coffin-Lowry Syndrome. Cell **117:** 387–398.
4. ZHOU, G. *et al.* 2006. Dominance of SOX9 function over RUNX2 during skeletogenesis. Proc. Natl. Acad. Sci. USA **103:** 19004–19009.
5. OTTO, F. *et al.* 1997. Cbfa1, a candidate gene for cleidocranial dysplasia syndrome, is essential for osteoblast differentiation and bone development. Cell **89:** 765–771.
6. KOMORI, T. *et al.* 1997. Targeted disruption of Cbfa1 results in a complete lack of bone formation owing to maturational arrest of osteoblasts. Cell **89:** 755–764.
7. XIAO, G. *et al.* 1998. Role of the alpha2-integrin in osteoblast-specific gene expression and activation of the Osf2 transcription factor. J. Biol. Chem. **273:** 32988–32994.
8. ROCA, H. *et al.* 2005. Cooperative interactions between RUNX2 and homeodomain protein-binding sites are critical for the osteoblast-specific expression of the bone sialoprotein gene. J. Biol. Chem. **280:** 30845–30855.
9. ITO, Y. 2004. Oncogenic potential of the RUNX gene family: 'overview'. Oncogene **23:** 4198–4208.
10. XIAO, G. *et al.* 2005. Cooperative interactions between activating transcription factor 4 and Runx2/Cbfa1 stimulate osteoblast-specific osteocalcin gene expression. J. Biol. Chem. **280:** 30689–30696.
11. JIANG, D. *et al.* 2004. Parathyroid hormone induction of the osteocalcin gene. Requirement for an osteoblast-specific element 1 sequence in the promoter and involvement of multiple-signaling pathways. J. Biol. Chem. **279:** 5329–5337.
12. AFZAL, F. *et al.* 2005. Smad function and intranuclear targeting share a Runx2 motif required for osteogenic lineage induction and BMP2 responsive transcription. J. Cell. Physiol. **204:** 63–72.
13. BIALEK, P. *et al.* 2004. A twist code determines the onset of osteoblast differentiation. Dev. Cell. **6:** 423–435.
14. IMAI, Y. *et al.* 2004. The corepressor mSin3A regulates phosphorylation-induced activation, intranuclear location, and stability of AML1. Mol. Cell Biol. **24:** 1033–1043.
15. HYNES, R.O. 2002. Integrins: bidirectional, allosteric signaling machines. Cell **110:** 673–687.
16. YOU, J. *et al.* 2001. Osteopontin gene regulation by oscillatory fluid flow via intracellular calcium mobilization and activation of mitogen-activated protein kinase in MC3T3-E1 osteoblasts. J. Biol. Chem. **276:** 13365–13371.
17. ENGLER, A.J. *et al.* 2006. Matrix elasticity directs stem cell lineage specification [see comment]. Cell **126:** 677–689.
18. FRANCESCHI, R.T. 1992. The role of ascorbic acid in mesenchymal differentiation. Nutr. Rev. **50:** 65–70.
19. XIAO, G. *et al.* 1997. Ascorbic acid-dependent activation of the osteocalcin promoter in MC3T3-E1 preosteoblasts: requirement for collagen matrix synthesis and the presence of an intact OSE2 sequence. Mol. Endocrinol. **11:** 1103–1113.

20. XIAO, G. *et al.* 2000. MAPK pathways activate and phosphorylate the osteoblast-specific transcription factor, Cbfa1. J. Biol. Chem. **275:** 4453–4459.
21. FRENDO, J.L. *et al.* 1998. Functional hierarchy between two OSE2 elements in the control of osteocalcin gene expression in vivo. J. Biol. Chem. **273:** 30509–30516.
22. XIAO, G. *et al.* 2002. Fibroblast growth factor 2 induction of the osteocalcin gene requires MAPK activity and phosphorylation of the osteoblast transcription factor, Cbfa1/Runx2. J. Biol. Chem. **277:** 36181–36187.
23. KIM, H.J. *et al.* 2003. The protein kinase C pathway plays a central role in the fibroblast growth factor-stimulated expression and transactivation activity of Runx2. J. Biol. Chem. **278:** 319–326.
24. WANG, F.S. *et al.* 2002. Superoxide mediates shock wave induction of ERK-dependent osteogenic transcription factor (CBFA1) and mesenchymal cell differentiation toward osteoprogenitors. J. Biol. Chem. **277:** 10931–10937.
25. ZIROS, P.G. *et al.* 2002. The bone-specific transcriptional regulator Cbfa1 is a target of mechanical signals in osteoblastic cells. J. Biol. Chem. **277:** 23934–23941.
26. KANNO, T. *et al.* 2007. Mechanical stress-mediated Runx2 activation is dependent on Ras/ERK1/2 MAPK signaling in osteoblasts. J. Cell. Biochem. **101:** 1266–1277.
27. GE, C. *et al.* 2007. Critical role of the extracellular signal-regulated kinase-MAPK pathway in osteoblast differentiation and skeletal development. J. Cell. Biol. **176:** 709–718.
28. FRANCESCHI, R.T. *et al.* 2003. Multiple signaling pathways converge on the Cbfa1/Runx2 transcription factor to regulate osteoblast differentiation. Connect. Tissue Res. **44**(Suppl 1): 109–116.

Identification of Additional Dimerization Partners of FIAT, the Factor Inhibiting ATF4-Mediated Transcription

RENÉ ST-ARNAUD[a,b,c] AND BILAL ELCHAARANI[a,b]

[a]Genetics Unit, Shriners Hospital for Children, Montreal, Quebec, Canada

[b]Department of Human Genetics, McGill University, Montreal, Quebec, Canada

[c]Departments of Medicine and Surgery, McGill University, Montreal, Quebec, Canada

ABSTRACT: FIAT is a leucine zipper protein whose name was coined for its interaction with ATF4 and the subsequent blockage of ATF4-directed osteocalcin gene transcription. FIAT is a nuclear protein that lacks a basic DNA-binding domain but contains three identifiable leucine zipper domains. FIAT heterodimerizes with ATF4 through one of these zippers and thereby prohibits ATF4 from binding to its cognate DNA sequence. We tested whether FIAT also interacts with additional basic domain-leucine zipper transcriptional regulators of osteoblast activity, such as the Fos family member Fra-1 or one of its dimerization partners, c-Jun. Transient transfection assays in osteoblastic MC3T3-E1 cells with the heterologous AP-1-tk-luciferase reporter revealed that FIAT does not affect c-Jun-mediated transcription, even in the presence of the c-Jun coactivator αNAC. However, FIAT inhibited transcriptional activation by a c-Jun∼Fra-1 heterodimer. Thus FIAT specifically inhibits Fra-1 transcriptional activity. These data identify a second target of the FIAT transcriptional repressor activity and suggest that FIAT can modulate early osteoblast activity by interacting with ATF4, as well as regulate later osteoblast function through inhibition of Fra-1.

KEYWORDS: AP-1; c-jun; fra-1; FIAT; osteoblast; transcriptional repression

INTRODUCTION

Convincing genetic evidence has accumulated demonstrating that transcription factors belonging to the basic domain-leucine zipper (bZIP) class play a significant role in regulating the differentiation and function of osteoblasts.[1,2]

Address for correspondence: René St-Arnaud, Genetics Unit, Shriners Hospital for Children, 1529 Cedar Avenue, Montreal, Quebec, Canada H3G 1A6. Voice: 514-282-7155; fax: 514-842-5581. rst-arnaud@shriners.mcgill.ca

Ann. N.Y. Acad. Sci. 1116: 208–215 (2007). © 2007 New York Academy of Sciences.
doi: 10.1196/annals.1402.028

Mice lacking the bZIP factor ATF4 are runted,[3-5] and ATF4 was shown to directly control osteoblast differentiation and activity,[6] and to indirectly control osteoclastogenesis.[7] Upon activation by the ribosomal S6 kinase 2 (RSK2), ATF4 regulates the onset of osteoblast differentiation, the transcription of the terminal osteoblastic differentiation marker osteocalcin, and the synthesis of the most abundant protein secreted by osteoblasts, type I collagen.[6] When activated by protein kinase A (PKA) phosphorylation, ATF4 activates the transcription of the RANKL gene and thus affects osteoclast formation.[7] The role of ATF4 in regulating osteoblast function appears relevant in the molecular etiology of skeletal dysplasiae, as RSK2 is the kinase inactivated in Coffin–Lowry Syndrome.[8] Furthermore, ATF4 mediates neurofibromin signaling in osteoblasts, suggesting a role in the development of the skeletal abnormalities of patients with neurofibromatosis type I.[9]

Members of the Fos family of proteins have also been shown to control osteoblast activity postnatally. Ectopic expression of Fra-1 leads to a progressive increase in bone mass due to enhanced osteoblast differentiation.[10,11] Targeted inactivation of Fra-1 produces a reciprocal osteopenic phenotype,[12] and the overexpression or ablation of Fra-1 affects matrix Gla protein (MGP) transcription,[12] suggesting that MGP is a specific target of Fra-1. Overexpression of ΔFosB, a naturally occurring truncated form of FosB generated by alternative splicing, leads to a continuous postnatal increase in bone mass accrual, leading to osteosclerosis.[13-16] These results suggest that both Fra-1 and ΔFosB promote osteoblast function by regulating target gene transcription in the osteoblast lineage.

The leucine zipper motif consists of heptad repeats of leucine residues, which align along one face of an alpha helix. When aligned in parallel, the hydrophobic faces of two complementary helices form a coiled coil.[17] Leucine zipper dimerization serves to juxtapose adjacent regions of each of the dimer's partners that are rich in basic amino acid residues and that serve as the DNA-binding domain of the dimer.[18,19] Thus bZIP factors act as dimers, either partnering with themselves to form homodimers, or interacting with a partner to form a heterodimer. ATF4 can form homodimers[20,21] but can also heterodimerize with a variety of partners.[20-22] Fos family members, including Fra-1 and ΔFosB, are obligate heterodimers and their partners include the members of the c-jun family, c-Jun, JunB, and JunD.[23,24]

The activity of bZIP factors, such as ATF4 and Fra-1, is under strict control in a number of cell types, including osteoblasts. Several levels of regulation have been identified that impact on the activity of these molecules. These include transcriptional regulation,[25] stability,[26,27] posttranslational modifications, such as phosphorylation,[6,7,28] as well as interaction with dimerization partners.[24,29] The dimerization partner appears to influence specificity of DNA binding[20] as well as transcriptional activity.[30,31] Another mode of bZIP factor regulation involves interaction with leucine zipper partners to form inactive heterodimers. We have cloned and characterized factor inhibiting ATF4-mediated transcription (FIAT), whose name was coined based on its ability to repress ATF4-dependent osteocalcin gene transcription both *in vitro*[32] and *in vivo*.[33] FIAT, whose expression can be detected in the nucleus of osteoblasts,[33] is a 62 kDa protein containing leucine zippers but no identifiable DNA-binding basic domain.[33] FIAT cannot homodimerize[33] but interacts with ATF4 to form inactive dimers that cannot bind DNA, leading to repression of the expression of ATF4 transcriptional targets.[32,33] Transgenic mice overexpressing FIAT are osteopenic[32–34] due to reduced osteoblast activity.[33] Inhibition of FIAT expression by RNA knockdown results in a higher transcriptional activity from the osteocalcin promoter and enhanced mineralization (Yu, Gauthier, and St-Arnaud, unpublished). These results support the role of FIAT in regulating ATF4 activity and confirm the importance of FIAT in the modulation of early osteoblast activity.

It also remains a formal possibility that FIAT may disturb bone formation by other mechanisms. FIAT could heterodimerize with additional bZip transcription factors, such as Fra-1 or ΔFosB. Alternatively, FIAT could sequester dimerization partners away from these bone mass controlling-bZip factors. A likely candidate remains c-Jun, a well-established dimerization partner of Fos family members.[23] We thus tested the ability of FIAT to modulate c-Jun- or Fra-1-dependent transcription.

MC3T3-E1 osteoblastic cells[35] were transfected with a heterologous AP-1 reporter target, consisting of one copy of the canonical AP-1 target sequence (recognized by c-Jun homodimers or Fra-1/c-Jun heterodimers) that was subcloned upstream of the minimal TK promoter sequence driving the luciferase gene (AP-1-tk-luc).[36] Expression vectors for FIAT,[33] c-Jun,[37] the c-Jun coactivator αNAC,[36,37] or the tethered c-Jun~Fra-1 dimer[38] were cotransfected in various combinations with the reporter construct. FIGURE 1 shows that expression of FIAT by itself had no effect on transcription from the AP-1-tk-luc reporter. Transcription of the reporter allele was strongly induced by c-Jun, and this high transcriptional activity was not affected by increasing amounts of the FIAT expression vector (FIG. 1). Since the transcriptional activity of the c-Jun homodimer is potentiated by the αNAC coactivator in osteoblasts,[39] we tested whether FIAT affects αNAC-stimulated c-Jun activity. FIGURE 2 shows the characteristic increase in c-Jun-mediated transcription of the reporter gene when coexpressed with αNAC. FIAT did not repress αNAC-potentiated c-Jun-dependent transcription (FIG. 2). Similar results were obtained when a tethered c-Jun~c-Jun dimer[38] was used (data not shown). Taken together, these results show that the c-Jun protein is not a dimerization target of the FIAT repressor.

Fra-1 does not contain a transcriptional activation domain and induction of transcription by Fra-1 requires interaction with a dimerization partner.[40,41] Since we showed that FIAT does not interact with c-Jun (FIGS. 1 and 2), we used the tethered c-Jun~Fra-1 dimer[38] to determine whether FIAT can block Fra-1 activity. The tethered dimer was formed by joining the c-Jun and Fra-1 molecules via a flexible polypeptide tether that forces specific pairing.[38]

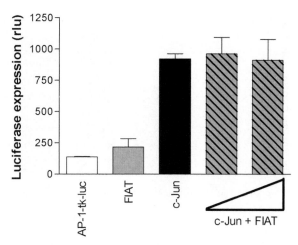

FIGURE 1. FIAT does not inhibit c-Jun-mediated transcription. MC3T3-E1 osteoblasts were transfected with the heterologous AP-1 reporter, AP-1-tk-luc, and expression vectors for c-Jun or FIAT, alone or in combination. Cell extracts were prepared 24 h post transfection and luciferase activity measured. Increasing amounts of the FIAT expression vector (depicted by the triangle) did not affect c-Jun-dependent transcription. Rlu = relative light units.

The c-Jun~Fra-1 dimer induced robust transcription of the luciferase reporter from the AP-1-tk-luc construct (FIG. 3). Coexpression of increasing amounts of the FIAT expression vector dose-dependently inhibited the activity of the c-Jun~Fra-1 dimer (FIG. 3). Since FIAT does not interact with c-Jun (FIGS. 1 and 2), we interpret these data to mean that FIAT can interact with the Fra-1 moiety of the c-Jun~Fra-1 dimer and block Fra-1-dependent transcription.

DISCUSSION

We have shown that FIAT can inhibit the transcriptional activity of the c-Jun~Fra-1 dimer, but not transcription induced by the c-Jun homodimer. Experiments are in progress to confirm protein–protein interactions between FIAT and Fra-1, and to examine the structure–function relationships mediating this interaction.

FIAT was previously shown to interact with the c-Jun transcriptional coactivator αNAC.[32,42] However, our results show that FIAT could not influence αNAC-potentiated c-Jun-mediated transcription (FIG. 2). Thus, the physiological relevance of the FIAT–αNAC interaction remains to be determined.

The c-Jun protein can dimerize with a wide range of partners.[23] Therefore, if FIAT interacted with c-Jun, a broad array of transcriptional responses might have been affected. The inability of FIAT to block c-Jun activity restricts the scope of FIAT-mediated repression and suggests that the inhibitory role of

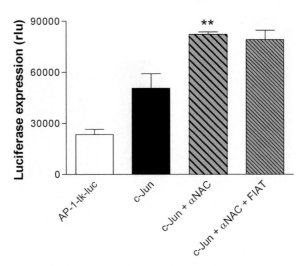

FIGURE 2. FIAT does not repress coactivator-potentiated c-Jun-mediated transcription. MC3T3-E1 osteoblasts were transfected with the heterologous AP-1 reporter, AP-1-tk-luc, and expression vectors for c-Jun, the c-Jun coactivator αNAC, or FIAT, alone or in combination. Cell extracts were prepared 24 h post transfection and luciferase activity measured. FIAT did not inhibit the αNAC-potentiated-c-Jun-dependent transcription. Rlu = relative light units; ** = $P < 0.01$ versus c-Jun alone.

FIAT is targeted to specific bZIP proteins. To date, we have identified two targets of FIAT-mediated repression: ATF4 and Fra-1.[33] Both targets are key regulators of osteoblastic differentiation and function.[6,7,12] Interestingly, ATF4 regulates osteoblast biology during development,[6] while the control of Fra-1 upon osteoblast function is exerted after birth.[10,12] Our findings that FIAT inhibits the activity of both factors suggest that it may affect gene transcription in osteoblasts both pre- and postnatally. It will be interesting to examine the expression patterns of each molecule to determine if they are restricted to particular stages of the osteoblastic differentiation sequence.

At any rate, our results establish that FIAT interacts with two key regulators of osteoblastic gene expression. This observation may explain why a modest perturbation in the level of FIAT expression translates into a significant bone phenotype.[32–34] FIAT may thus represent an interesting drug development target since even a partial modulation of its activity would presumably have a significant impact on bone mass.

ACKNOWLEDGMENTS

We thank Dr. L. Bakiri for providing the tethered c-Jun~c-Jun and c-Jun~Fra-1 expression vectors. This work was supported by NIH-NIAMS (1R01AR053287-01A1 to R.St-A.).

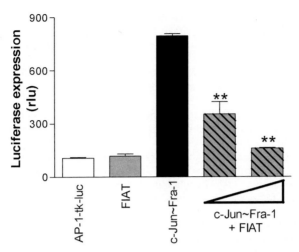

FIGURE 3. FIAT inhibits Fra-1-dependent transcription. MC3T3-E1 osteoblasts were transfected with the heterologous AP-1 reporter, AP-1-tk-luc, and expression vectors for the tethered c-Jun~Fra-1 dimer or FIAT, alone or in combination. Cell extracts were prepared 24 h post transfection and luciferase activity measured. Increasing amounts of the FIAT expression vector (depicted by the triangle) repressed c-Jun~Fra-1-mediated transcription. Since FIAT does not affect c-Jun activity, these data show that FIAT targets the Fra-1 moeity of the c-Jun~Fra-1 dimer. Rlu = relative light units; ** = $P < 0.01$ versus c-Jun~Fra-1.

REFERENCES

1. WAGNER, E.F. 2002. Functions of AP1 (Fos/Jun) in bone development. Ann. Rheum. Dis. **61**(Suppl 2): ii40–ii42.
2. YANG, X. & G. KARSENTY. 2002. Transcription factors in bone: developmental and pathological aspects. Trends Mol. Med. **8**: 340–345.
3. HETTMANN, T., K. BARTON & J.M. LEIDEN. 2000. Microphthalmia due to p53-mediated apoptosis of anterior lens epithelial cells in mice lacking the CREB-2 transcription factor. Dev. Biol. **222**: 110–123.
4. MASUOKA, H.C. & T.M. TOWNES. 2002. Targeted disruption of the activating transcription factor 4 gene results in severe fetal anemia in mice. Blood **99**: 736–745.
5. TANAKA, T. et al. 1998. Targeted disruption of ATF4 discloses its essential role in the formation of eye lens fibres. Genes Cells **3**: 801–810.
6. YANG, X. et al. 2004. ATF4 is a substrate of RSK2 and an essential regulator of osteoblast biology; implication for Coffin-Lowry Syndrome. Cell **117**: 387–398.
7. ELEFTERIOU, F. et al. 2005. Leptin regulation of bone resorption by the sympathetic nervous system and CART. Nature **434**: 514–520.
8. TRIVIER, E. et al. 1996. Mutations in the kinase Rsk-2 associated with Coffin-Lowry syndrome. Nature **384**: 567–570.
9. ELEFTERIOU, F. et al. 2006. ATF4 mediation of NF1 functions in osteoblast reveals a nutritional basis for congenital skeletal dysplasiae. Cell Metab. **4**: 441–451.
10. JOCHUM, W. et al. 2000. Increased bone formation and osteosclerosis in mice overexpressing the transcription factor Fra-1. Nat. Med. **6**: 980–984.

11. ROSCHGER, P. *et al.* 2004. Normal mineralization and nanostructure of sclerotic bone in mice overexpressing Fra-1. Bone **34:** 776–782.
12. EFERL, R. *et al.* 2004. The Fos-related antigen Fra-1 is an activator of bone matrix formation. EMBO J. **23:** 2789–2799.
13. KVEIBORG, M. *et al.* 2002. The increased bone mass in deltaFosB transgenic mice is independent of circulating leptin levels. Endocrinology **143:** 4304–4309.
14. KVEIBORG, M. *et al.* 2004. DeltaFosB induces osteosclerosis and decreases adipogenesis by two independent cell-autonomous mechanisms. Mol. Cell Biol. **24:** 2820–2830.
15. SABATAKOS, G. *et al.* 2000. Overexpression of DeltaFosB transcription factor(s) increases bone formation and inhibits adipogenesis. Nat. Med. **6:** 985–990.
16. SIMS, N.A. *et al.* 2002. Regulating DeltaFosB expression in adult Tet-Off-DeltaFosB transgenic mice alters bone formation and bone mass. Bone **30:** 32–39.
17. LANDSCHULZ, W.H., P.F. JOHNSON & S.L. MCKNIGHT. 1988. The leucine zipper: a hypothetical structure common to a new class of DNA binding proteins. Science **240:** 1759–1764.
18. GENTZ, R. *et al.* 1989. Parallel association of Fos and Jun leucine zippers juxtaposes DNA binding domains. Science **243:** 1695–1699.
19. GLOVER, J.N. & S.C. HARRISON. 1995. Crystal structure of the heterodimeric bZIP transcription factor c-Fos- c-Jun bound to DNA. Nature **373:** 257–261.
20. VALLEJO, M. *et al.* 1993. C/ATF, a member of the activating transcription factor family of DNA- binding proteins, dimerizes with CAAT/enhancer-binding proteins and directs their binding to cAMP response elements. Proc. Natl. Acad. Sci. USA **90:** 4679–4683.
21. HAI, T. & T. CURRAN. 1991. Cross-family dimerization of transcription factors Fos/Jun and ATF/CREB alters DNA binding specificity. Proc. Natl. Acad. Sci. USA **88:** 3720–3724.
22. CHEVRAY, P.M. & D. NATHANS. 1992. Protein interaction cloning in yeast: identification of mammalian proteins that react with the leucine zipper of Jun. Proc. Natl. Acad. Sci. USA **89:** 5789–5793.
23. HARTL, M., A.G. BADER & K. BISTER. 2003. Molecular targets of the oncogenic transcription factor jun. Curr. Cancer Drug Targets **3:** 41–55.
24. VINSON, C. *et al.* 2002. Classification of human B-ZIP proteins based on dimerization properties. Mol. Cell Biol. **22:** 6321–6335.
25. MATSUO, K. *et al.* 2000. Fosl1 is a transcriptional target of c-Fos during osteoclast differentiation. Nat. Genet. **24:** 184–187.
26. VIAL, E. & C. J. MARSHALL. 2003. Elevated ERK-MAP kinase activity protects the FOS family member FRA-1 against proteasomal degradation in colon carcinoma cells. J. Cell Sci. **116:** 4957–4963.
27. YANG, X. & G. KARSENTY. 2004. ATF4, the osteoblast accumulation of which is determined post-translationally, can induce osteoblast-specific gene expression in non-osteoblastic cells. J. Biol. Chem. **279:** 47109–47114.
28. GRUDA, M.C. *et al.* 1994. Regulation of Fra-1 and Fra-2 phosphorylation differs during the cell cycle of fibroblasts and phosphorylation *in vitro* by MAP kinase affects DNA binding activity. Oncogene **9:** 2537–2547.
29. CHINENOV, Y. & T.K. KERPPOLA. 2001. Close encounters of many kinds: Fos-Jun interactions that mediate transcription regulatory specificity. Oncogene **20:** 2438–2452.

30. FAWCETT, T.W. *et al.* 1999. Complexes containing activating transcription factor (ATF)/cAMP-responsive-element-binding protein (CREB) interact with the CCAAT/enhancer-binding protein (C/EBP)-ATF composite site to regulate Gadd153 expression during the stress response. Biochem. J. **339:** 135–141.

31. LIM, C. *et al.* 2000. Latency-associated nuclear antigen of Kaposi's sarcoma-associated herpesvirus (human herpesvirus-8) binds ATF4/CREB2 and inhibits its transcriptional activation activity. J. Gen. Virol. **81:** 2645–2652.

32. YU, V.W.C., C. GAUTHIER & R. ST-ARNAUD. 2006. Inhibition of ATF4 transcriptional activity by FIAT/g-taxilin modulates bone mass accrual. Ann. N. Y. Acad. Sci. **1068:** 131–142.

33. YU, V.W. *et al.* 2005. FIAT represses ATF4-mediated transcription to regulate bone mass in transgenic mice. J. Cell Biol. **169:** 591–601.

34. ST-ARNAUD, R. & V.W. YU. 2005. Inhibition of bone mass accrual by the FIAT transcriptional repressor. Med. Sci. (Paris). **21:** 1020–1021.

35. SUDO, H. *et al.* 1983. *In vitro* differentiation and calcification in a new clonal osteogenic cell line derived from newborn mouse calvaria. J. Cell Biol. **96:** 191–198.

36. MOREAU, A. *et al.* 1998. Bone-specific expression of the alpha chain of the nascent polypeptide- associated complex, a coactivator potentiating c-Jun-mediated transcription. Mol. Cell Biol. **18:** 1312–1321.

37. QUELO, I., M. HURTUBISE & R. ST-ARNAUD. 2002. alphaNAC requires an interaction with c-Jun to exert its transcriptional coactivation. Gene Expr. **10:** 255–262.

38. BAKIRI, L. *et al.* 2002. Promoter specificity and biological activity of tethered AP-1 dimers. Mol. Cell Biol. **22:** 4952–4964.

39. AKHOUAYRI, O., I. QUELO & R. ST-ARNAUD. 2005. Sequence-specific DNA binding by the {alpha}NAC coactivator is required for potentiation of c-jun-dependent transcription of the osteocalcin gene. Mol. Cell Biol. **25:** 3452–3460.

40. COHEN, D.R. *et al.* 1989. The product of a fos-related gene, fra-1, binds cooperatively to the AP-1 site with Jun: transcription factor AP-1 is comprised of multiple protein complexes. Genes Dev. **3:** 173–184.

41. WISDON, R. & I.M. VERMA. 1993. Transformation by Fos proteins requires a C-terminal transactivation domain. Mol. Cell Biol. **13:** 7429–7438.

42. YOSHIDA, K. *et al.* 2005. Interaction of the taxilin family with the nascent polypeptide-associated complex that is involved in the transcriptional and translational processes. Genes Cells **10:** 465–476.

(A)

(B)

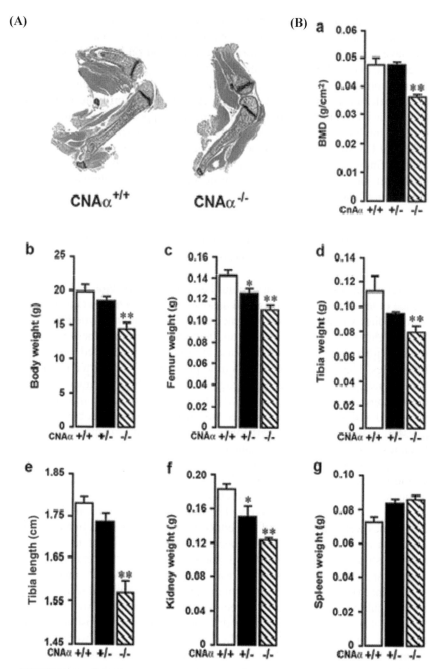

FIGURE 1. Gross phenotypic parameters (**A, B**) as shown of 10 week-old calcineurin Aα$^{+/+,+/-}$, and $^{-/-}$ mice. BMD = bone mineral density by Piximus (Lunar). *$P < 0.05$, **$P < 0.01$.

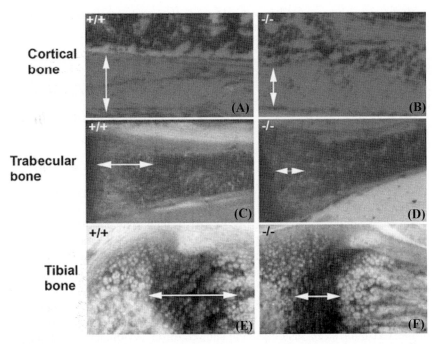

FIGURE 2. Hematoxylin and eosin sections of cortical (**A**, **B**) and trabecular bone (**C**, **D**) and of the head (**E**, **F**) in calcineurin Aα (+/+) (**A**, **C**, **E**) and Aα ($^{-/-}$) mice (**B**, **D**, **F**).

Results from RNAase protection assay performed on RNA isolated from whole muscle and bone tissue revealed the absence of CNAα isoform, while both β and γ isoforms were present (FIG. 5A). As CN isoforms interact with ryanodine receptors, which we have previously identified in both osteoclasts and osteoblasts,[6] we examined if RyR1 levels were altered upon the deletion of CNAα. While in muscle, RyR1 protein expression was reduced, it appeared to be enhanced in bone (FIG. 5B). These experiments were performed with protein extracts from whole bone and muscle; thus, the cellular origin of the elevated RyR1 in bone remains unclear.

FORCED OVEREXPRESSION OF CNAα IN OSTEOBLASTS

To examine the effect of CNAα on osteoblast differentiation, we used a novel protein transduction system that is mediated by a 12 amino acid-long, Arg-rich, HIV sequence, TAT. When fused to any protein (between 18 and 100 kDa), TAT causes the receptor-less movement of that protein into a cell within minutes. Unlike traditional transfection methods, TAT-mediated transduction can result in up to 90% delivery of an intact protein into the cell cytoplasm.[7] An anti-TAT antibody can then be used in immunodetection

TABLE 1. Comparisons of histological measures of diaphyseal cortical thickness and diaphyseal, metaphyseal, epiphyseal, and growth plate diameters (all in μm)

Parameter	CNAα$^{+/+}$	CNAα$^{-/-}$
Diameter		
Diaphysis	600	600
Metaphysis	180	150
Epiphysis	200	210
Growth plate	180	160
Cortical thickness		
Diaphysis	112	70

studies to differentiate the TAT fusion protein from the endogenously expressed protein.

To study cellular localization, MC3T3.E1 osteoblasts were incubated with TAT-CNAα (10 min at 37°C) and stained with both anti-TAT and anti-CNAα antibodies. Cells transduced with vehicle alone showed only anti-CNAα immune labeling representing endogenous CNAα. Transduction with TAT-CNAα resulted in a precise co-localization of the red and green immunostains within the same cells consistent with the influx of exogenously applied fusion protein that was detected by both antibodies.[3,5] Laser scanning confocal microscopy of the TAT-transduced fusion protein showed diffuse cytoplasmic staining in all image planes with negligible nuclear localization.[3]

Following transduction, we measured the expression of CN isoforms and the differentiation markers Runx2, osteocalcin, and bone sialoprotein (BSP) (FIG. 6). Whereas there was minimal change in the expression of endogenous CNAα, both CNAβ and CNAγ were elevated in TAT-CNAα-transduced cells. This was accompanied by a time-dependent and marked elevation in the expression of BSP and osteocalcin and a less marked effect on Runx2, confirming previous results.[3]

INHIBITION OF CNAα IN OSTEOBLASTS

We have already demonstrated that bone marrow-derived osteoblasts from CNAα$^{-/-}$ mice displayed a reduced propensity to differentiate to a mature phenotype as assessed by CFU-F formation.[3] These studies further indicated a significant concentration-dependent decrease in all osteoblast differentiation markers, namely Runx2, BSP, and osteocalcin, with both calcineurin inhibitors cyclosporine A and FK506.[3] However, when we used calvarial osteoblasts instead of bone marrow-derived cells, the effect of FK506 on differentiation converted from inhibitory to stimulatory (FIG. 7).

FIGURE 7 thus shows a profound increase in alkaline phosphatase staining of calvarial osteoblasts with 1 and 10 mg/mL FK506. Real time polymerase chain reaction (PCR) of RNA isolated from cells exposed to 1 mg/mL FKF06

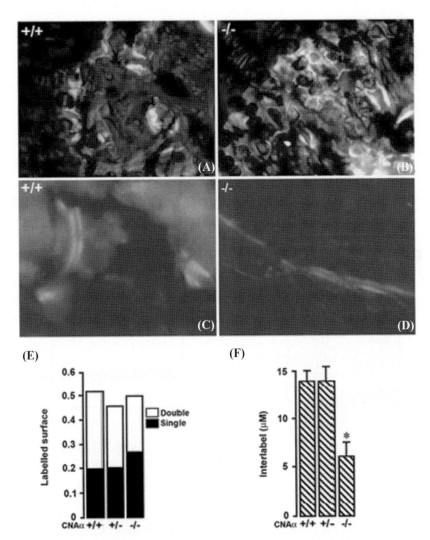

FIGURE 3. Bone sections following double tetracycline labeling applied at 8 day intervals shown at low (**A, B**) and high (**C, D**) magnifications in calcineurin Aα (+/+) (**A, C**) and Aα ($^{-/-}$) bone (**B, D**). Extent of labeling between experimental groups, or following single or double labeling (**E**). Spacings between successive labeled regions observed in the A$\alpha^{+/+}$ and $^{-/-}$ mice (**F**). *$P < 0.05$.

showed a further enhancement of osteoblast differentiation in CNA$\alpha^{-/-}$ cultures compared to FK-506-treated CNA$\alpha^{+/+}$ cells. It appears therefore that (a) calvarial osteoblasts respond differently to calcineurin inhibition than bone marrow-derived cells and (b) the stimulatory responses are exaggerated, rather

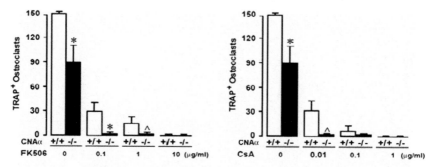

FIGURE 4. Tartrate-resistant acid phosphatase (TRAP) positive osteoclast formation in bone marrow cell cultures from calcineurin $A\alpha^{+/+}$ and $A\alpha^{-/-}$ mice treated with various concentrations of cyclosporine and FK506. $*P < 0.05$; $\wedge P < 0.07$.

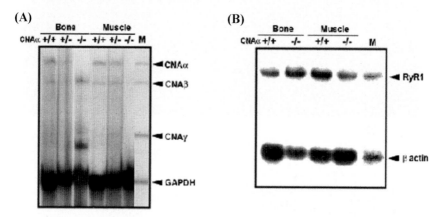

FIGURE 5. RNAse protection assay (**A**) and Western blot (**B**) comparing bones and muscle for calcineurin isoform expression (**A**) and ryanodine receptor expression (**B**) from calcineurin $A\alpha^{+/+}$, $A\alpha^{+/-}$, and $A\alpha^{-/-}$ mice, respectively.

than reduced, in the absence of CNAα, suggesting that the effect of FK506 was not exerted via CN inhibition. The molecular mechanism underlying the latter observations is currently under investigation.

ROLE OF CALCINEURIN IN SKELETAL REMODELING

We concur with others that CN is the Ca^{2+}-sensitive enzyme upstream of the transcriptional regulator NFAT2, which has been shown, in both osteoclasts and osteoblasts, to be critical to cell differentiation.[4,8–10] Dephosphorylation of NFAT2 by CN leads to its nuclear localization and binding together with AP1 transcription factors to gene promoters in the osteoclast precursor,[8–10] and together with osterix to gene promoters in the

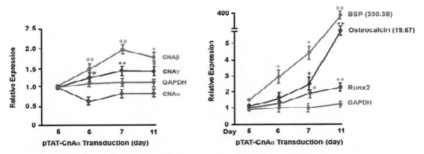

FIGURE 6. Real time quantitative RT-PCR using SYBR green to demonstrate differences in the expression of calcineurin isoforms Aα, Aβ, and Aγ, as well as markers of osteoblastic differentiation, namely Runx2, bone sialoprotein (BSP), osteocalcin, or GAPDH (control gene) 11 days after a one-time transduction with TAT-calcineurin Aα (200 nM for 10 min). $*P < 0.05$; $**P < 0.01$.

osteoblast precursor[11] to ultimately govern the differentiation of the respective cell types.

While it is clear that CN and its downstream signal NFAT2 is permissive to the differentiation of osteoblasts to a more mature, mineralizing phenotype, the role of CN in the osteoclast formation and function appears two-pronged. It stimulates osteoclastogenesis via NFAT2, as a primary mechanism.[4,8–10] However, CN activation in mature osteoclasts tends to lower resorptive activity,[5] a mechanism through which extracellular Ca^{2+} likely exerts its feedback role in inhibiting excessive resorption.[12] It is therefore not unexpected that deletion of CNAα causes low-turnover osteoporosis primarily due to low bone formation. It is also not surprising that osteoclast surfaces are normal in these mice, (TABLE 2), as inhibited osteoclastogenesis is likely balanced by stimulated resorption, with no net increase in TRAP-labeled surfaces.

CLINICAL RELEVANCE

Acute, rapid, and severe bone loss[13] follows the administration of either cyclosporine A or FK506 to organ transplant patients. The initial bone loss is of a high-turnover variety, characterized by elevated bone resorption and formation.[14] Rates of resorption and formation subsequently drop, but fracture rates can be as high as 65%, for example, after a liver transplant.[13] The human osteoporosis is not recapitulated fully in the CNAα mouse, particularly as the latter does not display increased resorption. Instead, overall resorption is normal and formation is low.

Several reasons could underlie this phenotypic mismatch. First, the high-turnover bone loss in calcineurin inhibitor-induced human osteoporosis may result from noncalcineurin effects of drugs, such as FK506. Calcineurin-independent effects on osteoblast differentiation are indeed obvious in this

(A)

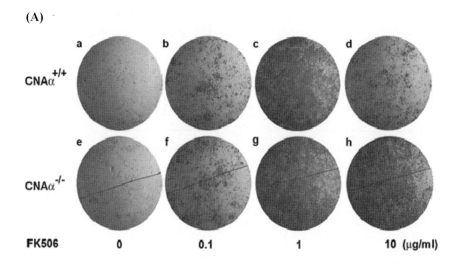

(B)

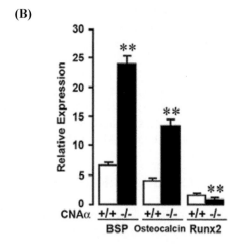

FIGURE 7. Effect of various concentrations of FK506 on alkaline phosphatase staining (**A**) and the expression of osteoblast differentiation markers Runx2, BSP, and OG1 (**B**) in calvarial osteoblasts derived from calcineurin $A\alpha^{+/+}$ and $A\alpha^{-/-}$ mice. $**P < 0.01$.

study (above). Second, acute calcineurin inhibition by drug therapy may cause initial osteoclastogenesis, which may be so profound as to overcome the inhibitory effects of calcineurin inhibition on resorptive function. Finally, T cells may mediate the acute effects of calcineurin inhibition,[15] an action that is not seen in the face of chronic deficiency.

Whatever the mechanism, it is obvious that the suppressive effects of calcineurin inhibition on bone formation, also seen with long-term therapy in humans,[13] results from reduced osteoblastic differentiation. Indeed, this may confound the effects of glucocorticoids, which are commonly coadministered with calcineurin inhibitors. Further studies are likely to unravel the detailed action of "calcineurin inhibitors" on bone, which may involve effects other than those exerted through a reduction in enzyme activity.

ACKNOWLEDGMENTS

These studies were supported by grants to MZ and HCB from the National Institutes of Health, and the Department of Veterans Affairs.

REFERENCES

1. ARAMBURU, J., J. HEITMAN & G.R. CRABTREE. 2004. Calcineurin: a central controller of signaling in eukaryotic cells. EMBO Rep. **4:** 343–348.
2. STANKUNAS, K., I.A. GRAEF, J.R. NEILSON, *et al.* 1999. Signaling through calcium, calcineurin, and NFAT in lymphocyte activation and development. Cold Spring Harbour Symp. Quant. Biol. **64:** 505–516.
3. SUN, L., H.C. BLAIR, Y. PENG, *et al.* 2005. Calcineurin regulates bone formation by the osteoblast. Proc. Natl. Acad. Sci. USA **102:** 17130–17135.
4. SUN, L., Y. PENG, N. ZAIDI, *et al.* 2006. Evidence that calcineurin is required for the genesis of bone-resorbing osteoclasts. Am. J. Physiol. Renal Physiol. **292:** F285–F291.
5. SUN, L., B.S. MOONGA, M. LU, *et al.* 2003. Molecular cloning, expression and function of osteoclastic calcineurin α. Am. J. Physiol. Renal Physiol. **284:** F575–F583.
6. ADEBANJO, O.A., H.K. ANANDATHREETHAVARADA, A.P. KOVAL, *et al.* 1999. A new function for CD38/ADP-ribosyl cyclase in nuclear Ca^{2+} homeostasis Nat. Cell Biol. (article) **7:** 409–414.
7. DOLGILEVICH, S., N. ZAIDI, J. SONG, *et al.* 2002. Transduction of TAT fusion proteins into osteoclasts and osteoblasts. **299:** 505–509.
8. KIM, Y., K. SATO, M. ASAGIRI, *et al.* 2005. Contribution of nuclear factor of activated T cells c1 to the transcriptional control of immunoreceptor osteoclast-associated receptor but not triggering receptor expressed by myeloid cells-2 during osteoclastogenesis. J. Biol. Chem. **280:** 32905–32913.
9. ASAGIRI, M., K. SATO, T. USAMI, *et al.* 2005. Autoamplification of NFATc1 expression determines its essential role in bone homeostasis. J. Exp. Med. **202:** 1261–1269.
10. MATSUO, K., D.L. GALSON, C. ZHAO, *et al.* 2004. Nuclear factor of activated T-cells (NFAT) rescues osteoclastogenesis in precursors lacking c-Fos. J. Biol. Chem. **279:** 26475–26480.
11. KOGA, T., Y. MATSUI, M. ASAGIRI, *et al.* 2005. NFAT and Osterix cooperatively regulate bone formation. Nat. Med. **11:** 880–885.

12. ZAIDI, M., B.S. MOONGA & C.L.-H. HUANG. 2003. Ca^{2+} sensing and cellular signaling processes in the local regulation of osteoclastic bone resorption. Biol. Rev. **79:** 79–100.
13. EPSTEIN, S., A.M. INZERILLO, J. CAMINIS & M. ZAIDI. 2003. Disorders associated with acute rapid and severe bone loss. J. Bone Miner. Res. **18:** 2083–2094.
14. EPSTEIN, S., E. SHANE & J.P. BILEZIKIAN. 1995. Organ transplantation and osteoporosis. Curr. Opin. Rheumatol. **7:** 255–261.
15. ZAIDI, M., J. IQBAL, A.M. INZERILLO, *et al.* 2002. The cellular and molecular aspects of immunosuppressant osteoporosis. *In* Osteoporosis: Pathophysiology and Clinical Management. Contemporary Endocrinology Series. Ed: E.S. Orwoll, M. Bliziotes: 523–535. Humana Press. Totowa, NJ.

The Role of NFAT in Osteoclast Formation

HIROSHI TAKAYANAGI

Department of Cell Signaling, Graduate School, Tokyo Medical and Dental University, Bunkyo-ku, Tokyo

Center of Excellence Program for Frontier Research on Molecular Destruction and Reconstruction of Tooth and Bone, Tokyo Medical and Dental University, Bunkyo-ku, Tokyo

ABSTRACT: Osteoclasts are cells of monocyte–macrophage origin that degrade bone matrix. Receptor activator of NF-κB ligand (RANKL) induces osteoclast formation in the presence of macrophage-colony-stimulating factor (M-CSF) and costimulatory signals. RANKL induces activation of the TNF receptor-associated factor 6 (TRAF6) and c-Fos pathways, which lead to the osteoclast-specific event, that is, autoamplification of nuclear factor of activated T cells (NFAT)c1, the master transcription factor for osteoclast differentiation. Autoamplification of NFATc1 is dependent on the calcium signaling of immunoglobulin-like receptors associated with immunoreceptor tyrosine-based activation motif (ITAM)-harboring adaptors. In addition to the calcineurin–NFATc1 axis, calcium signaling activates the calmodulin-dependent kinase pathway, which also plays a critical role in osteoclast formation. Such advances in the understanding of the molecular mechanism of osteoclast differentiation are expected to lead to novel therapeutic approaches to bone diseases.

KEYWORDS: osteoclast; RANKL; NFATc1; osteoimmunology

INTRODUCTION

Bone homeostasis depends on a balance between bone formation and bone resorption.[1] Disruption of the balance leads to various bone diseases. For example, bone loss in postmenopausal osteoporosis and bone destruction in inflammatory diseases, such as rheumatoid arthritis, are caused by excessive bone resorption by osteoclasts relative to the bone formation by osteoblasts. In osteopetrosis, a defect in osteoclast differentiation or function results in

Address for correspondence: Hiroshi Takayanagi, Department of Cell Signaling, Graduate School, Tokyo Medical and Dental University, Yushima 1-5-45, Bunkyo-ku, Tokyo 113-8549, Japan. Voice: +81-3-5803-5471; fax: +81-3-5803-0192.

taka.csi@tmd.ac.jp

Ann. N.Y. Acad. Sci. 1116: 227–237 (2007). © 2007 New York Academy of Sciences.
doi: 10.1196/annals.1402.071

abnormally high bone density and fragility.[2] Therefore, understanding the mechanism of osteoclast formation is important for the development of new therapeutic strategies against bone diseases.

Osteoclasts are cells of hematopoietic origin that decalcify and degrade the bone matrix. They are large, multinucleated cells formed by the fusion of precursor cells of monocyte–macrophage lineage.[3] Osteoclast differentiation is supported by bone-marrow stromal cells or osteoblasts through cell–cell contact. The molecules that are expressed by osteoblasts and are involved in the induction of osteoclastogenesis have been extensively studied.[4] It is widely accepted that supporting cells express RANKL and ligands for costimulatory signals, in addition to M-CSF.[5] Here we summarize recent advances in the understanding of the mechanism of osteoclast formation by focusing on the role of NFATc1.

MEDIATORS OF RANKL SIGNALING

RANKL is a TNF superfamily cytokine and was originally identified as an activator of dendritic cells expressed by activated T cells.[6,7] Subsequent studies revealed RANKL to be a key cytokine for osteoclastogenesis.[8,9] Targeted disruption of *RANKL* results in the defective formation of lymph nodes as well as severe osteopetrosis due to impaired formation of osteoclasts,[10] indicating that this molecule is critical for both the immune and bone systems. The receptor for RANKL, RANK, is expressed in osteoclast precursor cells as well as in mature osteoclasts and dendritic cells. Targeted disruption of *RANK* also results in severe osteopetrosis,[11] suggesting the RANKL/RANK interaction plays an essential role in osteoclast formation *in vivo*. Osteoprotegerin (OPG), a soluble decoy receptor of RANKL, acts as a physiological inhibitor of RANKL,[12] and the dynamic balance between RANKL and OPG regulates the levels of bone resorption.

The signals immediately activated by RANKL have been extensively studied. TRAF family proteins are adapter molecules for various cytokine receptors including TNF receptor superfamily and Toll/IL-1 receptor family members. RANK contains three TRAF6-binding sites[13] and TRAF6-deficient mice exhibit severe osteopetrosis, indicating that TRAF6 is an essential mediator of RANK signaling during osteoclastogenesis.[14] RANKL induces rapid activation of JNK and p38 as well as that of NF-κB, while such activation is severely abrogated in TRAF6-deficient cells.[14] NF-κB is a dimeric transcription factor complex composed of p65 (RelA), c-Rel, RelB, NF-κB1 (p50/p105), and NF-κB2 (p52/p100). Although *p50*- or *p52*-deficient mice have no obvious bone phenotype, mice doubly deficient in p50 and p52 develop osteopetrosis due to a defect in osteoclast differentiation, suggesting a critical role for NF-κB in osteoclastogenesis.[15,16] The AP-1 transcription factor is composed of the Fos, Jun, and activating transcription factor (ATF) family proteins. RANK activates

AP-1 through an induction of its critical component c-Fos in a relatively early phase, and c-Fos-deficient mice develop severe osteopetrosis due to a lack of osteoclasts.[17]

These studies indicate that the TRAF6-NF-κB and c-Fos pathways comprise essential downstream signaling events activated by RANKL.

THE SEARCH FOR OSTEOCLAST-SPECIFIC REGULATORS

NF-κB and AP-1 are activated by RANKL in the early phase and thus play an essential role in osteoclastogenesis, but these transcription factors are also activated by other cytokines, such as IL-1, which are not capable of inducing osteoclast differentiation. These observations suggest that RANKL has an as yet unknown target gene that is specifically linked to osteoclast differentiation. Therefore, many groups took part in efforts to identify such genes that were extremely active around 2001. In a genome-wide search for such genes, we found that NFATc1 is the transcription factor most strongly induced by RANKL.[18] The NFAT transcription factor family was originally identified in T cells and now comprises five members including NFATc1 (NFAT2), NFATc2 (NFAT1), NFATc3 (NFAT4), NFATc4 (NFAT3), and NFAT5. The activation of NFAT is mediated by a specific phosphatase, calcineurin, which is activated by calcium/calmodulin signaling. Consistent with this finding, calcineurin inhibitors, such as FK506 and cyclosporin A, have been shown to potently inhibit osteoclastogenesis. Another Japanese group also reported that NFATc1 is important for osteoclast differentiation,[19] and it is a perhaps surprising fact that no fewer than four Japanese groups were working at the same time on the role of NFAT in osteoclasts independently.[20,21]

THE ESSENTIAL AND SPECIFIC ROLE OF NFATc1

The next important task was to demonstrate that NFATc1 is essential for osteoclast differentiation. NFATc1-deficient mice have been generated by two groups, but the mice have a defect in cardiac valve formation, which leads to embryonic death at around E14.5, so the function of NFATc1 in bone tissue has not received much attention.[22,23] It has not proven to be easy to provide convincing genetic evidence that *NFATc1* is essential for osteoclast formation. Therefore, we developed a new culture system of embryonic stem cells that differentiate into osteoclasts in response to RANKL and M-CSF. This system revealed the essential and sufficient role of *NFATc1* in *in vitro* osteoclastogenesis by making it evident that NFATc1-deficient embryonic stem cells are unable to differentiate into osteoclasts, and that the ectopic expression of NFATc1 induced the BMMs to undergo osteoclast differentiation in the absence of RANKL.[18]

However, as mentioned above, *in vivo* analysis of NFATc1-deficient mice has been hampered by the constraint of embryonic lethality. To overcome this problem, we applied chimeric mouse approaches to osteoclast biology: (1) adoptive transfer of hematopoietic stem cells to osteoclast-deficient c-Fos-deficient mice, and (2) c-Fos-deficient blastocyst complementation. These methods were originally developed in the field of immunology to analyze lymphocytes in embryonically lethal mice. These techniques successfully demonstrated the critical function of NFATc1 in osteoclastogenesis *in vivo*.[24] Since NFATc1-deficient mice die owing to the defect in the cardiac valve formation, it is possible to overcome this lethal defect if NFATc1 is expressed in the heart. Winslow *et al.* reported that NFATc1-deficient mice crossed with transgenic mice expressing Tie2-promoter-driven NFATc1 are viable and, importantly, develop osteopetrosis.[25] These results collectively provide clear evidence that *NFATc1* is essential for osteoclast differentiation.

AUTOAMPLIFICATION OF NFATc1 DETERMINES ITS SPECIFICITY

Since NFATc1 and NFATc2 play a redundant role in the immune system, the question arises as to how NFATc1 plays such an exclusive function in osteoclastogenesis. An interesting observation was obtained from rescue experiments using *NFATc1*$^{-/-}$ osteoclast precursor cells: osteoclast formation in *NFATc1*$^{-/-}$ cells was recovered not only by forced expression of NFATc1 but also by that of NFATc2. How can we reconcile the indispensable *in vivo* role of NFATc1 in osteoclastogenesis with the observation that *NFATc1* deficiency is compensated for by forced expression of NFATc2? After a considerable amount of effort, we came up with the following putative explanation: NFATc1 and NFATc2 proteins have a similar function to induce osteoclastogenesis if and when they are ectopically expressed at a similarly high level. Therefore, the essential role of NFATc1 is not achieved by the unique function of the protein, but by an *NFATc1*-specific gene regulatory mechanism.

Accordingly, we analyzed the mRNA expression of *NFATc1* and *NFATc2* genes during osteoclastogenesis. The mRNA of *NFATc1* is induced selectively and potently by RANKL while *NFATc2* mRNA is expressed constitutively in precursor cells at a low level. Importantly, FK506, which suppresses the activity of NFAT through an inactivation of calcineurin, downregulates the induction of *NFATc1,* but not *NFATc2*. This suggests that *NFATc1* is selectively autoregulated by NFAT during osteoclastogenesis. As expected, chromatin immunoprecipitation (ChIP) experiments revealed that NFATc1 is recruited to the *NFATc1* but not the *NFATc2* promoter 24 h after RANKL stimulation, and this occupancy persists during the terminal differentiation of osteoclasts, indicating the autoamplification mechanism by NFATc1 is specifically operative in the *NFATc1* promoter.

Why does the autoamplification occur only in the case of *NFATc1* gene regulation? If NFAT-binding sites are only found in the *NFATc1* promoter, the selective recruitment would be easily explained. However, NFAT-binding sites are found in both the *NFATc1* and the *NFATc2* promoters: the promoter sequence thus cannot explain the difference. Histone acetylation is a marker of the transcriptionally active chromatin structure, and transcriptional coactivators, such as CBP and PCAF, have histone acetylase activity. Investigation of the recruitment of CBP and PCAF to the *NFATc1* promoter yielded positive results. The rate of histone acetylation in the *NFATc1* promoter increased gradually after RANKL stimulation, and methylation of histone H3 lysine 4, which is characteristic of a transcriptionally active locus, is also upregulated exclusively in the *NFATc1* promoter, while this was not observed in the *NFATc2* promoter. Conversely, the *NFATc2* promoter is constantly associated with methylated DNA-binding proteins, such as methyl-CpG-binding protein 2 (MeCP2), suggesting epigenetic modification of the *NFATc2* promoter is responsible for the muted pattern of gene expression. Thus, contrasting epigenetic modification of the *NFATc1* and the *NFATc2* promoters might explain their unique spatiotemporal induction pattern during osteoclastogenesis. We have documented these findings in more details elsewhere.[24] In conclusion, the essential role of the *NFATc1* gene is determined not only by the function of the encoded protein but also by an NFATc1-specific gene regulatory mechanism. It remains an issue to be pursued in future work to determine how such a specific form of epigenetic regulation arose specifically in osteoclasts.

ITAM AND OSTEOCLAST COSTIMULATION

Despite the evident importance of the calcium–NFATc1 pathway, it remained unclear how RANKL specifically activates calcium signals leading to the induction of *NFATc1*, since RANK belongs to the TNF receptor family, which is not directly related to calcium signaling. The screening of osteoclast-specific genes has shed light on a novel type of receptor. Osteoclast-associated receptor (OSCAR) is an immunoglobulin-like receptor expressed by osteoclasts, which is involved in the cell–cell interaction between osteoblasts and osteoclasts.[26] OSCAR associates with an adaptor molecule, Fc receptor common γ subunit (FcRγ), which is required for the cell-surface expression of OSCAR and its signal transduction.[27] FcRγ harbors an immunoreceptor tyrosine-based activation motif (ITAM), which is critical for the activation of calcium signaling in immune cells. Another ITAM-harboring adaptor, DNAX-activating protein 12 (DAP12), is also involved in the formation and function of osteoclasts. It is noteworthy that mice doubly deficient in FcRγ and DAP12 exhibit severe osteopetrosis owing to a differentiation blockade of osteoclasts, demonstrating that the immunoglobulin-like receptors associated with FcRγ and DAP12 are essential for osteoclastogenesis.[27] These receptors include OSCAR, triggering receptor expressed in myeloid cells-2 (TREM-2), signal-regulatory protein β1 (SIRPβ1), and paired immunoglobulin-like receptor-A (PIR-A), although the ligands and exact functions of each of these receptors remain to be elucidated. The triggering of either receptor by cross-linking with an antibody accelerates RANKL-induced osteoclast differentiation, indicating the activating role of these immunoglobulin-like receptors in osteoclastogenesis. However, in the absence of RANKL, the stimulation of these receptors alone is unable to induce osteoclast differentiation, suggesting that these receptor-mediated signals act cooperatively with RANKL but cannot substitute for the signal. Therefore, ITAM-mediated signals can be appropriately designated costimulatory signals for RANK.[5]

ITAM was initially recognized as a common sequence in the cytoplasmic tails of the signaling chains associated with the T cell receptor, B cell receptor, and certain Fc receptors. In immune cells, the ITAM is typically tyrosine phosphorylated by Src family kinases upon stimulation, resulting in the docking of the Syk tyrosine family kinase and other adapter molecules, which contain the Src homology 2 domain. Deficiency of Syk or inhibition of Syk by its specific inhibitor results in the impairment of osteoclast differentiation *in vitro*, suggesting the importance of Syk as a mediator of ITAM signaling in osteoclastogenesis.[27,28]

Downstream of ITAM signaling, PLCγ activates calcium signaling and the NFAT family of transcription factors in immune cells. In osteoclast differentiation, PLCγ is also phosphorylated in response to RANKL downstream of ITAM signals.[27] PLC cleaves the membrane phospholipid phosphatidylinositol-4,5-bisphosphate (PIP2) into the second messenger molecules inositol-1,4,5-trisphosphate (IP3) and diacylglycerol (DAG). IP3 directly increases intracellular calcium levels by inducing the release of endoplasmic reticulum calcium stores, while DAG activates PKC at the plasma membrane. The PLCγ family consists of two members, PLCγ1 and PLCγ2, and both isoforms require phosphorylation on specific tyrosine residues for their catalytic activity. A recent study showed that targeted deletion of *PLCγ2* in mice results in an osteopetrotic phenotype owing to a defect in osteoclastogenesis and that PLCγ2-deficient cells do not highly express NFATc1 after RANKL stimulation,[29] suggesting an important role of PLCγ2 in ITAM-stimulated calcium signaling.

CaMK AND CREB PLAY AN IMPORTANT ROLE IN OSTEOCLASTOGENESIS

In the course of transmitting a variety of signals, intracellular Ca^{2+} binds to calcium-binding proteins, the most important of which is calmodulin. Calmodulin contains four Ca^{2+}-binding EF hands arranged in pairs at both its N and C termini, with a helix linking the two. In response to transient increases in intracellular Ca^{2+}, Ca^{2+} interaction with the four EF hands leads to a dramatic conformational shift, creating an open form of calmodulin. This Ca^{2+}-dependent conformational change induces activation of its downstream effector proteins, the calcium/calmodulin-activated kinases (CaMKs), as well as the phosphatase calcineurin. Recently, we demonstrated that CaMKIV is critically involved in osteoclast differentiation and function.[30] Pharmacological inhibition of CaMKs as well as the genetic ablation of *Camk4* reduced CREB phosphorylation and downregulated the expression of c-Fos, which is required for the induction of NFATc1. Furthermore, the CaMK–CREB pathway regulates the expression of osteoclast-specific genes in cooperation with NFATc1. It is worth mentioning that two calcium-regulated pathways mediated by CaMK and calcineurin converge on the transcriptional control of osteoclast-specific genes.

TRANSCRIPTIONAL TARGETS OF NFATc1

Accumulating evidence suggests that a number of osteoclast-specific genes are directly regulated by NFATc1. Based on promoter analyses, the *TRAP*,[18] *calcitonin receptor*,[18] *cathepsin K*,[31] and *β3 integrin*[32] genes are regulated by NFATc1. OSCAR is also regulated by NFATc1.[33,34] Promoter deletion and ChIP studies in addition to an analysis using knockout cells convincingly showed that the *OSCAR* promoter is the direct target of NFATc1. The AP-1 complex is known to be a transcriptional partner of NFAT in lymphocytes, and crystal structure analysis revealed the formation of the NFAT:AP-1 complex to be crucial for DNA binding. Similarly, an NFAT:AP-1 complex is important for the induction of the *TRAP* and *calcitonin receptor* genes as well as the robust autoamplification of NFATc1.[18] It has also been shown that NFATc1 cooperates with PU.1 and MITF in the effect on the *cathepsin K* and the *OSCAR* promoters.[31] It is noteworthy that both PU.1 and MITF, which are thought to be important for the survival of osteoclast precursor cells, also participate in osteoclast-specific gene induction at the terminal stage of differentiation. Thus, NFATc1 forms an osteoclast-specific transcriptional complex containing AP-1 (Fos/Jun), PU.1, and MITF for the efficient induction of osteoclast-specific genes. Components of the NFATc1 transcriptional complex are not always the same: the cooperation between NFATc1 and PU.1/MITF was not observed in the case of the *calcitonin receptor* promoter,[34] suggesting that the differential composition of the transcriptional complex may contribute to the spatiotemporal expression of each gene during osteoclastogenesis.

CONCLUSIONS

Discovery of the RANKL–RANK system has brought about rapid progress in the understanding of the regulatory mechanism of osteoclast

234

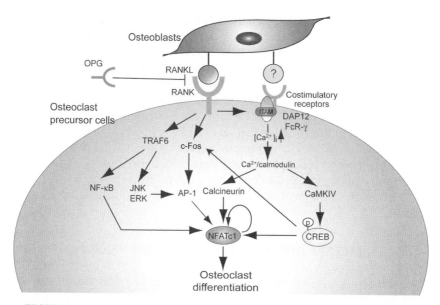

FIGURE 1. Schematic of signaling cascades in osteoclast differentiation. Osteoclastogenesis is supported by osteoblasts or bone marrow stromal cells, which provide RANKL, M-CSF, and poorly characterized ligands for costimulatory receptors. RANKL binding to RANK results in the recruitment of TRAF6, which activates NF-κB and MAPKs. The induction of NFATc1, a key transcription factor for osteoclastogenesis, is dependent on the transcription factors AP-1 (containing c-Fos) and NF-κB. Costimulatory signals for RANK: immunoreceptors associated with ITAM-harboring adaptors stimulate Ca^{2+} signaling. NFATc1 is translocated to the nucleus after the dephosphorylation by calcineurin that is activated by calcium (Ca^{2+}) signaling. Ca^{2+}/calmodulin kinases IV is a main kinase that activates cAMP response element-binding protein (CREB), which is also important for osteoclast differentiation. Induction of c-Fos is partly mediated by CREB. In the nucleus, NFATc1 works together with other transcription factors, such as AP-1, PU.1, MITF, and CREB, to induce various osteoclast-specific genes.

differentiation. The knowledge obtained from osteopetrotic mouse models is now being adapted to the context of the RANKL-stimulated signaling network (FIG. 1). As discussed above, much has been understood, but several questions remain. For some of the genes activated by RANKL, including the MAPKs, the *in vivo* significance has not been established yet. In ITAM-mediated costimulatory signaling, most of the ligands for the immunogloblin-like receptors have not been identified. The molecules that connect RANK with ITAM signaling are also unknown. Among the NFATc1 target genes, the genes that directly promote the differentiation process remain to be elucidated. In this regard, the reports on the role of DC-STAMP and v-ATPase V0 subunit d2 in osteoclast fusion are of interest.[35,36] Further detailed studies will be necessary to obtain a complete understanding of the osteoclast signaling pathways. Molecular understanding of these pathways is expected to provide perhaps unprecedented opportunities to find novel therapeutic strategies for the treatment of bone disease.

ACKNOWLEDGMENTS

The work was supported in part by Grants-in-Aid for Creative Scientific Research from Japan Society for the Promotion of Science, Grants-in-Aid for the 21st century COE program and Genome Network Project from Ministry of Education, Culture, Sports, Science, and Technology of Japan, Health Sciences Research Grants from the Ministry of Health, Labour and Welfare of Japan, and grants from Kanae Foundation for Life & Socio-Medical Science, Tokyo Biochemical Research Foundation, Yokoyama Foundation for Clinical Pharmacology.

REFERENCES

1. ZAIDI, M. 2007. Skeletal remodeling in health and disease. Nat. Med. **13:** 791–801.
2. TEITELBAUM, S.L. & F.P. ROSS. 2003. Genetic regulation of osteoclast development and function. Nat. Rev. Genet. **4:** 638–649.
3. BOYLE, W.J., W.S. SIMONET & D.L. LACEY. 2003. Osteoclast differentiation and activation. Nature **423:** 337–342.
4. SUDA, T. et al. 1999. Modulation of osteoclast differentiation and function by the new members of the tumor necrosis factor receptor and ligand families. Endocrine Rev. **20:** 345–357.
5. TAKAYANAGI, H. 2007. Osteoimmunology: shared mechanisms and crosstalk between the immune and bone systems. Nat. Rev. Immunol. **7:** 292–304.
6. ANDERSON, D.M. et al. 1997. A homologue of the TNF receptor and its ligand enhance T-cell growth and dendritic-cell function. Nature **390:** 175–179.
7. WONG, B.R. et al. 1997. TRANCE is a novel ligand of the tumor necrosis factor receptor family that activates c-Jun N-terminal kinase in T cells. J. Biol. Chem. **272:** 25190–25194.
8. YASUDA, H. et al. 1998. Osteoclast differentiation factor is a ligand for osteoprotegerin/osteoclastogenesis-inhibitory factor and is identical to TRANCE/RANKL. Proc. Natl. Acad. Sci. USA **95:** 3597–3602.
9. LACEY, D.L. et al. 1998. Osteoprotegerin ligand is a cytokine that regulates osteoclast differentiation and activation. Cell **93:** 165–176.
10. KONG, Y.Y. et al. 1999. OPGL is a key regulator of osteoclastogenesis, lymphocyte development and lymph-node organogenesis. Nature **397:** 315–323.
11. DOUGALL, W.C. et al. 1999. RANK is essential for osteoclast and lymph node development. Genes Dev. **13:** 2412–2424.

12. Bucay, N. *et al.* 1998. Osteoprotegerin-deficient mice develop early onset osteoporosis and arterial calcification. Genes Dev. **12:** 1260–1268.
13. Gohda, J. *et al.* 2005. RANK-mediated amplification of TRAF6 signaling leads to NFATc1 induction during osteoclastogenesis. EMBO J. **24:** 790–799.
14. Naito, A. *et al.* 1999. Severe osteopetrosis, defective interleukin-1 signalling and lymph node organogenesis in TRAF6-deficient mice. Genes Cells **4:** 353–362.
15. Franzoso, G. *et al.* 1997. Requirement for NF-κB in osteoclast and B-cell development. Genes Dev. **11:** 3482–3496.
16. Iotsova, V. *et al.* 1997. Osteopetrosis in mice lacking NF-κB1 and NF-κB2. Nat. Med. **3:** 1285–1289.
17. Grigoriadis, A.E. *et al.* 1994. c-Fos: a key regulator of osteoclast-macrophage lineage determination and bone remodeling. Science **266:** 443–448.
18. Takayanagi, H. *et al.* 2002. Induction and activation of the transcription factor NFATc1 (NFAT2) integrate RANKL signaling for terminal differentiation of osteoclasts. Dev. Cell **3:** 889–901.
19. Ishida, N. *et al.* 2002. Large scale gene expression analysis of osteoclastogenesis in vitro and elucidation of NFAT2 as a key regulator. J. Biol. Chem. **277:** 41147–41156.
20. Ikeda, F. *et al.* 2004. Critical roles of c-Jun signaling in regulation of NFAT family and RANKL-regulated osteoclast differentiation. J. Clin. Invest. **114:** 475–484.
21. Matsuo, K. *et al.* 2004. Nuclear factor of activated T-cells (NFAT) rescues osteoclastogenesis in precursors lacking c-Fos. J. Biol. Chem. **279:** 26475–26480.
22. de la Pompa, J.L. *et al.* 1998. Role of the NF-ATc transcription factor in morphogenesis of cardiac valves and septum. Nature **392:** 182–186.
23. Ranger, A.M. *et al.* 1998. The transcription factor NF-ATc is essential for cardiac valve formation. Nature **392:** 186–190.
24. Asagiri, M. *et al.* 2005. Autoamplification of NFATc1 expression determines its essential role in bone homeostasis. J. Exp. Med. **202:** 1261–1269.
25. Winslow, M.M. *et al.* 2006. Calcineurin/NFAT signaling in osteoblasts regulates bone mass. Dev. Cell **10:** 771–782.
26. Kim, N., M. Takami, J. Rho, *et al.* 2002. A novel member of the leukocyte receptor complex regulates osteoclast differentiation. J. Exp. Med. **195:** 201–209.
27. Koga, T. *et al.* 2004. Costimulatory signals mediated by the ITAM motif cooperate with RANKL for bone homeostasis. Nature **428:** 758–763.
28. Mocsai, A. *et al.* 2004. The immunomodulatory adapter proteins DAP12 and Fc receptor γ-chain (FcRγ) regulate development of functional osteoclasts through the Syk tyrosine kinase. Proc. Natl. Acad. Sci. USA **101:** 6158–6163.
29. Mao, D., H. Epple, B. Uthgenannt, *et al.* 2006. PLCγ2 regulates osteoclastogenesis via its interaction with ITAM proteins and GAB2. J. Clin. Invest. **116:** 2869–2879.
30. Sato, K. *et al.* 2006. Regulation of osteoclast differentiation and function by the CaMK-CREB pathway. Nat. Med. **12:** 1410–1416.
31. Matsumoto, M. *et al.* 2004. Essential role of p38 mitogen-activated protein kinase in cathepsin K gene expression during osteoclastogenesis through association of NFATc1 and PU.1. J. Biol. Chem. **279:** 45969–45979.
32. Crotti, T.N. *et al.* 2006. NFATc1 regulation of the human β3 integrin promoter in osteoclast differentiation. Gene **372:** 92–102.

33. KIM, K. *et al*. 2005. Nuclear factor of activated T cells c1 induces osteoclast-associated receptor gene expression during tumor necrosis factor-related activation-induced cytokine-mediated osteoclastogenesis. J. Biol. Chem. **280:** 35209–35216.

34. KIM, Y. *et al*. 2005. Contribution of nuclear factor of activated T cells c1 to the transcriptional control of immunoreceptor osteoclast-associated receptor but not triggering receptor expressed by myeloid cells-2 during osteoclastogenesis. J. Biol. Chem. **280:** 32905–32913.

35. YAGI, M. *et al*. 2005. DC-STAMP is essential for cell-cell fusion in osteoclasts and foreign body giant cells. J. Exp. Med. **202:** 345–351.

36. LEE, S.H. *et al*. 2006. v-ATPase V0 subunit d2-deficient mice exhibit impaired osteoclast fusion and increased bone formation. Nat. Med. **12:** 1403–1409.

Mechanisms of Osteoclastic Secretion

HAIBO ZHAO AND F. PATRICK ROSS

Department of Pathology and Immunology, Washington University School of Medicine, St. Louis, Missouri, USA

ABSTRACT: A tight balance between bone resorption by osteoclasts and bone formation by osteoblasts is required for the maintenance of bone mass and integrity. A net increase in bone resorption over formation results in osteoporosis, a disease associated with significantly morbidity and mortality. Following attachment via the integrin $\alpha v\beta 3$, osteoclasts degrade bone by generation of the ruffled border, the unique resorptive organelle of the cell. The adherent cell then secretes into the subcellular space protons and acidic proteases. We review here the concepts relating to the mechanisms of regulated secretion and provide preliminary data on the role of one protein important for secretion by osteoclasts.

KEYWORDS: osteoclasts; ruffled border; secretion

MECHANISM OF BONE RESORPTION

The skeleton is a dynamic organ that undergoes renewal throughout life by resorption of old bone by osteoclasts followed by new bone formation by osteoblasts. A tight balance between bone resorption and bone formation is required for the maintenance of bone mass and integrity. Enhanced osteoclastic bone resorption leads to reduced bone mass and compromised bone strength, leading to an increased risk of fractures, as seen in osteoporosis.

Osteoclasts are multinucleated cells formed by fusion of mononuclear precursors of the monocyte/macrophage lineage. Under the action of M-CSF and RANKL, the two major osteoclastogenic cytokines, precursors undergo transcriptional programming leading to the expression of a number of functionally important proteins that endow the cells with efficient and unique machinery for bone resorption.[1] Attachment of osteoclasts to bone induces formation of the "actin ring" at the sealing zone. This structure surrounds a specialized plasma membrane domain called the ruffled border, the unique resorptive organelle of the osteoclast that is generated by fusion of secretory vesicles with the bone-apposing plasma membrane. As a result, protons and lysosomal enzymes are

Address for correspondence: Haibo Zhao, Washington University School of Medicine, Department of Pathology and Immunology, Campus Box 8118, 660 South Euclid Avenue, St. Louis, MO 63110. Voice: 314-454-8588. fax: 314-454-5088.
hzhao@wustl.edu

Ann. N.Y. Acad. Sci. 1116: 238–244 (2007). © 2007 New York Academy of Sciences.
doi: 10.1196/annals.1402.058

vectorially secreted into the resorption lacuna to dissolve bone mineral and digest organic matrix, respectively.[2]

During the past few years, genetic studies in man and rodents have unveiled a number of molecules that mediate osteoclastic bone resorption. The integrin $\alpha v \beta 3$ and c-src are critical for osteoclast cytoskeleton organization and actin-ring formation.[3,4] Carbonic anhydrase II, the a3 subunit of vacuolar proton pump, the CLC-7 chloride channel and OSTM1 (also called gray lethal and identified recently as the β subunit/chaperone of CLC-7), are components of essential machinery that regulates secretion of hydrochloric acid, thus dissolving the inorganic components of bone.[5-7] Cathepsin K is a major lysosomal enzyme critical for bone degradation, with its preferred target being type I collagen, the major protein in bone.[1] Importantly, the a3 subunit of the vacuolar H^+ ATPase, CLC-7, and OSTM1 are all lysosomal membrane proteins that translocate to the ruffled border in resorbing osteoclasts. Taken together, these data indicate that the secretory vesicles fusing with the plasma membrane of the osteoclast to form its ruffled border are of lysosomal origin or represent lysosome-related organelles. Consistent with this assumption, inhibition of the expression of the small GTPase Rab7, which regulates specifically late endosomal/lysosomal vesicular trafficking, impairs ruffled border formation and osteoclast secretion.[8] In a related report, mice lacking Rab3D as a result of genetic deletion, exhibit increased bone mass; furthermore, localization of GFP-labeled Rab3D reveals specific intracellular vesicles of unknown composition or function.[9] Despite this progress in understanding how the osteoclast functions, the molecules modulating the transportation and fusion of secretory vesicles with the plasma membrane have not been identified.

MECHANISMS OF EXOCYTOSIS

In eukaryotic cells, exocytosis is a fundamental process through which cells release hydrophilic secretory products (such as neurotransmitters, hormones, and enzymes) to the extracellular space, or translocate specific functional proteins (including receptors, channels, transporters, and pumps) into the plasma membrane. There are two major types of exocytosis, constitutive versus regulated. In constitutive exocytosis, which requires no cell stimulation, secretory proteins are delivered to the endoplasmic reticulum, sorted and packaged in the Golgi complex, and ultimately discharged at the cell surface.[10] In regulated exocytosis, secretory proteins are stored in the cytoplasmic organelles after leaving the Golgi. Upon stimulation, these vesicles move toward and fuse with the plasma membrane.[11]

In spite of obvious difference in their regulatory mechanisms, the common final step in both secretory pathways is the fusion of exocytotic vesicles with the plasma membrane. Recent advances indicate that intracellular membrane fusion is mediated by v- (vesicular) and t- (target) SNAREs (soluble

N-ethylmaleimide-sensitive fusion protein (NSF) attachment protein (SNAP) receptors). These proteins assemble to form a four-helix bundle that closely juxtaposes the vesicle and target membranes.[12] Although SNARE proteins are sufficient to catalyze membrane fusion, additional proteins are required to achieve the specificity and special/temporal control of SNARE-mediated fusion. One of these regulators is the synaptotagmin (Syt) family of vesicular trafficking proteins.

To date 15 Syt isoforms have been identified in mammalian cells. These proteins are characterized by a short intraluminal region, a single transmembrane domain followed by a large cytoplasmic portion containing two copies of a C2 domain, which are homologous to the regulatory C2 motif that confers calcium and phospholipid binding in protein kinase C. Each Syt family member has a distinct tissue distribution and different calcium- and phospholipid-binding affinities. Thus, while Syts I, II, III, V, and X are expressed predominantly in the nervous system and neuronal endocrine cells, others are ubiquitous.[13,14]

The function of Syt family proteins in the regulation of exocytosis has been well characterized in synaptic vesicle secretion in neurons, large dense-core vesicle secretion in neuronal endocrine PC12 cells, and insulin secretion in pancreatic β cells. In all three models, and probably in all other cells, more than one Syt isoform is localized in the same vesicle. Although several isoforms have been shown to associate with synaptic vesicles, a recent functional study found out that only Syt I, II, and IX are essential for fast, synchronized neuronal transmission.[15] Similarly, Syt I, VII, and IX have been reported to localize at and regulate large dense-core vesicle secretion in PC12 cells.[16] Finally, Syt III, V, VII are localized at the insulin-containing vesicles in β cells.[17]

Unlike Syt I, which is restricted to the membrane of synaptic vesicles, Syt VII has a broad tissue distribution.[18] It is targeted to and regulates calcium-dependent exocytosis of lysosomes in fibroblasts and neurons, dense-core vesicles in neuroendocrine PC12 cells, insulin-containing secretory granules in pancreatic islet β-cells, lytic granules/secretory lysosomes in cell toxic lymphocytes, and lysosomes/phagosomes in macrophages.[19] Since macrophages are osteoclast precursors, Syt VII is a candidate to regulate secretory activity in the exclusive bone resorptive cell.

ROLE OF SYT IN OSTEOCLAST BIOLOGY

Given its expression pattern and role in other cells types, we hypothesized that Syt VII plays an important role in osteoclast function. To test this hypothesis, we characterized the bone phenotypes of Syt VII$^{-/-}$ mice and find that the knockouts have a trend to smaller size at all ages but with no overt developmental abnormality in the skeleton. Dual energy X-ray absorptiometry

(DEXA) analysis of 2-month-old wild-type and Syt VII$^{-/-}$ male mice revealed that Syt VII$^{-/-}$ mice have a consistent decrease in bone mineral density and that serum levels of CTx, a biomarker of global bone resorption, were significantly decreased in knockout animals. Taken together, these data indicate that the observed lower bone mass in Syt VII$^{-/-}$ mice results from a decrease of turnover in bone remodeling. To confirm these findings we performed detailed histomorphometric analyses of sections of tibiae from 2-month-old male mice. There was an approximately twofold reduction in trabecular volume in Syt VII$^{-/-}$ mice relative to wild-type littermate controls. These data suggest but do not prove that Syt VII specifically regulates secretion by the osteoclast.

To determine whether the bone phenotype observed in Syt VII$^{-/-}$ mice is due to an autonomous defect in osteoclasts we turned to *in vitro* culture models. Bone marrow macrophages (BMMs) from wild-type and Syt VII$^{-/-}$ mice were cultured with RANKL and M-CSF for 5 days. At different stages, cells were fixed and stained for tartrate-resistant acid phosphatase (TRAP), a widely used marker of osteoclast differentiation. The expression of several osteoclast marker genes was also detected by RT-PCR and Western blot analysis at each stage. The total number of multinucleated TRAP$^+$ osteoclasts, mRNA expression of cathepsin K, MMP-9, and β_3-integrin and protein levels of c-src and cathepsin K are all indistinguishable in wild-type and Syt VII$^{-/-}$ cells as they undergo osteoclastogenesis. These data demonstrate that Syt VII does not regulate osteoclast differentiation.

Next, we examined several aspects of osteoclast function. Like wild-type cells, Syt VII$^{-/-}$ osteoclasts cultured on bone slices form characteristic actin-rings, suggesting that the activating signals and cytoskeleton organization are intact in knockout cells. However, the secretion of cathepsin K into the resorption lacuna, shown by dual staining of cathepsin K and F-actin, is greatly reduced in Syt VII$^{-/-}$ osteoclasts compared with wild-type cells. While the resorption pits in wild-type cultures are deep and well defined, those in Syt VII$^{-/-}$ osteoclast cultures are shallow and irregular. Consistent with these morphologic findings, the level of CTx, the type I collagen fragments cleaved by cathepsin K during bone resorption, is markedly decreased in Syt VII$^{-/-}$ culture medium.

We then asked whether Syt VII regulates lysosome secretion in osteoclasts as it does in other cell types. We first examined the localization of GFP-tagged Syt VII in osteoclasts by laser confocal microscopy. In osteoclasts cultured on glass coverslips, Syt VII is co-localized with lysosomal membrane protein, LAMP2, in perinuclear lysosomes. In contrast, on bone Syt VII is found mainly at the ruffled border membrane circumscribed by actin-rings, where it also costained with LAMP2 and cathepsin K. Next, we fractionated the cytosol of mature osteoclasts by self-generated iodoxanol density gradient ultracentrifugation, followed by Western blot analysis to determine protein distribution. The peak of active cathepsin K (the 27 kDa isoform) is located in the lysosomal fraction containing LAMP-2, whereas the 36 kDa proform of cathepsin K comigrates with the trans-Golgi network marker, syntaxin 6. Interestingly, Syt VII and Rab7 and TI-VAMP, all lysosomal proteins, were in the same fraction. Last, coimmunoprecipitation experiments, demonstrated that Syt VII and TI-VAMP form a complex in osteoclasts. Therefore, Syt VII regulates the SNARE-mediated fusion of secreted lysosomes with the plasma membrane in osteoclasts.

To confirm the cell autonomous effects observed in our earlier studies, we constructed a retroviral vector expressing N-terminal triple-FLAG tagged Syt VII. Transduction of recombinant virus into Syt VII$^{-/-}$ BMMs, and their use as osteoclast precursors, demonstrated that reintroduction of the protein rescued bone resorption completely, as demonstrated by cathepsin K secretion, pit formation, and medium CTx levels.

In this study, we have provided evidence suggesting that Syt VII is an important player regulating lysosome secretion and insertion of lysosomal membrane proteins at the ruffled border in osteoclasts. At the moment, it is unclear whether a3, CLC7-7/OSTM1, cathepsin K, and Rab7 reach the ruffled border via the same or distinct populations of secretory lysosomes, nor is it known which of these vesicles may be associated with and modulated by Syt VII.

An RT-PCR screening of mRNA expression of all 15 isoforms in osteoclasts revealed that in addition to Syt VII a number of isoforms are expressed (Zhao H, unpublished data). It is likely that some of these isoforms are also localized at lysosomes and are partially functionally redundant with Syt VII in lysosomal secretion by osteoclasts. Indeed, Syt VII has been shown to cluster different Syt isoforms by its C2A and C2B domains in response to Ca^{2+}. The characterization of subcellular localization and functions of other Syt isoforms in osteoclasts merit further investigation.

Osteoclast activity is regulated systemically and locally by hormones, growth factors, cell–matrix and cell–cell interactions, and mechanical stress. This raises the question as to how Syt VII is modulated by signaling pathways downstream of these factors. Syts act as calcium sensors in triggering exocytosis and their C2A and C2B domains have been shown to bind calcium and are functionally important for Syt-mediated exocytosis. Ligation of the integrin $\alpha v \beta 3$, which is essential for osteoclast activation and function, increases cytosolic calcium to μM level,[20] which is within the scope of calcium-binding affinity of Syt VII. In addition, it is only when cultured on mineralized matrix but not on plastic or cartilage that osteoclasts undergo dramatic secretion and ruffled border formation. Therefore, Syt VII may sense and bind increased intracellular calcium and promote exocytosis upon osteoclast activation. Accumulating evidence indicates that the phosphorylation of Syts also modulates their function.[21] The activation of osteoclasts by RANKL, M-CSF, and the integrin $\alpha v \beta 3$ triggers a series of phosphorylation events. Thus, it will be interesting to determine if Syt VII is phosphorylated in osteoclasts.

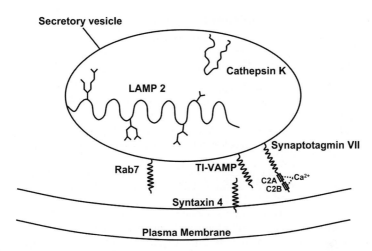

FIGURE 1. Model for regulated secretion by osteoclasts. Osteoclasts bound to bone generate a signal arising from activation of a cytokine receptor or integrin (not shown), leading to an increase in intracellular calcium. The raised levels of calcium bind to the C-terminal C2 domains of synaptotagmin VII, stimulating interactions between v-SNARES, such as TI-VAMP and t-SNARES, including syntaxin 4. The consequent close approximation of vesicular and plasma membranes leads to their fusion and hence secretion of cathepsin K into the resorptive space. Rab7 and perhaps Rab3D[9] play a role in targeting secretory vesicles to the bone-apposed membrane surface.

SUMMARY AND CONCLUSION

In summary, we have provided the novel insight that the secretory vesicles used by osteoclasts for bone resorption, are lysosomes or lysosome-related organelles. Syt VII is associated with these vesicles and regulates their exocytosis. Lack of synaptotagmin VII inhibits cathepsin K secretion in osteoclasts and therefore Syt VII plays an important role in bone biology (FIG. 1).

REFERENCES

1. TEITELBAUM, S.L. & F.P. ROSS. 2003. Genetic regulation of osteoclast development and function. Nat. Rev. Genet. **4:** 638–649.
2. VAANANEN, H.K., H. ZHAO, M. MULARI, *et al.* 2000. The cell biology of osteoclast function. J. Cell Sci. **113:** 377–381.
3. FENG, X., D.V. NOVACK, R. FACCIO, *et al.* 2001. A Glanzmann's mutation in b3 integrin specifically impairs osteoclast function. J. Clin. Invest. **107:** 1137–1144.
4. MIYAZAKI, T., A. SANJAY, L. NEFF, *et al.* 2004. Src kinase activity is essential for osteoclast function. J. Biol. Chem. **279:** 17660–17666.
5. SLY, W.S., D. HEWETT-EMMETT, M.P. WHYTE, *et al.* 1983. Carbonic anhydrase II deficiency identified as the primary defect in the autosomal recessive syndrome

of osteopetrosis with renal tubular acidosis and cerebral calcification. Proc. Natl. Acad. Sci. USA **80:** 2752–2756.

6. LI, Y.P., W. CHEN, Y. LIANG, *et al.* 1999. Atp6i-deficient mice exhibit severe osteopetrosis due to loss of osteoclast-mediated extracellular acidification. Nat. Genet. **23:** 447–451.

7. LANGE, P.F., L. WARTOSCH, T.J. JENTSCH, *et al.* 2006. ClC-7 requires Ostm1 as a beta-subunit to support bone resorption and lysosomal function. Nature **440:** 220–223.

8. ZHAO, H., T. LAITALA-LEINONEN, V. PARIKKA, *et al.* 2001. Downregulation of small GTPase Rab7 impairs osteoclast polarization and bone resorption. J. Biol. Chem. **276:** 39295–39302.

9. PAVLOS, N.J., J. XU, D. RIEDEL, *et al.* 2005. Rab3D regulates a novel vesicular trafficking pathway that is required for osteoclastic bone resorption. Mol. Cell. Biol. **25:** 5253–5269.

10. HOWELL, G.J., Z.G. HOLLOWAY, C. COBBOLD, *et al.* 2006. Cell biology of membrane trafficking in human disease. Int. Rev. Cytol. **252:** 1–69.

11. CHIEREGATTI, E. & J. MELDOLESI. 2005. Regulated exocytosis: new organelles for non-secretory purposes. Nat. Rev. Mol. Cell Biol. **6:** 181–187.

12. JAHN, R. & R.H. SCHELLER. 2006. SNAREs–engines for membrane fusion. Nat. Rev. Mol. Cell Biol. **7:** 631–643.

13. CHAPMAN, E.R. 2002. Synaptotagmin: a Ca2+ sensor that triggers exocytosis? Nat. Rev. Mol. Cell Biol. **3:** 498–508.

14. SUDHOF, T.C. 2002. Synaptotagmins: why so many? J. Biol. Chem. **277:** 7629–7632.

15. XU, J., T. MASHIMO & T.C. SUDHOF. 2007. Synaptotagmin-1, -2, and -9: Ca(2+) sensors for fast release that specify distinct presynaptic properties in subsets of neurons. Neuron **54:** 567–581.

16. TUCKER, W.C., J.M. EDWARDSON, J. BAI, *et al.* 2003. Identification of synaptotagmin effectors via acute inhibition of secretion from cracked PC12 cells. J. Cell Biol. **162:** 199–209.

17. GUT, A., C. KIRALY, M. FUKUDA, *et al.* 2001. Expression and localisation of synaptotagmin isoforms in endocrine (β)-cells: their function in insulin exocytosis. J. Cell Sci. **114:** 1709–1716.

18. ANDREWS, N.W. & S. CHAKRABARTI. 2005. There's more to life than neurotransmission: the regulation of exocytosis by synaptotagmin VII. Trends Cell Biol. **15:** 626–631.

19. CZIBENER, C., N.M. SHERER, S.M. BECKER, *et al.* 2006. Ca2+ and synaptotagmin VII-dependent delivery of lysosomal membrane to nascent phagosomes. J. Cell Biol. **174:** 997–1007.

20. ZIMOLO, Z., G. WESOLOWSKI, H. TANAKA, *et al.* 1994. Soluble avb3-integrin ligands raise [Ca2+]i in rat osteoclasts and mouse-derived osteoclast-like cells. Am. J. Physiol. **266:** C376–C381.

21. LEE, B.H., X. MIN, C.J. HEISE, *et al.* 2004. WNK1 phosphorylates synaptotagmin 2 and modulates its membrane binding. Mol. Cell. **15:** 741–751.

New Roles for Osteoclasts in Bone

BRENDAN F. BOYCE,[a,b] ZHENQIANG YAO,[a] QIAN ZHANG,[a]
ROULIN GUO, [a] YAN LU,[a] EDWARD M. SCHWARZ,[b]
AND LIANPING XING[a,b]

[a]Department of Pathology and Laboratory Medicine, University of Rochester
Medical Center, Rochester, New York, USA

[b]Center for Musculoskeletal Research, University of Rochester Medical Center,
Rochester, New York, USA

ABSTRACT: Osteoclasts have long been recognized as the cells that resorb bone in normal bone remodeling and in pathologic conditions in
which bone resorption is increased. They are derived from precursors in
the mononuclear phagocyte lineage, which arise in the bone marrow and
fuse with one another to form the multinucleated cells that resorb calcified matrixes under the influence of osteoblastic cells in bone marrow.
There is growing evidence that osteoclast precursors (OCPs) and osteoclasts have functions in and around bone other than bone resorption.
For example, they modulate the differentiation of other cells, including
osteoblastic cells; they regulate hematopoietic stem cell movement from
the bone marrow to the bloodstream; and they are secretory cells that
participate in immune responses. In this article, we review these findings,
which support new roles for osteoclasts and OCPs in the growing field of
osteoimmunology and in common pathologic conditions affecting bones
and joints.

KEYWORDS: osteoclast; cytokine; osteoblast; bone resorption; inflammatory arthritis

INTRODUCTION

Understanding of the molecular mechanisms that regulate osteoclast formation, differentiation, activity, and survival, has advanced significantly in
the past decade since the discovery of receptor activator of NF-κB lig-
and (RANKL) and osteoprotegerin (OPG) (reviewed in Ref. 1). Osteoblas-
tic/stromal cells express these cytokines in response to cytokines, hormones, growth factors, and mechanical stresses and are the major cell type
regulating osteoclasts positively and negatively during bone modeling and

Address for correspondence: Dr. Brendan F. Boyce, Department of Pathology and Laboratory
Medicine, University of Rochester Medical Center, 601 Elmwood Ave, Box 626, Rochester, NY 14642.
Voice: 585-275-5837; fax: 585-273-3637.
Brendan_Boyce@urmc.rochester.edu

Ann. N.Y. Acad. Sci. 1116: 245–254 (2007). © 2007 New York Academy of Sciences.
doi: 10.1196/annals.1402.084

remodeling. RANKL signals through its receptor, RANK, in osteoclast precursors (OCPs)[1] and sequentially activates NF-κB, c-Fos, and NFATc1[2] to induce osteoclast formation and activation. This role for osteoblastic cells had been anticipated for many years since Rodan and Martin[3] proposed it, based on *in vitro* assays of bone marrow cells. However, recent reports have added an unexpected twist to the relationship between osteoclastic and osteoblastic cells. Consequently, osteoclastic cells are now being recognized not only as bone-resorbing cells, but also as cells with additional functions. These include regulation of the function of other cell types, such as osteoblastic cells, secretion of cytokines, and roles as immune cells in inflammatory bone diseases. These findings suggest that inhibition of some osteoclast functions could have previously unexpected detrimental or beneficial effects on bone mass and strength.

RECENTLY IDENTIFIED OSTEOCLASTIC–OSTEOBLASTIC INTERACTIONS

Previous studies have determined that direct osteoclastic–osteoblastic interaction is required *in vitro* for most cytokine or hormone-induced osteoclast formation (reviewed in Ref. 4) through upregulation of RANKL expression on the surface of cytoplasmic extensions of osteoblastic cells. This was considered previously to be a one-way regulatory relationship between these two cell types, with osteoblastic cells having a dominant role. However, recent studies indicate that osteoclastic cells are not entirely enslaved by osteoblastic cells and that they can directly regulate osteoblastic cells.[5] For example, studies by Zhao *et al.*[6] showed that in addition to the direct contact between these cell types through RANKL–RANK interaction, there is also direct contact through ephrinB2, a ligand expressed by OCPs, and its receptor, EphB4, on osteoblastic cells. Ephrin/Eph signaling is required during embryonic development for numerous processes, including arterial–venous link up and neuronal axon pathfinding in which cell processes can extend over relatively large distances (reviewed in Ref. 7).

Osteoblastic stromal cells, like osteocytes within bone, have extensive dendritic processes extending from their cell bodies and thus they are able to interact with many hematopoietic cells, including OCPs, within the bone marrow cavity. Zhao *et al.* found that during interaction between osteoclastic and osteoblastic cells, typical ligand–receptor signaling through the EphB4 receptor in osteoblastic cells promoted osteoblast differentiation from precursors, while so-called reverse signaling through the ephrinB2 ligand back into OCPs inhibited OCP differentiation.[6] This inhibition occurs through a mechanism that downregulates c-Fos activation of NFATc1. The stimulatory pathway in osteoblastic cells has yet to be identified. These findings raise the possibility that activation of these pathways could lead to increased bone mass in individuals with diseases, such as rheumatoid arthritis and osteoporosis, which are characterized by increased bone resorption and decreased bone formation, by inhibiting bone resorption and stimulating bone formation.

More recently, Lee *et al.*[8] reported an unexpected negative regulatory role for osteoclasts in osteoblast differentiation. Using comparative mRNA expression profiling to examine how osteoclasts differ from macrophages, they identified a cDNA fragment that encodes a protein,[8] which is almost identical to Atp6v0d2. Atp6v0d2 is a subunit of v-ATPase, a component of the V-type H^+ ATP6i proton pump complex that secretes H^+ from osteoclasts.[9] Cl^- flow passively through the chloride channel, ClC-7, and along with H^+ forms HCl, which dissolves the mineral in bone matrix. Atp6v0d2 is highly expressed around the sealing zone associated with the ruffled borders of osteoclasts during resorption, but it is not expressed by osteoblasts. Lee *et al.* generated Atp6v0d2-deficient mice and found that the mice had increased bone mass as a consequence of two distinct mechanisms: the mice had defective fusion of OCPs associated with a reduction in both osteoclast size and bone resorption; however, but they also had increased numbers of osteoblasts and enhanced bone formation. Expression of DC-STAMP is required in OCPs for precursor fusion,[10] but the fusion defect was not due to reduced expression of DC-STAMP. Atp6v0d2 is not expressed by osteoblasts, and it is not known how it inhibits bone formation. However, its effects appear to be mediated by osteoblast-extrinsic factors. These findings suggest that osteoclasts or their precursors negatively regulate osteoblasts through the function of Atp6v0d2, perhaps through the effects of a protein secreted by them. The results also indicate that this subunit of the protein pump plays a positive role in OCP fusion. Like the studies by Zhao *et al.*, they also raise the possibility that a single therapeutic agent could be developed to inhibit osteoclast activity and increase bone formation, presumably by different mechanisms.

OSTEOCYTE REGULATION OF OSTEOCLASTS

Osteocytes are the most abundant osteoblastic cells, comprising >90% of bone cells. It has long been speculated that osteocytes play important roles in mineral homeostasis by controlling movement of ions and cations into and from the matrix around them into the fluid that fills the canalicular network within bone. Osteocyte dendritic processes extend throughout the bone around them and they interact with those of other osteocytes and with stromal cells within the adjacent bone marrow (reviewed in Ref. 11). Previous studies have suggested that viable osteocytes can inhibit osteoclast activation[12] and that apoptotic osteocytes may send signals to osteoblastic stromal cells in marrow to induce osteoclast formation.[13]

More recently, studies by Tatsumi *et al.*[14] to investigate the role of osteocytes in bone remodeling have provided more definitive evidence that these cells regulate osteoclast formation and activation through upregulation of RANKL expression. They generated transgenic mice that express the receptor for diphtheria toxin (DT) specifically in osteocytes and found that RANKL mRNA levels increased in their bone marrow cells 2 days after injecting the mice with DT. Six days later, 70–80% of osteocyte lacunae in bone were empty as a result of DT-induced apoptosis. The authors concluded that dying osteocytes sent osteoclastogenic signals through their dendritic processes that lead to increased RANKL expression by osteoblastic cells in the bone marrow. Dramatic bone loss was observed in the mice 40 days after DT injection, and this was followed within 90 days of injection by new bone formation by osteoblasts to replace the lost bone. Furthermore, DT injected 1 day before mice were unloaded by tail suspension prevented the bone loss seen in control mice, providing further evidence that viable osteocytes negatively regulate osteoclastic bone resorption. In this experiment, it appears that DT quickly induced osteocyte apoptosis or at least prevented osteocytes from sending unloading-induced signals to induce osteoclast formation. The mechanism(s) whereby dying osteocytes signal presumably to osteoblastic cells to induce osteoclast activation and viable osteocytes limit osteoclast formation have yet to be established. Nevertheless, these elegant studies by Tatsumi et al. support a growing role proposed for osteocytes in the regulation of osteoblastic cells and bone mass[11] and strengthen the proposal that osteocytes are potential targets for therapeutic intervention in the prevention of bone loss and the maintenance of bone mass.

OSTEOCLASTS AS IMMUNE CELLS

Understanding of the mechanisms regulating osteoclastogenesis has been facilitated by the development of in vitro techniques to obtain relatively large and pure populations of OCPs from bone marrow macrophages or splenocytes (reviewed in Ref. 15). These techniques have facilitated the identification of OCPs in bone marrow, blood, and spleens of mice and humans using antibodies to CD11b and Gr-1, and to other precursor markers, such as c-fms (the receptor for c-Fos) and RANK.[16–18] Using these techniques, others and we have demonstrated that OCP numbers are increased in the peripheral blood of humans[19] and mice[16–20] with inflammatory arthritis and high blood levels of TNF and that treatment with anti-TNF therapy reverses this increase. OCPs express CXCR4, the receptor for stromal cell-derived growth factor (SDF-1), a chemokine that regulates the movement of hematopoietic cells in and out of bone marrow and other sites (reviewed in Ref. 21). Cytokines, such as TNF, regulate the expression of SDF-1 by bone marrow stromal cells (our unpublished observations) and in this way can promote OCP egression from the marrow into the bloodstream from where they can concentrate at sites of inflammation in and around bones.[21] OCPs not only respond to TNF, but they and osteoclasts also secrete TNF and other cytokines, such as IL-1 and IL-6, in response to TNF.[22] These direct effects of TNF on OCPs occur *in vitro* in the absence of RANKL. They could provide a mechanism in addition to TNF's established indirect action to increase osteoclastogenesis through RANKL expression by accessory cells.[23] This could further induce osteoclast formation at sites in bone where TNF levels are increased, such as in rheumatoid joints and around affected teeth of patients with periodontal disease. Thus, through an autocrine mechanism, TNF could induce an autoamplifying vicious cycle at such sites to enhance osteoclast formation.[21]

TNF also induces expression in OCPs of c-Fos,[21] a transcription factor whose expression is required for osteoclast differentiation downstream of RANKL/RANK/NF-κB signaling.[21] Furthermore, treatment of OCPs overexpressing retroviral c-Fos with TNF induces increased resorptive activity of osteoclasts formed from them more than from cytokine-treated OCPs expressing retroviral GFP.[21] Thus, TNF could directly increase not only osteoclast formation, but also osteoclastic resorptive activity to more aggressively resorb bone at sites where TNF levels are increased.

We have found that osteoclast activity can be increased by other mechanisms in addition to those described above. For example, at sites of TNF-induced inflammation in TNF-transgenic mice, osteoclasts increase their expression of the lymphangiogenic factor, VEGF-C,[24] a member of the VEGF family of angiogenic proteins.[25] This increase in VEGF-C expression is associated with increased formation of lymphatic vessels around affected joints,[24,26] which likely enhances the immune response by increasing the flow of lymph and other inflammatory mediators from the affected joints. We have also found that VEGF-C enhances the resorptive activity, but not the formation of osteoclasts *in vitro*,[24] an effect that is not observed with VEGF-A or VEGF-D. The precise roles of VEGF-C secreted by OCPs will require further investigation, and it will be important to determine if inhibition of this function of OCPs in inflammatory arthritis is beneficial or detrimental to affected joints.

OSTEOCLAST PRECURSOR INTERACTION WITH BONE MATRIX INCREASES INTERLEUKIN-1 SECRETION

In studies examining the effects of c-Fos overexpression in OCPs, we observed that these cells differentiated spontaneously into osteoclasts when cultured on bone slices in the presence of M-CSF alone, but not on plastic (Ref. 27 and manuscript submitted). Furthermore, GFP-expressing OCPs cultured on bone slices in the upper chamber of a transwell culture system induced the formation of osteoclasts from c-Fos-expressing osteoclasts cultured below on plastic. This effect was inhibited by an IL-1 receptor antagonist, indicating that osteoclast formation was induced by IL-1 released by the OCPs in response to their interaction with bone matrix. IL-1 is known to increase osteoclast activation[28] in part at least by prolonging osteoclast survival[29] and higher concentrations of bisphosphonates are required to induce osteoclast apoptosis when the cells are cultured on bone slices compared to plastic.[30] These findings provide further evidence that osteoclasts can enhance their activity by interacting with bone matrix and that inhibition of IL-1 signaling in osteoclasts and OCPs is another therapeutic approach to reduce bone loss.

Osteoclast formation and activity are regulated negatively predominantly by OPG secreted by osteoblastic stromal cells (reviewed in Ref. 23) and the RANKL/OPG ratio is a major determinant of bone resorption and bone mass. However, osteoclasts and OCPs can also negatively regulate their formation directly in response to RANKL. For example, although RANKL activation of c-Fos induces NFATc1 activation in OCPs, it also induces OCP secretion of interferon-β, which binds to its receptor on OCPs.[31] This leads to degradation of tumor necrosis factor receptor associated protein (TRAF) 6, which is an adaptor molecule that mediates RANKL/RANK-induced signaling because RANK does not have intrinsic kinase activity (reviewed in Ref. 23).

TNF can also negatively regulate osteoclast formation directly, but it appears to do this through a different mechanism from RANKL. Both RANKL and TNF activate NF-κB, c-Fos, and NFATc1 in OCPs to induce osteoclast formation directly, but TNF induces fewer osteoclasts than RANKL *in vitro.*[21] Both of these cytokines activate the canonical and noncanonical NF-κB pathways in OCPs. The former pathway is activated by phosphorylation and subsequent degradation of inhibitory kappa kinase β (reviewed in Ref. 32), which frees NF-κB p65 and p50 dimers to translocate to the nucleus where they induce expression of osteoclastogenic genes. The noncanonical pathway is activated by phosphorylation and degradation of the inhibitory NF-κB protein, p100. When not activated, p100 is bound to the NF-κB protein, RelB. Upon activation, p100 is degraded. A p52 fragment of p100 is released as a degradation product and it binds to RelB. p52/RelB dimers then translocate to the nucleus to induce gene expression.

We have found that TNF upregulates expression of NF-κB p100 in OCPs, while RANKL does not.[33] We speculated that this increase in p100 expression was responsible for the reduced induction of osteoclast formation by TNF compared to RANKL. To examine this hypothesis, we treated OCPs from wild-type (wt) and NF-κB p100-deficient mice with RANKL and TNF. We found that TNF induced similar numbers of osteoclasts from the p100-/- OCPs as RANKL, but significantly fewer from wt cells than RANKL. These findings suggest that TNF induction of osteoclast formation is limited directly in OCPs by this p100-mediated mechanism. Thus, NF-κB p100 may function as a negative regulator of osteoclastogenesis by controlling proliferation of OCPs to limit excessive bone resorption mediated by TNF in various bone diseases and is a potential target for therapeutic intervention.

OSTEOCLASTS AS REGULATORS
OF HEMATOPOIETIC CELL FUNCTION

Osteoblastic cells on endosteal surfaces in mice are known to support hematopoietic stem cell (HSCs), through signals that maintain self-renewal potential in undifferentiated stem cells (reviewed in Ref. 34). Some HSCs reside at the endosteal surfaces in so-called niches from where they leave the marrow and enter the blood. Interestingly, PTH positively regulates this process[35] through a cyclic AMP-Jagged-1-dependent mechanism[36] that promotes the egression of hematopoietic stem and progenitor cells (HSPCs) from the bone marrow into the bloodstream.

Osteoclasts are hematopoietic cells derived from myeloid lineage mononuclear precursors. Failure of osteoclast formation or function leads to the development of osteopetrosis in humans and other mammals (reviewed in Ref. 37). Osteopetrosis is characterized by the failure of osteoclasts to resorb bone matrix formed in the metaphyses of long bones during endochondral ossification. This failure results in variable degrees of filling of the marrow cavity of long bones by unresorbed bone and cartilage, associated with a reduction in the volume of hematopoietic marrow and extramedullary hematopoiesis. Consequently, subjects with osteopetrosis are typically anemic and have impaired immune responses with an increased propensity to develop infections, which can be fatal. However, a direct role for osteoclasts in HSC mobilization had not been anticipated.

A recent study by Kollet *et al.*[38] has identified an unexpected role for osteoclasts in this HSC egression. They knew that stress situations, such as inflammation and chemotherapy, or treatment with granulocyte colony-stimulating factors, induce massive stem cell mobilization from the marrow through activation of proteolytic enzymes, which break adhesions between stem cells and their bone marrow microenvironment (reviewed in Ref. 39). They noted that HSC mobilization was associated with increased numbers of osteoclasts on bone surfaces and found that RANKL-induced increased osteoclastogenesis *in vivo* was accompanied by stem and progenitor cell mobilization along with increased osteoclast expression of MMP-9 and cathepsin K. These enzymes not only degrade proteins in calcified matrices, they also cleave membrane-bound kit ligand, a growth and adhesion factor for HSCs. In addition, RANKL decreased the expression of osteoblast kit ligand and osteopontin, which also affects stem cell numbers. The authors examined mice deficient in the protein tyrosine phosphatase-epsilon (PTP ε), which have dysfunctional osteoclasts and mild osteopetrosis and found that RANKL did not mobilize stem and progenitor cells in these mice. Thus, these findings define a role for osteoclast activation in HSC mobilization.

These data now link osteoblasts and HSCs with osteoclasts in normal and pathologic bone remodeling. As noted earlier, RANKL expression increased in response to inflammatory cytokines in inflammatory and other bone disease in which bone resorption is increased. As outlined above, OCP numbers increase in blood in response to TNF and can give rise to osteoclasts at sites of inflammation in bone. Why HSCs leave the marrow cavity and circulate in the blood remains poorly understood, particularly when many remain *in situ*. Further studies will be required to determine their function and the role of normal and particularly increased bone remodeling in this process.

SUMMARY

Cells in the mononuclear phagocyte lineage have multiple functions in immune and other responses, and specific subsets of them, such as Kupffer cells, microglia, and pulmonary macrophages, reside in particular organs where they have specific functions. Until recently, osteoclasts and OCPs have been considered to have a very restricted function to resorb bone during bone remodeling. The reports of multiple additional functions for osteoclastic cells described above, some of them similar to those of monocytes, raise the possibility that they may have even more similarities and more as yet undiscovered functions. While profound suppression of bone resorption with increasingly potent bisphosphonates has been the goal of many clinical studies and pharmaceutical company chemists, the findings that osteoclasts and OCPs have additional roles in and around bone suggest that complete or near complete suppression of osteoclast formation or some of their functions could have unwanted adverse affects on osteoblastic cells or immune responses.

ACKNOWLEDGMENTS

Some of the work reported in this review is supported by NIH Grants AR43510 to BFB and AR48697 to LX and AR54041. We thank Ildiko Nagy for help with preparation of the article and Xiaoyun Zhang for technical assistance.

REFERENCES

1. ASAGIRI, M. & H. TAKAYANAGI. 2007. The molecular understanding of osteoclast differentiation. Bone **40:** 251–264.
2. YAMASHITA, T. *et al.* 2007. NF-kappaB p50 and p52 regulate receptor activator of NF-kappaB ligand (RANKL) and tumor necrosis factor-induced osteoclast precursor differentiation by activating c-Fos and NFATc1. J. Biol. Chem. **282:** 18245–18253.
3. RODAN, G.A. & T.J. MARTIN. 1981. Role of osteoblasts in hormonal control of bone resorption-a hypothesis. Calcif. Tiss. Int. **33:** 349–351.
4. MARTIN, T.J. & N.A. SIMS. 2005. Osteoclast-derived activity in the coupling of bone formation to resorption. Trends Mol. Med. **11:** 76–81.
5. BOYCE, B.F. & L. XING. 2006. Osteoclasts, no longer osteoblast slaves. Nat. Med. **12:** 1356–1358.

6. ZHAO, C. *et al.* 2006. Bidirectional ephrinB2-EphB4 signaling controls bone home-ostasis. Cell Metab. **4:** 111–121.

7. DAVY, A. & P. SORIANO. 2005. Ephrin signaling *in vivo*: look both ways. Dev. Dyn. **232:** 1–10.

8. LEE, S.H. *et al.* 2006. v-ATPase V0 subunit d2-deficient mice exhibit impaired osteoclast fusion and increased bone formation. Nat. Med. **12:** 1403–1409.

9. SUN-WADA, G.H. *et al.* 2003. Diversity of mouse proton-translocating ATPase: presence of multiple isoforms of the C, D and G subunits. Gene **302:** 147–153.

10. YAGI, M. *et al.* 2005. DC-STAMP is essential for cell-cell fusion in osteoclasts and foreign body giant cells. J. Exp. Med. **202:** 345–351.

11. BONEWALD, L.F. 2007. Osteocytes as dynamic, multifunctional cells. Ann. N.Y. Acad. Sci.

12. TOMKINSON, A. *et al.* 1998. The role of estrogen in the control of rat osteocyte apoptosis. J. Bone Miner. Res. **13:** 1243–1250.

13. VERBORGT, O., G.J. GIBSON, & M.B. SCHAFFLER. 2000. Loss of osteocyte integrity in association with microdamage and bone remodeling after fatigue *in vivo*. J. Bone Miner. Res. **15:** 60–67.

14. TATSUMI, S. *et al.* 2007. Targeted ablation of osteocytes induces osteoporosis with defective mechanotransduction. Cell Metab. **5:** 464–475.

15. SUDA, T. *et al.* 1999. Modulation of osteoclast differentiation and function by the new members of the tumor necrosis factor receptor and ligand families. Endocr. Rev. **20:** 345–357.

16. LI, P. *et al.* 2004. Systemic tumor necrosis factor alpha mediates an increase in peripheral CD11bhigh osteoclast precursors in tumor necrosis factor alpha-transgenic mice. Arthritis Rheum. **50:** 265–276.

17. JACQUIN, C. *et al.* 2006. Identification of multiple osteoclast precursor populations in murine bone marrow. J. Bone Miner. Res. **21:** 67–77.

18. HOROWITZ, M.C. & J.A. LORENZO. 2004. The origins of osteoclasts. Curr. Opin. Rheumatol. **16:** 464–468.

19. RITCHLIN, C.T. *et al.* 2003. Mechanisms of TNF-alpha- and RANKL-mediated osteoclastogenesis and bone resorption in psoriatic arthritis. J. Clin. Invest. **111:** 821–831.

20. YAO, Z. *et al.* 2006. Tumor necrosis factor-alpha increases circulating osteoclast precursor numbers by promoting their proliferation and differentiation in the bone marrow through up-regulation of c-Fms expression. J. Biol. Chem. **281:** 11846–11855.

21. BOYCE, B.F., E.M. SCHWARZ & L. XING. 2006. Osteoclast precursors: cytokine-stimulated immunomodulators of inflammatory bone disease. Curr. Opin. Rheumatol. **18:** 427–432.

22. O'KEEFE, R.J. *et al.* 1997. Osteoclasts constitutively express regulators of bone re-sorption: an immunohistochemical and *in situ* hybridization study. Lab. Investig. **76:** 457–465.

23. TAKAYANAGI, H. 2005. Mechanistic insight into osteoclast differentiation in os-teoimmunology. J. Mol. Med. **83:** 170–179.

24. GUO, R. *et al.* 2007. RANKL stimulates osteoclasts to release the lymphatic growth factor, VEGF-C, and enhances osteoclastic bone resorption through an autocrine mechanism J. Bone Miner. Res. **22:** S30.

25. TAMMELA, T. *et al.* 2005. The biology of vascular endothelial growth factors. Cardiovasc. Res. **65:** 550–563.

26. PROULX, S.T. *et al*. 2007. MRI and quantification of draining lymph node function in inflammatory arthritis. Ann. N.Y. Acad. Sci.
27. YAO, Z. *et al*. 2006. Osteoclast precursors induce their differentiation to osteoclasts by interacting with bone matrix and secreting cytokines. J. Bone Miner. Res. **21:** S262.
28. JIMI, E. *et al*. 1999. Interleukin 1 induces multinucleation and bone-resorbing activity of osteoclasts in the absence of osteoblasts/stromal cells. Exp. Cell Res. **247:** 84–93.
29. JIMI, E. *et al*. 1998. Activation of NF-κB is involved in the survival of osteoclasts promoted by Interleukin-1. J. Biol. Chem. **273:** 8799–8805.
30. HUGHES, D.E. *et al*. 1995. Bisphosphonates promote apoptosis in murine osteoclasts *in vitro* and *in vivo*. J. Bone Miner. Res. **10:** 1478–1487.
31. TAKAYANAGI, H. *et al*. 2002. RANKL maintains bone homeostasis through c-Fos-dependent induction of interferon-beta. Nature **416:** 744–749.
32. PERKINS, N.D. 2007. Integrating cell-signalling pathways with NF-kappaB and IKK function. Nat. Rev. Mol. Cell Biol. **8:** 49–62.
33. YAO, Z. 2007. NF-κB2 limits TNF-induced osteoclastogenesis through precursor cell cycle regulation. J. Bone Miner. Res. **22:** S42.
34. ADAMS, G.B. & D.T. SCADDEN. 2006. The hematopoietic stem cell in its place. Nat. Immunol. **7:** 333–337.
35. CALVI, L.M. *et al*. 2003. Osteoblastic cells regulate the haematopoietic stem cell niche. Nature **425:** 841–846.
36. WEBER, J.M. *et al*. 2006. Parathyroid hormone stimulates expression of the Notch ligand Jagged1 in osteoblastic cells. Bone **39:** 485–493.
37. TOLAR, J., S.L. TEITELBAUM & P.J. ORCHARD. 2004. Osteopetrosis. N. Engl. J. Med. **351:** 2839–2849.
38. KOLLET, O. *et al*. 2006. Osteoclasts degrade endosteal components and promote mobilization of hematopoietic progenitor cells. Nat. Med. **12:** 657–664.
39. COTTLER-FOX, M.H. *et al*. 2003. Stem cell mobilization. Hematol. Am. Soc. Hematol. Educ. Prog. 419–437.

Similarities and Contrasts in Ryanodine Receptor Localization and Function in Osteoclasts and Striated Muscle Cells

CHRISTOPHER L.-H. HUANG,[a] LI SUN,[b] JAMES A. FRASER,[a] ANDREW A. GRACE,[c] AND MONE ZAIDI[b]

[a]Physiological Laboratory, University of Cambridge, Cambridge, United Kingdom

[b]Mount Sinai Bone Program and Department of Medicine, Mount Sinai School of Medicine, New York, New York, USA

[c]Department of Biochemistry, University of Cambridge, Cambridge, United Kingdom

ABSTRACT: This review compares ryanodine receptor (RyR)-mediated Ca^{2+} signaling processes in muscle and osteoclast cells. In muscle, RyR-mediated release of an intracellularly stored, sarcoplasmic reticular (SR), Ca^{2+} is triggered by voltage-sensitive dihydropyridine receptor (DHPR)-L-type Ca^{2+} channels either through an allosteric coupling with the RyR in skeletal muscle or a Ca^{2+}-induced Ca^{2+} release initiated by extracellular Ca^{2+} entry in cardiac muscle. Both cell subtypes are nevertheless capable of Ca^{2+}-induced SR Ca^{2+} release with cardiac muscle additionally showing a store overload-induced Ca^{2+} release (SOICR) driven by SR luminal Ca^{2+} under some pathological conditions. Osteoclasts similarly show cytosolic Ca^{2+} elevations driven by release of intracellular Ca^{2+} stores that culminate in motile activity in turn modifying bone resorptive activity. However, such triggering is controlled by ambient Ca^{2+} rather than membrane potential with features strongly suggestive of control by a surface membrane Ca^{2+} receptor. Yet common actions of the RyR-specific agents perchlorate, dantrolene Na, ryanodine, caffeine, adenosine 3′,5′-cyclic diphosphate ribose (cADPr) and ruthenium red implicate RyR in signaling in all these cell types. These findings were reconciled by reports confirming and uniquely localizing a cell surface rather than microsomal osteoclastic RyR that might itself detect ambient Ca^{2+} possibly through its otherwise intraluminal positioned low-affinity Ca^{2+}-binding site in parallel with the SOICR mechanism in cardiac muscle. Such a mechanism could interact with other osteoclast processes transferring Ca^{2+} between cytosol, intracellular stores and extracellular space and be integrated with systemic processes regulating Ca^{2+} homeostasis.

Address for correspondence: Dr. Christopher Huang, Physiological Laboratory, University of Cambridge, Downing Street, Cambridge CB2 3EG, UK. Voice: +44-0-1223-333822; fax: +44-0-1223-333840.

clh11@cam.ac.uk

Ann. N.Y. Acad. Sci. 1116: 255–270 (2007). © 2007 New York Academy of Sciences.
doi: 10.1196/annals.1402.064

KEYWORDS: ryanodine receptor; skeletal muscle; cardiac muscle; osteo-clasts; calcium

THE RYANODINE RECEPTOR-CA^{2+} RELEASE CHANNEL

Ryanodine receptors (RyRs) are large homotetrameric Ca^{2+} release chan-nels each consisting of ~565 kDa polypeptides that normally occur in muscle microsomal membranes in which they gate release of intracellularly stored Ca^{2+}.[1,2] Each component monomer associates with a range of accessory pro-teins, some potentially regulatory; these include calmodulin, FKPB12.6, pro-tein kinase A, the protein phosphatases 1 and 2A, and calmodulin-dependent protein kinase II.[3] It has been suggested that FKBP12.6 binding promotes a functionally important coupling of component RyR subunits during excitation–contraction coupling.[3–6] In addition, both junctin and triadin on the luminal side of the sarcoplasmic reticulum (SR) interact with calsequestrin to extents dependent on Ca levels within the SR.[4,7]

In skeletal and cardiac muscle, RyRs initiate cellular activation through releasing intracellularly stored SR Ca^{2+} in response to triggering by trans-verse tubular voltage sensor-dihydropyridine receptors (DHPRs) that also act as voltage-gated L-type Ca^{2+} channels.[8] The precise characteristics of this ac-tivation depend on the specific DHPR and RyR types expressed. Skeletal and cardiac muscles express different RyR subtypes in turn controlled by distinct, DHPRs. Of known isoforms, RyR1 occurs in the terminal cisternal membranes of skeletal muscle SR[9] whereas RyR2 and RyR3 typically occur in cardiac and brain microsomal membranes.[2]

CYTOSOLIC TRIGGERING OF RYR-MEDIATED CA^{2+} RELEASE IN STRIATED MUSCLE

Both skeletal and cardiac muscle RyR molecules can thus be triggered through influences exerted from their cytosolic aspects. The RyR1 found in skeletal muscle is normally allosterically coupled to and gated by its related DHPR independently of extracellular Ca^{2+}[10–14] (FIG. 1A, B). Electrophysio-logical and pharmacological evidence now strongly supports a voltage-sensing role for the DHPR, which generates a steeply voltage-dependent electrical sig-nature or charge movement with depolarizing voltage change directly driving RyR gating resulting in release of SR Ca^{2+}.[8] It is now thought that the complex kinetics of this gating, q$_\gamma$ signal, specific to skeletal and absent in cardiac mus-cle (FIG. 2A), directly reflects the component reciprocal allosteric interactions between the RyR-Ca^{2+} release channels and surface membrane tubular DHPR-voltage sensors. Furthermore, RyR modification by the inhibitors ryanodine, daunorubicin, and micromolar tetracaine modified the kinetics but preserved

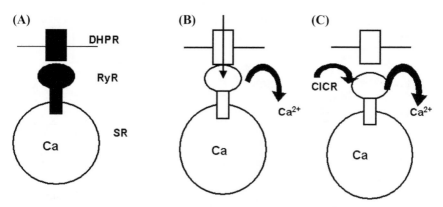

FIGURE 1. RyR signaling in skeletal muscle illustrating (**A**) geometrical relationship between cell surface DHPRs and microsomal RyRs that form (**B**) an allosteric activation scheme in which configurational changes driven by alterations in membrane potential in the DHPR directly lead to activation of SR Ca^{2+} release by allosteric coupling. However, uncoupling of the DHPR and RyR permit a CICR culminating in propagated Ca^{2+} waves under some pathological conditions (**C**).

the steady-state voltage dependence and pharmacological identity of the q_γ signal consistent with such an expected cross-talk between the RyRs and the DHPR-voltage sensors.[12–14] Conversely, the RyR-specific twitch potentiator perchlorate both shifted the activation voltages for this activation process and reversed both the steady-state effects of tetracaine and the kinetic effects of ryanodine and daunorubicin.[14]

In contrast, gating of cardiac RyR2s is triggered by the Ca^{2+} influx through cardiac DHPRs in response to membrane depolarization through a Ca^{2+}-induced Ca^{2+}-release (CICR) mechanism[3,15–20] (FIG. 2A). Excitation–contraction coupling (ECC) characteristics and slow L-type C^{2+} currents–typical of skeletal muscle recover with expression of skeletal DHPR α-subunits in dysgenic murine skeletal myotubes.[21] In contrast, cardiac DHPR expression leads to a cardiac-type ECC associated with fast, large amplitude, L-type Ca^{2+} currents with both dependent on extracellular Ca^{2+}.[22] A chimeric cardiac DHPR containing the cytoplasmic loops of the skeletal DHPR produces a skeletal muscle-type ECC but associated with large L-type Ca^{2+} currents.[22] Conversely, dyspedic skeletal myotubes lacking RyR1 lack an ECC[23] specifically recovering with RyR1 but not RyR2 transfection.[24,25] The latter instead results in spontaneous Ca^{2+} oscillations and the Ca^{2+} waves resembling those observed in Ca^{2+}-overloaded cardiac myocytes,[26,27] reflecting a possible greater sensitivity of RyR2 to a CICR activated by cytosolic Ca^{2+}.

Nevertheless, skeletal muscle exposed to hypertonic solutions can be driven into a CICR[28] manifest both as discrete foci of elevated cytosolic $[Ca^{2+}]$[29] or as regenerative, propagated Ca^{2+} waves that closely resembled those observed

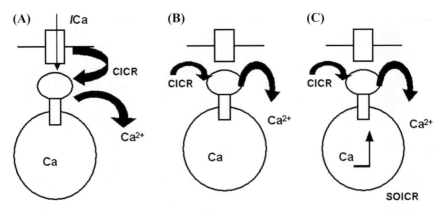

FIGURE 2. RyR signaling mechanisms in cardiac muscle involving RyR gating by a CICR from the SR by an initial Ca^{2+} entry, I_{Ca}, through voltage-gated DHPRs (**A**). Under certain conditions, this can lead to CICR culminating in propagated Ca^{2+} waves under some pathological conditions (**B**), or a release of SR Ca^{2+} by triggering at the luminal aspect of the RyR leading to an SOICR (**C**).

in dyspedic skeletal myotubes expressing RyR2[24,25] in models for the clinical condition of malignant hyperthermia. Similar, possibly pathological events, often implicated in cardiac arrhythmogenesis in whole hearts, occur in cardiac myocytes, following overload of their SR with Ca^{2+}[26,27] (FIGS. 1C, 2B). Furthermore, both processes were blocked by RyR inhibitors and enhanced by the RyR potentiator caffeine.[24,25]

LUMINAL TRIGGERING OF RyR-MEDIATED Ca^{2+} RELEASE IN CARDIAC MUSCLE

In cardiac cells, increasing evidence implicates altered intracellular Ca^{2+} homeostasis that involves the RyRs, in cardiac arrhythogenesis whether triggered by early or delayed afterdepolarizing activity, through such spontaneous release of SR Ca^{2+} in turn triggering these cystolic Ca^{2+} waves and oscillations.[3] One possible scheme for such actions, which has invoked SR *luminal* mechanisms for triggering of Ca^{2+} release, provides a contrast with mechanisms of actions involving regulatory agents directed at the *cystolic* aspect of the RyR (FIG. 2C). This hypothesis suggested an existence of a store overload-induced Ca^{2+} release (SOICR) mechanism that is triggered once a threshold level of SR Ca^{2+} content is exceeded.[30] This would predict that normal myocytes would have a threshold for SOICR that is greater than the SR-free Ca^{2+} levels either in resting conditions or in cells undergoing catecholaminergic stimulation. The result would be little or no spontaneous Ca^{2+} release. In contrast, cardiac RyR2-Ca^{2+} release channels that contain mutations thought to cause the clinical condition of catecholaminergic polymorphic ventricular tachycardia (CPVT)[31] might be abnormally sensitive to activation by luminal Ca^{2+} suggesting a reduced SOICR threshold.[30] In such a situation, whereas the SR Ca^{2+} may remain below the SOICR threshold in resting cells, a SR Ca^{2+} overloading that follows catecholaminergic stimulation may now elevate SR Ca levels to above this critical threshold. The result would be a triggering of spontaneous SR Ca^{2+} release. Similar arrhythmogenic Ca^{2+} release mechanisms have also been implicated in the arrhythmogenesis associated with cardiac failure. Such SOICR phenomena have been reported in HEK293 cells expressing *WT* RyR2 where Ca^{2+} stores were elevated by increased extracellular $[Ca^{2+}]_o$.[30] Such a mechanism has also been suggested for the clinical condition of CPVT[32–34] through its associated RyR gene mutations increasing SR Ca^{2+} release into the cytosol in turn leading to membrane potential changes responsible for triggered arrhythmogenesis.[31,35–37] SOICR has indeed been reported in HEK293 cells expressing *RyR2* containing the *N4104K*, *R4496C*, and *N4895D* mutations associated with CPVT with mutant RyR2 showing increased sensitivities to activation by luminal Ca^{2+} of 300 μM compared to *WT*. Such findings were corroborated in studies of arrhythmogenic properties at the whole heart level in a *RyR2-R4496C* (+/−) knockin murine CPVT mouse model.[38]

OSTEOCLAST REGULATION BY LOCAL [Ca^{2+}]

To what extent do analogies from relatively simple activation systems found in muscle apply to cells with more complex functions, such as bone resorbing osteoclasts? Osteoclasts show sharp contrasts from striated muscle in structure and function, yet both cell types share Ca^{2+}-driven signaling phenomena. Recent studies have demonstrated an involvement of the RyR within their particular, both geometrically and functionally distinct, signaling schemes. They also implicate RyRs in a more complex set of cellular regulatory actions than has been apparent from the evidence from striated muscle alone.

First, Ca^{2+} exerts regulatory actions on osteoclast activity, but acts from the extracellular space, rather than from the cytosol. Osteoclasts become exposed to high extracellular $[Ca^{2+}]$ (8–20 mM[39]) during hydroxyapatite dissolution and early evidence suggested that this in turn inhibits their bone resorptive activity and associated release of enzyme, suggestive of local feedback mechanisms for their functional control.[40,41] Second, triggering of such activation mechanisms similarly results in elevations of cytosolic Ca^{2+}, as well as altered motile activity. Thus, cellular physiological studies demonstrate rapid, sustained alterations in cytosolic $[Ca^{2+}]$ accompanied by a cell retraction following applications of millimolar extracellular $[Ca^{2+}]$.[42–44] Third, such cytosolic $[Ca^{2+}]$ elevations could reflect release of intracellularly stored Ca^{2+} in addition to extracellular Ca^{2+} entry, in common with the situation in striated muscle. Experiments applying ionomycin to isolated osteoclasts elicited cytosolic $[Ca^{2+}]$ transients that fully returned to base line in osteoclasts bathed in Ca^{2+}-free external solution, but these disappeared with repeated ionophore application suggesting depletion of a finite Ca store. In contrast, such Ca^{2+} peaks decayed to sustained levels and persisted with repeated ionomycin applications, suggestive of a replenishing extracellular Ca^{2+} entry in osteoclasts bathed in Ca^{2+}-containing solutions.

In other respects, the regulatory system in osteoclasts differed from features found in striated muscle. First, the cytosolic $[Ca^{2+}]$ elevations differed from those shown by striated muscle in their insensitivity to the Ca^{2+} channel-specific reagents, nifedipine, BAYK 8644, verapamil, and diltiazem all of which also spared the cell retraction and changes in bone resorption following elevations of extracellular $[Ca^{2+}]$.[45]

Second, cellular mechanisms directly sensing ambient Ca^{2+} rather than membrane potential in turn inducing either allosterically coupled SR Ca^{2+} release in skeletal muscle or extracellular Ca^{2+} entry initiating CICR have been implicated in triggering these effects of extracellular $[Ca^{2+}]$. Ca^{2+} sensing processes of this kind occur in numerous cell types including thyroid parafollicular cells, parathyroid chief cells,[46] gastrointestinal enterocytes,[47,48] renal juxtaglomerular and proximal tubular cells,[49] neurons,[50] cytotrophoblasts,[51] keratinocytes,[44,52] and testicular Leydig cells.[53]

Nevertheless, such sensing appears similarly to involve mediation by a membrane-resident regulatory molecule, albeit in this situation acting as a Ca^{2+} receptor (CaR). However, to be effective in osteoclasts, such divalent ion, Ca^{2+}, receptors (CaR) would sense high, millimolar, Ca^{2+} levels.[40,54-57] Such a regulatory mechanism involving Ca^{2+} sensing, rather than Ca^{2+} entry was consistent with results using other, cell-impermeant, divalent, Ni^{2+}, Cd^{2+} or trivalent, La^{3+}, ions, which triggered similar cytosolic $[Ca^{2+}]$ changes, even in cells studied in Ca^{2+}-free, extracellular media. Furthermore, quantification of these responses when Ni^{2+} acted as surrogate agonist yielded graded activation and inactivation curves modulated toward lower agonist concentrations by reduced $[Ca^{2+}]$ and $[Mg^{2+}]$.[56] Furthermore, such cytosolic signaling events appeared to result in motile events in the osteoclast in common with a mechanical activation in striated muscle. Cell retraction and reductions in both acid phosphatase release and bone resorptive activity followed these actions on cytosolic Ca^{2+},[58-61] with similar orders of potency: $La^{3+} > Cd^{2+} > Ni^{2+} > Ca^{2+} > Ba^{2+} = Sr^{2+} > Mg^{2+}$ among agonist ions.[54,56] In addition, the processes concerned shared an albeit smaller sensitivity to the membrane voltage: these agonist-like actions were modulated by membrane potential manipulations produced by K^+ and its ionophore, valinomycin.[54]

EVIDENCE FOR A RyR INVOLVEMENT IN LOCAL CONTROL OF OSTEOCLASTIC ACTIVITY

Explorations of the effect of physiologically effective agents added to these analogies in the cellular activation between cell types in demonstrating pharmacological parallels between the processes of osteoclast activation and the RyR-dependent signaling processes observed in striated muscle. First, perchlorate ions, known to facilitate SR Ca^{2+} release by skeletal muscle RyR-Ca^{2+} release channels,[62,63] produced transient elevations of cytosolic $[Ca^{2+}]$, sustained cell retraction and reduced quantitative *in vitro* indicators of osteoclastic bone resorptive activity.[64] Second, the RyR-antagonist dantrolene Na inhibited such changes.[65] Third, ryanodine itself produced voltage-dependent inhibition of effects of Ni^{2+} on cytosolic $[Ca^{2+}]$.[60] Finally, the RyR agonist caffeine, even at concentrations considerably lower than active levels in muscle, produced releases of intracellularly stored Ca^{2+} through bell-shaped concentration-response curves as well as occlusive interactions with Ni^{2+} and extracellular $[Ca^{2+}]$.[66]

LOCAL CONTROL OF OSTEOCLASTIC ACTIVITY INVOLVES A UNIQUELY SITED CELL SURFACE RyR

Labeling studies then confirmed and uniquely localized such an osteoclastic RyR or RyR-like CaR molecule. However, they show important contrasts to the situation in striated muscle in which such RyRs are localized to the microsomal fraction, while being controlled by signaling events by DHPRs at the cell surface. Instead, they demonstrate a unique localization of the osteoclastic RyR at the cell surface (FIG. 3). A specific [^{3}H]-ryanodine binding to freshly isolated rat osteoclasts was displaced by ryanodine itself, the membrane-impermeant substitute agonist Ni^{2+} and the RyR antagonist ruthenium red. Immunostaining studies went on to localize the normally intraluminal (SR) aspect of the putative RyR to the extracellular, and its cytosolic aspect to the intracellular compartment of the osteoclasts in which they occurred. Thus, antiserum raised to an epitope located within the channel-forming domain of RyRs potentiated cytosolic Ca^{2+} responses to Ni^{2+}. Serial confocal sections and immunogold scanning electron microscopy localized the antibody binding to the plasma membrane in intact, unfixed, osteoclasts. In contrast, antiserum, Ab[34] directed to a putative intracellular epitope neither potentiated CaR activation nor stained live osteoclasts but successfully stained permeabilized cells in a distinctive cytoplasmic pattern.[67]

Pharmacological findings also favored this plasma membrane localization.[68] Extracellular ruthenium red and adenosine 3',5'-cyclic diphosphate ribose (cADPr) both triggered voltage-sensitive elevations in cytosolic $[Ca^{2+}]$ and these maneuvers attenuated subsequent cytosolic $[Ca^{2+}]$ responses to external Ni^{2+} compatible with common sites acted upon by extracellularly applied Ni^{2+}, cADPr, and ruthenium red. Further details of possible mechanisms by which this unique cell surface RyR-CaR acts in the transduction of extracellular $[Ca^{2+}]$ into alterations of intracellular $[Ca^{2+}]$ remain to be explored. One might invoke the intraluminal low-affinity Ca^{2+}-binding site,[69] which in the case of the osteoclast would assume a uniquely extracellular position, in parallel

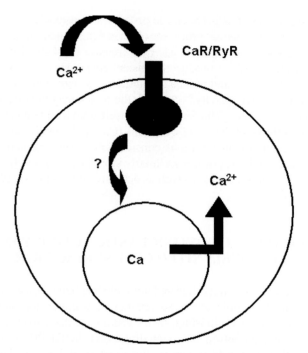

FIGURE 3. Possible scheme illustrating osteoclastic signaling by a surface RyR whose cytosolic aspect continues to face the intracellular compartment, with a luminal aspect occurring on the extracellular side of the RyR.

with the situation in cardiac RyR2 actions in CPVT. Alternatively, the RyR might be coupled to a CaR made from conventional 7-pass G-protein-coupled molecules types described elsewhere.[70] There also remains the question as to how the surface event induces release of intracellularly stored Ca^{2+}.[9]

CaR FUNCTION WITHIN THE CONTEXT OF OSTEOCLASTIC Ca^{2+} HOMEOSTASIS

Preliminary evidence suggests that osteoclasts do possess the complementary processes by which Ca^{2+} is transferred between cytosol, intracellular stores, and extracellular space in common with similar functions in striated muscle and other nonexcitable cells for such a scheme to work. First, store refilling in osteoclasts similarly depends upon thapsigargin-sensitive Ca^{2+}-ATPase additionally including a capacitative Ca^{2+} influx. Thapsigargin-elevated cytosolic $[Ca^{2+}]$ in osteoclasts in Ca^{2+}-free extracellular solutions but reduced cytosolic $[Ca^{2+}]$ signals elicited by subsequent application of extracellular Ni^{2+}. Restoration of extracellular $[Ca^{2+}]$ then resulted in a cytosolic [Ca^{2+}] overshoot. Similarly, applications of ionomycin induced cytosolic [Ca^{2+}] elevations in cells bathed in Ca^{2+}-free solutions, which then failed to respond to subsequent applications of the agonist Ni^{2+} consistent with a Ca^{2+} store depletion. However, the cells then showed similar cytosolic [Ca^{2+}] overshoots with subsequent restorations of the extracellular Ca^{2+} again consistent with a capacitative Ca^{2+} influx from the extracellular space through a cytosolic route.[71] Second, although the Ca^{2+} transport mechanisms that restore basal cytosolic [Ca^{2+}] levels remain unclear, there is evidence for a Ca^{2+}-ATPase on the osteoclast dorsal surface[72] and functional, electrophysiological, and immunochemical evidence for a Na^+/Ca^{2+} exchanger comparable to the surface membrane exchanger in cardiac muscle that could be linked to the proton extrusion that is a primary determinant of the rate and extent of bone resorption.[73]

MODULATION OF CaR/RyR FUNCTION

There is also preliminary evidence for an integration of this local extracellular Ca^{2+} sensing system within remaining processes regulating osteoclast cellular physiology. First, CD38/ADP-ribosyl cyclase has been implicated in control of bone resorption through cADPr.[74] RT-PCR demonstrated cytosolic mRNA for the ADP-ribosyl cyclase, CD38, that cyclizes NAD^+ to cADPr; the latter gates RyR-mediated Ca^{2+} release. Confocal microscopy and Western blotting localized this CD38 to plasma membrane. Furthermore, CD38 triggering by agonist antibody in the presence of NAD^+ substrate triggered cytosolic Ca^{2+} signals attenuated by ryanodine and caffeine. Conversely, anti-CD38 agonist antibody inhibited bone resorption and elevated interleukin-6 (IL-6) secretion whereas IL-6 enhanced CD38 mRNA expression, findings that together suggested an organization of these phenomena into a functioning feedback loop.

Second, there is evidence that the CaR sensing processes are integrated with systemic mechanisms including those related to calcitonin (CT) function, and including its effects upon cAMP-mediated signaling. Physiological, femtomolar, CT, amylin, calcitonin gene-related peptide (CGRP), cholera toxin, and dibutyryl-cAMP, all thought to influence the cAMP signaling system, inhibited Ca^{2+} responses to both Ca^{2+} and Ni^{2+}. These effects were reversed by the inhibition of protein kinase A phosphorylation by IP-20.

Third, cytokine systems involving IL-6 interact with the CaR-mediated signaling. IL-6 may be involved in an autocrine–paracrine loop that sustains osteoclastic activity despite locally generated and otherwise inhibitory elevations in local extracellular [Ca^{2+}] resulting from bone resorption.[75] It inhibited the cytosolic [Ca^{2+}] signals triggered by extracellular Ca^{2+} or Ni^{2+}. Nonpermeabilized osteoclasts incubated with anti-IL-6 receptor (IL-6R) show a peripheral

TABLE 1. Similarities and contrasts between RyR- mediated signaling in striated muscle and the osteoclast

	Striated muscle	Osteoclasts
Primary trigger	Voltage sensor detection of membrane potential	Ca^{2+} receptor (CaR) detection of extracellular $[Ca^{2+}]$
Cell surface DHPR	Present	Absent
RyR localization	Microsomal	Cell surface
Intracellular change	Elevation of cytosolic $[Ca^{2+}]$	Elevation of cytosolic $[Ca^{2+}]$
Source of cytosolic $[Ca^{2+}]$ elevation	Intracellular Ca^{2+} store	Intracellular Ca^{2+} store; with possibly extracellular Ca^{2+} entry
Physiological end-result of activation	Troponin activation and generation of tension	Cell retraction and inhibition of bone resorptive activity
Pharmacological sensitivity to DHPR-specific agents	Sensitivity to Ca^{2+} channel agonists and blockers	No action
Pharmacological sensitivity to RyR-specific agents	Sensitive to extracellular applications of ryanodine, perchlorate, caffeine, ruthenium red, and Ni^{2+}	Sensitive to extracellular applications of ryanodine, perchlorate, caffeine
Ca^{2+} homeostatic mechanisms	Demonstrable Ca^{2+}-ATPase, Na^{+}-Ca^{2+} exchange (cardiac muscle) activity,	Demonstrable Ca^{2+}-ATPase, Na^{+}-Ca^{2+} exchange activity, as well as capacitative Ca^{2+} entry
Integration with systemic homeostatic mechanisms	Catecholaminergic mechanisms	Sensitivity to interleukin-6, calcitonin, amylin, CGRP and H^{+}
Translational importance	Malignant hyperthermia (striated muscle); cardiac arrhythmogenesis, including catecholaminergic polymorphic ventricular tachycardia (cardiac muscle)	Osteoclastic and bone mass disorders

plasma membrane fluorescence on confocal microscopy. Applied IL-6 but not IL-11 prevented the inhibition of osteoclastic bone resorption induced by high extracellular Ca^{2+}. This effect was reversed by excess soluble IL-6 receptor (sIL6-R). Conversely, elevations of extracellular $[Ca^{2+}]$ or culturing on bone induced increased osteoclast release of IL-6 and enhanced their IL-6 and IL-6R mRNA histostaining.

Fourth, vitamin D-binding protein (DBP) and the macrophage-activating factor (DBP-MAF) left following removal of its sialic acid or galactose residue both inhibit extracellular Ca^{2+} sensing.[76]

Finally, extracellular pH also influences Ca^{2+} sensing: decreased pH (7.8 to 4.0) increased cytosolic Ca^{2+} signals elicited by extracellular Ni^{2+} or Ca^{2+} (5 mM). This effect persisted in cells studied in Ca^{2+}-free solutions consistent with altered intracellularly stored Ca^{2+} release as opposed to entry of extracellular Ca^{2+}.[77]

OTHER POSSIBLE ROLES FOR OSTEOCLASTIC RyRs

TABLE 1 summarizes some of the similarities and differences drawn out by the present article. Finally, the osteoclast may provide further examples of RyR-based signaling systems with unexpected configurations. There have been suggestions for an existence of RyRs within their inner nuclear membrane activated through a CD38/ADP-ribosyl cyclase with nucleoplasmic catalytic sites that would drive an intranuclear NAD^+ cyclization to a cADPr that in turn might trigger nucleoplasmic Ca^{2+} influx.[78–80] Elevated nuclear $[Ca^{2+}]$ enhances the expression of a number of early response genes such as c-fos, c-jun, jun-B, fos-B, nur-77, and zif-268, as well as certain late genes such as those for IL-2, IL3,[75] and IL6,[81,82] and might thereby contribute to regulation of longer-term processes, such as gene expression and apoptosis.[83]

ACKNOWLEDGMENTS

C.L.-L.H. and A.A.G. thank the Medical Research Council, the Wellcome Trust, and the British Heart Foundation for their generous support. M.Z. is supported by grants from the National Institutes of Health, and the Department of Veterans Affairs. J.A.F. thanks Gonville and Caius College for Research Fellowship support.

REFERENCES

1. BERRIDGE, M.J. 1993. Inositol trisphosphate and calcium signalling. Nature **361:** 315–325.

2. MEISSNER, G. 1994. Ryanodine receptor/Ca^{2+} release channels and their regulation by endogenous effectors. Annu. Rev. of Physiol. **56:** 485–508.
3. BERS, D.M. 2002. Cardiac excitation-contraction coupling. Nature **415:** 198–205.
4. WEHRENS, X.H., S.E. LEHNART, S.R. REIKEN, *et al.* 2004. Protection from cardiac arrhythmia through ryanodine receptor-stabilizing protein calstabin2. Science **304:** 292–296.
5. YANO, M., K. ONO, T. OHKUSA, *et al.* 2000. Altered stoichiometry of FKBP12.6 versus ryanodine receptor as a cause of abnormal Ca^{2+} leak through ryanodine receptor in heart failure. Circulation **102:** 2131–2136.
6. YANO, M., Y. IKEDA & M. MATSUZAKI. 2005. Altered intracellular Ca^{2+} handling in heart failure. J. Clin. Invest. **115:** 556–564.
7. ZHANG, L., J. KELLEY, G. SCHMEISSER, *et al.* 1997. Complex formation between junctin, triadin, calsequestrin, and the ryanodine receptor. Proteins of the cardiac junctional sarcoplasmic reticulum membrane. J. Biol. Chem. **272:** 23389–23397.
8. HUANG, C.L.-H. 1990. Voltage-dependent block of charge movement components by nifedipine in frog skeletal muscle. J. Gen. Physiol. **96:** 535–558.
9. HUANG, C.L.-H. 1993. Intramembrane Charge Movements in Skeletal Muscle. Oxford University Press. Oxford.
10. SCHNEIDER, M.F. & W.K. CHANDLER. 1973. Voltage-dependent charge in skeletal muscle: a possible step in excitation-contraction coupling. Nature **342:** 244–246.
11. NAKAI, J., R.T. DIRKSEN, H.T. NGUYEN, *et al.* 1996. Enhanced dihydropyridine receptor channel activity in the presence of ryanodine receptor. Nature **380:** 72–75.
12. HUANG, C.L.-H. 1996. Kinetic isoforms of intramembrane charge in intact amphibian striated muscle. J. Gen. Physiol. **107:** 515–534.
13. HUANG, C.L.-H. 1997. Dual actions of tetracaine on intramembrane charge in amphibian skeletal muscle. J. Physiol. **501:** 589–606.
14. HUANG, C.L.-H. 1998. The influence of caffeine on intramembrane charge movements in intact frog striated muscle. J. Physiol. **512:** 707–721.
15. FABIATO, A. 1985. Time and calcium dependence of activation and inactivation of calcium-induced release of calcium from the sarcoplasmic reticulum of a skinned canine cardiac Purkinje cell. J. Gen. Physiol. **85:** 247–289.
16. CANNELL, M. B, H. CHENG & W.J. LEDERER. 1995. The control of calcium release in heart muscle. Science **268:** 1045–1049.
17. WEHRENS, X.H., S.E. LEHNART, T. HUANG, *et al.* 2003. FKBP12.6 deficiency and defective calcium release channel (ryanodine receptor) function linked to exercise-induced sudden cardiac death. Cell **113:** 829–840.
18. KONTULA, K., P.J. LAITINEN, A. LEHTONEN, *et al.* 2005. Catecholaminergic polymorphic ventricular tachycardia: recent mechanistic insights. Cardiovasc. Res. **67:** 379–387.
19. YANO, M., T. YAMAMOTO, N. IKEMOTO & M. MATSUZAKI. 2005. Abnormal ryanodine receptor function in heart failure. Pharmacol. Ther. **107:** 377–391.
20. PHROMMINTIKUL, A. & N. CHATTIPAKORN. 2006. Roles of cardiac ryanodine receptor in heart failure and sudden cardiac death. Int. J. Cardiol. **112:** 142–152.
21. TANABE, T., K.G. BEAM, J.A. POWELL & S. NUMA. 1988. Restoration of excitation contraction coupling and slow Ca current in dysgenic muscle by dihydropyridine receptor complementary DNA. Nature **336:** 134–139.

22. TANABE, T., A. MIKAMI, S. NUMA & K.G. BEAM. 1990. Cardiac-type excitation-contraction coupling in dysgenic skeletal muscle injected with cardiac dihydropyridine receptor cDNA. Nature **344:** 451–453. Tanabe, T., K.G. Beam, B.A. Adams, *et al.* 1990. Regions of the skeletal muscle dihydropyridine receptor critical for excitation-contraction coupling. Nature **346:** 567–569.

23. TAKESHIMA, H., M. IINO, H. TAKEKURA, *et al.* 1994. Excitation-contraction uncoupling and muscular degeneration in mice lacking functional skeletal muscle ryanodine-receptor gene. Nature **369:** 556–559.

24. YAMAZAWA, T., H. TAKESHIMA, T. SAKURAI, *et al.* 1996. Subtype specificity of the ryanodine receptor for Ca^{2+} signal amplification in excitation-contraction coupling. EMBO J. **15:** 6172–6177.

25. NAKAI, J., T. OGURA, F. PROTASI, *et al.* 1997. Functional nonequality of the cardiac and skeletal ryanodine receptors. Proc. Natl. Acad. Sci. USA **94:** 1019–1022.

26. LIPP, P. & E. NIGGLI. 1993. Microscopic spiral waves reveal positive feedback in subcellular calcium signalling. Biophys. J. **65:** 2272–2276.

27. CHENG, H., M.R. LEDERER, W.J. LEDERER & M.B. CANNELL. 1996. Calcium sparks and $[Ca^{2+}]_i$ waves in cardiac myocytes. Am. J. Physiol. **270:** C148–C159.

28. CHAWLA, S., J.N. SKEPPER, A.R. HOCKADAY & C.L.-H. HUANG. 2001. Calcium waves induced by hypertonic solutions in intact frog skeletal muscle fibres. J. Physiol. **536:** 351–360.

29. KLEIN, M.G., H. CHENG, L.F. SANTANA, *et al.* 1996. Two mechanisms of quantized calcium release in skeletal muscle. Nature **379:** 455–458.

30. JIANG, D., B. XIAO, D. YANG, *et al.* 2004. RyR2 mutations linked to ventricular tachycardia and sudden death reduce the threshold for store-overload-induced Ca^{2+} release (SOICR). Proc. Natl. Acad. Sci. USA **101:** 13062–13067.

31. PRIORI, S.G., C. NAPOLITANO, N. TISO, *et al.* 2001. Mutations in the cardiac ryanodine receptor gene (hRyR2) underlie catecholaminergic polymorphic ventricular tachycardia. Circulation **103:** 196–200.

32. WEHRENS, X.H. & A.R. MARKS. 2003. Altered function and regulation of cardiac ryanodine receptors in cardiac disease. Trends Biochem. Sci. **28:** 671–678.

33. FRANCIS, J., V. SANKAR, V.K. NAIR & S.G. PRIORI. 2005. Catecholaminergic polymorphic ventricular tachycardia. Heart Rhythm **2:** 550–554.

34. GEORGE, C.H., H. JUNDI, N. L THOMAS, *et al.* 2007. Ryanodine receptors and ventricular arrhythmias: emerging trends in mutations, mechanisms and therapies. J. Mol. Cell Cardiol. **42:** 34–50.

35. ROSEN, M.R. & P. DANILO JR. 1980. Effects of tetrodotoxin, lidocaine, verapamil, and AHR-2666 on ouabain-induced delayed afterdepolarizations in canine Purkinje fibers. Circ. Res. **46:** 117–124.

36. LEENHARDT, A., V. LUCET, I. DENJOY, *et al.* 1995. Catecholaminergic polymorphic ventricular tachycardia in children. A 7-year follow-up of 21 patients. Circulation **91:** 1512–1519.

37. KUMMER, J.L., R. NAIR & S.C. KRISHNAN. 2006. Images in cardiovascular medicine. Bidirectional ventricular tachycardia caused by digitalis toxicity. Circulation **113:** e156–e157.

38. CERRONE, M., B. COLOMBI, M. SANTORO, *et al.* 2005. Bidirectional ventricular tachycardia and fibrillation elicited in a knock-in mouse model carrier of a mutation in the cardiac ryanodine receptor. Circ. Res. **96:** e77–e82.

39. SILVER, I.A., R.J. MURRILLS & D.J. ETHERINGTON. 1988. Microelectrode studies on the acid microenvironment beneath adherent macrophages and osteoclasts. Exp. Cell Res. **175:** 266–276.

40. MALGAROLI, A., J. MELDOLESI, A.Z. ZALLONE & A. TETI. 1989. Control of cytosolic free calcium in rat and chicken osteoclasts. The role of extracellular calcium and calcitonin. J. Biol. Chem. **264:** 14342–14347.

41. ZAIDI, M., H. K DATTA, A. PATCHELL, *et al.* 1989. 'Calcium-activated' intracellular calcium elevation: a novel mechanism of osteoclast regulation. Biochem. Biophys. Res. Commun. **163:** 1461–1465.

42. DATTA, H.K., I. MACINTYRE & M. ZAIDI. 1989. The effect of extracellular calcium elevation on morphology and function of isolated rat osteoclasts. Biosci. Reports **9:** 747–751.

43. MOONGA, B.S., D.W. MOSS, A. PATCHELL & M. ZAIDI. 1990. Intracellular regulation of enzyme secretion from rat osteoclasts and evidence for a functional role in bone resorption. J. Physiol. **429:** 29–45.

44. ZAIDI, M. 1990. "Calcium receptors" on eukaryotic cells with special reference to the osteoclast. Biosci. Reports **10:** 493–507.

45. DATTA, H.K., I. MACINTYRE & M. ZAIDI. 1990. Intracellular calcium in the control of osteoclast function. I. Voltage-insensitivity and lack of effects of nifedipine, BAYK8644 and diltiazem. Biochem. Biophys. Res. Commun. **167:** 183–188.

46. BROWN, E.M., G. GAMBA, I.D. RICCARDI, *et al.* 1993. Cloning and characterization of an extracellular calcium sensing receptor from bovine parathyroid. Nature **366:** 575–579.

47. GAMA, L., L.M. BAXENDALE-COX & G.E. BREITWIESER. 1997. Ca^{2+}-sensing receptors in intestinal epithelium. Am. J. Physiol. **273:** C1168–C1175.

48. PAZIANAS, M., O.A. ADEBANJO, V.S. SHANKAR, *et al.* 1995. Extracellular cation sensing by the enterocyte. Prediction of a novel divalent cation "receptor". Biochem. Biophys. Res. Commun. **210:** 948–953.

49. RICCARDI, D., J. PARK, W.S. LEE, *et al.* 1995. Cloning and functional expression of a rat kidney extracellular calcium/polyvalent cation-sensing receptor. Proc. Natl. Acad. Sci. USA **92:** 131–135.

50. QUINN, S.J., C.P. YE, R. DIAZ, *et al.* 1997. The Ca^{2+}-sensing receptor: a target for polyamines. Am. J. Physiol. **273:** C1315–C1323.

51. LUNDGREN, S., G. HJALM, P. HELLMAN, *et al.* 1994. A protein involved in calcium sensing of the human parathyroid and placental cytotrophoblast cells belongs to the LDL-receptor protein superfamily. Exp. Cell Res. **212:** 344–350.

52. BROWN, E.M. 1991. Extracellular Ca^{2+} sensing, regulation of parathyroid cell function, and role of Ca^{2+} and other ions as extracellular (first) messengers. Physiol. Rev. **71:** 371–411.

53. ADEBANJO, O.A., J. IGIETSEME, C. L-H. HUANG & M. ZAIDI. 1998. The effect of extracellularly applied divalent cations on cytosolic Ca^{2+} in murine leydig cells: evidence for a Ca^{2+}-sensing receptor. J. Physiol. **513:** 399–410.

54. ZAIDI, M., J. KERBY, C.L.-H. HUANG, *et al.* 1991. Divalent cations mimic the inhibitory effects of extracellular ionized calcium on bone resorption by isolated rat osteoclasts: further evidence for a "calcium receptor". J. Cell. Physiol. **149:** 422–427.

55. ZAIDI, M., V.S. SHANKAR, C.M.R. BAX, *et al.* 1992. Characterization of the osteoclast calcium receptor. *In* Calcium Regulation and Bone Metabolism. D.V. Cohn & A.R. Tashjian Jr., Eds.: 170–174. Elsevier. Amsterdam.

56. SHANKAR, V.S., A.S. ALAM, C.M. BAX, *et al.* 1992. Activation and inactivation of the osteoclast Ca^{2+} receptor by the trivalent cation, La^{3+}. Biochem. Biophys. Res. Commun. **187:** 907–912.

57. SHANKAR, V.S., C.M. BAX, A.S. ALAM, *et al.* 1992. The osteoclast Ca^{2+} receptor is highly sensitive to activation by transition metal cations. Biochem. Biophys. Res. Commun. **187:** 913–918.

58. BAX, C.M., V.S. SHANKAR, B.S., MOONGA, *et al.* 1992. Is the osteoclast calcium "receptor" a receptor-operated calcium channel? Biochem. Biophys. Res. Commun. **183:** 619–625.

59. BAX, B.E., V.S. SHANKAR, C.M.R. BAX, *et al.* 1993. Functional consequences of the interaction of Ni^{2+} with the osteoclast Ca^{2+} receptor. Exp. Physiol. **78:** 517–529.

60. ZAIDI, M., V.S. SHANKAR, A.S. TOWHIDUL ALAM, *et al.* 1992. Evidence that a ryanodine receptor triggers signal transduction in the osteoclast. Biochem. Biophys. Res. Commun. **188:** 1332–1336.

61. SHANKAR, V.S., C.M.R. BAX, B.E. BAX, *et al.* 1993. Activation of the Ca^{2+} 'receptor' on the osteoclast by Ni^{2+} elicits cytosolic Ca^{2+} signals: evidence for receptor activation and inactivation, intracellular Ca^{2+} redistribution and divalent cation modulation. J. Cell. Physiol. **155:** 120–129.

62. HUANG, C.L.-H. 1986. The differential effects of twitch potentiators on charge movements in frog skeletal muscle. J. Physiol. **380:** 17–33.

63. HUANG, C.L.-H. 1987. "Off" tails of intramembrane charge movements in frog skeletal muscle in perchlorate-containing solutions. J. Physiol. **384:** 492–509.

64. MOONGA, B.S., H.K. DATTA, P.J.R. BEVIS, *et al.* 1991. Correlates of osteoclast function in the presence of perchlorate ions in the rat. Exp. Physiol. **76:** 923–933.

65. MIYAUCHI, M., K.A. HRUSKA, E.M. GREENFIELD, *et al.* 1990. Osteoclast cytosolic calcium, regulated by voltage-gated calcium channels and extracellular calcium, controls podosome assembly and bone resorption. J. Cell Biol. **111:** 2543–2552.

66. SHANKAR, V.S., M. PAZIANAS, C.L-H. HUANG, *et al.* 1995. Caffeine modulates Ca^{2+} receptor activation in isolated rat osteoclasts and induces intracellular Ca^{2+} release. Am. J. Physiol. **268:** F447–F454.

67. ZAIDI, M., V.S. SHANKAR, R.E. TUNWELL, *et al.* 1995. A ryanodine receptor-like molecule in the osteoclast plasma membrane is a functional component of the osteoclast Ca^{2+} sensor. J. Clin. Investig. **96:** 1582–1590.

68. ADEBANJO, O.A., V.S. SHANKAR, M. PAZIANAS, *et al.* 1996. Extracellularly applied ruthenium red and cADP ribose elevate cytosolic Ca^{2+} in isolated rat osteoclasts. Am. J. Physiol. **270:** F469–F475.

69. ANDERSON, K., F.A. LAI, Q.Y. LIU, *et al.* 1989. Structural and functional characterization of the purified cardiac ryanodine receptor-Ca^{2+} release channel complex. J. Biol. Chem. **264:** 1329–1335.

70. KAMEDA, T., H. MANO, Y. YAMADA, *et al.* 1998. Calcium-sensing receptor in mature osteoclasts, which are bone resorbing cells. Biochem. Biophys. Res. Commun. **245:** 419–422.

71. ZAIDI, M., V.S. SHANKAR, C.M.R. BAX, *et al.* 1993. Linkage of extracellular and intracellular control of cytosolic Ca^{2+} in rat osteoclasts in the presence of thapsigargin. J. Bone Min. Res. **8:** 961–967.

72. ZAIDI, M., A.S.M.T. ALAM, V.S. SHANKAR, *et al.* 1993. Cellular biology of bone resorption. Biol. Rev. Camb. Philos. Soc. **68:** 197–264.

73. MOONGA, B.S., R. DAVIDSON, L. SUN, *et al.* 2001. Identification and characterization of a sodium/calcium exchanger, NCX-1, in osteoclasts and its role in bone resorption. Biochem. Biophys. Res. Commun. **283:** 770–775.

74. SUN, L., O.A. ADEBANJO, B.S. MOONGA, *et al.* 1999. CD38/ADP-ribosyl cyclase: a new role in the regulation of osteoclastic bone resorption. J. Cell Biol. **146:** 1161–1171.

75. ADEBANJO, O.A., B.S. MOONGA, T. YAMATE, *et al.* 1998. Mode of action of interleukin-6 on mature osteoclasts. Novel interactions with extracellular Ca^{2+} sensing in the regulation of osteoclastic bone resorption. J. Cell Biol. **142:** 1347–1356.

76. ADEBANJO, O.A., B.S. MOONGA, J.G. HADDAD, *et al.* 1998. A possible new role for vitamin D-binding protein in osteoclast control. Inhibition of Ca^{2+} sensing at low physiological concentrations. Biochem. Biophys. Res. Commun. **249:** 668–671.

77. ADEBANJO, O.A., V.S. SHANKAR, M. PAZIANAS, *et al.* 1994. Modulation of the osteoclast Ca^{2+} receptor by extracellular protons. Possible linkage between Ca^{2+} sensing and extracellular acidification. Biochem. Biophys. Res. Commun. **194:** 742–747.

78. ADEBANJO, O.A., H.K. ANANDATHREETHAVARADA, A.P. KOVAL, *et al.* 1999. A new function for CD38/ADP-ribosyl cyclase in nuclear Ca^{2+} homeostasis. Nat. Cell Biol. **7:** 409–414.

79. GERASIMENKO, O.V., J.V. GERASIMENKO, A.V. TEPIKIN & O.H. PETERSEN. 1995. ATP-dependent accumulation and inositol trisphosphate- or cyclic ADP-ribose-mediated release of Ca^{2+} from the nuclear envelope. Cell **80:** 439–444.

80. SANTELLA, L. & E. CARAFOLI. 1997. Calcium signaling in the cell nucleus. FASEB J. **11:** 1091–2109.

81. BELLIDO, T., N. STAHL, T.J. FARRUGGELLA, *et al.* 1996. Detection of receptors for interleukin-6, interleukin-11, leukemia inhibitory factor, oncostatin M, and ciliary neurotrophic factor in bone marrow stromal/osteoblastic cells. J. Clin. Investig. **97:** 431–437.

82. FRANCHIMONT, N., S. RYDZIEL & E. CANALIS. 1997. Interleukin-6 is autoregulated by transcriptional mechanisms in cultures of rat osteoblastic cells. J. Clin. Investig. **100:** 1797–1803.

83. NICOTERA, P. & A.D. ROSSI. 1994. Nuclear Ca^{2+}: physiological regulation and role in apoptosis. Mol. Cell. Biochem. **135:** 89–98.

Posttranslational Regulation of Bim by Caspase-3

HIDETOSHI WAKEYAMA, TORU AKIYAMA, YUHO KADONO,
MASAKI NAKAMURA, YASUSHI OSHIMA, KOZO NAKAMURA,
AND SAKAE TANAKA

*Department of Orthopaedic Surgery, Faculty of Medicine, The University
of Tokyo, Tokyo, Japan*

ABSTRACT: Bim is a proapoptotic BH3-domain-only member of the Bcl-2
family, and its expression is regulated both transcriptionally and post-
translationally. We developed an *in vitro* system examining the posttranslational regulation of Bim. Since Bim is a strong mediator of apoptosis, it
has been quite difficult to establish cell lines stably overexpressing Bim.
Coexpression of Bcl-2 enabled us to obtain mouse embryonic fibroblasts
(MEFs) in which Bim is overexpressed and Bcl-2 expression is regulated by Tet-off system. Reduction of Bcl-2 levels by doxycycline treatment induced caspase-3 and caspase-7 activation, which was followed by
Bim degradation. Bim degradation was suppressed by gene knockdown
of *caspase-3*, but not by *caspase-7* knockdown. The same posttranslational regulation of Bim was observed in osteoclasts. These results suggest that caspase-3 negatively regulates Bim expression by stimulating its
degradation, thus creating a negative feedback loop in the Bim–caspase
axis.

KEYWORDS: osteoclasts; Bim; caspase-3; degradation

INTRODUCTION

Apoptosis is genetically programmed cell death to remove the unwanted
cells.[1] The abnormalities of apoptosis regulation induce various sicknesses,
such as cancers, autoimmune diseases, and degenerative disorders.[2] Recent
studies have revealed that the apoptosis of osteoblasts and osteoclasts is strictly
regulated, and plays important roles in maintaining the skeletal integrity. However, the molecular events implicated in osteoblast and osteoclast apoptosis
have not been fully elucidated. There are two types of apoptosis pathways; the
death receptor pathway and the mitochondrial pathway.[3] Bim is a proapoptotic BH (Bcl-2 homology) 3-only member of Bcl-2 family[4] and induces the

Address for correspondence: Sakae Tanaka, Department of Orthopaedic Surgery, Faculty of
Medicine, The University of Tokyo, 7-3-1 Hongo, Bunkyo-ku, Tokyo 113-0033, Japan. Voice: +81-3-
3815-5411; ext: 33376; fax: +81-3-3818-4082.
    TANAKAS-ORT@h.u-tokyo.ac.jp

Ann. N.Y. Acad. Sci. 1116: 271–280 (2007). © 2007 New York Academy of Sciences.
doi: 10.1196/annals.1402.001

mitochondrial apoptosis pathway through cytochrome c release from mitochondria.[5,6] Released cytochrome c interacts with Apaf-1 and caspase-9 to form the apoptosome.[7] Caspase-9 in the apoptosome activates effector caspases (caspase-3 and caspase-7) that lead to apoptosis.[8] Bim is expressed in hematopoietic, epithelial, neuronal, and germ cells,[9] and Strasser and coworkers generated *bim*-deficient mice and demonstrated that Bim is essential for apoptosis of T-lymphocytes, B-lymphocytes, myeloid cells, neurons, and osteoclasts.[10–14] Accumulating evidence has revealed that the expression of Bim is regulated at both transcriptional and posttranslational levels.[15] It has been reported that Bim is regulated at the transcriptional level in hematopoietic progenitors and neurons,[12,13,16] and the forkhead-like transcription factor FOXO3A (forkhead box O3A; also known as FKHRL1) is involved in the transcriptional regulation of Bim in several types of cells.[17–24] The other important regulation of Bim is the posttranslational regulation that includes phosphorylation and ubiquitination. Bim is phosphorylated by extracellular-regulated kinase (ERK),[25–30] c-Jun N-terminal kinase (JNK),[31,32] and AKT,[33] and affects the expression level or the proapoptotic function of Bim. We and other groups previously reported that Bim is regulated via the ubiquitin–proteasome degradation process.[14,25,28,34,35] We found that Bim expression was markedly upregulated in the course of osteoclast apoptosis without changing its transcriptional level and was downregulated by M-CSF treatment. M-CSF maintained the protein level of Bim at low levels by inducing its ubiquitination, which was at least partly mediated by an E3 ubiquitin ligase c-Cbl.[14]

POSTTRANSLATIONAL REGULATION OF BIM IN MEF/BCL-2 CELLS

To investigate the molecular mechanisms underlying the degradation of Bim in detail, we analyzed the protein dynamics of Bim in mouse embryonic fibroblasts (MEFs). Since we found it difficult to overexpress Bim_{EL} in MEF cells, probably due to its strong proapoptotic activity, we generated MEF cells in which the expression level of Bcl-2 can be regulated by the Tet-off system (MEF/Bcl-2 cells) (FIG. 1). Bim_{EL} can be stably overexpressed in the cells without causing cell death. Doxycycline treatment reduced mRNA and protein levels of Bcl-2 in a time-dependent manner (FIG. 2A, B). After 12 h of doxycycline treatment, activation of caspase-3 was observed as shown by cleaved caspase-3 immunoblotting, which was associated with the decrease in Bim protein levels (FIG. 2A) without affecting its mRNA level (FIG. 2B). The protein levels of other Bcl-2 family members did not appear to alter during this period (FIG. 2A). To further investigate the mechanism of Bim degradation in MEF/Bcl-2 cells, we examined the effect of various proteinase inhibitors. The degradation of Bim was not affected by aprotinin, calpain inhibitor V, or cathepsin inhibitor E64, but was inhibited by a broad-spectrum

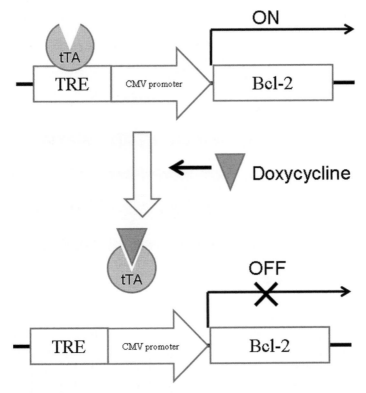

FIGURE 1. Schematic representation of Bcl-2 Tet-off system. Tet-off system is based on a tetracycline-regulatable transactivator (tTA), which induces transcription in the absence of tetracycline or its analog doxycycline (Dox). This promoter is composed of a tetracycline-responsive element (TRE) followed by a minimal promoter of the human cytomegalovirus (hCMV) immediate early gene. The tTA protein is a fusion protein composed of the TRE-binding domain of Tc repressor protein and the herpes simplex virus VP16 activation domain. In the presence of doxycycline, the binding of tTA to TRE was suppressed and the expression of Bcl-2 was shut down (500 ng/mL).[37]

caspase inhibitor zVAD-fmk and a caspase-3-specific inhibitor zDEVD-fmk (FIG. 3).

REGULATION OF BIM DEGRADATION BY CASPASE-3 BUT NOT BY CASPASE-7

To analyze the distinct role of caspases in the degradation of Bim in MEF/Bcl-2 cells, we performed gene silencing of caspases using the RNA interference technique. MEF/Bcl-2 cells overexpressing Bim were transduced

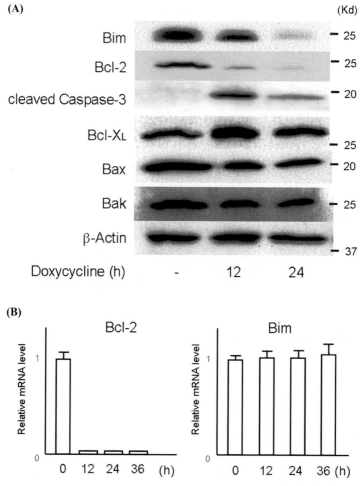

FIGURE 2. Regulation of Bcl-2 and Bim in MEF/Bcl-2 cells. (**A**) Time course of change in the expression of Bim, Bcl-2, cleaved caspase-3, and other Bcl-2 family members in Bim$_{EL}$ overexpressing MEF Bcl-2 Tet-off cells treated with doxycycline. Decrease in Bim expression levels was observed following Bcl-2 downregulation, while the expression of Bcl-xL, Bax, and Bak expression levels did not change. (**B**) Transcriptional regulation of *bcl-2* and *bim* in MEF/Bcl-2 cells. Doxycycline efficiently downregulated *bcl-2* mRNA levels, while *bim* mRNA level did not change in real-time PCR. The y-axis indicates relative mRNA levels.

with small hairpin RNA (shRNA) of caspases-3 (shRNA/caspases-3) using retrovirus vectors, and the cells were treated with doxycycline as described above. Caspase-3 expression was efficiently suppressed by shRNA/caspase-3 (FIG. 4A, upper panel), and active caspase-3 was barely detectable in shRNA/caspase-3-transduced cells even after doxycycline treatment. Bim

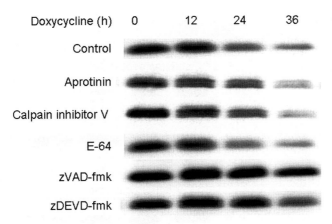

FIGURE 3. Regulation of Bim degradation by caspase(s). Bim degradation in MEF/Bcl-2 cells overexpressing Bim was suppressed by a pan caspase inhibitor zVAD-fmk and a caspase-3-specific inhibitor zDEVD-fmk, but not by aprotinin, calpain inhibitor V, or E-64.

degradation was markedly decreased compared to the control cells (FIG. 4A, lower panel). Although shRNA/caspase-7 introduction efficiently reduced caspase-7 expression in the cells, it did not affect caspase-3 activation or doxycycline-dependent Bim degradation (FIG. 4B). These data indicate that caspase-3 but not caspase-7 is involved in Bim degradation.

REGULATION OF BIM IN OSTEOCLASTS

We finally investigated whether similar regulation of Bim is observed in osteoclasts. Osteoclasts generated from mouse bone marrow cells in the presence of recombinant human M-CSF (10 ng/mL) and soluble RANKL (100 ng/mL) underwent cell death within 48 h after removal of these cytokines.[36] The protein level of Bim increased after 12 h of the cytokine removal and reduced again after 24 h (FIG. 5A). The reduction of Bim levels was associated with the activation of caspase-3 and caspase-7, and zVAD-fmk but no other proteinase inhibitors maintained Bim at high levels after 24 h of the cytokine removal (FIG. 5B). To further confirm the role of caspase-3 on Bim degradation in OCs, we generated OCs from *caspase-3*-deficient mouse bone marrow cells. The degradation of Bim was much reduced in *caspase-3 -/-* OCs, and a high level of Bim was maintained 24 h after the cytokine withdrawal (FIG. 5B). These results suggest that caspase-3 is critically involved in the degradation of Bim in osteoclasts as well.

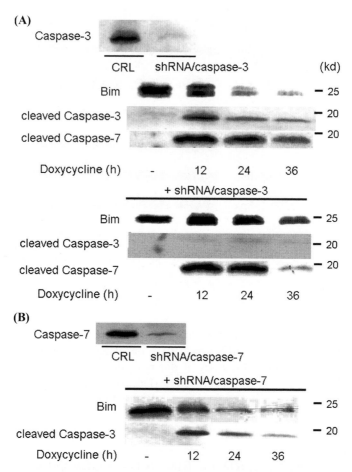

FIGURE 4. Involvement of caspase-3 in Bim degradation. (**A**) (*Upper*) Retrovirus vector-mediated small hairpin RNA of caspase-3 (shRNA/caspase-3) transduction to MEF/Bcl-2 cells efficiently downregulated caspase-3 expression. Knockdown of caspase-3 by shRNA suppressed Bim degradation in MEF/Bcl-2 cells. (*Middle*) In the control cells caspase-3 was activated after 12 h of doxycycline stimulation, and Bim expression level decreased in a time-dependent manner. (*Lower*) In shRNA/caspase-3 transduced cells, caspase-3 activation was barely observed, and degradation of Bim was markedly suppressed. (**B**) Caspase-7 does not affect Bim degradation. (*Upper*) Caspase-7 was reduced in shRNA/caspase-7-transduced cells. (*Lower*) ShRNA/caspase-7 introduction did not affect caspase-3 activation or Bim degradation by doxycycline in MEF/Bcl-2 cells.

INVOLVEMENT OF CASPASE-3 IN BIM DEGRADATION

In this study, we demonstrated the involvement of caspase-3 in the post-translational regulation of Bim in MEF cells and osteoclasts. Bim activates caspase cascades by inducing cytochrome *c* release from the mitochondria, and

(A)

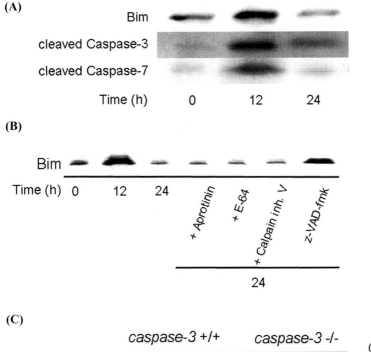

(B)

(C)

FIGURE 5. Involvement of caspase-3 in Bim degradation in osteoclasts. (A) Time course of change in Bim expression during osteoclast apoptosis. Osteoclasts were generated from bone marrow cells in the presence of soluble RANKL (100 ng/mL) and recombinant human M-CSF (10 ng/mL). They underwent apoptosis after 24 h of cytokine withdrawal. Expression of Bim, cleaved caspase-3, and cleaved caspase-7 was detected by Western blotting. Bim expression levels were increased 12 h after the cytokine deprivation, and then decreased after 24 h in association with caspase-3 and caspase-7 activation. (B) Bim degradation was suppressed by a pan-caspase inhibitor zVAD-fmk but not by aprotinin, calpain inhibitor V, or cathepsin inhibitor E64. (C) Bim expression in *caspase-3* -/- osteoclasts. Osteoclasts were generated from bone marrow cells of *caspase-3* -/- mice or their normal littermates, and subjected to apoptosis assay. Bim degradation was not observed 24 h after the cytokine removal in *caspase-3* -/- OCs. Caspase-7 activation was observed in *caspase-3* -/- OCs at the comparable level as control OCs.

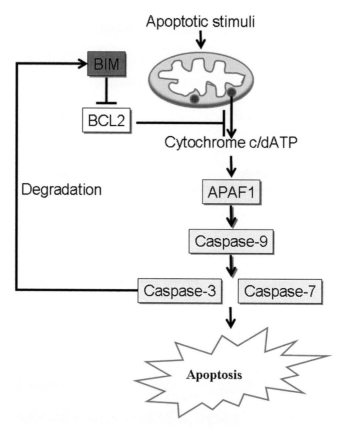

FIGURE 6. Possible role of caspase-3 on Bim degradation. Apoptotic stimuli induce cytochrome *c* release from mitochondria, which is considered to be induced by the inhibition of Bcl-2 by Bim, and subsequently cause caspase-3 and caspase-7 activation. Activated caspase-3 in turn causes degradation of Bim, thus creating a negative feedback loop.

induces apoptosis of the cells. Caspase-3 in turn induces Bim degradation (FIG. 6), although the molecular events underlying still remain elusive. Since osteoclasts are primary cells for bone resorption, the regulation of their apoptosis is critical for maintaining skeletal homeostasis and can be a potential therapeutic target. Further study is required to elucidate the detailed mechanisms of Bim regulation and its importance in osteoclast biology.

ACKNOWLEDGMENTS

This work was supported in part by Grants-in-Aid from the Ministry of Education, Culture, Sports, Science, and Technology of Japan and the Health

Science research grants from the Ministry of Health, Labor, and Welfare of Japan to S.T.

REFERENCES

1. KERR, J.F., A.H. WYLLIE & A.R. CURRIE. 1972. Apoptosis: a basic biological phenomenon with wide-ranging implications in tissue kinetics. Br. J. Cancer **26:** 239–257.
2. THOMPSON, C.B. 1995. Apoptosis in the pathogenesis and treatment of disease. Science **267:** 1456–1462.
3. OPFERMAN, J.T. & S.J. KORSMEYER. 2003. Apoptosis in the development and maintenance of the immune system. Nat. Immunol. **4:** 410–415.
4. GROSS, A., J.M. MCDONNELL & S.J. KORSMEYER. 1999. BCL-2 family members and the mitochondria in apoptosis. Genes Dev. **13:** 1899–1911.
5. O'CONNOR, L. *et al.* 1998. Bim: a novel member of the Bcl-2 family that promotes apoptosis. EMBO J. **17:** 384–395.
6. PUTHALAKATH, H. *et al.* 1999. The proapoptotic activity of the Bcl-2 family member Bim is regulated by interaction with the dynein motor complex. Mol. Cell **3:** 287–296.
7. ACEHAN, D. *et al.* 2002. Three-dimensional structure of the apoptosome: implications for assembly, procaspase-9 binding, and activation. Mol. Cell **9:** 423–432.
8. SIEGEL, R.M. 2006. Caspases at the crossroads of immune-cell life and death. Nat. Rev. Immunol. **6:** 308–317.
9. O'REILLY, L.A. *et al.* 2000. The proapoptotic BH3-only protein bim is expressed in hematopoietic, epithelial, neuronal, and germ cells. Am. J. Pathol. **157:** 449–461.
10. BOUILLET, P. *et al.* 1999. Proapoptotic Bcl-2 relative Bim required for certain apoptotic responses, leukocyte homeostasis, and to preclude autoimmunity. Science **286:** 1735–1738.
11. BOUILLET, P. *et al.* 2002. BH3-only Bcl-2 family member Bim is required for apoptosis of autoreactive thymocytes. Nature **415:** 922–926.
12. PUTCHA, G.V. *et al.* 2001. Induction of BIM, a proapoptotic BH3-only BCL-2 family member, is critical for neuronal apoptosis. Neuron **29:** 615–628.
13. WHITFIELD, J. *et al.* 2001. Dominant-negative c-Jun promotes neuronal survival by reducing BIM expression and inhibiting mitochondrial cytochrome c release. Neuron **29:** 629–643.
14. AKIYAMA, T. *et al.* 2003. Regulation of osteoclast apoptosis by ubiquitylation of proapoptotic BH3-only Bcl-2 family member. Bim. Embo. J. **22:** 6653–6664.
15. HUANG, D.C. & A. STRASSER. 2000. BH3-Only proteins-essential initiators of apoptotic cell death. Cell **103:** 839–842.
16. SHINJYO, T. *et al.* 2001. Downregulation of Bim, a proapoptotic relative of Bcl-2, is a pivotal step in cytokine-initiated survival signaling in murine hematopoietic progenitors. Mol. Cell Biol. **21:** 854–864.
17. ESSAFI, A. *et al.* 2005. Direct transcriptional regulation of Bim by FoxO3a mediates STI571-induced apoptosis in Bcr-Abl-expressing cells. Oncogene **24:** 2317–2329.
18. GILLEY, J., P.J. COFFER & J. HAM. 2003. FOXO transcription factors directly activate bim gene expression and promote apoptosis in sympathetic neurons. J. Cell Biol. **162:** 613–622.

19. MOLLER, C. *et al.* 2005. Stem cell factor promotes mast cell survival via inactivation of FOXO3a-mediated transcriptional induction and MEK-regulated phosphory-lation of the proapoptotic protein Bim. Blood **106:** 1330–1336.

20. ROSAS, M. *et al.* 2005. Cytokine mediated suppression of TF-1 apoptosis requires PI3K activation and inhibition of Bim expression. FEBS Lett. **579:** 191–198.

21. STAHL, M. *et al.* 2002. The forkhead transcription factor FoxO regulates transcrip-tion of p27Kip1 and Bim in response to IL-2. J. Immunol. **168:** 5024–5031.

22. SUNTERS, A. *et al.* 2003. FoxO3a transcriptional regulation of Bim controls apopto-sis in paclitaxel-treated breast cancer cell lines. J. Biol. Chem. **278:** 49795–49805.

23. URBICH, C. *et al.* 2005. FOXO-dependent expression of the proapoptotic protein Bim: pivotal role for apoptosis signaling in endothelial progenitor cells. FASEB J. **19:** 974–976.

24. DIJKERS, P.F. *et al.* 2002. FKHR-L1 can act as a critical effector of cell death induced by cytokine withdrawal: protein kinase B-enhanced cell survival through maintenance of mitochondrial integrity. J. Cell Biol. **156:** 531–542.

25. LEY, R. *et al.* 2003. Activation of the ERK1/2 signaling pathway promotes phos-phorylation and proteasome-dependent degradation of the BH3-only protein. Bim. J. Biol. Chem. **278:** 18811–18816.

26. LEY, R. *et al.* 2005. Regulatory phosphorylation of Bim: sorting out the ERK from the JNK. Cell Death Differ. **12:** 1008–1014.

27. LEY, R. *et al.* 2004. Extracellular signal-regulated kinases 1/2 are serum-stimulated "Bim(EL) kinases" that bind to the BH3-only protein Bim(EL) causing its phos-phorylation and turnover. J. Biol. Chem. **279:** 8837–8847.

28. LUCIANO, F. *et al.* 2003. Phosphorylation of Bim-EL by Erk1/2 on serine 69 pro-motes its degradation via the proteasome pathway and regulates its proapoptotic function. Oncogene **22:** 6785–6793.

29. HARADA, H. *et al.* 2004. Survival factor-induced extracellular signal-regulated kinase phosphorylates BIM, inhibiting its association with BAX and proapoptotic activity. Proc. Natl. Acad. Sci. USA **101:** 15313–15317.

30. BISWAS, S.C. & L.A. GREENE. 2002. Nerve growth factor (NGF) down-regulates the Bcl-2 homology 3 (BH3) domain-only protein Bim and suppresses its proapop-totic activity by phosphorylation. J. Biol. Chem. **277:** 49511–49516.

31. LEI, K. & R.J. DAVIS. 2003. JNK phosphorylation of Bim-related members of the Bcl2 family induces Bax-dependent apoptosis. Proc. Natl. Acad. Sci. USA **100:** 2432–2437.

32. PUTCHA, G.V. *et al.* 2003. JNK-mediated BIM phosphorylation potentiates BAX-dependent apoptosis. Neuron **38:** 899–914.

33. QI, X.J., G.M. WILDEY & P.H. HOWE. 2006. Evidence that Ser87 of BimEL is phosphorylated by Akt and regulates BimEL apoptotic function. J. Biol. Chem. **281:** 813–823.

34. STYLES, N.A., W. ZHU & X. LI. 2005. Phosphorylation and down-regulation of Bim by muscarinic cholinergic receptor activation via protein kinase C. Neurochem. Int. **47:** 519–527.

35. MELLER, R. *et al.* 2006. Rapid degradation of Bim by the ubiquitin-proteasome pathway mediates short-term ischemic tolerance in cultured neurons. J. Biol. Chem. **281:** 7429–7436.

36. MIYAZAKI, T. *et al.* 2000. Reciprocal role of ERK and NF-kappaB pathways in survival and activation of osteoclasts. J. Cell Biol. **148:** 333–342.

37. MASUI, S. *et al.* 2005. An efficient system to establish multiple embryonic stem cell lines carrying an inducible expression unit. Nucleic Acids Res. **33:** e43.

Osteocytes as Dynamic Multifunctional Cells

LYNDA F. BONEWALD

University of Missouri, School of Dentistry, Kansas City, Missouri, USA

ABSTRACT: The target of bone systemic factors and therapeutics has been assumed to be primarily osteoblasts and/or osteoclasts and their precursors. All the action with regard to bone modeling or remodeling has been assumed to take place on the bone surface. In this scenario, cells below the bone surface, that is, osteocyte, are considered to be inactive placeholders in the bone matrix. New data show osteocytes are involved. In addition to the function of osteocytes translating mechanical strain into biochemical signals between osteocytes and cells on the bone surface to affect (re)modeling, new functions are emerging. Osteocytes are exquisitely sensitive to mechanical strain in the form of shear stress compared to osteoblasts or osteoclasts and communicate with each other, with cells on the bone surface, and with marrow cells. Osteocytes are able to move their cell body and their dendritic processes and appear to be able to modify their local microenvironment. A novel function now attributed to osteocytes includes regulation of phosphate metabolism. Therefore, in addition to osteoblasts and osteoclasts, osteocytes are also important for bone health.

KEYWORDS: osteocytes; mechanical load; E11/gp38; perilacunar matrix; Dmp1; Pex; FGF23

OSTEOCYTES AS ORCHESTRATORS OF BONE ADAPTATION TO LOAD

Galileo in 1638 first suggested that the shape of bones is related to loading, but it was Julius Wolff in 1892 who proposed that bone accommodates or responds to strain (for review see Ref. 1).[1] Harold Frost proposed "Four Windows" of mechanical strain: (1) disuse or lack of strain that results in bone loss as occurs in astronauts, immobilized patients, and in the jaw bones of edentulous patients; (2) homeostasis, where resorption equals formation and is therefore necessary for the normal maintenance of bone mass; (3) modeling, where the application of load, such as exercise, results in an increase in

Address for correspondence: Lynda F. Bonewald, Ph.D., University of Missouri, School of Dentistry, 650 E 25th St., Kansas City, MO 64108. Voice: 816-235-2068; fax: 816- 235-5524.
bonewaldl@umkc.edu

Ann. N.Y. Acad. Sci. 1116: 281–290 (2007). © 2007 New York Academy of Sciences.
doi: 10.1196/annals.1402.018

bone mass essentially through new bone formation; and (4) pathologic overload where extreme load is applied, such as occurs with new military recruits and race horses where tremendous resorption is followed by formation.[2] So for decades, (if not centuries), the question has been asked as to how bone responds to load or to unloading. A central theory for the past 30 to 40 years is that the osteocyte is ideally located in bone to sense mechanical strain and translate strain (or lack of) into biochemical signals to cells on the bone surface. However, this theory has been difficult to prove as osteocytes have been relatively inaccessible.

The osteocyte is derived from the osteoblast. The osteoblast progenitor is recruited to the bone surface where it differentiates into a polygonal matrix-producing cell. It is not clear why some of these cells are designated to become osteocytes,[3,4] but they make connections with existing embedded cells and then are engulfed in osteoid where they are referred to as osteoid–osteocytes.[5] Once the matrix around them becomes mineralized they are referred to as mature osteocytes. It is this cell embedded in mineralized matrix that has been referred to as a passive, inactive cell acting as a placeholder in bone.

Osteocytes make up more than 90–95% of all bone cells in the adult skeleton, whereas osteoblasts compose less than 5% and osteoclasts less than 1%. Osteocytes are viable for years, even decades, whereas osteoblasts live lifetimes of weeks and osteoclasts of days. The unique feature of osteocytes is the formation of long dendritic processes that travel through small tunnels in the bone matrix called canaliculi that connect osteocytes within their caves or lacunae with cells on the bone surface. These processes have been shown to extend into the bone marrow.[6] Osteocytes are thought to send signals of both bone resorption and bone formation, but this has not been validated. It has been suggested that dying or dead osteocytes send signals of resorption[7,8] and recently it has been shown that a protein highly expressed in osteocytes called sclerostin can target osteoblasts to inhibit bone formation.[9] Osteocytes may act as conductors with regard to directing both osteoclast and osteoblast activity, as orchestrators of bone remodeling (see FIG. 1).

A major question in the field is how does the osteocyte sense mechanical strain? It is thought that cells on the bone surface (lining cells, osteoblasts) are most likely to be subjected to substrate strain, whereas osteocytes are more likely to sense mechanical strain due to fluid flow shear stress. Osteocytes as compared to osteoblasts are more responsive to fluid flow shear stress than to other forms of mechanical strain, such as substrate stretching.[10] It has been proposed that osteocytes sense shear stress mainly along their dendritic processes, or along their dendritic processes and the cell body. Recently, it has also been proposed that cilia may play a role in osteocyte mechanosensation.[11] Preliminary data show that in vitro, osteocyte cell deformation correlates with magnitude of shear stress, which in turn correlates with a biological response, that of prostaglandin release (data not shown). It remains to be shown if this occurs in vivo. PKD1 and 2, components of cilia, which are known to have

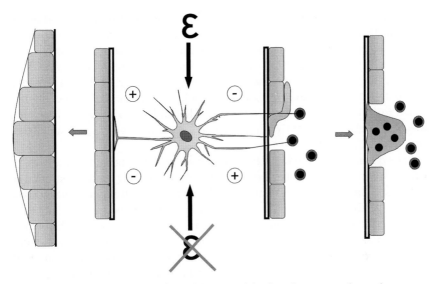

FIGURE 1. Simplified diagram of the potential roles of osteocytes in mechanotransduction. The osteocyte is in direct contact with cells on the bone surface and can extend processes into the marrow space. The role of the lining cell in mechanotransduction remains a mystery. The lining cells could be an intermediary for signaling or could be bypassed by the osteocyte. ε = Normal Loading. With normal loading of the skeleton that maintains normal homeostasis, the osteocyte sends signals inhibiting osteoclast activity.[46,47] Expression of Dmp1, necessary for mineralization of osteoid, is increased in response to load.[43] Osteocytes may also send signals to mesenchymal stem cells to differentiate into osteoblasts.[48] $X\varepsilon$ = Unloading. Upon immobilization, the osteocyte sends signals of resorption.[46] These signals may come from dying, apoptotic osteocytes[7,49,50] or from viable cells.[51] It has also been shown that sclerostin expression is increased in response to unloading, potentially inhibiting the activity of osteoblasts.[52]

a mechanosensory function in the kidney, are expressed in bone. Deletion of PKD1 function results in animals with a bone defect.[12] However, even though cilia most likely do play an important role in osteocytes, it is not clear how and if cilia are responsible for mechanotransduction.

FORMATION OF THE OSTEOCYTE LACUNOCANALICULAR SYSTEM

We have asked the question, how is the lacunocanalicular network formed and what is its function? The dogma has been that osteocytes form their dendrites through a passive process once the cell becomes surrounded by osteoid. This concept arose because the osteocyte has 30% of the cytoplasm of a matrix producing osteoblast, therefore, the cytoplasm shrinks as the dendritic processes are left behind in the matrix. In a search for markers highly expressed on osteocytes, the E11/gp38 molecule was found first in MLO-Y4 osteocyte-like cells and also in early embedding osteocytes in bone but not in cells on the bone surface.[13,14] E11/gp38 is a 40 kDa transmembrane protein thought to play a role in the formation of dendritic processes in various cell types. Cells with extensive cellular projections, such as podocytes and type 1 alveolar lung cells, etc., express high amounts of E11/gp38. This membrane molecule appears to play a role in dendrite elongation, as MLO-Y4 cells subjected to fluid flow shear stress elongate their processes and this elongation was blocked by siRNA.[14] Conditional deletion of this gene is a neonatal lethal due to lung defects.[15]

In vivo loading resulted in elevation in both gene and protein expression of E11/gp38, not only near the bone surface, but in deeply embedded bone in response to loading.[14] It was not clear why a molecule proposed to have a role in dendrite formation would be increased in deeply embedded osteocytes—cells thought to have their dendrites stationary and tethered to the walls of their canaliculi.[16,17] However, dynamic imaging of viable calvaria has shown that osteocytes can extend and retract their dendritic processes.[18] This suggests that E11/gp38 could be involved in the extension of dendrites in osteocytes embedded in bone in response to load. Observations using static data limit our thinking and ability to form more accurate and novel hypotheses, whereas dynamic imaging has opened a whole new area for investigation.

OSTEOCYTE MODIFICATION OF THEIR MICROENVIRONMENT

Early pioneers in osteocyte biology noted from histological sections of bone that there appeared to be increases of osteocyte lacunar size with various diseases or conditions.[19,20] This led to the highly controversial theory of "osteocytic osteolysis" where it was proposed that osteocytes could remove bone.[21] This theory fell out of favor because it was proposed that osteocytes could use the same mechanisms as osteoclasts to remove bone. Recently, with the advent of new technology, such as atomic force microscopy and nanoindentation, Raman spectrometry, and Scanning Acoustic Microscopy, data are emerging to suggest that the perilacunar matrix surrounding the osteocyte is distinct from the rest of the bone matrix. It appears that the perilacunar matrix has a lower elastic modulus or is hypomineralized or "softer" than the surrounding matrix (data not published). Changes in perilacunar matrix could alter the type or magnitude of strain sensed by the osteocyte and therefore modify their responsiveness to load.[22]

The osteocyte can modify its microenvironment in response to environmental factors. It has been shown that administration of glucocorticoids appears to enlarge or increase this perilacunar matrix and also leads to a significant increase in lacunar size.[23] Glucocorticoids not only appear to induce apoptosis of osteoblasts and osteocytes,[24] but appear to inhibit osteoclast activity.[25] It has been proposed that osteocytes send signals to the osteoclast to initiate re-modeling. If the osteoclast is prevented from responding to the osteocyte call to resorb, then the osteocyte may become compromised and begin to remove mineral from its lacunae and from its surrounding matrix. It is not clear at this time what molecular mechanisms would be responsible. As the osteocyte can remove mineral from its lacunae and perilacunar matrix, this cell may also be able to modify the diameter of its canaliculi. Any change in canalicular diameter would have an effect on the flow of the bone fluid. An increase in canalicular diameter would decrease shear stress and a decrease in diameter would increase shear stress. It has been shown that the number of canaliculi per osteocyte increase with age.[26,27] It is not known if the already embedded osteocyte can generate new canaliculi or with bone remodeling with new bone deposition, the new osteocytes have greater numbers of dendrites/canaliculi. This could be one of the reasons why the aging skeleton is less responsive to load.

ROLE OF OSTEOCYTES IN MINERAL METABOLISM

One can ascribe function to a cell type by identification of specific or se-lective markers of known function. Whereas E11/gp38 is a marker for early osteocytes, *sost*/sclerostin is a marker for late embedded osteocytes.[28,29] Dele-tion of Sost in mice results in increased bone formation and mutation of *Sost* in humans results in Sclerostosis.[30] Sclerostin is a direct inhibitor of the Wnt pathway through binding to Lrp5.[31–33] The anabolic effect of PTH may be through inhibition of Sost expression.[34] These data support the hypothesis that the osteocyte can regulate bone formation and mineralization by targeting the osteoblast.

There are also three key molecules expressed in osteocytes that play a role in phosphate homeostasis; highly expressed in osteocytes and an avian osteo-cyte specific marker, Dentin Matrix Protein 1, (Dmp1), and *Pex*/Phex, (Phos-phate Regulating Neutral Endopeptidase on Chromosome X) both highly ex-pressed in osteocytes,[35,36] and FGF23, also expressed in osteocytes, but at much lower levels.[37] Deletion or mutation of either Pex or Dmp1 results in hypophosphatemic rickets resulting from a dramatic elevation of FGF23 in osteocytes.[37,38] Hypophosphatemic rickets in humans is caused by inactivat-ing mutations of *Pex* and autosomal recessive hypophosphatemia in humans is due to mutations in Dmp1, both resulting in elevated circulating levels of FGF23.[38–40] FGF23 is a phosphaturic factor that prevents reabsorption of Pi by the kidney leading to hypophosphatemia (see Fig. 2).

Dmp1 appears to have functions in addition to regulation of FGF23 expres-sion. Dmp1 has been shown to have a nuclear localization sequence,[41] the secreted protein is highly phosphorylated, and recombinant Dmp1 nucleates apatite.[42] This suggests that Dmp1 could function intracellulary to regulate transcription and extracellulary to regulate mineralization of osteoid. Dmp1

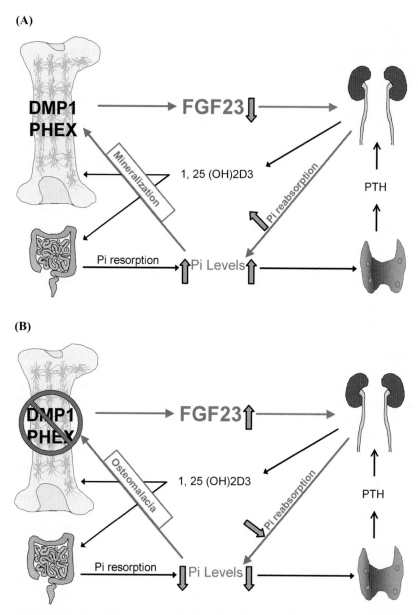

FIGURE 2. Simplified diagram of the interactions between Dmp1, Phex, and FGF23, all shown to be expressed in osteocytes. Both Dmp1 and Phex appear to downregulate FGF23 expression, which in turn allows reabsorption of phosphate by the kidney thereby maintaining sufficient circulating phosphate to maintain bone mineral content (**A**). In the absence of either Dmp1 or Phex, FGF23 is highly elevated in osteocytes leading to phosphate excretion by the kidney thereby lowering circulating phosphate leading to osteomalacia and rickets (**B**). Osteocytes appear to play a major role in mineral homeostasis.

null mice fed a high phosphate diet showed rescue of the length of the long bones, but not a full rescue of the osteomalacia.[38] Unmineralized osteoid surrounding osteocytes was still present in these animals. This suggests that Dmp1 has both a systemic and a local function. As the length of the bones in the Dmp1 null is rescued by high phosphate diet, this suggests that Dmp1 regulates FGF23 expression, which in turn systemically regulates phosphate. As the osteomalacia is not completely corrected, this suggests that Dmp1 regulates mineralization of bone osteoid. Based on these observations we have proposed that the osteocyte network can function as an endocrine gland to regulate phosphate homeostasis. As both Dmp1 and Phex are regulated by mechanical loading,[43–45] it will be important to determine if skeletal loading can play a role in mineral and phosphate metabolism.

ACKNOWLEDGMENT

This work was supported by NIH NIAMS PO1 AR46798.

REFERENCES

1. BONEWALD, L. 2007. *In* The Osteocyte. D.F.R. Marcus, C. Rosen, Eds.: Ch.8 Osteoporosis. Elsevier.
2. FROST, H.M. 1992. Perspectives: bone's mechanical usage windows. Bone Mineral **19:** 257–271.
3. MAROTTI, G., V. CANE, S. PALAZZINI & C. PALUMBO. 1990. Structure-function relationships in the osteocyte. Ital. J. Min. Electrolyte Metab. **4:** 93–106.
4. FRANZ-ODENDAAL, T.A., B.K. HALL & P.E. WITTEN. 2006. Buried alive: how osteoblasts become osteocytes. Dev. Dyn. **235:** 176–190.
5. PALUMBO, C. 1986. A three-dimensional ultrastructural study of osteoid-osteocytes in the tibia of chick embryos. Cell Tissue Res. **246:** 125–131.
6. KAMIOKA, H., T. HONJO & T. TAKANO-YAMAMOTO. 2001. A three-dimensional distribution of osteocyte processes revealed by the combination of confocal laser scanning microscopy and differential interference contrast microscopy. Bone **28:** 145–149.
7. VERBORGT, O., N.A. TATTON, R.J. MAJESKA & M.B. SCHAFFLER. 2002. Spatial distribution of Bax and Bcl-2 in osteocytes after bone fatigue: complementary roles in bone remodeling regulation? J. Bone Miner. Res. **17:** 907–914.
8. MANOLAGAS, S.C. 2006. Choreography from the tomb: An emerging role of dying osteocytes in the purposeful, and perhaps not so purposeful, targeting of bone remodeling. BoneKEy **3:** 5–14.
9. VAN BEZOOIJEN, R.L., P. TEN DIJKE, S.E. PAPAPOULOS & C.W. LOWIK. 2005. SOST/sclerostin, an osteocyte-derived negative regulator of bone formation. Cytokine Growth Factor Rev. **16:** 319–327.
10. WESTBROEK, I., N.E. AJUBI, M.J. ALBLAS, *et al.* 2000. Differential stimulation of prostaglandin G/H synthase-2 in osteocytes and other osteogenic cells by pulsating fluid flow. Biochem. Biophys. Res. Commun. **268:** 414–419.

11. BONEWALD, L.F. 2006. Mechanosensation and transduction in osteocytes. Bonekey Osteovision **3:** 7–15.

12. XIAO, Z., S. ZHANG, J. MAHLIOS, *et al.* 2006. Cilia-like structures and polycystin-1 in osteoblasts/osteocytes and associated abnormalities in skeletogenesis and Runx2 expression. J. Biol. Chem. **281:** 30884–30895.

13. SCHULZE, E., M. WITT, M. KASPER, *et al.* 1999. Immunohistochemical investigations on the differentiation marker protein E11 in rat calvaria, calvaria cell culture and the osteoblastic cell line ROS 17/2.8. Histochem. Cell Biol. **111:** 61–69.

14. ZHANG, K., C. BARRAGAN-ADJEMIAN, L. YE, *et al.* 2006. E11/gp38 selective expression in osteocytes: regulation by mechanical strain and role in dendrite elongation. Mol. Cell Biol. **26:** 4539–4552.

15. RAMIREZ, M.I., G. MILLIEN, A. HINDS, *et al.* 2003. T1alpha, a lung type I cell differentiation gene, is required for normal lung cell proliferation and alveolus formation at birth. Dev. Biol. **256:** 61–72.

16. WEINBAUM, S., S.C. COWIN & Y. ZENG. 1994. A model for the excitation of osteocytes by mechanical loading-induced bone fluid shear stresses. J. Biomech. **27:** 339–360.

17. YOU, L.D., S. WEINBAUM, S.C. COWIN & M.B. SCHAFFLER. 2004. Ultrastructure of the osteocyte process and its pericellular matrix. Anat. Rec. A Discov. Mol. Cell. Evol. Biol. **278:** 505–513.

18. VENO, P., D.P. NICOLELLA, P. SIVAKUMAR, *et al.* 2006. Live imaging of osteocytes within their lacunae reveals cell body and dendrite motions. J. Bone. Min. Res. **21** (Suppl 1): 538.

19. BONUCCI, E. & G. GHERARDI. 1977. Osteocyte ultrastructure in renal osteodystrophy. Virchows Arch. A Pathol. Anat. Histol **373:** 213–231.

20. MARIE, P.J. & F.H. GLORIEUX. 1983. Relation between hypomineralized periosteocytic lesions and bone mineralization in vitamin D-resistant rickets. Calcif. Tissue Int. **35:** 443–448.

21. BELANGER, L.F. 1968. Osteocytic osteolysis. Calcif. Tissue Res. **4:** 1–12.

22. RATH BONIVTCH, A., L.F. BONEWALD & D.P. NICOLELLA. 2006. Tissue strain amplification at the osteocyte lacuna: a microstructural finite element analysis. J. Biomech. **40:** 2199–2206.

23. LANE, N.E., W. YAO, M. BALOOCH, *et al.* 2006. Glucocorticoid-treated mice have localized changes in trabecular bone material properties and osteocyte lacunar size that are not observed in placebo-treated or estrogen-deficient mice. J. Bone Miner. Res. **21:** 466–476.

24. WEINSTEIN, R.S., R.L. JILKA, A.M. PARFITT & S.C. MANOLAGAS. 1998. Inhibition of osteoblastogenesis and promotion of apoptosis of osteoblasts and osteocytes by glucocorticoids. Potential mechanisms of their deleterious effects on bone. J. Clin. Invest. **102:** 274–282.

25. KIM, H.J., H. ZHAO, H. KITAURA, *et al.* 2006. Glucocorticoids suppress bone formation via the osteoclast. J. Clin. Invest **116:** 2152–2160.

26. OKADA, S., S. YOSHIDA, S.H. ASHRAFI & D.E. SCHRAUFNAGEL. 2002. The canalicular structure of compact bone in the rat at different ages. Microsc. Microanal. **8:** 104–115.

27. HOLMBECK, K., P. BIANCO, I. PIDOUX, *et al.* 2005. The metalloproteinase MT1-MMP is required for normal development and maintenance of osteocyte processes in bone. J. Cell Sci. **118**(Pt 1): 147–156.

28. VAN BEZOOIJEN, R.L., B.A. ROELEN, A. VISSER, *et al.* 2004. Sclerostin is an osteocyte-expressed negative regulator of bone formation, but not a classical BMP antagonist. J. Exp. Med. **199:** 805–814.

29. POOLE, K.E., R.L. VAN BEZOOIJEN, N. LOVERIDGE, *et al.* 2005. Sclerostin is a delayed secreted product of osteocytes that inhibits bone formation. FASEB J. **19:** 1842–1844.
30. BALEMANS, W., M. EBELING, N. PATEL, *et al.* 2001. Increased bone density in sclerosteosis is due to the deficiency of a novel secreted protein (SOST). Hum. Mol. Genet. **10:** 537–543.
31. LI, X., Y. ZHANG, H. KANG, *et al.* 2005. Sclerostin binds to LRP5/6 and antagonizes canonical Wnt signaling. J. Biol. Chem. **280:** 19883–19887.
32. ELLIES, D.L., B. VIVIANO, J. MCCARTHY, *et al.* 2006. Bone density ligand, Sclerostin, directly interacts with LRP5 but not LRP5G171V to modulate Wnt activity. J. Bone Miner. Res. **21:** 1738–1749.
33. SEMENOV, M., K. TAMAI & X. HE. 2005. SOST is a ligand for LRP5/LRP6 and a Wnt signaling inhibitor. J. Biol. Chem. **280:** 26770–26775.
34. BELLIDO, T. 2006. Downregulation of SOST/sclerostin by PTH: a novel mechanism of hormonal control of bone formation mediated by osteocytes. J. Musculoskelet. Neuronal Interact. **6:** 358–359.
35. TOYOSAWA, S., S. SHINTANI, T. FUJIWARA, *et al.* 2001. Dentin matrix protein 1 is predominantly expressed in chicken and rat osteocytes but not in osteoblasts. J. Bone Miner. Res. **16:** 2017–2026.
36. WESTBROEK, I., K.E. DE ROOIJ & P.J. NIJWEIDE. 2002. Osteocyte-specific monoclonal antibody MAb OB7.3 is directed against Phex protein. J. Bone Miner. Res. **17:** 845–853.
37. LIU, S., J. ZHOU, W. TANG, *et al.* 2006. Pathogenic role of Fgf23 in Hyp mice. Am. J. Physiol. Endocrinol. Metab. **291:** E38–E49.
38. FENG, J.Q., L.M. WARD, S. LIU, *et al.* 2006. Loss of DMP1 causes rickets and osteomalacia and identifies a role for osteocytes in mineral metabolism. Nat. Genet. **38:** 1310–1315.
39. [no authors listed] 1995. A gene (PEX) with homologies to endopeptidases is mutated in patients with X-linked hypophosphatemic rickets. The HYP Consortium. Nat. Genet. **11:** 130–136.
40. LORENZ-DEPIEREUX, B., M. BASTEPE, A. BENET-PAGES, *et al.* 2006. DMP1 mutations in autosomal recessive hypophosphatemia implicate a bone matrix protein in the regulation of phosphate homeostasis. Nat. Genet. **38:** 1248–1250.
41. NARAYANAN, K., A. RAMACHANDRAN, J. HAO, *et al.* 2003. Dual functional roles of dentin matrix protein 1. Implications in biomineralization and gene transcription by activation of intracellular Ca2+ store. J. Biol. Chem. **278:** 17500–17508.
42. HE, G., T. DAHL, A. VEIS & A. GEORGE. 2003. Nucleation of apatite crystals *in vitro* by self-assembled dentin matrix protein 1. Nat. Mater. **2:** 552–558.
43. GLUHAK-HEINRICH, J., L. YE, L.F. BONEWALD, *et al.* 2003. Mechanical loading stimulates dentin matrix protein 1 (DMP1) expression in osteocytes *in vivo*. J. Bone Miner. Res. **18:** 807–817.
44. GLUHAK-HEINRICH, J., D. PAVLIN, W. YANG, *et al.* 2007. MEPE expression in osteocytes during orthodontic tooth movement. Arch. Oral Biol. **52:** 684–690.
45. YANG, W., Y. LU, I. KALAJZIC, *et al.* 2005. Dentin matrix protein 1 gene cis-regulation: use in osteocytes to characterize local responses to mechanical loading *in vitro* and *in vivo*. J. Biol. Chem. **280:** 20680–20690.
46. TATSUMI, S., K. ISHII, N. AMIZUKA, *et al.* 2007. Targeted ablation of osteocytes induces osteoporosis with defective mechanotransduction. Cell Met. **5:** 464–475.

47. Gu, G., M. Mulari, Z. Peng, *et al.* 2005. Death of osteocytes turns off the inhibition of osteoclasts and triggers local bone resorption. Biochem. Biophys. Res. Commun. **335:** 1095–1101.
48. Heino, T.J., T.A. Hentunen & H.K. Vaananen. 2004. Conditioned medium from osteocytes stimulates the proliferation of bone marrow mesenchymal stem cells and their differentiation into osteoblasts. Exp. Cell Res. **294:** 458–468.
49. Noble, B.S., H. Stevens, N. Loveridge & J. Reeve. 1997. Identification of apoptotic changes in osteocytes in normal and pathological human bone. Bone **20:** 273–282.
50. Aguirre, J.I., L.I. Plotkin, S.A. Stewart, *et al.* 2006. Osteocyte apoptosis is induced by weightlessness in mice and precedes osteoclast recruitment and bone loss. J. Bone Miner. Res. **21:** 605–615.
51. Zhao, S., Y.K. Zhang, S. Harris, *et al.* 2002. MLO-Y4 osteocyte-like cells support osteoclast formation and activation. J. Bone Miner. Res. **17:** 2068–2079.
52. Robling, A.G., T. Bellido & C.H. Turner. 2006. Mechanical stimulation *in vivo* reduces osteocyte expression of sclerostin. J. Musculoskelet. Neuronal Interact. **6:** 354.

A Chromosomal Inversion within a Quantitative Trait Locus Has a Major Effect on Adipogenesis and Osteoblastogenesis

CHERYL L. ACKERT-BICKNELL,[a] JESSE L. SALISBURY,[a,b]
MARK HOROWITZ,[c] VICTORIA E. DeMAMBRO,[a]
LINDSAY G. HORTON,[a] KATHRYN L. SHULTZ,[a]
BEATA LECKA-CZERNIK,[d] AND CLIFFORD J. ROSEN[a]

[a] The Jackson Laboratory, Bar Harbor, Maine, USA

[b] Graduate School for Biomedical Science, University of Maine, Orono, Maine, USA

[c] Yale University School of Medicine, New Haven, Connecticut, USA

[d] University of Arkansas Medical Center, Little Rock, Arkansas, USA

ABSTRACT: We mapped a quantitative trait locus (QTL) for BMD to mid-distal chromosome (Chr) 6 in a cross between C57BL/6J (B6) and C3H/HeJ (C3H). The B6.C3H-6T (6T) congenic was developed to map candidate genes in this QTL. Recently, a 25 cM paracentric inversion was discovered on Chr 6 in C3H/HeJ; we found 6T also carries this inversion. Microarrays from the liver of B6 and 6T uncovered two narrow bands of decreased gene expression in close proximity to the predicted locations of the inversion breakpoints. Changes in specific gene expression in 6T were consistent with its phenotype of low trabecular bone volume and marrow adipogenesis. The BXH recombinant inbred (RI) strains do not carry the C3H/HeJ inversion. To test if the inversion, or allelic effects, were responsible for the 6T phenotype, we made a new congenic, B.H-6, developed by introgressing a 30 Mb region of C3H genomic sequence from BXH6 onto a B6 background. While genetically identical to 6T, this new congenic had a distinct metabolic and skeletal phenotype, with more body fat and greater trabecular BV/TV compared to B6 or 6T. We conclude that the phenotype of 6T cannot be explained by simple allelic differences in one or more genes from C3H. Rather, 6T demonstrates that disordered regulation of gene expression by genomic rearrangement can have a profound effect on a complex trait, such as BMD, and that genomic rearrangement can supersede the effects of various alleles.

Address for correspondence: Dr. Clifford J. Rosen, The Jackson Laboratory, 600 Main Street, Bar Harbor, ME, USA, 04609. Voice: 207-288-6787; fax: 207-288-6073.
rofe@aol.com

Ann. N.Y. Acad. Sci. 1116: 291–305 (2007). © 2007 New York Academy of Sciences.
doi: 10.1196/annals.1402.010

KEYWORDS: bone mineral density; bone architecture; chromosomal inversion; congenic strain; C57BL/6J inbred mouse strain; C3H/HeJ inbred mouse strain

INTRODUCTION

Osteoporosis is a disorder associated with increased skeletal fragility.[1,2] Low bone mineral density (BMD), a characteristic feature of this disease, represents the most important risk factor for development of osteoporotic fractures. BMD is a polygenic trait with estimates of heritability ranging up to 80%.[3,4] Polymorphisms in numerous genes, including the Vitamin D receptor (*VDR*), collagen-type 1 (*COL1A1*), estrogen receptor *(ER)*, parathyroid hormone *(PTH)*, and insulin-like growth factor 1 (*IGF1*) have been associated with BMD in humans, although each account for a small percentage of the variance in this trait.[5-7] Despite extensive studies with large population cohorts, we are only beginning to understand how genes responsible for the heritable component of this syndrome contribute to fracture risk.[4] Not surprisingly, the mouse has become critical in understanding this process and for hypothesis testing. We identified a quantitative trait locus (QTL) for the phenotype of total femoral volumetric BMD and serum IGF-1 on mid-distal Chromosome (Chr) 6 of the mouse, in a cross between the C3H/HeJ (C3H) and C57BL/6J (B6) inbred mouse strains.[8,9] The B6.C3H-6T (6T) congenic strain of mice was developed for the purpose of gaining insight into the biology underlying this QTL and was made by introgressing a region of Chr 6 from C3H onto a B6 background by 10 generations of selective backcrossing. This was followed by several generations of intercrossing to generate mice that were homozygous for B6 alleles for the entire genome, except for the region between *D6Mit93* and *D6Mit216*, which was homozygous for the C3H alleles.[10] Female 6T mice have lower vBMD than either the B6 background strain, or the C3H donor strain mice, and have a smaller periosteal circumference, slightly shorter femurs, and low serum IGF-1. 6T mice also exhibit a decrease in trabecular bone volume fraction (Bone Volume/Total Volume, BV/TV%) of the distal femur that is coincident with an increase in marrow adipocytes and an impairment in osteoblast differentiation.[10,11]

During the development of the B6.C3H-6T congenic strain, we observed that no recombination events occurred between *D6Mit124* and *D6Mit150*. These markers had been mapped in other strains to be greater than 20 cM apart; subsequently we found these two markers to be located more than 45 Mb apart on Chr 6 (http://www.informatics.jax.org/). Further experimentation determined there was a 25 cM paracentric inversion on mid-distal Chr 6 in the C3H/HeJ strain and that the Foundation Stocks for this strain, kept at the Jackson Laboratory (Bar Harbor, ME, USA), are homozygous for this inversion. More specifically, it was found that the *D6Mit124* and *D6Mit150* markers were both located within the inverted region, but that *D6Mit93* was not. To identify candidate genes related to the QTL of interest, we asked whether the inversion per se, or alleles within the QTL, or both, were responsible for the development of a unique skeletal and metabolic phenotype that included a profound change in stromal cell allocation from preosteoblasts into adipocytes.

MATERIALS AND METHODS

Animal

All strains used for the studies reported herein were obtained from our research colonies at The Jackson Laboratory, Bar Harbor, Maine. All mice were produced by pair matings, with progeny weaned at 22–25 days of age and housed in groups of 2–5 of the same sex in polycarbonate cages (324 cm^2) with sterilized White Pine shavings. Colony environmental conditions included 14:10-h light:dark cycles, with free access to acidified water (pH 2.5 with HCl to retard bacterial growth) that contains 0.4 mg/mL of vitamin K (menadione Na bisulfite), and irradiated NIH31 diet containing 6% fat, 19% protein, Ca:P of 1.15:0.85, plus vitamin and mineral fortification (Purina Mills International, Brentwood, MO, USA). All procedures involving mice were reviewed and approved by the Institutional Animal Care and Use Committee of The Jackson Laboratory.

Microarray

Liver was collected from three 8-week-old B6 and three 6T female mice. Mice were fasted for 5 h before tissue collection. Tissue samples were stored in RNA later (Ambion, Austin, TX, USA) following dissection and later homogenized in TRIzol (Invitrogen, Carlsbad, CA, USA). Total RNA was isolated by standard TRIzol methods according to the manufacturer's protocols, and quality was assessed using a 2100 Bioanalyzer instrument and RNA 6000 Nano LabChip assay (Agilent Technologies, Palo Alto, CA, USA). Following reverse transcription with an oligo(dT)-T7 primer (Affymetrix, Santa Clara, CA, USA), double-stranded cDNA was synthesized with the superscript double-stranded cDNA synthesis custom kit (Invitrogen). In an *in vitro* transcription (IVT) reaction with T7 RNA polymerase, the cDNA was linearly amplified and labeled with biotinylated nucleotides (Enzo Diagnostics, Farmingdale, NY, USA). Fifteen micrograms of biotin-labeled and fragmented cRNA was then hybridized onto MOE430v2.0∗ GeneChip™ arrays (Affymetrix) for 16 h at 45°C. Posthybridization staining and washing were performed according to the manufacturer's protocols using the Fluidics Station 450 instrument (Affymetrix). Finally, the arrays were scanned with a GeneChip™ Scanner 3000 laser confocal slide scanner. The images were quantified using GeneChip^{TM} Operating Software(GCOS) v1.2. Probe level data were imported into the R software environment and expression values were summarized using the Robust MultiChip Average (RMA) function in the R/affy package as previously described.[12] Using the R/maanova package, an analysis of variance (ANOVA) model was applied to the data, and F1, F2, F3, and Fs test statistics were constructed along with their permutation *P*-values. False discovery rate was then assessed using the R/q value package to estimate q-values from calculated test statistics.[12] The data discussed in this publication have been deposited in NCBI's Gene Expression Omnibus (GEO, http://www.ncbi.nlm.nih.gov/geo/) and are accessible through GEO Series accession number GSE5959.

Genotyping

Mice were genotyped by preparing genomic DNA from digestion of 1 mm tail tips in 0.5 mL of 50 mM NaOH for 10 min at 95°C, then pH was adjusted to 8.0 with 1M Tris-HCl. Genotyping of individual mouse DNAs was accomplished by polymerized chain reaction (PCR) using oligonucleotide primer pairs (Mit markers, www-genome.wi.mit.edu/cgi-bin/mouse/index) from several sources (Research Genetics, Birmingham, AL, USA; Invitrogen; IDT, Coralville, IA; Qiagen, Valencia, CA, USA). These primer pairs amplify simple dinucleotide repeated sequences of anonymous genomic DNA that are of different sizes and, via gel electrophoresis, can uniquely discriminate between B6 and C3H genomes. Details of standard PCR reaction conditions have been described previously.[13] PCR products from B6, C3H, and F1 hybrids were used as electrophoretic standards in every gel to identify the genotypes of mice (i.e., *b6/b6, b6/c3, c3/c3*). In addition, three single nucleotide polymorphisms (SNPs) were assayed to further determine the ends of the C3H-like congenic region. Assays for SNPs were performed at KBiosciences, Herts, UK (www.KBioscience.co.uk).

Sequencing

Primers for sequencing were designed using the MacVector sequence analysis package (Version 7.1.1, Accelrys Inc). Cycle sequencing of PCR-DNA templates was performed using Applied Biosystems' BigDye Terminator v3.1 cycle sequencing kit. Sequencing reactions were then purified using Agencourt's CleanSEQ magnetic bead purification system. Purified reactions were run on Applied Biosystems 3730xl DNA Analyzer using POP 7 polymer. Raw data files were analyzed using Applied Biosystems DNA Sequencing Analysis Software, Version 5.2.

Development of the B.H-6 Congenic Strain

BHX6 mice were mated with B6 progenitor strain mice and the offspring mated back to B6. Mice were typed at each generation for the markers *D6Mit102, D6Mit320, D6Mit108, D6Mit366,* and *D6Mit150* (Mit markers, www.genome.wi.mit.edu/cgi-bin/mouse/index) and mice that were *c3/b6* for all of these markers were backcrossed to B6. This was repeated until eight generations of backcrossing had been achieved. The N8F1 mice were intercrossed producing N8F2 progeny. Mice homozygous for the C3H-like alleles were further intercrossed for two more generations, yielding N8F4. This new congenic was named B.H-6.

Body Composition

Areal BMD and body composition were assessed using peripheral dual-energy X-ray absorptiometry (PIXImus, GE-Lunar, Madison, WI, USA). Whole body (exclusive of the head) composition measures of lean mass, fat mass, and a percentage of body fat were obtained.

Necropsy for Femur Collection

All bones for density and architecture analysis were collected from female, 16-week-old mice. This is the age at which mice have acquired their adult femoral mass.[14] Mice were necropsied, whole body weights recorded, and tissue samples collected. Skeletal preparations (lumbar vertebral columns, pelvis, and attached hind limbs) were placed in 95% EtOH for a period of not less than 2 weeks. Femurs were dissected free of remaining muscle and connective tissue, and placed in 95% EtOH for storage until subsequent analyses were conducted.

pQCT for Volumetric (v)BMD Bone Densitometry

Femur lengths were measured with digital calipers (Stoelting, Wood Dale, IL, USA) and then measured for density using the SA Plus densitometer (Orthometrics, Stratec SA Plus Research Unit, White Plains, NY, USA). Calibration of the SA Plus instrument was established with hydroxyapatite standards of known density (50–1,000 mg/mm^3) with cylindrical diameters 2.4 mm and length 24 mm that approximate mouse femurs. Daily quality control of the SA Plus instrument's operation was checked with a manufacturer-supplied phantom. The bone scans were analyzed with threshold settings to separate bone from soft tissue. Thresholds of 710 and 570 mg/cm^3 were used to determine cortical bone areas and surfaces that yielded area values consistent with his-tomorphometrically derived values. To determine mineral content, a second analysis was carried out with thresholds of 220 and 400 mg/cm^3 selected so that mineral from most partial voxels (0.07 mm) was included in the analysis. Density values were calculated from the summed areas and associated mineral contents. Precision of the SA Plus for repeated measurement of a single femur was found to be 1.2–1.4%. Isolated femurs were scanned at seven locations at 2 mm intervals, beginning 0.8 mm from the distal ends of the epiphyseal condyles. Total vBMD values were calculated by dividing the total mineral content by the total bone volume and expressed as mg/mm^3.

Micro-CT40 for Distal Trabecular Bone

Femurs were scanned using a Micro-CT40 microcomputed tomographic in-strument (Scanco Medical AG, Bassersdorf, Switzerland) to evaluate trabec-ular bone volume fraction and microarchitecture in the secondary spongiosa of the distal femur. Daily quality control of the instrument's operation was checked with a manufacturer-supplied phantom. The femurs were scanned at low-resolution energy level of 55KeV, and intensity of 145 μA. Approxi-mately 100 slices were measured just proximal to the distal growth plate, with an isotropic pixel size of 12 μm and slice thickness of 12 μm. Trabecular bone volume fraction (%BV/TV) and microarchitecture properties were evaluated in the secondary spongiosa, starting ∼ 0.6 mm proximal to the growth plate, and extending proximally 1.5 mm.

Statistical Assessment

Data are expressed as mean ± SEM in tables and figures. Statistical eval-uation of bone and body composition was conducted using JMP version 6 software (SAS, Cary, NC, USA). To account for differences in body size be-tween strains, a stepwise ANCOVA approach was used for PIXI, pQCT, and μCT data using body weight and femur length as covariates. Nonsignificant covariates and interactions were removed in a stepwise fashion until the final model was obtained.

RESULTS

Determining the End Points of the Inversion

We previously published that the upper breakpoint of the inversion was prox-imal to D6Mit124 (at 71.3 Mb, NCBI Build 36) and the lower breakpoint was distal to D6Mit150 (at 116.1 Mb, NCBI Build 36).[15] Recombination between

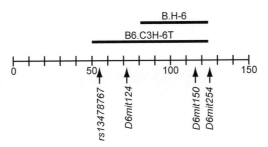

FIGURE 1. Relationship between the 6T and B.H-6 congenics with the inversion breakpoints are shown. Vertical ticks indicate 10 Mb intervals along Chr 6. The upper breakpoint of the inversion is located between *rs13478767 and D6Mit124* and the lower breakpoint is located between *D6Mit150* and *D6Mit254*. The 6T congenic was made by introgressing the region between *D6Mit93* (at 52 Mb, not shown) and *D6Mit150* from C3H on the B6 background by selective backcrossing.[10] Further mapping has shown that the C3H-like region extends to *D6Mit216* (at 121.1 Mb), but not to *Dmit254*. The new B.H-6 congenic carries C3H-like alleles from *D6Mit102* to rs3727110 at 122.0 Mb.

C3H and strains not carrying the inversion is not possible within the inverted region. To further resolve the location of the inversion breakpoints, we examined recombination patterns in the original B6XC3H F2 mapping cross in which the BMD QTL on Chr 6 was discovered. This cross is described in detail elsewhere.[8,9] The markers *D6Mit93* and *D6Mit124* were typed in the original mapping analysis. *D6Mit124* was clearly located within the inverted region, whereas *D6Mit93* was not. We then found that 21 mice in the F2 cross had had a recombination event between these two markers. We typed these mice for the following markers, *D6Mit183*, *D6Mit175*, and *D6Mit17*. In four of the 21 mice, a recombination event that had occurred between *D6Mit183* and *D6Mit175* was found, but there were no recombination events distal to *D6Mit175*. We then sequenced *rs13478767*, a SNP located between *D6Mit183* and *D6Mit175* and found that in one mouse, recombination events distal to *rs13478767* had occurred. We therefore concluded that the upper inversion breakpoint must be distal to *rs13478767*, located at 55.2 Mb (NCBI Build 36), but proximal to *D6Mit124* at 71.3 Mb.

We employed a similar strategy to determine the location of the lower breakpoint. We knew from the FISH analysis that *D6Mit150* was located within the inverted region.[15] The next marker typed in the original F2 mapping cross was *D6Mit59*, located at 138.8 Mb. To narrow the possible interval in which the breakpoint could be located, we then typed *D6Mit254* (at 125.3 Mb), *D6Mit135* (at 128.8 Mb), and *D6Mit219* (132.4 Mb) in DNA from 11 of the original F2 mapping cross mice. We found that a recombination event had occurred between the markers *D6Mit150* and *D6Mit254* in 2 of the 11 mice examined. Therefore, we concluded that the lower inversion breakpoint was distal to *D6Mit150*, but proximal to *D6Mit254*. The locations of these markers, relative to the C3H-like congenic region of the 6T strain, are shown in FIGURE 1.

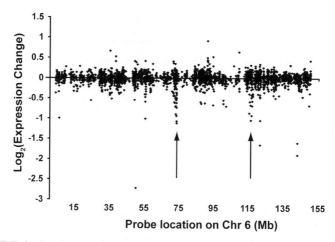

FIGURE 2. Gene expression levels in the liver of female 6T mice, relative to age-matched female B6 mice, was examined by microarray. We plotted the genomic location, in Mb (NCBI Build 36), to which each probe set mapped versus relative expression level. As can be seen here, there is a clear pattern of decreased gene expression at ~72 Mb and at ~116 Mb, as indicated by the black arrows. These zones of decreased gene expression are located in close proximity to the predicted breakpoints for the paracentric inversion found in C3H/HeJ. This inversion is carried in the 6T congenic strain.

Gene Expression Along Chromosome 6 in B6.C3H-6T

We next examined the pattern of differential gene expression on Chr 6 in the liver of 6T and B6 by microarray. Relative expression change for each probe set was calculated as described above. We then plotted expression changes in 6T versus the chromosomal location of the probe set. As can be seen in FIGURE 2, there are scattered genes along Chr 6 that show significantly high or low gene expression, as expected. Expression changes in these genes are likely, either directly or indirectly, due to a consequence of allelic differences between these two strains. Interestingly, at approximately 72 Mb and 116 Mb, a clear pattern gene of expression emerges. In other words, in these two locations, that is, the breakpoints of the inversion, the majority of genes are clearly downregulated in 6T liver.

Development of the B.H-6 Congenic Strain

To test the hypothesis that the inversion was responsible for the differential gene expression, we looked to another set of inbred strains. The BXH recombinant inbred (RI) strains were generated from crosses between the B6 and the C3H/HeJ strains.[16] These RI strains were developed in the early 1970s, and must have been made either before the inversion arose on Chr 6

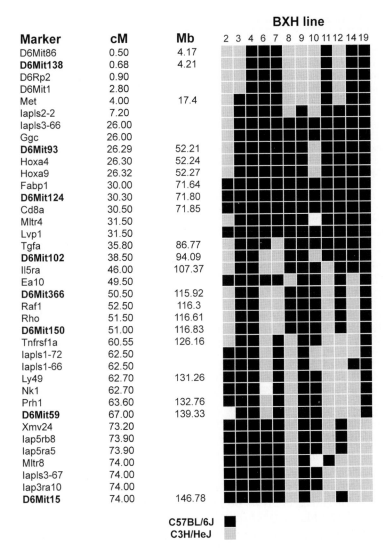

FIGURE 3. Mapping data for 12 of the BXH RI strains were obtained from the mouse genome database. The map location in CentiMorgans (CM) and the genomic region (if available) were also determined for all markers. Eight additional markers (*in bold*) were typed for all 12 lines. C3H-like alleles are shown in gray and B6-like alleles are shown in black. Data were not obtainable for all strains for all markers (denoted by white squares).

in C3H/HeJ or before this inversion became fixed, as none of the resulting BXH RI lines are *c3/c3* for the entire inversion region (FIG. 3). To determine if the low BMD and low serum IGF-1 phenotypes of the 6T congenic mouse were a function of allelic differences from C3H, or due entirely to the Chr

TABLE 1. Anthropomorphic aspects of three strains of mice

	C57BL/6J	B6.C3H-6T	B.H-6
Body weight (g) [a]	20.12 ± 0.34	19.32 ± 0.62	21.44 ± 0.66
Lean mass (g) [b,c]	14.37 ± 0.10	14.76 ± 0.19	13.70 ± 0.20*,**
% Total body fat [b,c,d]	19.60 ± 0.33	15.48 ± 0.91*	22.02 ± 0.87*,**

[a] The covariate strain was not significant.
[b] Strain was a significant covariate.
[c] Body weight was a significant covariate.
[d] Strain X Body weight was a significant covariate.
*Significantly different than C57BL/6J ($p = 0.05$).
**Significantly different than B6.C3H-6T ($p = 0.05$).

6 inversion, we then generated a new congenic strain, B.H-6. The allele distribution patterns of the 12 remaining BXH RI lines are publicly available at (http://www.informatics.jax.org/searches/riset˙form.shtm). We obtained DNA from these 12 strains and further genotyped them for 8 additional markers (*D6Mit138, D6Mit93, D6mit124, D6Mit102, D6Mit366, D6mit150, D6Mit59,* and *D6Mit15*). Two lines carried C3H-like alleles for the majority of the genomic region that is inverted in the present day C3H/HeJ strain: BXH2 and BXH6 (FIG. 3). BXH6 was chosen as the donor strain for the congenic because: (1) we were concerned that the C3H-like region of BXH2 did not extend distal enough, and (2) the proximal end of the region of interest in BXH2 appears to have undergone extensive recombination, resulting in short alternating stretches of B6-like and C3H-like sequences.

The B.H-6 congenic was generated by introgressing an approximately 30 Mb region of Chr 6 from the BXH6 RI line onto a B6 background, by eight generations of backcrossing. The B.H-6 congenic mice are *c3/c3* for the region between *D6Mit102* and *D6Mit150*, but are otherwise *b6/b6* for the remainder of the genome. Additional SNP testing was done on the N8F4 generation mice to further determine the ends of the C3H-like congenic region. It was determined that the B.H-6 congenic is *b6/b6* like at *rs3708822* at 89.3 Mb (NCBI build 36) but is *c3/c3* like at *D6Mit102* at 93.4 Mb (NCBI build 36). At the distal end the B.H-6 congenic is *c3/c3* as far distal as *rs3727110* at 122.0 Mb (NCBI build 36).

Phenotype of the B.H-6 Congenic

At 16 weeks of age, female B.H-6 congenic weigh more than both the B6 progenitor and the 6T congenic stain, although the covariate of strain was not significant in the ANCOVA fit model. Female B.H-6 mice had higher absolute fat mass (data not shown), and a higher percentage of total body fat than either B6 or 6T. In addition, this strain had a lower lean mass, as noted in TABLE 1. Unlike the 6T congenic, 16-week-B.H-6 female mice had similar total femoral

TABLE 2. Differences in the cortical bone mass for three strains of mice

	C57BL/6J	B6.C3H-6T	B.H-6
Total vBMD (mg/mm^3) ^{a,b,d}	0.585 ± 0.004	0.530 ± 0.006*	0.587 ± 0.007**
Total cortical vBMD (mg/mm^3) ^{a,b}	1.099 ± 0.002	1.085 ±0.003*	1.086 ± 0.003*
Cortical thickness (mm)^{a,b}	0.190 ± 0.001	0.174 ± 0.001*	0.181±0.002*,**
Periosteal circumference (mm) ^{a,b,c}	4.88 ± 0.02	4.77 ± 0.02*	4.89 ± 0.02**
Endosteal circumference (mm) ^{a,c}	3.68 ± 0.02	3.67 ± 0.01	3.76 ± 0.02*,**

^{a}Strain was a significant covariate.
^{b}Femur length was a significant covariate.
^{c}Body weight was a significant covariate.
^{d}Strain X Femur length was a significant covariate.
*Significantly different than C57BL/6J ($p = 0.05$).
**Significantly different than B6.C3H-6T ($p = 0.05$).

TABLE 3. Differences in the distal trabecular region of the femur in three strains

	C57BL/6J	B6.C3H-6T	B.H-6
BV/TV (%) ^{a}	8.17 ± 0.3	6.02 ± 0.4*	9.46 ±0.4*,**
Trabecular number (/mm) ^{a,b,c,d,e}	3.80 ± 0.05	3.53 ± 0.06*	4.04 ±0.05*,**
Trabecular thickness (μm)^f	0.0471 ± 0.0007	0.0465 ± 0.001	0.0493 ± 0.0009
Connectivity density (mm^{-3}) ^{a}	61.86 ± 3.21	34.69 ± 4.53*	70.96 ± 3.98**
SMI ^{a}	2.90 ± 0.04	3.27 ± 0.04*	2.81 ± 0.05**

^{a}Strain was a significant covariate.
^{b}Body weight was a significant covariate.
^{c}Femur length was a significant covariate.
^{d}Femur length X Body weight was a significant covariate.
^{e}Strain X Femur length was a significant covariate.
^{f}The covariate strain was not significant.
*Significantly different than C57BL/6J ($p = 0.05$).
**Significantly different than B6.C3H-6T ($p = 0.05$).

volumetric BMD (vBMD) compared to the B6 progenitor strain. Interestingly, like 6T, B.H-6 mice had reduced cortical vBMD and had thinner cortices at the midshaft of the femur than B6 (TABLE 2). While there was no difference between B6 and B.H-6 for periosteal circumference, the endosteal circumference at the midshaft of the femur was larger in B.H-6 (TABLE 2). Most impressively, the B.H-6 mice had a striking increase in BV/TV% of the distal femur, compared to either the B6 or the 6T strains. This increase in trabecular bone volume was primarily a function of more trabeculae, rather than a significant change in trabecular thickness (TABLE 3). The structure model index (SMI) of the distal femoral trabecular bone was lower in B.H-6, indicating a more plate-like appearance to the trabeculae, as can be seen in FIGURE 4.

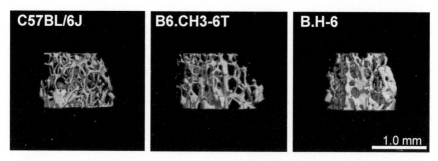

FIGURE 4. Trabecular bone architecture was examined in the distal femur of female, 16-week-old C57BL/6J (B6), B6.C3H-6T (6T), and B.H-6 mice. The BV/TV% was greatly increased in the B.H-6 strain, compared to either the B6 progenitor strain, or the 6T congenic, which carried the inversion from C3H/HeJ. In addition, the trabecular number was greatly increased in B.H-6. There was an increase in trabecular thickness, but this was not statistically significant.

DISCUSSION

The arrangement of genes within the eukaryotic genome is not random. Studies have shown that functionally related genes can be found clustered within discrete blocks. These blocks can be as large as several megabases in mammals (reviewed in Hurst *et al.*[17]). One such block exists in mice at the distal end of the 6T congenic region. Between 115 and 125 Mb, several genes involved in adipocyte maturation and fatty acid metabolism, as well as osteoblast differentiation, can be found, including *Pparg*, *Alox5*, *Adipor2,* and *Tpi1*. Disruption in the organization and order of genes within the genome, without disturbing the structure of a gene unit, can cause a variety of human diseases.[18] In this article we show that the 6T and B.H-6 congenic strains, despite having the same *C3H* alleles on distal Chr 6 and identical B6 backgrounds, have a drastically different anthropomorphic and skeletal phenotype; that is, 6T is a small lean mouse with very low BMD but marrow adipogenesis, whereas B.H-6 is fatter with higher BMD. These data support the thesis that the chromosomal inversion, not the allelic differences between C3H and B6, is responsible for the metabolic and skeletal phenotypes of 6T.

The most likely candidate gene to be disrupted by this inversion because of its genomic location is *Pparg*, a nuclear receptor that is essential for adipogenesis, and also a negative regulator of osteoblastogenesis when activated by specific ligands. We previously showed that *Pparg* expression is increased in calvarial osteoblasts from 6T, compared to B6, although in the liver, *Pparg* expression was significantly downregulated in 6T versus B6.[11] To determine if *Pparg* was involved in the 6T phenotype, we recently identified two SNPs in the promoter region of exon B, adjacent to a Cebpa-binding site, which codes for

TABLE 4. Assessment of the effects of allelic differences found within the C3H/HeJ *Pparg2* promoter on promoter function. (WT — B6-like, Mut — C3H-like, *$p < 0.001$)

Treatment	48 h	% Decrease	72 h	% Decrease
WT	16.64 ± 0.58		28.40 ± 1.63	
Mut	11.1 ± 0.48	−32*	16.5 ± 0.6	−42*
WT + CEBPα	47.07 ± 8.31		54.4 ± 0.96	
Mut + CEBPα	36.0 ± 1.1	−23*	39.0 ± 4.5	−26*
No transfection	0.18 ± 0.04		0.45 ± 0.06	
CEBPα alone	0.02 ± 0.01		0.04 ± 0	

The B6 (WT) *Pparg2* promoter was cloned into a pGL3 Luciferase expression vector (Promega, Madison, WI). Two polymorphisms from the C3H/HeJ strain were introduced using the QuikChange Site Directed Mutagenesis kit (Stratagene, La Lolla, CA) and this construct was referred to as "Mut." Constructs were transfected into UAMS33 cells and cells were treated with either vehicle or vehicle + CEBPA. Luciferase expression is presented at 48 and 72 h. Transfection efficiency was controlled by co-transfection with a Renilla expression vector (pRL-SV40, Promega, Madison, WI).

expression of the PPARG2 protein that is fat specific. Transient transfection of this promoter polymorphism containing a C3H sequence into UAMS stromal cells resulted in significant downregulation of *Pparg* expression, consistent with a hypomorphic polymorphism in the parental C3H mouse (see TABLE 4). Interestingly, homozygous *Pparg*[−/−] mice (*Pparg*[tm1Tka]) die at embryonic day 10.5 to 11 *post coitum* due to placental insufficiency but the *Pparg*[+/−] mouse is viable and appears to have normal development of all major organs.[19] *Pparg*[+/−] mice have higher areal BMD by DXA and markedly increased trabecular bone volume (BV/TV%) at 8 weeks of age compared to wildtype.[20] Moreover, it appears these mice maintain a higher BV/TV% than controls through 52 weeks of age, suggesting resistance to age-related trabecular bone loss.[20] The increase in bone volume fraction is strikingly similar to the skeletal phenotype of the B.H-6 congenic strain and most likely represents a significant genetic contribution from C3H (i.e., a hypomorphic polymorphism), which also has increased bone volume fraction in the distal femur. On the other hand, 6T mice have reduced bone volume fraction, despite the allelic effects of the C3H polymorphism, strongly suggesting that the inversion has disrupted the regulation of this gene, resulting in marrow adipogenesis, impaired osteoblast differentiation, and low bone mass.

In conclusion, we have created two mouse strains with nearly identical alleles for genes in the distal region of Chr 6, but with widely different metabolic and skeletal phenotypes. The chromosomal inversion in distal Chr 6 clearly determines the effects of several closely related genes on fat and bone. Further studies of the regulatory region 3′ to the inversion breakpoint will help determine the evolutionary significance of this genomic region and provide us with a better understanding of the relationship between adipogenesis and osteoblastogenesis.

ACKNOWLEDGMENT

The authors would like to thank Mr. Jesse Hammer for his assistance in the preparation of this manuscript. This work was funded by a grant from NIAMS AR45433.

REFERENCES

1. COMPSTON, J. & C.J. ROSEN. 2006. *Fast Facts: Osteoporosis*. Fifth edition. Health Press. Oxford.
2. 1994. WHO assessment of fracture risk and its application to screening for post-menopausal osteoporosis. WHO technical reports series **834:**
3. CUSACK, S. & K. CASHMAN. 2003. Impact of genetic variation on metabolic response of bone to diet. Proc. Nutr. Soc. **62:** 901–912.
4. RALSTON, S. & B. DE CROMBRUGGHE. 2006. Genetic regulation of bone mass and susceptibility to osteoporosis. Genes Dev. **20:** 2492–2506.
5. THAKKINSTIAN, A., C. D'ESTE, J. EISMAN, *et al*. 2004. Meta-analysis of molecular association studies: vitamin D receptor gene polymorphisms and BMD as a case study. J. Bone Miner. Res. **19:** 419–428.
6. BUSTAMANTE, M., X. NOGUES, A. ENJUANES, *et al*. 2007. COL1A1, ESR1, VDR and TGFB1 polymorphisms and haplotypes in relation to BMD in Spanish post-menopausal woman. Osteoporos. Int. **18:** 235–243.
7. RIVADENEIRA, F., J. HOUWING-DUISTERMAAT, N. VAESSEN, *et al*. 2003. Association between an insulin-like growth factor 1 gene promoter polymorphism and bone mineral density in the elderly: the Rotterdam Study. J. Clin. Endocrinol. Metab. **88:** 3878–3884.
8. BEAMER, W.G., K.L. SHULTZ, L.R. DONAHUE, *et al*. 2001. Quantitative trait loci for femoral and lumbar vertebral bone mineral density in C57BL/6J and C3H/HeJ inbred strains of mice. J. Bone Miner. Res. **16:** 1195–1206.
9. ROSEN, C.J., G. CHURCHILL, L.R. DONAHUE, *et al*. 2000. Mapping quantitative trait loci for serum insulin-like growth factor-I levels in mice. Bone **27:** 521–528.
10. BOUXSEIN, M.L., C.J. ROSEN, C.H. TURNER, *et al*. 2002. Generation of a new congenic mouse strain to test the relationship among serum insulin-like growth factor I, bone mineral density and skeletal morphology *in vivo*. J. Bone Miner. Res. **17:** 570–579.
11. ROSEN, C.J., C. ACKERT-BICKNELL, M.L. ADAMO, *et al*. 2004. Congenic mice with low serum IGF-1 have increased body fat, reduced bone mineral density, and an altered osteoblast differentiation program. Bone **35:** 1046–1058.
12. SHOCKLEY, K. & G.A. CHURCHILL. 2006. Gene expression analysis of mouse chromosome substitution strains. Mammalian Genome **17:** 598–614.
13. SVENSON, K., Y.-C. CHEAH, K.L. SHULTZ, *et al*. 1995. Strain distribution pattern for the SSLP markers in the SWXJ recombinant inbred strain set: chromosomes 1 to 6. Mammalian Genome **6:** 867–872.
14. BEAMER, W.G., L.R. DONAHUE, C.J. ROSEN, *et al*. 1996. Genetic variability in adult bone density among inbred strains of mice. Bone **18:** 397–403.
15. AKESON, E.C., L.R. DONAHUE, W.G. BEAMER, *et al*. 2006. Chromosomal inversion discovered in C3H/HeJ mice. Genomics **87:** 311–313.

16. TAYLOR, B.A., H.G. BEDIGIAN & H. MEIER. 1977. Genetic studies of the Fv-1 locus in mice: linkage with Gpd-1 in recombinant inbred lines. J. Virol. **23:** 106–109.
17. HURST, L., C. PAL & M. LERCHER. 2004. The evolutionary dynamics of eukaryotic gene order. Nat. Rev. Genet. **5:** 299–310.
18. KLEINJAN, D. & V. VAN HEYNINGEN. 2005. Long-range control of gene expression: emerging mechanisms and disruption in disease. Am. J. Hum. Genet. **76:** 8–32.
19. BARAK, Y., M.C. NELSON, E.S. ONG, *et al.* 1999. PPAR gamma is required for placental, cardiac, and adipose tissue development. Mol. Cell **4:** 585–595.
20. AKUNE, T., S. OHBA, S. KAMEKURA, *et al.* 2004. PPARgamma insufficiency enhances osteogenesis through osteoblast formation from bone marrow progenitors. J. Clin. Invest. **113:** 846–855.

Osteoblasts Cultured on Three-Dimensional Synthetic Hydroxyapatite Implanted on a Chick Allantochorial Membrane Induce Ectopic Bone Marrow Differentiation

LUCIA MANCINI, ROBERTO TAMMA, MARIA SETTANNI,
CLAUDIA CAMERINO, NICOLA PATANO, GIOVANNI GRECO,
MAURIZIO STRIPPOLI, AND ALBERTA ZALLONE

*Department of Human Anatomy and Histology, School of Medicine,
University of Bari, Bari, Italy*

ABSTRACT: Osteoblast (OB) activities have been studied on hydroxyapatite three-dimensional (3D) scaffolds in comparison with traditional planar substrata. OBs cultured on 3D displayed increased proliferation, differentiation, and matrix protein synthesis, when compared to 2D cultures on the same substrata. Confluent cultures, however, could not be maintained for long, due to insufficient fluid diffusion within 3D scaffolds that impaired cell viability. Thus, confluent OB 3D cultures were implanted on the allantochorial membrane of chick embryos. Vessels from the embryo colonized the bone-like network giving rise in the presence of OBs to an ectopic bone marrow formation in the intratrabecular spaces. In the absence of OBs, when the biomaterial alone was implanted, blood vessels were still present but hematopoietic marrow was absent. In both cases osteoclasts (OCs) derived from the host were found on the implant surface. These results indicated that scaffolds with cells can be easily vascularized and confirmed the role of OBs in the definition of the microenvironment that induce blood marrow differentiation in the intratrabecular spaces.

KEYWORDS: osteoblast; bone implants; allantochorion; hematopoietic niche

INTRODUCTION

Bone injuries, genetic malformations, and several diseases often require implantation of grafts. Three different grafting materials are currently used:

Address for correspondence: Alberta Zallone, Department of Human Anatomy and Histology, School of Medicine, University of Bari, Piazza G. Cesare 11 Bari, Italy 70124. Voice: 39-080-5478307; fax: 39-080-5478308.
a.zallone@anatomia.uniba.it

Ann. N.Y. Acad. Sci. 1116: 306–315 (2007). © 2007 New York Academy of Sciences.
doi: 10.1196/annals.1402.008

autografts, allografts, and synthetic materials. Although autogenous bone is the "gold standard" graft material, it is limited in supply and necessitates traumatic harvesting procedures. Tissue engineering is a new concept involving three-dimensional (3D) autologous tissue growth before implantation into patients. Ideally, bone tissue engineering will provide unlimited autologous bone tissue without the need of tissue harvesting procedures other than the relatively simple one of harvesting marrow cells from the patient. Scaffolds for tissue engineering have been prepared using a variety of materials, including synthesized hydroxyapatite, molded in porous structures that serve as scaffolds for bone marrow-derived cells, having open pores with a high degree of interconnectivity.[1]

Three-dimensional culture systems have been successfully used in the investigation of complex biological processes, such as angiogenesis, wound healing, tumor invasion, and metastasis. Natural bone tissue has itself a very peculiar 3D organization and it is easily hypothesized that bone cells will behave differently on a 3D matrix compared to the already available data generated in traditional cultures.[2,3] Our knowledge about cells in culture, including complex functions as adhesion, migration, signaling, and cytoskeletal function, is primarily derived from studies on planar 2D tissue culture substrates. However, the importance of 3D ECM is recognized for epithelial cells, where 3D environments promote normal epithelial polarity and differentiation.[4] Fibroblastic cells have been studied mainly in 2D cell cultures, yet culturing them on flat substrates induces an artificial polarity between the lower and upper surfaces of these normally nonpolar cells; not surprisingly, fibroblast morphology and migration differ once suspended in collagen gels.[5,6] Three-dimensional matrix interactions also display enhanced cell biological activities and narrowed integrin usage.[7] Three-dimensional cultures are, therefore, a useful tool for studying cell activities in more physiological condition. A report about osteoblasts (OBs) cultured on 3D PLGA scaffolds shows that after 24 h there is an increased expression of osteopontin and VEGF, while expression of Col I did not increase with time as it did when cells were cultured on a 2D film.[8] These 3D cultures are also a potential tool for bioengineering.[9]

In the first part of our study we compared the molecular properties of OBs cultured on planar substrata or in 3D scaffolds; subsequently, to overcome impaired cell viability due to limited fluid diffusion within the scaffold and to study if and how these scaffolds could be invaded by blood vessels, we implanted the 3D cultures on the chick chorioallantoic membrane (CAM), thus used as a nourishing support for a 3D OB cultures developed on synthetic hydroxyapatite supports.[10]

Our results demonstrate that not only vessels migrated within the trabecular meshes, but that a bone marrow of chick origin developed in this unusual site only in the presence of OBs in the implants.

MATERIALS AND METHODS

Biomaterial

We used Skelite™ (Millenium Biologix Inc., Ontario, Canada), which is a porous biphasic material composed of $65 \pm 15\%$ hydroxyapatite (HA) and $35 \pm 15\%$ tricalcium phosphate with less than 15% of other phosphates. The pore size is 400 ± 150 μm and the Ca/P ratio is 2 ± 0.2. The Skelite was crushed into fragments 4–5 mm^3 in size and directly transferred to culture dishes.

Human OB Cell Preparation

Human OBs were harvested from bone fragments obtained during hip replacement surgery on adult patients after their informed consents. Specimens were cleaned of soft tissues, reduced to fragments approximately 2 mm in size, and digested with 0.5 mg/mL *Clostridium histolyticum* neutral collagenase (Sigma Chemical Co., St. Louis, MO, USA) in phosphate-buffered saline (PBS) for 30 min at 37°C. According to Gehron–Robey and Termine protocol,[11,12] neutral collagenase is able to remove fibroblasts and blood cells from the bone fragments.

The bone chips were then washed vigorously three times with Iscove-modified Dulbecco's medium (IMDM, Gibco Ltd., Uxbridge, UK) containing 3.02 g/L of sodium bicarbonate and cultured in IMDM supplemented with 10% fetal calf serum (FSC, Gibco Ltd.), 100 IU/mL of penicillin, 100 μg/mL of streptomycin, 2.5 μg/mL of amphotericin B and 50 IU/mL of mycostatin (Gibco Ltd.) at 37°C, in a water saturated atmosphere containing 5% CO_2. The media were changed every 3 days. Under these conditions OBs in the explants proliferated and migrated to the culture dishes, reaching confluence within 4 weeks. Cells were then trypsinized and transferred to appropriate dishes for characterization and experiments.

OB Characterization

Human OBs were characterized according to the well-established parameters of alkaline phosphatase activity (3), production of cAMP in response to PTH 10^{-8} mol/L (Sigma Chemical Co.), and synthesis of osteocalcin in response to 1,25-dihydroxy-vitamin D_3 10^{-8} mol/L (Sigma Chemical Co.).

RNA Isolation and Reverse-Transcriptase Polymerase Chain Reaction (RT-PCR) Amplification

Human OBs were subjected to mRNA extraction, using spin columns (RNasy, Qiagen, Hilden, Germany) according to the manufacturer's

instructions, to detect the expression of RUNX-2, COLL-1, Osteopontin, osteocalcin, and SDF-1; GAPDH was used as housekeeping gene. By using Invitrogen kit, the resulting cDNA (20 ng) was subjected to PCR amplification using
Platinum Taq DNA Polymerase (Invitrogen Life Technologies, Milan, Italy)
and appropriate primers. PCR products were analyzed by 1.5% Agarose gel
electrophoresis containing 0.01% ethidium bromide, and the resulting bands
were detected by a light-sensitive CCD video system (BioDocAnalyze, What-
man Biometra Goettingen, Germany).

Implants

Scaffolds were first seeded with OBs obtained from primary cultures and
cultured for 1 week, to allow confluence and invasion of the scaffolds by
OBs. Eventually 12 scaffolds with OBs and 12 control scaffolds without cells
were implanted, after gentle scraping, on allantochorial membrane of chick
embryos at day 6 of incubation at 37°C, near to a vessel branch. Embryos
viability and development of scaffold vascularization were monitored daily by
a stereomicroscope. After different times of incubation (3, 5, 7 days) they were
explanted and fixed in paraformaldehyde 4% in PBS for 30 min at 4°C.

Microscopy

Specimens were dehydrated in a graded series of alcohol-acetone and embedded in Epon-Araldite (Durcupan, Sigma Chemical Co.). Semithin sections
were cut by a Leica RM2155 Rotary Microtome (Leica Microsystems, Milan,
Italy). The sections were stained with Toluidine Blue 1%. The coverslips were
mounted with Entellan (Merck, Germany) before examining the slides in a
Nikon Y-FL Microscope (Nikon, Italy).

RESULTS

The OBs seeded on Skelite fragments underwent a rapid proliferation and
within a week by phase contrast microscopy cells within the meshes were
observed (FIG. 1A). Sections prepared from samples at this stage show OBs
on the biomaterial surfaces and within the cavities in the Skelite trabeculae
(FIG. 1B). We compared the proliferation rate of the same population of OBs
cultured on planar or 3D Skelite substrate. The results, evaluated by titer-tek
technique, show a much more rapid proliferation rate on the 3D scaffolds
from 24 to 72 h (FIG. 1C). We then compared the gene expression of the
OB master gene, RUNX-2, and of other genes coding bone matrix proteins
as collagen type 1, osteopontin (OPN), and osteocalcin (OC). The results

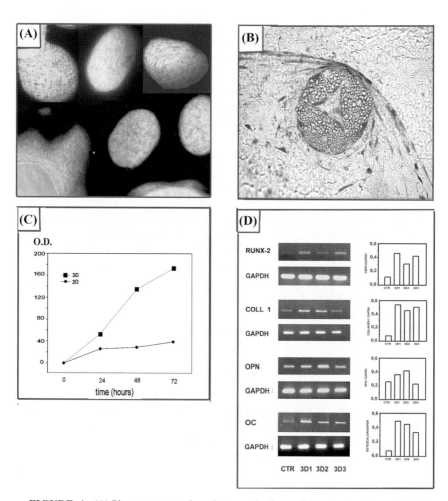

FIGURE 1. (A) Phase contrast microphotograph of a scaffold colonized by osteoblasts that filled most of the intratrabecular space (50×). **(B)** Microphotograph of a sectioned scaffold; osteoblasts line the surface and the inner channel, surrounded by less differentiated cells and loose matrix (200×). **(C)** The diagram shows, measured by optical density of the prestained cell lysate, the different proliferation rate on 2D and on 3D. **(D)** Osteoblasts cultured on 3D scaffolds show, compared to 2D, an enhanced gene expression of RUNX-2, collagen type 1 (COLL 1), osteopontin (OPN), and osteocalcin (OC). Histograms show the ratio with housekeeper gene GAPDH.

obtained in three different cultures indicated an increased gene expression for all the genes. Osteopontin only was increased in two cultures out of three (FIG. 1D).

In these culture conditions the cells within the third week undergo degeneration, starting from the inner part of the biomaterial (not shown). To overcome this problem, due to impaired diffusion of fluids as the cells progressively fill the scaffold meshes, with the aim of prolonging the culture time, we decided to search for a physiological incubator capable of furnishing vascularization and nutrients to the implants. The aim was also to verify if bone formation could be observed in the scaffold colonized by OBs in longer time laps. We decided to implant the Skelite fragments on the allantochorial membrane of chick embryos at day 6 of incubation, when the membrane is sufficiently developed with well-organized vascularization.

Twelve fragments with OBs, as well as 12 fragments without cells as controls, were implanted on the embryo membranes, after a gentle scraping of the surface, in close proximity of a vessel branch. *In vivo* observations demonstrated that blood vessel in a few days invaded the scaffolds (FIG. 2A). The fragments were fixed and sectioned after 3, 5, and 7 days, respectively.

Microscopic observations showed a good vascularization in all the fragments implanted. OBs were observed on the Skelite trabeculae as well as OCs, derived from the host, as only OBs were present on the scaffolds before implantation (FIG. 2B, F). Allantochorial cells and vessels were present between the Skelite trabeculae in all cases, but with striking differences: if the implant was Skelite alone, in all the specimens only loose connective tissue and blood vessel were visible in the trabecular spaces (FIG. 2G, H), while, if the Skelite had been already colonized by OBs before the implantation, a well-developed bone marrow was always observed between the trabeculae (FIG. 2C, D), indicating an inductive effect exerted by these cells in the ectopic differentiation of hematopoietic cells.

To explain these results we wanted to investigate if OBs cultured on hydroxyapatite were capable of synthesizing SDF-1, the stromal-derived growth factor identified as responsible for the differentiation of blood marrow and by RT-PCR we were able to confirm that, indeed, this cytokine was expressed in our culture conditions (FIG. 2E).

DISCUSSION

Our experiments give several indications. First they show, as supposed, that OBs in 3D perform much better than in 2D. Proliferation and specific differentiation are increased and the bone matrix genes appear expressed at a greater extent. However, when the meshes of the scaffold are filled with cells and matrix, the diffusion becomes insufficient and cells die. The implant on the allantochorion is a solution that permits to maintain the culture for at least another week. These results could also have been obtained utilizing a bioreactor with forced fluid circulation, but we also wanted to verify if the implant could be invaded by blood vessels. Two completely different, but both interesting results derived from the implants observation: the first is that allantochorial membrane can be used as a source of vascularization to follow

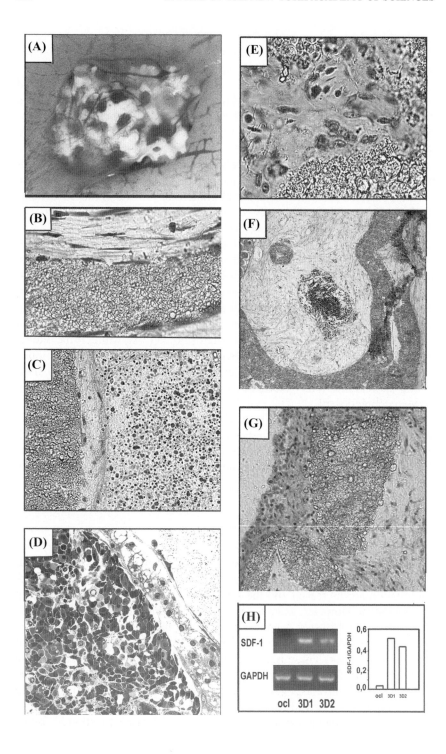

the development of a tridimensional, more differentiated state of OBs, not possible in normal *in vitro* cultures for the difficulty of fluid diffusion. Allantochorial membrane can offer the fresh nutrients and the oxygen not readily available in culture media. The second and more interesting finding is the differentiation of bone marrow between the trabecular meshes, which occurs only when OBs are present in the implant (100% of the specimen observed). Although embryonic blood vessels first differentiate in the allantochorion during embryogenesis, marrow in normal conditions does not differentiate in this site. The development of the bone marrow is a coordinated process in which blood precursors migrate and colonize spaces carved out of embryonic bone and cartilage. In adults, the formed elements of the blood are produced in the bone marrow. Thus, an intimate physical association between bone cells and blood cells is established early in life. Many investigators have demonstrated that primary and transformed murine OB cell lines secrete cytokines that could modulate marrow stem cell function. The list includes granulocyte-colony-stimulating factor (G-CSF), M-CSF, GM-CSF, interleukin-1 (IL-1), IL-6, and others.[13,14] Primary human OBs have been less well characterized, but they express many molecules known to modulate hematopoiesis. For example, G-CSF, GM-CSF, M-CSF, IL-1, IL-6, IL-7, leukemia inhibitory factor (LIF), OPG, receptor activator of NF-κB ligand (RANKL), stromal-derived factor (SDF-1), tumor necrosis factor-alpha (TNF-α), and vascular endothelial growth factor (VEGF) have all been detected using human cells.[15–17] Several animal models strongly implicate OBs and marrow endothelial cells in the expression of SDF-1 in localizing human hematopoietic progenitor cells into the marrow.[18] These and other findings strongly suggest a reciprocal relationship between OBs and hematopoietic cells,[19] but the dimension of this interaction has yet to be completely defined. It has been postulated that OBs maintain hematopoiesis by virtue of creating a niche. This idea has been supported by the fact that in RUNX-2-deficient mice the lack of osteoblastic maturation also results in a total lack of bone marrow throughout the entire skeleton, however, the number of hematopoietic precursors in the yolk sac at embryonic day 10.5 and in the liver at E12.5 is normal.[20,21] Recently,[22] a similar result was obtained

---

FIGURE 2. (**A**) Scaffold on the allantochorial membrane: several vessels invaded the intertrabecular spaces (20×). (**B**) Osteoblasts on the scaffold display an active plump aspect and are surrounded by layers of less differentiated cells (300×). (**C**) Bone marrow is already present after 5 days on the membrane (200×). (**D**) After 7 days the bone marrow is well differentiated between the trabeculae of the scaffold (250×). (**E**) Mature osteoclast on scaffold surface (300×). (**F**) (120×) and (**G**) (200×). When scaffolds without osteoblasts were transplanted, the embryonic tissue with vessels invaded the scaffolds, but bone marrow never developed. (**H**) Osteoblasts cultured on 3D scaffolds express stromal-derived factor 1 (SDF-1), cytokine responsible for marrow homing, as verified by RT-PCR. Negative control is a primary culture of osteoclasts.

implanting subcutaneously in nude mice synthetic scaffolds loaded with osteogenic cells: the development of a bone marrow was observed. Our results give another experimental support to the hypothesized role of OBs in creating a niche for hematopoiesis.[23]

In conclusion, we demonstrated the enhanced proliferative and synthetic activity of human OBs, when cultured on a 3D macroporous hydroxyapatite, the good vascularization capability of these cultures, once transplanted *in vivo* on allantochorial membrane and the induction of bone marrow differentiation by OBs in an ectopic site.

ACKNOWLEDGMENTS

We wish to acknowledge the expert technical support of Antonella Grano. This work has been financed by grants from Europeans Space Agency (ESA): MAP program ESA AO-LS-99-MED-024 (ERISTO), Italian Space Agency (ASI), OSMA project and Fondazione Cassa di Risparmio di Puglia.

REFERENCES

1. ZAMBONIN, G. & GRANO M. 1995. Biomaterials in orthopaedic surgery: effects of different hydroxyapatites and demineralized bone matrix on proliferation rate and bone matrix synthesis by human osteoblasts. Biomaterials **16:** 397–402.
2. BANCROFT, G.N., V.I. SIKAVITSAS, J. VAN DEN DOLDER, et al. 2002. Fluid flow increases mineralized matrix deposition in 3D perfusion culture of marrow stromal osteoblasts in a dose-dependent manner. Proc. Natl. Acad. Sci. USA **99:** 12600–12605.
3. SIKAVITSAS, V.I., G.N. BANCROFT, H.L. HOLTORF, et al. 2003. Mineralized matrix deposition by marrow stromal osteoblasts in 3D perfusion culture increases with increasing fluid shear forces. Proc. Natl. Acad. Sci. USA **100:** 14683–14688.
4. ROSKELLEY, C.D. & M.J. BISSELL. 1995. Dynamic reciprocity revisited: a continuous, bidirectional flow of information between cells and the extracellular matrix regulates mammary epithelial cell function. Biochem. Cell Biol. **73:** 391.
5. ELSDALE, T. & J. BARD. 1972. Collagen substrata for studies on cell behaviour. J. Cell Biol. **54:** 626.
6. FRIEDL, P. & E.B. BROCKER. 2000. The biology of cell locomotion within three-dimensional extracellular matrix. Cell Mol. Life Sci. **57:** 41.
7. CUKIERMAN, E., R. PANKOV, D.R. STEVENS & K.M. YAMADA. 2001. Taking cell-matrix adhesions to the third dimension. Science **294:** 1708–1712.
8. HUANG, W., B. CARLSEN, I. WULUR, et al. 2004. BMP-2 exerts differential effects on differentiation of rabbit bone marrow stromal cells grown in two-dimensional and three-dimensional systems and is required for *in vitro* bone formation in a PLGA scaffold. Exp. Cell Res. **299:** 325–334.
9. MAENO, S., Y. NIKI, H. MATSUMOTO, et al. 2005. The effect of calcium ion concentration on osteoblast viability, proliferation and differentiation in monolayer and 3D culture. Biomaterials **26:** 4847–4855.

10. TISCHER, T., M. SCHIEKER, M. STENGELE, *et al.* 2004. 3D-culturing of human osteoblastic cells with vessel like nutrient supply. Z. Orthop. Ihre. Grenzgeb. **142:** 344–349.

11. GEHRON-ROBEY, P.G. & J.D. TERMINE. 1985. Human bone cells *in vitro*. Calcif. Tissue Int. **37:** 453–460.

12. ROBEY, P.G., M.F. YOUNG, K.C. FLANDERS, *et al.* 1987. Osteoblasts synthesize and respond to transforming growth factor-type beta (TGF-beta) *in vitro*. J. Cell Biol. **105:** 457–463.

13. DECKERS, M.M., R.L. VAN BEZOOIJEN, G. VAN DER HORST, *et al.* 2002. Bone morphogenetic proteins stimulate angiogenesis through osteoblast-derived vascular endothelial growth factor A. Endocrinology **143:** 1545–1553.

14. SHEN, F., M.J. RUDDY, P. PLAMONDON & S.L. GAFFEN. 2005. Cytokines link osteoblasts and inflammation: microarray analysis of interleukin-17- and TNF-alpha-induced genes in bone cells. J. Leukoc. Biol. **77:** 388–399.

15. BILBE, G., E. ROBERTS, M. BIRCH & D.B. EVANS. 1996. PCR phenotyping of cytokines, growth factors and their receptors and bone matrix proteins in human osteoblast-like cell lines. Bone **19:** 437–445.

16. BLAIR, H.C., B.A. JULIAN, X. CAO, *et al.* 1999. Parathyroid hormone-regulated production of stem cell factor in human osteoblasts and osteoblast-like cells. Biochem. Biophys. Res. Commun. **255:** 778–784.

17. TAICHMAN, R.S. 2005. Blood and bone: two tissues whose fate are intertwined to create the hematopoietic stem-cell niche. Blood **105:** 2631–2639.

18. KORTESIDIS, A., A. ZANNETTINO, S. ISENMANN, *et al.* 2005. Stromal-derived factor-1 promotes the growth, survival, and development of human bone marrow stromal stem cells. Blood **105:** 3793–3801.

19. BALDUINO, A., S.P. HURTADO, P. FRAZAO, *et al.* 2005. Bone marrow subendosteal microenvironment harbours functionally distinct haemosupportive stromal cell populations. Cell Tissue Res. **319:** 255–266.

20. DUCY, P., R. ZHANG, V. GEOFFROY, *et al.* 1997. Osf2/Cbfa1: a transcriptional activator of osteoblast differentiation. Cell **89:** 747–754.

21. WANG, Q., T. STACY, M. BINDER, *et al.* 1996. Disruption of the Cbfa2 gene causes necrosis and hemorrhaging in the central nervous system and blocks definitive hematopoiesis. Proc. Natl. Acad. Sci. USA **93:** 3444–3449.

22. GOMI, K., M. KANAZASHI, D. LICKORISH, *et al.* 2004. Bone marrow genesis after subcutaneous delivery of rat osteogenic cell-seeded biodegradable scaffolds into nude mice. J. Biomed. Mater. Res. **71A:** 602–607.

23. CALVI, L.M., G.B. ADAMS, K.W. WEIBRECHT, *et al.* 2003. Osteoblastic cells regulate the haematopoietic stem cell niche. Nature **425:** 841–846.

TRAIL Is Involved in Human Osteoclast Apoptosis

GIACOMINA BRUNETTI,[a] ANGELA ORANGER,[a] GIORGIO MORI,[b]
ROBERTO TAMMA,[a] ADRIANA DI BENEDETTO,[a] PAOLO PIGNATARO,[a]
FELICE R. GRASSI,[c] ALBERTA ZALLONE,[a] MARIA GRANO,[a] AND
SILVIA COLUCCI[a]

[a]Department of Human Anatomy and Histology, University of Bari, Bari, Italy

[b]Department of Biomedical Science, University of Foggia, Italy

[c]Department of Oral Science, University of Bari, Bari, Italy

ABSTRACT: Control of osteoclast (OC) apoptosis has been recognized as
a critical regulatory factor in bone remodeling. TRAIL, a member of
the TNF superfamily, induces apoptosis in neoplastic and normal cells.
However, few data are available on the effects of TRAIL on bone cells,
thus in the present study we investigated TRAIL role on the apoptosis
of human mature OCs. We show that TRAIL treatment causes reduced
cell viability, loss of nuclei integrity, and derangement of the actin mi-
crofilament in OCs. We also demonstrated that the death receptor DR5,
upregulated by TRAIL, could be the mediator of TRAIL-induced OC
apoptosis.

KEYWORDS: TRAIL; TRAIL receptors; osteoclasts; apoptosis

INTRODUCTION

Net bone mass represents the relative activities of osteoblasts and osteoclasts
(OCs). Bone remodeling is an ever-occurring event characterized by sequential tethering of the activities of OCs and osteoblasts. In fact, the majority of
acquired, systemic diseases of the skeleton reflects imbalance between OC and
osteoblast activity in the remodeling process. Remodeling units are initiated
by the appearance of OCs that degrade a packet of bone. These resorptive
cells are replaced, at the same location, by osteoblast precursors, which synthesize new bone. Bone resorption reflects the sum of OC recruitment and
apoptosis and the rate at which the average cell degrades matrix. Considerable
progress has been made in understanding the mechanisms of OC apoptosis

Address for correspondence: Silvia Colucci, Ph.D., Department of Human Anatomy and Histology,
University of Bari, Medical School, Piazza Giulio Cesare, 11, 70124 Bari, Italy. Voice: +39-080-
5478361; fax: +39-080-5478308.

s.colucci@anatomia.uniba.it

Ann. N.Y. Acad. Sci. 1116: 316–322 (2007). © 2007 New York Academy of Sciences.
doi: 10.1196/annals.1402.011

in rodents, but little is known about the human mechanism. The induction of apoptosis by extracellular signals involves ligands related to TNF superfamily, such as TNF-related apoptosis-inducing ligand (TRAIL), also known as Apo2 ligand (Apo2L). To date, human receptors for TRAIL have been identified and characterized. Of these, DR4 and DR5 are able to transduce a death signal, whereas, DcR1 and DcR2 do not possess a complete death domain and are not able to mediate apoptosis.[1,2] TRAIL induces cell death in a wide variety of tumor cell lines. Although the selective cytotoxicity of TRAIL is not entirely clear it may be related to the balance between death and "decoy" receptors expressed by the cells. Even though *in vitro* studies reported by us and other investigators demonstrated that human OCs generated from two different systems, peripheral blood mononuclear cells (PBMCs) and cord blood monocytes, express TRAIL receptors, the TRAIL-induced apoptotic pathway has not been at present elucidated in human OCs.[3,4] Thus, in this study we show new data demonstrating that TRAIL induces apoptosis in human-differentiated OCs.

MATERIALS AND METHODS

Cell Cultures

OCs were obtained from heparinized human PBMCs collected from healthy donors. PBMCs were isolated by centrifugation over a Histopaque 1077 density gradient (Sigma Chemical Co., St. Louis, MO, USA), cultured in the presence of 25 ng/mL recombinant human macrophage colony-stimulating factor (rh-MCSF) and 30 ng/mL receptor activator of nuclear factor-kappa B ligand (rh-RANKL– R&D Systems Inc., MN, USA) for 18–21 days.

Cell Viability Assay

Cell viability was measured by 3-(4,5-dimethylthiazol-2-yl)-2,5-diphenyltetrazolium bromide (MTT) assay. PBMCs were cultured in 96-well tissue culture plates with 25 ng/mL of M-CSF and 30 ng/mL of RANKL. The cells, after removal of M-CSF and RANKL, were treated with different concentrations of rh-TRAIL (10 to 500 ng/mL) for 24 and 48 h to evaluate cell viability on preosteoclasts, on days 7 and 14, and mature OCs on day 21. In parallel, the effect of 500 ng/mL TRAIL on OC viability was evaluated in cells pretreated for 30 min with 5 μg/mL of antagonist anti-DR5-neutralizing monoclonal antibody (mAb) (Alexis, San Diego, CA, USA).

Detection of Apoptosis by Phalloidin and DAPI Staining

Cells were cultured on coverslips in the presence of M-CSF and RANKL. Once they had differentiated into mature OCs, after removal of M-CSF and RANKL, they were treated with rh-TRAIL at 100 and 500 ng/mL for 48 h. OCs were fixed and stained with 4,6-diamidino-2-phenylindole (DAPI, Sigma Aldrich, Milan, Italy) and with fluorescein-labeled phalloidin (F-PHD) (Sigma Aldrich). The photomicrographs were obtained using a Nikon Ellipse E400 microscope equipped with Nikon Plan Fluor 10x/0.30 dicl. The microscope was connected with a Nikon digital camera DxM 1200.

Western Blot Analysis

PBMCs were cultured in 6-well tissue culture plates with 25 ng/mL of M-CSF and 30 ng/mL of RANKL. After removal of M-CSF and RANKL, on day 21 mature OCs were treated with 500 ng/mL rh-TRAIL for 0–8 h to evaluate TRAIL receptor expression. Proteins were subjected to 12% SDS-PAGE gel, transferred to nitrocellulose membrane, and probed for various proteins (DcR1, DcR2, DR4, DR5, and β−actin).

RESULTS

Sensitivity to TRAIL in human OCs was investigated by MTT assay. As shown in FIGURE 1, the OC viability was reduced by TRAIL in a dose- and

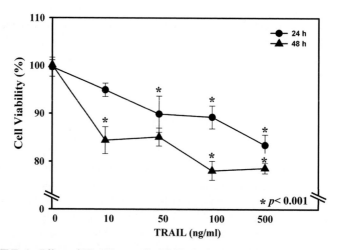

FIGURE 1. Effect of TRAIL on cell viability in human differentiated osteoclasts. Osteoclasts treated for 24 and 48 h with TRAIL at the indicated concentrations were analyzed by MTT assay to evaluate cell viability. Results are expressed as mean values of optical density at 570 nm ± standard error.

time-dependent manner. In particular, when OCs were exposed to 50 ng/mL rh-TRAIL for 24 h their viability was hardly but significantly decreased, and the effect was further enhanced at 500 ng/mL TRAIL respect to untreated OCs. After 48-h TRAIL treatment, the lowest concentration (10 ng/mL) of the molecule already induced a significant reduction of OC viability, and this effect was further marked at 100 and 500 ng/mL TRAIL. No effect on cell viability was found in OC precursors treated with increasing concentrations of TRAIL on days 7 and 14 (data not shown).

It is known that morphological changes correlate with cell death. Thus, the effect of TRAIL on the integrity of nuclei and the cytoskeletal organization of OCs was investigated. In particular, after 48 h, in control conditions OCs exhibited round nuclei (FIG. 2B) and actin microfilaments organized in

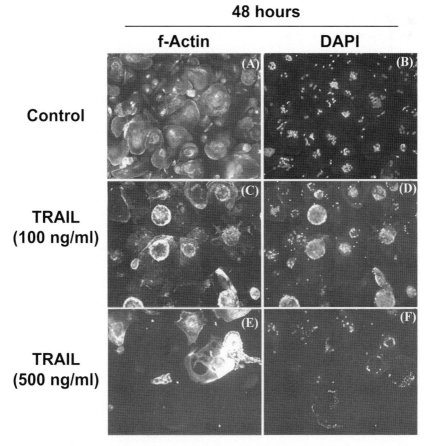

FIGURE 2. Morphology of osteoclasts after 48-h TRAIL treatment. Fluorescent DNA-binding dye, (DAPI), and fluorescein-labeled phalloidin (F-PHD) staining were performed to detect nuclear morphology (**B, D, F**) and actin microfilaments (**A, C, E**).

podosomes (FIG. 2A), the typical adhesion structures of these cells. The evidence of nuclei fragmentation and derangement in actin microfilaments was observed only in some of the OCs treated with 100 ng/mL rh-TRAIL after 48 h (FIG. 2D and C, respectively). However, picnotic nuclei and a complete derangement of actin organization, consisting in the loss of podosomes and a consequent detachment of the cells from the substrate, were found in the OCs treated with 500 ng/mL TRAIL for 48 h (FIG. 2E–F). In this condition, a marked reduction of the cell number could be observed, the shape of the residual cells was less round and their ruffled membranes suggested that the apoptotic process was taking place. Furthermore, in TRAIL-treated OC, the loss of nuclear integrity was biochemically demonstrated by DNA fragmentation (data not shown).

Additionally, the expression of TRAIL receptors in differentiated OCs after 500 ng/mL TRAIL treatment was investigated by Western blot. Results showed that the expression of DR4, DcR1, and DcR2 were not affected by TRAIL treatment (data not shown). Differently, the expression of DR5 was significantly increased by TRAIL, reaching the maximum level within 2 h and resulting still higher in respect to the untreated OCs up to 8 h of treatment (FIG. 3). Furthermore, to assess the specific involvement of DR5 in TRAIL-induced OC apoptosis we investigated OC viability in the presence of TRAIL using an anti-DR5 neutralizing antibody. We demonstrated that the reduction of OCs viability induced by TRAIL after 48 h was completely abolished only in the presence of anti-DR5-neutralizing antibody (FIG. 4).

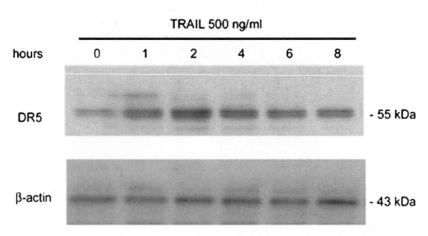

FIGURE 3. TRAIL upregulates DR5 expression in human differentiated osteoclasts. Human mature osteoclasts, treated with 500 ng/mL TRAIL at the indicated times, were lysed and analyzed by Western blot analysis to detect the protein levels of DR5.

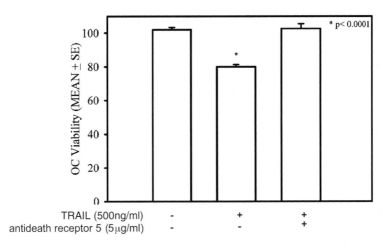

FIGURE 4. Effect of antideath receptor 5-neutralizing antibody on osteoclast viability. Osteoclasts were cultured in control conditions (−), in the presence of TRAIL (+) with (+) or without (−) anti-DR5-neutralizing antibodies and cell viability was assessed by MTT assay. Osteoclasts were pretreated for 30 min with the indicated antibody before the addition of TRAIL.

DISCUSSION

OC life span is the keystone to bone remodeling and could provide a useful target for the treatment of skeletal disorders. Although it has been demonstrated that OC apoptosis is mainly regulated by the Fas/FasL system,[5] other apoptotic molecules, such as TRAIL, cannot be excluded in the regulation of this process. Our findings demonstrate that TRAIL affects human OC apoptosis mainly through DR5. With particular regard to the bone cells, the expression of TRAIL receptors has been described in normal and transformed osteoblasts[6] as well as in OCs in pathological and physiological conditions,[3,4] suggesting that TRAIL could play a role in bone cell apoptosis.[4,7] In this article, we have demonstrated that human OCs, once differentiated, were sensitive to TRAIL-induced apoptosis. This evidence lies in the reduction of cell viability and increased number of apoptotic OCs induced by TRAIL. Furthermore, TRAIL treatment induced actin filament disruption, podosome loss, and OC detachment from the substrate. Additionally, in fully differentiated human OCs we have herein documented for the first time that TRAIL caused upregulation of the levels of DR5 but no modulation of either DR4 or the two-decoy receptor expression. These findings suggest that DR5 could be the major receptor involved in TRAIL-mediated apoptosis of human mature OCs. This hypothesis is in agreement with the demonstration that only anti-DR5-neutralizing antibody completely restored the TRAIL-induced reduction of OC viability. In conclusion, this study shows that TRAIL-induced OC apoptosis, mainly mediated by DR5, could be an additional and/or alternative mechanism inducing apoptosis of bone-resorbing cells

ACKNOWLEDGMENT

This investigation was supported by Agenzia Spaziale Italiana (ASI-OSMA grant to Maria Grano).

REFERENCES

1. SHERIDAN, J.P., S.A. MARSTERS, R.M. PITTI, *et al*. 1997. Control of TRAIL-induced apoptosis by a family of signaling and decoy receptors. Science **277:** 818–821.
2. DEGLI-ESPOSTI, M.A., P.J. SMOLAK, H. WALCZAK, *et al*. 1997. Cloning and characterization of TRAIL-R3, a novel member of the emerging TRAIL receptor family. J. Exp. Med. **186:** 1165–1170.
3. COLUCCI, S., G. BRUNETTI, R. RIZZI, *et al*. 2004. T cells support osteoclastogenesis in an *in vitro* model derived from human multiple myeloma bone disease: the role of the OPG/TRAIL interaction. Blood **104:** 3722–3730.
4. ROUX, S., P. LAMBERT-COMEAU, C. SAINT-PIERRE, *et al*. 2005. Death receptors, Fas and TRAIL receptors, are involved in human osteoclast apoptosis. Biochem. Biophys. Res. Commun. **333:** 42–50.
5. WU, X., M.A. MCKENNA, X. FENG, *et al*. 2003. Osteoclast apoptosis: the role of Fas *in vivo* and *in vitro*. Endocrinology **144:** 5545–5555.
6. BU, R., C.W. BORYSENKO, Y. LI, *et al*. 2003. Expression and function of TNF-family proteins and receptors in human osteoblasts. Bone **33:** 760–770.
7. TINHOFER, I., R. BIEDERMANN, M. KRISMER, *et al*. 2006. A role of TRAIL in killing osteoblasts by myeloma cells. FASEB J. **20:** 759–761.

Proteomic Analysis of Hydroxyapatite Interaction Proteins in Bone

HAI-YAN ZHOU

Laboratory for the Study of Skeletal Disorders and Rehabilitation, Department of Oral and Developmental Biology, Harvard School of Dental Medicine and Children's Hospital, Boston, Massachusetts, USA

ABSTRACT: Biomineralization involves proteins and/or other macro-molecules directly in controlling the mineral crystal nucle-ation/induction, growth, and maturation. To identify these proteins in bone, bovine bone EDTA/NaCl extract was passed through a hydrox-yapatite column followed by washing with 0.5 M NaCl and the bound proteins were collected and analyzed by mass spectrometry. More than 30 proteins were identified. While as described previously, albumin, α2-HS glycoprotein, decorin, biglycan, osteoadherin, osteonectin, etc. were included, collagen α2 (I), matrix extracellular phosphoglycopro-tein, secreted phosphoprotein 24, chondroadherin, lumican, perlecan, thrombospondin 1, nucleobindin, etc. were for the first time shown to directly interact with calcium phosphate mineral.

KEYWORDS: biomineralization; hydroxyapatite; collagen α2 (I); matrix extracellular phosphoglycoprotein; secreted phosphoprotein 24; chon-droadherin; lumican; perlecan; thrombospondin 1; nucleobindin

INTRODUCTION

Biomineralization is the process by which vertebrates, many invertebrates, some bacteria, and many plants form inorganic minerals for ion storage, sup-port, protection/defense, feeding, sound reception, gravity perception, toxic waste disposal, orientation in the earth's magnetic field, etc. The process in-volves organic proteins and/or other macromolecules directly in controlling the mineral crystal nucleation/induction, growth, and maturation. To identify these proteins in bone, bone EDTA/NaCl extract obtained from fresh bovine bone powders by demineralization with 0.5 M EDTA/1.0M NaCl (pH 7.4) was passed through a hydroxyapatite (HAP, 40 μm particle size, Bio-Rad, Hercules, CA, USA) column followed by washing with 0.5 M NaCl, 10 mM Tris-HCl,

Address for correspondence: Hai-Yan Zhou, Laboratory for the Study of Skeletal Disorders and Rehabilitation, Enders Building, Room 914, Children's Hospital, 300 Longwood Avenue, Boston, MA 02115. Voice: 617-919-2035; fax: 617-730-0122.

hai-yan.zhou@childrens.harvard.edu

Ann. N.Y. Acad. Sci. 1116: 323–326 (2007). © 2007 New York Academy of Sciences.
doi: 10.1196/annals.1402.023

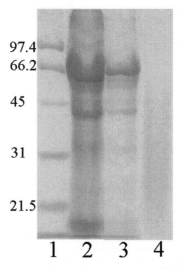

FIGURE 1. SDS-polyacrylamide gel electrophoresis of HAP interaction proteins. All materials were separated on 10% gels under reducing conditions and stained with GelCode Blue Stain Reagent (Pierce Biotechnology, Rockford, IL, USA). Lane 1, molecular mass standards (molecular weights are indicated on the left in Kilodalton); lane 2, EDTA/NaCl extract; lane 3, HAP-bound materials; lane 4, HAP-unbound materials.

pH 7.4, and the bound proteins were collected, reduced with dithiothreitol, alkylated with iodoacetamide, and digested with trypsin prior to tandem mass spectrometry (MS/MS), which was performed at the Proteomics Center at Children's Hospital, Boston, MA.

RESULTS AND DISCUSSION

The HAP interaction proteins were extracted by demineralization with EDTA/NaCl (FIG. 1, lane 3). However, not all the extracted proteins were HAP-binding (FIG. 1, lane 4). More than 30 proteins were systematically identified (TABLE 1). Most of them are extracellular matrix proteins. Although the interactions of HAP with histones may be nonspecific, the interaction with nuclear Ca^{2+}-binding nucleobindin is physiological because it has been found in bone extracellular matrix.[1] While as described previously, several proteins, such as α2-HS glycoprotein, albumin, decorin, biglycan, osteoadherin, and osteonectin, were found in the HAP-bound fraction, collagen α2 (I), matrix extracellular phosphoglycoprotein, secreted phosphoprotein 24, chondroadherin, lumican, perlecan, thrombospondin 1, nucleobindin, etc. for the first time and/were shown to directly interact with calcium phosphate mineral.

It is known that mineralization starts when it is associated with type I collagen and at the specific sites within the collagen fibrils.[2,3] The interaction between

TABLE 1. Hydroxyapatite (HAP) interaction proteins

Collagen	Collagen α2 (I)
Phosphoproteins	Secreted phosphoprotein 24, matrix extracellular phosphoglyco protein
Proteoglycans	Biglycan, decorin, osteoadherin/osteomodulin, chondroadherin, lumican, perlecan
Matricellular proteins	Osteonectin, thrombospondin-1
Blood-derived proteins	Albumin, α2-HS glycoprotein, hemoglobin β-A
Nuclear proteins	Nucleobindin, histone H2B, histone H2A.1, histone H1.1
Others	Myosin-9/nonmuscle IIa, etc.

HAP and collagen α2 (I) suggests that collagen α2 (I) is able to induce mineral crystal formation and/or direct the size, shape, and orientation of the crystals. *In vitro* calcification studies show that biglycan and osteonectin serve as apatite nucleators whereas albumin and α2-HS glycoprotein inhibit the formation and growth of mineral crystals.[4] It is clear from theoretical observations and data already available in these *in vitro* systems that phosphorylated proteins are apt to be among the most important in influencing the biomineralization process in calcium phosphate systems.[2,4] At least three phosphorylated proteins, secreted phosphoprotein 24, matrix extracellular phosphoglycoprotein and osteonectin, were found in the HAP-bound proteins. However, two other major bone extracellular phosphoproteins osteopontin and bone sialoprotein, both of which affected *in vitro* and *in vivo* calcification,[4,5] could not be detected. This may be due to low MS/MS spectrum quality and/or simultaneous fragmentation of multiple peptide ions having similar m/z values, both of which can lead to failure of assigning correct peptide sequences to the experimental MS/MS spectra. To solve these problems, the study incorporating ion exchange chromatography in preparation of the HAP-bound proteins is being undertaken. Also, this modification will help detect some missed minor components.

In conclusion, this study provides new insight into the function of the newly identified proteins and thereby the mechanism for the mineral crystal nucleation, growth, and maturation.

ACKNOWLEDGMENTS

This work was supported by NIH grant 2RO1AG014701–17A2 and Departmental Orthopaedic Research Fund.

REFERENCES

1. PETERSSON, U. *et al.* 2004. Nucleobindin is produced by bone cells and secreted into the osteoid, with a potential role as a modulator of matrix maturation. Bone **34:** 366–376.

2. GLIMCHER, M.J. 1992. The nature of the mineral component of bone and the mechanism of calcification. *In* Disorders of Bone and Mineral Metabolism. F.L. Coe
   & M.J. Favus, Eds.: 265–286. Raven Press, New York.
3. LEE, D.D. & M.J. GLIMCHER. 1991. Three-dimensional spatial relationship between
   the collagen fibrils and the inorganic calcium phosphate crystals of pickerel
   (*Americanus americanus*) and herring (*Clupea harengus*) bone. J. Mol. Biol. **217:**
   487–501.
4. BOSKEY, A.L. 1998. Biomineralization: conflict, challenges, and opportunities. J.
   Cell. Biochem. Suppl. **30/31:** 83–91.
5. WANG, J. *et al.* 2006. Site-specific *in vivo* calcification and osteogenesis stimulated
   by bone sialoprotein. Calcif. Tissue Int. **79:** 179–189.

Nuclear Receptors and Bone

RONALD N. MARGOLIS

National Institute of Diabetes and Digestive and Kidney Diseases, Bethesda, Maryland, USA

ABSTRACT: Nuclear receptors (NRs) represent a class of ligand-dependent and -independent transcription factors with importance to the regulation of development, reproduction, and metabolism. The emergence of new understanding of the structure, function, and role in disease of NRs provides new insights into the interaction between genetics and the environment, with NRs representing new targets for the development of therapeutic agents. NRs play key roles in bone health and contribute to our understanding of diseases and disorders that result in osteopenia and osteoporosis. The Nuclear Receptor Signaling Atlas (www.nursa.org) is an online repository of information about NRs and provides a community-wide resource designed to help catalyze new advances in biology and medicine.

KEYWORDS: nuclear receptors; osteoporosis; Nuclear Receptor Signaling Atlas; NURSA

GENERAL CONCEPTUAL FRAMEWORK

Nuclear receptors (NRs) comprise a family of ligand-dependent and -independent transcription factors that regulate development, reproduction, and homeostasis. Many members of the NR superfamily are responsive to ligands that may include hormones (estrogen, thyroid), vitamins (D and A), and metabolites (dietary lipids, bile acids, xenobiotics, drugs). NRs are characterized by structural domains that are well conserved through mammalian evolution.[1] These domains enable binding of ligand and to DNA, homo- and hetero-dimerization, as well as binding of nuclear accessory proteins present as regulatory cofactors. NRs form labile complexes in the nucleus in the context of *trans*-acting complexes that bind to discrete motifs in chromatin found largely in or near the promoter regions of target genes. Ultimately the action of NRs is reflected in changes in expression of these target genes. The roles of NRs in bone and osteoporosis have long been explored and stemmed initially from the observation of decreased bone mineral density (BMD), development

Address for correspondence: Ronald Margolis, Ph.D., Division of Diabetes, Endocrinology and Metabolic Diseases, National Institute of Diabetes and Digestive and Kidney Diseases, NIH, 6707 Democracy Blvd., Room 693, Bethesda, MD 20892-5460. Voice: 301-594-8819; fax: 301-435-6047. rm76f@nih.gov

Ann. N.Y. Acad. Sci. 1116: 327–334 (2007). © 2007 New York Academy of Sciences.
doi: 10.1196/annals.1402.043

of osteopenia, and increased risk of fracture in hypogonadal/postmenopausal women.[2] Understanding of the role of estrogen in the female and emerging understanding of its role in men, led to a fundamental appreciation of the role(s) of signaling through the estrogen receptor and an overall appreciation of the role of steroid hormones in maintenance of proper bone health.[3] This review briefly summarizes the roles of the NR superfamily in regulation of homeostasis and highlights some novel actions in bone.

While the regulation of a major tissue, such as bone, is both complex and interrelated, at least one series of regulatory factors, the NRs, have gained significant importance as effectors of signals that regulate bone development and homeostasis. The importance of NRs to understanding of biology and medicine is reflected in the enormous growth in the literature since the identification of the first member of this family of receptors, the steroid hormone estrogen.[4] Subsequent discoveries at the level of mechanism of action, including purification and cloning of the receptors, have led to increased delineation of structure, function, and role in disease. As discussed above, NRs are characterized by highly conserved structural domains including those for binding of ligand, DNA, and coregulators as well as for homo- and hetero-dimerization.[3–5] Indeed, the identification of coregulators more than 10 years ago[6] further enriched the picture,[7] helping to increase our understanding of the molecular mechanism(s) of action. With these advances has come an appreciation that tissue selectivity and specificity of action of NRs is an important consideration for understanding how and when NRs exert their effects. Rapidly forming and labile associations of nuclear accessory proteins found as coactivators, corepressors, and chromatin-remodeling factors, as well as other parts of the transcriptional machinery, together with NRs form multi-subunit molecular complexes that bind to regions of chromatin known as hormone response elements found in or near the promoter regions of target genes. Once bound, the NR–coregulator complex affects gene expression through either activation or repression of transcription. Recent studies have shown that expression of NRs has both an anatomic pattern and a circadian overlay, with functional implications that are still being deciphered.[8–11]

Physiological roles for the vitamin D receptor (VDR) and the estrogen receptor in calcium homeostasis and maintenance of bone density have clear importance to bone, and reflect the underlying molecular mechanism(s) of action.[12] No less important are the glucocorticoid and thyroid receptors in the regulation of inflammation, carbohydrate and lipid metabolism, and cellular integrity, with effects on bone that have variably been described as direct and indirect. Indeed, each has been implicated in osteopenia and subsequent risk of fracture when present at chronic or inappropriate levels. Still other NRs previously not associated with bone have been suggested to have actions in bone. Examples include several orphan NRs, such as the estrogen-related receptor (ERR), acting to mediate the actions of estrogen on bone,[13] and the peroxisome proliferator-activated receptor (PPAR). While NRs represent one of a series

of important transcription factors that regulate the development and activities of bone cells, there have now emerged several interesting new roles for NRs with potential actions in bone homeostasis.[14] An example is the finding that the PPAR-γ isoform of the PPAR subfamily of NRs and the role it plays in the regulation of cell fate determination of mesenchymal stem cells (MSC) and subsequent decisions that determine the development of osteoblasts.[15] In the adult, synthetic ligands for PPAR-γ agonists are now a mainstay of therapeutic intervention in type 2 diabetes, but actions on the regulation of bone turnover have demonstrated differential effects, with some ligands causing decreased bone formation, others increased, and still others having no effect on bone formation, highlighting the need to better understand tissue-specific actions of synthetic ligands for NRs. Whether these puzzling actions are the result of tissue-specific differences in responsiveness or to steroid hormone receptor polymorphisms is yet unclear.[16] Nevertheless, these actions of ligands for NRs emphasize the importance to clinical practice. The use of NRs as targets in the development of therapeutic agents to combat a host of diseases and conditions has only amplified the need to better understand how they work. Examples include agonists or partial agonists, such as tamoxifen in the treatment of estrogen-receptor positive breast cancer, prednisone for inflammation, thyroid hormone for thyroid insufficiency, and raloxifene for osteopenia. Since we are still trying to understand NRs and their roles in pathophysiology in bone, there is a continuing need to evaluate the basic mechanism(s) of action of these hormones in acting through their respective NRs to regulate bone and mineral metabolism and the activities of cells that play key roles in bone. One way to exploit these new advances in understanding the basic molecular biology and physiology of NR structure and function is to use high throughput screens with small chemical compounds in assays of receptor function to identify new compounds that show both selectivity and specificity of action as the basis for the development of new therapeutic agents.[17]

With the emergence of increased understanding of the molecular mechanism of action of NRs and the consequent focus on NRs as targets for the development of new therapeutic agents, coalescing information about NR structure, function, and role in disease in an easily accessible online repository was felt to be an important goal. The objective was to provide a community-wide resource beneficial to all investigators in this, and other fields. But with greater complexity and larger volumes of information comes greater need for the means to collate and communicate this information in ways that are useful to the broader community of investigators. The recently developed Nuclear Receptor Signaling Atlas (NURSA), an NIH-funded consortium, brought together investigators and informaticists from multiple disciplines to develop, accrue, and communicate information on the NR superfamily.[18] The ultimate goal of NURSA is to provide a central repository for information on NRs together with conceptual advances that could catalyze new ideas to fully elucidate NR function. A web portal (www.nursa.org) was designed to enable users to gain

access to information on NRs, coregulators, ligands, and target genes, as well as key literature citations and raw data sets generated by NURSA investigators.[19] Through the development of a series of molecule pages it has become possible to choose a particular NR receptor, coregulator, or ligand and conduct detailed, *in silico* searches for additional information about that molecule. One such example is the VDR, long linked to bone health indirectly through roles in calcium absorption and reciprocal regulation of PTH levels, and perhaps directly through actions on bone cells.[20] For the VDR, the molecule page includes, e.g., links to expression profiling of anatomic distribution in tissue of the mouse (http://www.nursa.org/molecule.cfm?molType=receptor&molId=1I1). Such studies of anatomic distribution may help to reinforce what is already known from other studies, but may serve to identify new avenues for investigation. By hyperlinking the primary NURSA content in the molecule pages with community and NIH-based web sites (e.g., NCBI; www.ncbi.nih.gov) containing additional, even orthogonal, data and information from databases at many other sites users can mine data and conduct queries tailored to individual interests and needs.

One effort with relevance to a system-wide understanding of NR structure, function, and role in disease has been fostered by NURSA investigators through development of high throughput data gathering.[10,11] As noted above for the VDR, expression profiling of NRs in the mouse using quantitative polymerase chain reaction (qPCR)-based assays has revealed patterns of expression that extend beyond what was previously known, suggesting potential roles in tissues over and above those already identified. Further probing in mouse models has revealed physiological variation that includes a circadian periodicity in expression in specific tissues.[11,21] These findings suggest both an anatomic and a temporal pattern of expression contributing to and dependent upon homeostatic mechanisms, and including interactions with the environment. Through the use of associated bioinformatics tools it has become possible to analyze the large data sets to help reveal patterns not immediately evident from study of isolated data sets. The result of the application of unbiased cluster analysis, and other informatics tools, to analysis of the large quantities of data developed has been the formulation of an hypothesis termed the "Nuclear Receptor Ring of Physiology," suggested by the anatomic expression data and providing a blueprint for further probing studies.[10] This type of systems biology-based analysis fosters the generation of new hypotheses to be tested both in the lab and *in silico*. Indeed, evidence from "classical" hormonal signaling pathways suggests the need to understand a role for system-wide regulatory pathways in the regulation of bone. Recent evidence demonstrating that the key pituitary releasing hormones for thyroid hormone and estrogen, TSH and FSH, respectively, may play roles in altered bone function suggest important roles for both the hypothalamic-pituitary-thyroid and hypothalamic-pituitary-gonadal axes, and remind us that many of the classical steroid hormones exist in the context of system-wide, classical, physiological feedback loops, integrated to form whole

body networks. These studies lend credence to the need for a systems-wide appreciation for regulatory hormone networks to help elucidate mechanisms underlying both homeostasis and development of disease.[22,23]

In a recent New York Academy of Sciences symposium on Skeletal Development and Remodeling [24] in a session on Novel Targets and Therapeutics for Bone Loss[14] a number of receptors for peptide and steroid hormones were identified with potential roles either as agents with anticatabolic or anabolic actions in bone. A troubling limitation was apparent through lack of true tissue specificity with effects felt at the level of potential off target, or undesired/less than optimal, effects.[15] With the greater appreciation of the role of NRs in bone, an ongoing concern has been to better define the underlying mechanism of action of NRs in bone, and other tissues, and to identify tissue-specific actions that can lead to highly specific therapeutic agents. The goal is to elicit a desired response solely in bone without the off target effects dictated by the broader distribution of that receptor in other tissues. The current New York Academy of Sciences symposium (2nd Conference on Skeletal Biology and Medicine) followed up on the important issues discussed earlier and had a session devoted to NRs in Bone. While NURSA introduced an important new resource, two specific examples of actions of NRs with emerging importance to our understanding of bone and mineral homeostasis and roles in disease were discussed. With a specific focus on two of the "classic" steroid receptors, the glucocorticoid and VDR, respectively, new studies suggest that while much is known about these two key systemic hormone receptors, there is much that remains to be established with new findings providing evidence of exciting and novel functions. The VDR has a number of important roles in bone and mineral metabolism, not the least of which is the contribution signaling through which it maintains proper calcium balances. It has long been known that one important downstream target for ligand-activated VDR is the cellular calcium-binding protein, calbindin.[25] More recently the notion of calbindin as primarily a calcium-binding protein has been challenged with new data that suggest a role in the activation of an important calcium channel protein, TRPV. These findings have been interpreted to help explain how calcium reabsorption in kidney can result in calcium movement through and across cells without resulting in dangerous buildup of intracellular calcium levels. A classical VDR/retinoid X receptor (RXR) heterodimer partner formation with ligand (vitamin D) appears to recruit chromatin remodeling, and other coregulators, required for activation of calbindin gene expression. No less important to understanding homeostasis of bone is the receptor for adrenal glucocorticoids. The glucocorticoid receptor (GR) is important in the regulation of overall energy balances, as well as being an important modulator of immune cell function. In bone it has long been known that excess glucocorticoids, such as might be found in Cushing's disease or in chronic treatment of inflammation, have been held responsible for osteopenia and subsequent osteoporosis and fracture[26] with the explanation that excess steroids had an effect on osteoclast apoptosis and bone resorption. Now,

recent findings confirm glucocorticoid effects on osteoclasts but also suggest that the effect occurs within the osteoclast at the level of cytoskeletal elements required for formation of the tight seal between the osteoclast and the nascent resorption pit.[27] The result is an imbalance in resorption and since resorption and formation are tightly coupled, bone formation is compromised with resulting osteopenia. Thus, the indirect inhibition of bone formation may very well underlie the actions of chronic steroid administration, mediated through GR action on signaling and downstream gene expression in the osteoclast.[27] These vignettes serve to further highlight the need to push the boundaries of our current knowledge of NRs beyond the known and expected to encompass new visions of their role(s) in bone and mineral metabolism and actions on bone homeostasis.

FUTURE DIRECTIONS

There is a continuing need to forge a better understanding of the cell and tissue-specific actions of NRs and their associated coregulators and role(s) in maintenance of proper homeostatic balances. This is particularly relevant to bone, whose overall structure needs to be stable while its constituent components remain labile. Starting with the precursors for bone cells in the marrow, the roles of NRs in cell fate determination, differentiation, and function of mature osteoblasts and osteoclasts need to be better defined and understood. It may then become possible to consider how changes in anatomical and temporal expression of important effectors, such as NRs, interact with the environment in the development of pathology. The ultimate goal will be to identify potential targets for the development of novel and highly specific therapeutic agents.

REFERENCES

1. MANGELSDORF, D.J., C. THUMMEL, M. BEATO, *et al.* 1995. The nuclear receptor superfamily: the second decade. Cell **83:** 835–839.
2. LANE, N.E. 2006. Epidemiology, etiology, and diagnosis of osteoporosis. Am. J. Obstet. Gynecol. **194**(2 Suppl):s3–s11.
3. SYED, F.A., D.G. FRASER, T.C. SPELSBERG, *et al.* 2007. Effects of loss of classical estrogen response element signaling on bone in male mice. Endocrinology **148:** 1902–1910.
4. EVANS, R.M. 2005. The nuclear receptor superfamily: a rosetta stone for physiology. Mol. Endo. **19:** 1429–1438.
5. MOORE, D.D., S. KATO, W. XIE, *et al.* 2006. International Union of Pharmacology. LXII. The NR1H and NR1I Receptors: Constitutive Androstane Receptor, Pregnene X Receptor, Farnesoid X Receptor α, Farnesoid X Receptor β, Liver X Receptor α, Liver X Receptor β, and Vitamin D Receptor. Pharmacol. Rev. **58:** 742–759.

6. LONARD, D.M. & B.W. O'MALLEY 2006. The expanding cosmos of nuclear receptor coactivators. Cell **125**: 411–414.
7. ROSENFELD, M.G., V.V. LUNYAK & C.K. GLASS. 2006. Sensors and signals: a coactivator/corepressors/epigenetic code for integrating signaling-dependent programs of transcriptional response. Genes Dev. **20**: 1405–1428.
8. FU, M., T. SUN, A.L. BOOKOUT, *et al.* 2005. A nuclear receptor atlas: 3T3-LI adipogenesis. Mol. Endocrinol. **19**: 2437–2450.
9. BARISH, G.D., M. DOWNES, W.A. ALAYNICK, *et al.* 2006. A nuclear receptor atlas: macrophage activation. Mol. Endo. **19**: 2466–2477.
10. BOOKOUT, A.L., Y. JEONG, M. DOWNES, *et al.* 2006. Anatomical profiling of nuclear receptor expression reveals a hierarchical transcriptional network. Cell **126**: 789–799.
11. YANG, X., M. DOWNES, R.T. YU, *et al.* 2006. Nuclear receptor expression links the circadian clock to metabolism. Cell **126**: 801–810.
12. PIKE, J. Wesley, M.B. MEYER, M. WATANUKI, *et al.* 2007. Perspectives on mechanisms of gene regulation by 1,25-dihydroxyvitamin D3 and its receptors. J. Steroid Biochem. Mol. Biol. **103**: 389–95.
13. BONNELYE, E. & J.E. AUBIN. 2005. REVIEW: estrogen receptor-related receptor alpha: a mediator of estrogen response in bone. J. Biol. Chem. **280**: 3104–3111.
14. MARGOLIS, R.N. & S.J. WIMALAWANSA. 2006. Novel targets and therapeutics for bone loss. Ann. N. Y. Acad. Sci. 1068: 402–409.
15. GIAGINIS, C., A. TSANTILI-KAKOULIDOU & S. THEOCHRARIS 2007. Peroxisome proliferator-activated receptor-γ ligands as bone turnover modulators. Exp. Opin. Investig. Drugs **16**: 195–207.
16. GENNARI, L., V. DE PAOLA, D. MERLOTTI, *et al.* 2007. Steroid hormone receptor gene polymorphisms and osteoporosis: a pharmacogenomic review. Exp. Opin. Pharmacother. **8**: 537–553.
17. DOWNES, M., M.A. VERDECIA, A.J. ROECKER, *et al.* 2003. A chemical, genetic, and structural analysis of the nuclear bile acid receptor FXR. Mol. Cell **11**: 1079–1092.
18. MARGOLIS, R.N., R.M. EVANS & B.W. O'MALLEY. 2005. The Nuclear Receptor Signaling Atlas: development of a functional atlas of nuclear receptors. Mol. Endocrinol. **19**: 2433–2437.
19. LANZ, R.B., Z. JERICEVIC, W.J. ZUERCHER, *et al.* 2006. Nuclear Receptor Signaling Atlas (www.nursa.org): hyperlinking the Nuclear Receptor Signaling Community. Nucleic Acids Res. **34**: D221–D226.
20. BARTHEL, T.K., D.R. MATHERN, G.K. WHITFIELD, *et al.* 2007. 1,25-dihydroxyvitamin D3/VDR-mediated induction of FGF23 as well as transcriptional control of other bone anabolic and catabolic genes that orchestrate the regulation of phosphate and calcium mineral metabolism. J. Steroid Biochem. Mol. Biol. **103**: 381–388.
21. YIN, L., J. WANG, P.S. KLEIN & M.A. LAZAR. 2006. Nuclear Receptor Rev-erba is a critical lithium-sensitive component of the Circadian Clock. Science **311**: 1002–1005.
22. ABE, E., R.C. MARIANS, W. YU, *et al.* 2003. TSH is a negative regulator of skeletal remodeling. Cell **115**: 151–162.
23. SUN, L., Y. PENG, A.C. SHARROW, *et al.* 2006. FSH directly regulates bone mass. Cell **125**: 247–260.
24. ZAIDI, M. 2006. Skeletal development and remodeling in health, disease, and aging. Ann. N. Y. Acad. Sci. **1068**: xiii–575.

25. CHRISTAKOS, S., P. DHAWAN, X. PENG, *et al.* 2007. New insights into the function and regulation of vitamin D target proteins. J. Steroid Biochem. Mol. Biol. **103:** 405–410.

26. CANALIS, E., G. MASSIOTTI, A. GIUSTINA & J.P. BILEZIKIAN. 2007. Glucocorticoid-induced osteoporosis: pathophysiology and therapy. Osteoporos. Int. DOI10.1007/s00198-007-0394-0.

27. KIM, H.-J., H. ZHAO, H. KITAURA, *et al.* 2006. Glucocorticoids suppress bone formation via the osteoclast. J. Clin. Invest. **116:** 2152–2160.

Glucocorticoids and the Osteoclast

HYUN-JU KIM,[a] HAIBO ZHAO,[a] HIDEKI KITAURA,[a]
SANDIP BHATTACHARYYA,[b] JUDSON A. BREWER,[b]
LOUIS J. MUGLIA,[b] F. PATRICK ROSS,[a] AND STEVEN L. TEITELBAUM[a]

[a]Department of Pathology and Immunology, St. Louis, Missouri, USA

[b]Department of Pediatrics, Washington University School of Medicine,
St. Louis, Missouri, USA

ABSTRACT: Glucocorticoid (GC)-induced bone loss is the most common cause of secondary osteoporosis but its pathogenesis is controversial. GCs clearly suppress bone formation *in vivo* but the means by which they impact osteoblasts is unclear. Because bone remodeling is characterized by tethering of the activities of the two cells, the osteoclast is a potential modulator of the effect of GCs on osteoblasts. To address this issue we compared the effects of dexamethasone on wild-type (WT) osteoclasts with those derived from mice with disruption of the GC receptor in osteoclast lineage cells and found that the bone-degrading capacity of GC-treated WT cells is suppressed. The inhibitory effect of dexamethasone on bone resorption reflects failure of osteoclasts to organize their cytoskeleton in response to M-CSF. Dexamethasone specifically arrests M-CSF activation of RhoA, Rac, and Vav3, each of which regulate the osteoclast cytoskeleton. In all circumstances, mice lacking the GC receptor in osteoclast lineage cells are spared the impact of dexamethasone on osteoclasts and their precursors. Consistent with osteoclasts modulating the osteoblast-suppressive effect of dexamethasone, GC receptor-deficient mice are protected from the steroid's inhibition of bone formation.

KEYWORDS: glucocorticoids; osteoclasts; bone remodeling

Glucocorticoids (GCs) are among the most commonly prescribed drugs for inflammatory disorders including those that promote bone loss, such as periodontal disease and the periarticular osteolysis of rheumatoid arthritis. While efficiently suppressing inflammation, GCs also carry severe complications including a predisposition to osteoporosis, particularly early in the course of therapy. In fact, GC administration typically yields an annual average bone loss approximating 12% within the first year, which attenuates to about 3% thereafter.[1]

Address for correspondence: Steven L. Teitelbaum, M.D., Washington University School of Medicine, Department of Pathology and Immunology, Campus Box 8118, 660 South Euclid Avenue, St. Louis, MO 63110. Voice: 314-454-8463; fax: 314-454-5505.
teitelbs@wustl.edu

Ann. N.Y. Acad. Sci. 1116: 335–339 (2007). © 2007 New York Academy of Sciences.
doi: 10.1196/annals.1402.057

GCs impact bone directly and indirectly. The potential indirect skeletal effects of the steroids include suppressed synthesis and secretion of sex hormones. They also decrease calcium absorption by renal tubules and intestinal mucosa. This compendium of events theoretically promotes secondary hyperparathyroidism.[2] However, there is presently little proof that GC-treated patients develop meaningful hyperparathyroidism and, therefore, the body of evidence indicates that direct effects of the steroids on bone cells predominate. In fact, there are substantial data indicating GCs directly suppress the activity of bone-forming cells, *in vivo*.[1,3] Alternatively, precisely how these steroids impact osteoblasts *in vitro*, is controversial. While some investigators find that, similar to their effects *in vivo*, GCs suppress bone-forming cells *in vitro*,[4] others report they actually enhance mineralized nodule formation in culture.[5,6] The latter data raised the possibility, therefore, that the *in vivo* inhibitory effects of GCs on bone formation reflect suppression of an intermediary cell, which, in turn, secondarily dampens the osteoblast.

Bone remodeling is an ever-occurring event in mammals involving tethering of the activities of osteoclasts and osteoblasts. In this process, osteoclasts resorb a packet of bone followed by the appearance of osteoblasts, which deposit new osseous matrix at that remodeling site. The fact that osteoclast activity modulates that of osteoblasts in the remodeling process suggests the bone-resorbing cell may be the intermediary between the steroid and bone formation. In this scenario, GCs would directly suppress remodeling osteoclasts, leading to failure to effectively recruit and activate osteoblasts, thus suppressing bone formation.

To address this hypothesis, we generated mice in which the GC receptor (GCr) is conditionally deleted in osteoclast lineage cells.[7] This exercise involved mating GCr$^{flox/flox}$ mice to those expressing lysosome M Cre. These mice served as negative controls for all experiments involving the direct effects of GCs on osteoclasts.

We first asked if GCs affect proliferation of osteoclastogenic cells. To this end, we cultured wild-type (WT) or GCr-deficient osteoclast precursors, in the form of bone marrow macrophages (BMMs), in increasing amounts of dexamethasone and measured proliferation by incorporation of BrdU. We find that dexamethasone induces a dose-dependent inhibition of osteoclast precursor proliferation in WT BMMs, whereas, those cells lacking the GCr are unaffected by the steroid.

We next turned to the impact of dexamethasone on apoptosis of mature osteoclasts as Weinstein and colleagues have provided evidence, the steroid prolongs their life span.[8] Hence, we established osteoclastogenic cultures consisting of WT and GCr-/- BMMs in the presence of RANKL and M-CSF. Dexamethasone was added for the entire 5 days of culture, or the last 2 days, or 1 day. In agreement with Weinstein *et al.*,[8] GCs prolong the longevity of WT osteoclasts but do not impact those lacking its receptor. Thus, we have established that GCs both suppress osteoclast precursor proliferation and apoptosis of the mature resorptive cell, a circumstance that could sustain normal osteoclast number.

Our next exercise was to assess the impact of GCs on osteoclast differentiation. Once again, we produced 5-day osteoclastogenic cultures of WT and GCr-deficient BMMs and at the termination, measured expression of osteoclastogenic markers, MMP-9, TRAP, and Cathepsin-K mRNAs. GCs have no effect on expression of these hallmarks of osteoclast differentiation in either genotype.

We next asked if GCs suppress osteoclast function. To this end, we cultured WT and GCr-deficient osteoclasts on slices of whale dentin and assessed formation of resorption lacunae. Both WT and receptor-deficient cells generate abundant resorption pits in the absence of dexamethasone. In the presence of 100 nm of the steroid, WT bone resorption, but not the product of GCr-deficient osteoclasts, is markedly inhibited. Thus, we have determined that dexamethasone treatment of cells bearing the GCr substantially depresses bone resorption.

Organization of the osteoclast cytoskeleton is an essential component of its capacity to resorb bone.[9] To resolve if GCs disrupt the osteoclast cytoskeleton, we cultured cells in increasing amounts of dexamethasone and after 5 days, stained them for TRAP activity. Whereas, GCr-/- osteoclasts are unaffected by the corticosteroid, WT cells undergo dramatic cytoskeletal alterations at concentrations as low as 1 nm dexamethasone. These changes involve failure to spread and thus, characteristic osteoclasts are virtually absent in 10 nm of the steroid.

Osteoclasts undergo dramatic organization of their cytoskeleton during the resorptive cycle. Upon attachment to bone, the cell's actin organizes into a ring that isolates the resorptive microenvironment from the general extracellular space.[10] Thus, formation of the actin ring in osteoclasts is an important marker of their ability to resorb bone and the ring's presence correlates with resorptive activity. Thus, we asked if dexamethasone impacts actin ring formation by osteoclasts. Whereas the actin rings of GCr-deficient osteoclasts are unaffected by dexamethasone, those of WT cells are virtually eliminated by the steroid.

Cytoskeletal organization, in all cells, is under the aegis of the Rho family of GTPases, which include Rho A, Rac, and cdc42. We find that depletion of Rac, specifically in osteoclast lineage cells, prompts a substantial increase in bone mass due to osteoclast failure. Thus, we asked if dexamethasone exerts its disruptive effect on the osteoclast cytoskeleton by inhibiting Rho GTPases. To this end, we treated committed preosteoclasts with carrier or dexamethasone in the presence of M-CSF and measured activated Rho and Rac, as manifest by their GTP-association. M-CSF in the absence of dexamethasone substantially activates these two Rho GTPases within 5 min, an event which is obviated by dexamethasone.

Transit of Rho family GTPases from their inactive GDP-bound to their active GTP-bound form is under the influence of guanine nucleotide exchange factors (GEFs). Vav3 is an osteoclast-specific GEF critical for Rac activation and cytoskeletal organization.[11] We therefore asked if GCs suppress Rac activation by impacting Vav3. Vav3 activity was assessed as a manifestation of M-CSF-induced phosphorylation. Similar to its affect on Rho and Rac, dexamethasone inhibits M-CSF-mediated Vav3 activation. Thus, GCs disrupt the cytoskeleton by blocking Vav3 activation eventuating in failure to transit Rho GTPases to their GTP-bound form.

GCs exert their biological effects by both genomic and nongenomic means. To determine the mechanism whereby the steroid impacts osteoclasts, we measured phosphorylated Vav3 in the presence or absence of dexamethasone, in a time-dependent manner. Corticosteroid inhibition of the GEF requires a 16-h exposure, consistent with the conclusion that the disruptive effect of the steroid on the osteoclast cytoskeleton represents a genomic mechanism.

Having established that GCs inhibit osteoclast function *in vitro*, we asked if the same obtains *in vivo*. We treated dexamethasone-injected mice with parathyroid hormone to induce osteoclastogenesis and then prepared histological sections of calvariae to assess the morphology of these cells. In keeping with disruption of their cytoskeleton, WT, but not GCr-deficient mice, exposed to dexamethasone, generate osteoclasts, which are not attached to the bone surface and have a crenated, irregular appearance consistent with disruption of their cytoskeleton. This morphological suggestion of defective osteoclast function in WT cells, exposed to dexamethasone, is confirmed by serum TRACP5 levels. In contrast, those mice lacking the GCr continue to express high levels of the global bone resorption marker under the influence of parathyroid hormone despite the presence of dexamethasone. Furthermore, this distinction in total bone resorption between corticosteroid-treated WT and GCr-deficient mice is not reflected by differences in osteoclast number. Thus, the resorption-suppressing effect of dexamethasone, *in vivo*, reflects inhibition of the activity of osteoclasts and not their generation.

Having determined that GCs reduce osteoclast function, *in vitro* and *in vivo*, we turned to the impact of this phenomenon on remodeling, asking if this inhibited resorption translates to arrested bone formation. If so, corticosteroids administered to WT mice should suppress osteoclast activity, which in turn would inhibit osteoblast function. Alternatively, the steroids should not impact GCr-/- osteoclasts, *in vivo*, which would, in turn, spare bone formation.

We first assessed bone formation in WT and GCr-/- mice by kinetic histomorphometric measurements using time-spaced courses of tetracycline. While the mineral apposition and bone formation rates are suppressed in WT mice exposed to the steroid, dexamethasone has no such effect on those lacking the GCr exclusively in osteoclast lineage cells. These findings are substantiated by parallel changes in the bone formation markers, osteocalcin and alkaline phosphatase. We conclude, therefore, that GCs suppress bone formation by two distinct mechanisms. First, there is substantial evidence that the steroids directly inhibit osteoblast function. In addition, GCs suppress the osteoclast, which, in the context of remodeling, prompts additional blunting of osteoblast function.

These data, however, leave unanswered the observations that bone resorption is substantially increased during the first year of GC therapy in patients with inflammatory disorders. We posit that this phenomenon reflects continued presence of molecules, such as tumor necrosis factor-α (TNF-α) and RANK ligand (RANKL), which override the inhibitory effects of the steroids. In fact, we observed that both RANKL and TNF-α prevent the disruptive effects of dexamethasone on the osteoclast cytoskeleton. Thus, GCs may exert distinct effects on the skeleton in the presence and absence of inflammatory cytokines, the precise mechanisms of which remain to be determined.

REFERENCES

1. WEINSTEIN, R.S. 2001. Glucocorticoid-induced osteoporosis. Rev. Endocr. Metab. Disord. **2:** 65–73.
2. RUBIN, M.R. & J.P. BILEZIKIAN. 2002. Clinical review 151: the role of parathyroid hormone in the pathogenesis of glucocorticoid-induced osteoporosis: a reexamination of the evidence. J. Clin. Endocrinol. Metab. **87:** 4033–4041.
3. WEINSTEIN, R.S., R.L. JILKA, A.M. PARFITT, *et al.* 1998. Inhibition of osteoblastogenesis and promotion of apoptosis of osteoblasts and osteocytes by glucocorticoids. J. Clin. Invest. **102:** 274–282.
4. SMITH, E., R.A. REDMAN, C.R. LOGG, *et al.* 2000. Glucocorticoids inhibit developmental stage-specific osteoblast cell cycle. Dissociation of cyclin A-cyclindependent kinase 2 from E2F4-p130 complexes. J. Biol. Chem. **275:** 19992–20001.
5. AUBIN, J.E. 1999. Osteoprogenitor cell frequency in rat bone marrow stromal populations: role for heterotypic cell-cell interactions in osteoblast differentiation. J. Cell. Biochem. **72:** 396–410.
6. PURPURA, K.A., J.E. AUBIN & P.W. ZANDSTRA. 2004. Sustained in vitro expansion of bone progenitors is cell density dependent. Stem Cells **22:** 39–50.
7. KIM, H.J., H. ZHAO, H. KITAURA, *et al.* 2006. Glucocorticoids suppress bone formation via the osteoclast. J. Clin. Invest. **116:** 2152–2160.
8. WEINSTEIN, R.S., J.R. CHEN, C.C. POWERS, *et al.* 2002. Promotion of osteoclast survival and antagonism of bisphosphonate-induced osteoclast apoptosis by glucocorticoids. J. Clin. Invest. **109:** 1041–1048.
9. TEITELBAUM, S.L. 2000. Bone resorption by osteoclasts. Science **289:** 1504–1508.
10. TEITELBAUM, S.L. 2007. Osteoclasts: what do they do and how do they do it? Am. J. Pathol. **170:** 427–435.
11. FACCIO, R., S.L. TEITELBAUM, K. FUJIKAWA, *et al.* 2005. Vav3 regulates osteoclast function and bone mass. Nat. Med. **11:** 284–290.

Vitamin D

Molecular Mechanism of Action

SYLVIA CHRISTAKOS,[a] PUNEET DHAWAN,[a] BRYAN BENN,[a]
ANGELA PORTA,[a] MATTHIAS HEDIGER,[b] GOO T. OH,[c]
EUI-BAE JEUNG,[d] YAN ZHONG,[a] DARE AJIBADE,[a]
KOPAL DHAWAN,[a] AND SNEHA JOSHI[a]

[a]Department of Biochemistry and Molecular Biology, UMDNJ-New Jersey
Medical School, Newark, New Jersey, USA

[b]Institute for Biochemistry and Molecular Biology, University
of Bern, CH-3012 Berne, Switzerland

[c]Laboratory of Cardiovascular Genomics, Ewha Woman's University,
Seoul 120-750, Korea

[d]Laboratory of Veterinary Biochemistry and Molecular Biology, College of
Veterinary Medicine, Chungbuk National University, Chungbuk 361-763, Korea

ABSTRACT: Vitamin D maintains calcium homeostasis and is required
for bone development and maintenance. Recent evidence has indicated
an interrelationship between vitamin D and health beyond bone, including effects on cell proliferation and on the immune system. New developments in our lab related to the function and regulation of target
proteins have provided novel insights into the mechanisms of vitamin D
action. Studies in our lab have shown that the calcium-binding protein,
calbindin, which has been reported to be a facilitator of calcium diffusion, also has an important role in protecting against apoptotic cell death
in different tissues including protection against cytokine destruction of
osteoblastic and pancreatic β cells. These findings have important implications for the therapeutic intervention of many disorders including
diabetes and osteoporosis. Recent studies in our laboratory of intestinal calcium absorption using calbindin-D_{9k} null mutant mice as well as
mice lacking the 1,25-dihydroxyvitamin D_3 (1,25(OH)$_2$$D_3$) inducible epithelial calcium channel, TRPV6, provide evidence for the first time of
calbindin-D_{9k} and TRPV6 independent regulation of active calcium absorption. Besides calbindin, the other major target of 1,25(OH)$_2$$D_3$ in
intestine and kidney is 25(OH)D_3 24 hydroxylase (24(OH)ase), which is
involved in the catabolism of 1,25(OH)$_2$$D_3$. In our laboratory we have
identified various factors that cooperate with the vitamin D receptor in

Address for correspondence: Dr. Sylvia Christakos, Department of Biochemistry and Molecular
Biology, UMDNJ-New Jersey Medical School, 185 South Orange Ave., Newark, NJ 07103. Voice:
973-972-4033; fax: 973-972-5594.
    christak@umdnj.edu

Ann. N.Y. Acad. Sci. 1116: 340–348 (2007). © 2007 New York Academy of Sciences.
doi: 10.1196/annals.1402.070

regulating 24(OH)ase expression including C/EBP β, SWI/SNF (complexes that remodel chromatin using the energy of ATP hydrolysis) and the methyltransferases, CARM1 and G9a. Evidence is also presented for C/EBP β as a nuclear coupling factor that coordinates regulation of osteopontin by 1,25(OH)$_2$D$_3$ and PTH. Our findings define novel mechanisms that may be of fundamental importance in understanding how 1,25(OH)$_2$D$_3$ mediates its multiple biological effects.

KEYWORDS: calbindin; TRPV6 epithelial calcium channel; vitamin D; 1,25-dihydroxyvitamin D$_3$; osteopontin; parathyroid hormone; 25-hydroxyvitamin D$_3$ 24-hydroxylase

INTRODUCTION

The hormonally active form of vitamin D, 1,25-dihydroxyvitamin D$_3$ (1,25(OH)$_2$D$_3$), a principal factor that maintains calcium homeostasis, has also been reported to have numerous other physiological functions including inhibition of proliferation of certain cancer cells and preventing or partially protecting against certain autoimmune diseases. 1,25(OH)$_2$D$_3$ regulates gene expression in target cells by binding to the vitamin D receptor (VDR). The liganded VDR heterodimerizes with the retinoid X receptor (RXR), binds to vitamin D response elements (VDREs) in the promoter of target genes and, together with coactivators, affects target gene transcription.[1] Results of chromatin immunoprecipitation assays have indicated that the CBP/SRC coactivator complex (that has histone acetyltransferase activity) may be recruited first by 1,25(OH)$_2$D$_3$ followed by recruitment of the vitamin D receptor-interacting protein (DRIP) complex (that acts through recruitment of RNA polymerase II holoenzyme) (sequential model).[2] In addition, various other factors including Ras-activated Ets transcription factor, the SWI/SNF complex, which facilitates transcription by remodeling chromatin using the energy of ATP hydrolysis, and members of the CCAAT enhancer-binding protein family of transcription factors have also been reported to modulate VDR-mediated transcription.[3–7] This article will focus on research from our laboratory presented at the New York Academy of Sciences Second Conference on Skeletal Biology and Medicine related to novel mechanisms involved in VDR-mediated transcriptional activation as well as to the physiological significance of vitamin D target proteins as determined in studies using null mutant mice.

CALBINDIN AND TRPV6

One of the most pronounced effects of 1,25(OH)$_2$D$_3$ known is increased synthesis of the calcium-binding protein calbindin in the intestine.[8] There are two major subclasses of calbindin: calbindin-D$_{9k}$ (a 9,000 Mr calcium binding protein present in mammalian intestine and bovine, mouse and neonatal rat kidney) and calbindin-D_{28k} (a 28,000 Mr protein present in avian intestine, mammalian and avian kidney and pancreas, in mammalian osteoblastic cells, and in mammalian to molluskan brain).[9] Studies in our lab have shown that calbindin-D_{28k}, which has been reported to be a facilitator of calcium diffusion in avian intestine, has an important role in protecting against apoptotic cell death in different tissues including protection against excitotoxic cell death in brain and against cytokine destruction of osteoblastic and pancreatic β cells. Calbindin-D_{28k} blocks apoptosis by buffering calcium, preventing damage to the mitochondria and cell death that results from sustained elevations in intracellular calcium. We have shown that the antiapoptotic effects of calbindin-D_{28k} also involve inhibition of caspase 3.[10] The inhibition of caspase 3 by calbindin is not dependent on the ability of calbindin-D_{28k} to buffer calcium. GST pull down assays indicated that calbindin-D_{28k} binds directly to caspase 3.[10] Besides the inhibitor of apoptotic proteins, calbindin-D_{28k} is the only other natural endogenous inhibitor of caspase 3. These findings have important implications for the therapeutic intervention of many disorders including diabetes and osteoporosis. Further studies are needed to determine whether calbindin-D_{9k} also has an antiapoptotic function.

In the intestine, calbindin-D_{28k} in avian species and calbindin-D_{9k} in mammals have been localized primarily in the absorptive cells, which supports the role first proposed for calbindin as a facilitator of calcium diffusion through the cell interior toward the basolateral membrane. Also, in VDR knockout mice a major defect is in intestinal calcium absorption, which is accompanied by a 50% reduction in intestinal calbindin-D_{9k} mRNA, further supporting the proposed role of calbindin.[11] However, we and others have noted that the induction of calbindin does not always correlate with an increase in intestinal calcium absorption, suggesting that $1,25(OH)_2D_3$ has multiple effects at various control points in the intestinal cell.[12,13]

In addition to the role of $1,25(OH)_2D_3$ on transcellular movement of calcium, it is known that the rate of calcium entry can also be increased by $1,25(OH)_2D_3$. Recently, an apical calcium channel, TRPV6, which is induced by $1,25(OH)_2D_3$ and colocalized with calbindin in the duodenum, has been identified.[14] In VDR knockout mice TRPV6 mRNA was found to be more markedly decreased in the intestine than calbindin-D_{9k} mRNA.[15] Although these results are suggestive, the role of TRPV6 in vivo in $1,25(OH)_2D_3$-mediated active intestinal calcium absorption had not been addressed. The generation of calbindin-D_{9k} KO mice and TRPV6 KO mice made possible for the first time in vivo studies of mechanisms underlying $1,25(OH)_2D_3$-mediated intestinal calcium absorption.[16,17] We found that when fed a regular diet, TRPV6 KO mice and calbindin-D_{9k} KO mice have serum calcium levels similar to those of wild-type (WT) mice ($\sim$10 mg/Ca^{++}/dL). In the TRPV6 KO mice there is a threefold increase in serum PTH and a 2.4-fold increase in serum $1,25(OH)_2D_3$ levels.[17] Under low calcium conditions, serum $1,25(OH)_2D_3$ and PTH levels in the TRPV6 KO mice were not further increased but were

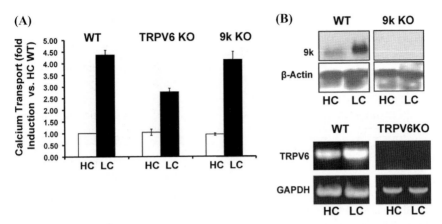

FIGURE 1. Active intestinal calcium transport in TRPV6 and calbindin-D_{9k} null mutant mice. **(A)** Active calcium transport was measured using everted duodenal sacs. $n = 6–18$/group. $P < 0.05$ for all groups for mice on the LC diet compared to mice on the HC diet. $P > 0.1$ for WT mice and calbindin-D_{9k} KO mice on the LC diet. $P < 0.05$ for TRPV6 KO mice on the LC diet compared to WT and calbindin-D_{9k} KO mice on the LC diet. **(B)** *Upper panel*: Representative Northern blot of calbindin-D_{9k} mRNA in intestine of WT mice on HC or LC diet and the absence of calbindin-D_{9k} mRNA in calbindin-D_{9k} KO mice. *Lower panel*: Representative RT-PCR analysis of TRPV6 mRNA in intestine of WT mice on HC or LC diet and the absence of TRPV6 mRNA in TRPV6 KO mice.

similar to high levels observed in WT mice. Active intestinal calcium absorption was measured by the everted gut sac assay. Under low dietary calcium conditions (mice were fed a low calcium diet (0.02%) from weaning for 4 weeks; LC) there was a 4.4-, 2.8-, and 4.2-fold increase in calcium absorption in the duodenum of WT, TRPV6 KO, and calbindin-D_{9k} KO mice, respectively (FIG. 1) compared to mice fed a high calcium diet (1.0%; HC). Duodenal calcium absorption was increased 2.6-fold in calbindin-D_{9k}/TRPV6 double KO mice fed the low calcium diet (not shown). Calcium uptake was not stimulated by low dietary calcium in the ileum of the WT or null mutant mice (not shown). This study provides evidence for the first time using null mutant mice for calbindin-D_{9k} and TRPV6-independent regulation of active intestinal calcium absorption, thus challenging the dogma for the need for calbindin-D_{9k} and TRPV6 for vitamin D-induced active intestinal calcium transport.

OTHER MAJOR TARGETS OF 1,25(OH)$_2$D$_3$ AND MOLECULAR MECHANISMS INVOLVED IN REGULATING TARGET GENE EXPRESSION

Besides calbindin and TRP channels, the other known pronounced effect of 1,25(OH)$_2$D$_3$ in intestine and kidney is increased synthesis of

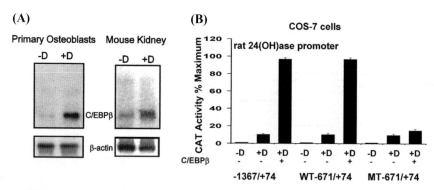

FIGURE 2. C/EBP β is a 1,25(OH)$_2$D$_3$ target in osteoblastic cells and in kidney and C/EBP β enhances 1,25(OH)$_2$D$_3$-induced 24(OH)ase transcription. **(A)** Northern blot analysis of the effect of 1,25(OH)$_2$D$_3$ on the expression of C/EBP β mRNA in primary osteoblasts (−D, vehicle; +D, 1,25(OH)$_2$D$_3$ treatment (10^{-8} M), 9 h) or mouse kidney (−D, vitamin D-deficient mice; +D, vitamin D-deficient mice treated with 1,25(OH)$_2$D$_3$ (30 ng 1,25(OH)$_2$D$_3$ for 48 h, 24 h, and 6 h prior to sacrifice). **(B)** CAT activity determined in extracts of COS-7 cells transfected with hVDR with or without cotransfection of 1 μg C/EBP β expression vector. Note the inhibition of the C/EBP β enhancement using the −671/+74 24(OH)ase promoter with a mutated C/EBP motif (C/EBP site at −395/−388).

24(OH)ase, the enzyme involved in metabolic inactivation of 1,25(OH)$_2$D$_3$. Thus 1,25(OH)$_2$D$_3$ regulates its own metabolism by inducing 24(OH)ase, protecting against hypercalcemia. 24(OH)ase is the most transcriptionally responsive vitamin D-inducible gene identified to date. Using Affymetrix gene chip array to identify new targets of 1,25(OH)$_2$D$_3$, we found 19 genes that were activated in mouse kidney by 1,25(OH)$_2$D$_3$ by a factor greater than 50%. They included 24(OH)ase (58-fold induction), calbindin-D$_{28k}$ (2.5-fold induction), and C/EBP β (4.4-fold induction). Since C/EBP family members were previously reported to be regulated by other steroids and since putative C/EBP sites were noted by sequence homology in the 24(OH)ase promoter, we focused on C/EBP β as a target of 1,25(OH)$_2$D$_3$. The induction of C/EBP β by 1,25(OH)$_2$D$_3$ was verified by Northern (FIG. 2A) and Western analysis (not shown). Induction of C/EBP β mRNA was observed not only in mouse kidney but also in primary osteoblasts (FIG. 2A). Transfection studies indicated that 1,25(OH)$_2$D$_3$ induction of 24(OH)ase transcription is enhanced a maximum of 10-fold by C/EBP β and a C/EBP β site was identified by mutatgenesis (FIG. 2B) and gel shift analysis (not shown) at −395/−388 in the rat 24(OH)ase promoter.[7] From this study a fundamental role was established for the first time for cooperative effects and cross talk between the C/EBP family of transcription factors and VDR-mediated transcription.

We found that PTH as well as 1,25(OH)$_2$D$_3$ can induce C/EBP β in osteoblastic cells (FIG. 3A), suggesting that C/EBPs may be mediators of PTH and 1,25(OH)$_2$D$_3$ actions that affect skeletal integrity and osteoblast function.

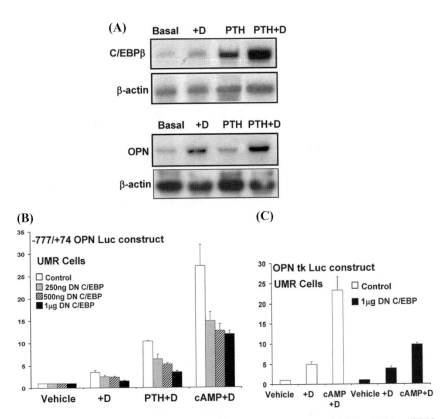

FIGURE 3. PTH enhances $1,25(OH)_2D_3$-induced C/EBP β and OPN mRNA and DN C/EBP inhibits PTH enhancement of OPN transcription. (**A**) Northern blot analysis was performed using total RNA from UMR 106 osteoblastic cells that had been treated with PTH (25 nM) for 4 h followed by treatment with $1,25(OH)_2D_3$ (10^{-8} M) for 16 h. (**B**) UMR cells were transfected with $-777/+74$ OPN promoter construct. Cells were treated with vehicle, $1,25(OH)_2D_3$ (10^{-8} M), PTH (25 nM) $+1,25(OH)_2D_3$ or 8 bromo cAMP (1 mM) $+ 1,25(OH)_2D_3$ in the presence or absence of increasing concentrations of C/EBP DN. (**C**) The OPN VDRE was sufficient to observe the inhibitory effect of DN CEBP.

Osteopontin (OPN), induced in response to $1,25(OH)_2D_3$ in osteoblasts, has been reported to modulate both resorption and mineralization. PTH and/or activation of PKA can enhance the $1,25(OH)_2D_3$ induction of OPN expression and transcription in osteoblastic cells (FIG. 3A, B). To understand regulatory mechanisms we asked whether C/EBP β, which is induced by $1,25(OH)_2D_3$ and PTH in osteoblastic cells, may play a role in the regulation of OPN transcription. In UMR106 osteoblastic cells cotransfected with a C/EBP dominant negative expression construct (DN-CEBP) and the mouse OPN promoter ($-777/+79$), the cAMP or PTH enhancement of $1,25(OH)_2D_3$-induced transcription was inhibited, suggesting the involvement of C/EBP in PTH-enhanced

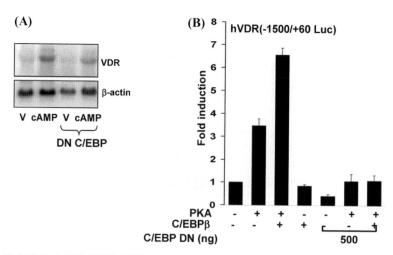

FIGURE 4. DN CEBP inhibits cAMP -induced VDR levels and PKA-mediated transcription of hVDR. (A) Northern blot analysis of VDR mRNA levels induced after cAMP treatment of UMR cells (1 mM 8 bromo cAMP, 9 h) in the presence or absence of prior transfection with DN C/EBP. (B) Inhibition of PKA-mediated transcription of VDR by DN C/EBP.

OPN transcription. The OPN VDRE was sufficient to observe the inhibitory effect of DN-C/EBP, indicating that C/EBP is not acting through a site in the OPN promoter (FIG. 3C). VDR mRNA levels, induced after cAMP treatment in UMR cells (1 mM 8 bromo cAMP, 9 h), were inhibited by prior transfection of DN-CEBP, suggesting that C/EBP may mediate its effect on the PTH and cAMP-mediated enhancement of OPN transcription through upregulation of VDR (FIG. 4A). Data using the hVDR promoter ($-1500/+60$) indicated that expression of PKA results in a 3.5 $+/-$ 0.5-fold induction of hVDR promoter activity that is further enhanced twofold in the presence of CEBP β. DN-C/EBP inhibited the C/EBP enhancement and the PKA induction of VDR transcription. (FIG. 4B). The findings presented provide evidence that C/EBP β is a nuclear coupling factor that coordinates regulation of osteoblast function by $1,25(OH)_2D_3$ and PTH, at least in part, by enhancing PKA-induced VDR transcription.

Various additional factors have been reported to modulate VDR-mediated transcription. It is well known that p160 coactivators, which have histone acetyltransferase activity, can bind to steroid receptors and enhance their activity. In addition to acetylation, methylation also occurs on core histones. Preliminary results in our laboratory have indicated that cooperativity between histone methyltransferases and p160 coactivators may also play a fundamental role in VDR-mediated transcriptional activation (FIG. 5).

In summary, findings from these studies define novel mechanisms that may be of fundamental importance in understanding how $1,25(OH)_2D_3$ mediates

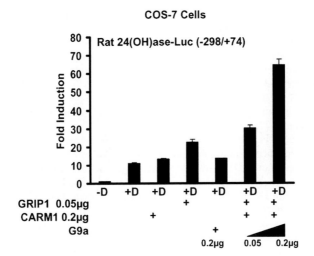

FIGURE 5. G9a, histone H3 lysine 9 methyltransferase, is a coactivator for VDR. COS-7 cells were transfected with rat 24(O)ase-luc ($-298/+74$) and expression vectors encoding VDR and CARM1 (coactivator-associated arginine methyltransferase 1), GRIP-1 and G9a as indicated. Cells were treated with vehicle or $1,25(OH)_2D_3$. 0.2 μg CARM1 or G9a alone had little effect of VDR-mediated transcription. However, transfection of G9a in combination with GRIP-1 and CARM1 resulted in a maximal sixfold greater enhancement of $1,25(OH)_2D_3$-induced transcription above the $1,25(OH)_2D_3$-induced response.

its pleiotropic effects. Understanding the function of target proteins as well as the multiple cofactors involved in VDR-mediated transcription may lead to the design of drugs that can selectively modulate specific $1,25(OH)_2D_3$ responses and thus may be used to treat bone loss disorders, autoimmune disorders, or various types of cancers.

REFERENCES

1. CHRISTAKOS, S., P. DHAWAN, Y. LIU, *et al.* 2003. New insights into the mechanisms of vitamin D action. J. Cell Biochem. **88:** 695–705.
2. KIM, S., N.K. SHEVDE & J.W. PIKE. 2005. 1,25-Dihydroxyvitamin D3 stimulates cyclic vitamin D receptor/retinoid X receptor DNA-binding, co-activator recruitment, and histone acetylation in intact osteoblasts. J. Bone Miner. Res. **20:** 305–317.
3. DWIVEDI, P.P., J.L. OMDAHL, I. KOLA, *et al.* 2000. Regulation of rat cytochrome P450C24 (CYP24) gene expression. Evidence for functional cooperation of Ras-activated Ets transcription factors with the vitamin D receptor in 1,25-dihydroxyvitamin D(3)-mediated induction. J. Biol. Chem. **275:** 47–55.
4. VILLAGRA, A., F. CRUZAT, L. CARVALLO, *et al.* 2006. Chromatin remodeling and transcriptional activity of the bone-specific osteocalcin gene require

CCAAT/enhancer-binding protein beta-dependent recruitment of SWI/SNF activity. J. Biol. Chem. **281:** 22695–22706.

5. CHRISTAKOS, S., P. DHAWAN, Q. SHEN, et al. 2006. New insights into the mechanisms involved in the pleiotropic actions of 1,25dihydroxyvitamin D3. Ann. N. Y. Acad. Sci. **1068:** 194–203.

6. GUTIERREZ, S., A. JAVED, D.K. TENNANT, et al. 2002. CCAAT/enhancer-binding proteins (C/EBP) beta and delta activate osteocalcin gene transcription and synergize with Runx2 at the C/EBP element to regulate bone-specific expression. J. Biol. Chem. **277:** 1316–1323.

7. DHAWAN, P., X. PENG, A.L. SUTTON, et al. 2005. Functional cooperation between CCAAT/enhancer-binding proteins and the vitamin D receptor in regulation of 25-hydroxyvitamin D3 24-hydroxylase. Mol. Cell Biol. **25:** 472–487.

8. CHRISTAKOS, S., Y. LIU, P. DHAWAN, et al. 2005. The calbindins: calbindin D9k and calbindin D28K. In Vitamin D, 2nd ed. Feldman, Pike & Glorieux, Eds.: 721–735. Chapter 42. Academic Press. San Diego, CA.

9. CHRISTAKOS, S. & Y. LIU. 2004. Biological actions and mechanism of action of calbindin in the process of apoptosis. J. Steroid Biochem. Mol. Biol. **89–90:** 401–404.

10. BELLIDO, T., M. HUENING, M. RAVAL-PANDYA, et al. 2000. Calbindin-D28k is expressed in osteoblastic cells and suppresses their apoptosis by inhibiting caspase-3 activity. J. Biol. Chem. **275:** 26328–26332.

11. LI, Y.C., M. AMLING, A.E. PIRRO, et al.1998. Normalization of mineral ion homeostasis by dietary means prevents hyperparathyroidism, rickets, and osteomalacia, but not alopecia in vitamin D receptor-ablated mice. Endocrinology **139:** 4391–4396.

12. KRISINGER, J., M. STROM, H.D. DARWISH, et al. 1991. Induction of calbindin-D 9k mRNA but not calcium transport in rat intestine by 1,25-dihydroxyvitamin D3 24-homologs. J. Biol. Chem. **266:** 1910–1913.

13. WANG, Y.Z., H. LI, M.E. BRUNS, et al. 1993. Effect of 1,25,28-trihydroxyvitamin D2 and 1,24,25-trihydroxyvitamin D3 on intestinal calbindin-D9K mRNA and protein: is there a correlation with intestinal calcium transport? J. Bone Miner. Res. **8:** 1483–1490.

14. PENG, J.B., E.M. BROWN & M.A. HEDIGER. 2003. Epithelial Ca2+ entry channels: transcellular Ca2+ transport and beyond. J. Physiol. **551:** 729–740.

15. VAN CROMPHAUT, S.J., M. DEWERCHIN & J.G. HOENDEROP. 2001. Duodenal calcium absorption in vitamin D receptor-knockout mice: functional and molecular aspects. Proc. Natl. Acad. Sci. USA **98:** 13324–13329.

16. LEE, G.-S., K.-Y. LEE, K.-C. CHOI, et al. 2007. A phenotype of a calbindin- D9k gene-knockout is compensated for by the induction of other calcium- transporter genes in a mouse model. J. Bone Miner. Res. 2007 Aug. 13 [Epub ahead of print].

17. BIANCO, S.D., J.B. PENG, H. TAKANAGA, et al. 2007. Marked disturbance of calcium homeostasis in mice with targeted disruption of the TRPV6 calcium channel gene. J. Bone Miner. Res. **22:** 274–285.

Parathyroid Hormone Regulates Histone Deacetylases in Osteoblasts

EMI SHIMIZU, [a] NAGARAJAN SELVAMURUGAN,[a]
JENNIFER J. WESTENDORF,[b] AND NICOLA C. PARTRIDGE[a]

[a]Department of Physiology and Biophysics, UMDNJ-Robert Wood Johnson
Medical School, Piscataway, New Jersey, USA

[b]Departments of Orthopedic Surgery and Molecular Biology and Biochemistry,
Mayo Clinic, Rochester, Minnesota, USA

ABSTRACT: Parathyroid hormone (PTH) functions as an essential regulator of calcium homeostasis and as a mediator of bone remodeling. We have already shown that PTH stimulates the expression of matrix metalloproteinase-13 (MMP-13), which is responsible for degrading components of extracellular matrix. We have hypothesized that histone deacetylases (HDACs) are involved with PTH-induced MMP-13 gene expression in the osteoblastic cell line, UMR 106–01. We have shown that PTH profoundly regulates HDAC4 in UMR 106–01 cells through a PKA-dependent pathway, leading to removal of HDAC4 from the MMP-13 promoter and its enhanced transcription. Understanding the mechanism of how HDACs affect osteoblast differentiation and mineralization will identify new theraupeutic methods for bone diseases, such as osteoporosis and multiple myeloma.

KEYWORDS: histone deacetylases; parathyroid hormone; osteoblasts

Parathyroid hormone (PTH) is secreted by the parathyroid as a polypeptide of 84 amino acids.[1] This hormone binds to PTH/PTH-related protein 1 receptor on osteoblast membranes[2] and acts to increase the concentration of calcium in the blood. PTH induces the expression of matrix metalloproteinase-13 by osteoblastic cells[3] (MMP-13, collagenase-3), which is responsible for degrading components of extracellular matrix.

In our previous data, we have shown that Runx2 binding to the runt domain (RD)-binding site and activator protein-1 (AP-1) binding to the AP-1 site are necessary for PTH-induced MMP-13 promoter activity and that the proteins interact with each other. Histone acetyltransferases (HATs) and histone deacetylases (HDACs) are directly associated with transcription factors in the promoter region and play a role regulating gene expression (FIG. 1).[4]

Address for correspondence: Nicola C. Partridge, Department of Physiology and Biophysics, University of Medicine and Dentistry of New Jersey-Robert Wood Johnson Medical School, 675 Hoes Lane, Piscataway, New Jersey 08854. Voice: 732-235-4552; fax: 732-235-3977.
partrinc@umdnj.edu

Ann. N.Y. Acad. Sci. 1116: 349–353 (2007). © 2007 New York Academy of Sciences.
doi: 10.1196/annals.1402.037

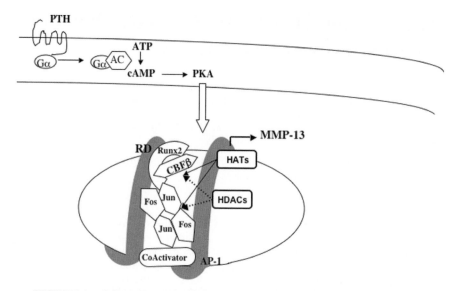

FIGURE 1. Schematic model of the promoter of rat MMP-13 gene. Runx2 binds to the runt domain (RD)-binding site and activator protein-1 (AP-1) binds to the AP-1 site. They are necessary for PTH-induced MMP-13 promoter activity and these proteins interact with each other. Histone acetylases (HATs) and histone deactylases (HDACs) are associated with these transcription factors in the MMP-13 promoter.

Runx2 (AML-3/Cbfa1) is an important transcription factor in bone cells and is required for osteoblast development, differentiation, and chondrocyte hypertrophy, and disruption of the Runx2 gene in mice induces skeletal defects.[5–7] We have examined how HDACs are associated with PTH-induced MMP-13 gene expression in the osteoblastic osteosarcoma cell line, UMR 106–01.

Regulation of gene expression is mediated by several mechanisms, such as DNA acetylation, methylation, ATP-dependent chromatin remodeling, and posttranslational modifications of histones. HDACs play a role in deacetylation of lysine residues in the tails of core histones. HDACs are classified into four groups. The class I HDACs (HDAC1, 2, 3, and 8) have high homology to the yeast global transcriptional regulator Rpd3 and mostly localize in the nucleus. In contrast, the class II HDACs (HDAC4, 5, 6, 7, 9, and 10) show homology to yeast Hda1. They have been shown to be actively maintained in the cytoplasm. They bind to an anchor protein, such as 14–3-3 proteins, and are imported into the nucleus when required.[8] The binding sites of HDAC4 are three serine residues, which are calcium–calmodulin-dependent kinase (CaMK) phosphorylation sites.[9,10] In UMR 106–01 cells, HDAC4 mostly localizes in the cytoplasm, and HDAC4 co-localizes and binds to 14–3-3 beta with or without PTH treatment. We found that HDAC4–14-3–3 beta binding is associated with CaMK phosphorylation by using mutant HDAC4 (point mutations of three serine residues) in osteoblastic cells. In addition, HDAC4 entry

to the nucleus is associated with protein kinase A since HDAC4 protein levels were decreased by the PKA inhibitor, H89, in the nucleus.

The class III HDACs (SIRT 1–7) are highly homologous to Sir (Silent information regulator) 2 and require NAD^+ for deacetylase activity. Sir2 protein is associated with chromatin silencing and has been found to influence cellular metabolism and aging.[11,12] Gao *et al.* cloned HDAC11 and classified it as a class IV and it associates with HDAC6.[13,14]

Previously, Schroeder *et al.* have shown HDAC3 interacts with Runx2 in osteoblastic cells. HDAC3 suppresses Runx2-mediated activation of the osteocalcin promoter.[15] Moreover, HDAC6 interacts with Runx2 and this contributes to $p21^{CIP1/WAF1}$ regulation in differentiating osteoblasts.[16] Recently, HDAC4 was shown to act as a negative regulator of chondrocyte hypertrophy due to constitutive Runx2 suppression by HDAC4.[17] Moreover, HDAC4 downregulates MEF2C (myocyte enhancer factor-2C). Endochondral bone formation is exquisitely sensitive to the balance between MEF2C and HDAC4.[18] We have shown that endogenous HDAC3, 4, and 6 interacts with Runx2 in UMR 106–01 cells. Thus, these HDACs may suppress Runx2 functions in UMR 106–01 cells.

To examine whether HDACs are involved in PTH-induced MMP-13 gene expression, we used the HDAC inhibitor, trichostatin A, in UMR 106–01 cells. It markedly stimulated basal transcription from the MMP-13 promoter in UMR 106–01 cells. However, this inhibitor blocks activities of both class I and class II HDACs. We examined which HDACs are expressed in UMR 106–01 or rat primary osteoblastic cells. HDAC3, 4, 6, and 7 are strongly expressed in both groups of cells. Interestingly, these HDACs are increased in the differentiation and mineralization stages compared to the proliferation stage. This result suggested that these HDACs might be associated with osteoblast differentiation and mineralization. On the other hand, HDAC1 and 5 were detected at lower levels than other HDACs in rat osteoblasts. Kang *et al.* have shown that class II HDACs, such as HDAC4 and HDAC5, act as corepressors for TGF-β-mediated transcriptional repression of Runx2 in differentiating osteoblasts.[19] They have shown that HDAC5 is expressed more strongly than HDAC4 in mouse primary osteoblasts. Moreover, enzymatic activities of class II HDACs, such as HDAC4, are suppressed by HDAC3-SMRT/N-CoR complex.[20] In addition, SIRT1 is not only associated with deacetylation of MEF2, but also forms complexes with HDAC4. The latter regulates sumoylation of MEF2, but not deacetylation.[21] We examined the presence of HDAC4 on the Runx2 binding site of the MMP-13 promoter by ChIP assay with UMR 106–01 cells with or without PTH treatment. The association of HDAC4 with the MMP-13 promoter on RD and AP-1 sites was decreased with PTH treatment. This reduction was restored to control levels in the presence of a PKA inhibitor. In addition, the precipitated Runx2 interacted with HDAC4 under basal conditions, and the interaction was decreased after PTH stimulation in osteoblastic cells.

Here, we have reported that PTH regulates the functions of HDACs in rat osteoblastic cells and this appears to be part of the overall regulation of MMP-13 transcription.

REFERENCES

1. STREWLER, G.J., P.H. STERN, J.W. JACOBS, et al. 1987. Parathyroid hormone like protein from human renal carcinoma cells. Structural and functional homology with parathyroid hormone. J. Clin. Invest. **80:** 1803–1807.
2. JUPPNER, H., A.B. ABOU-SAMRA, M. FREEMAN, et al. 1991. A G protein-linked receptor for parathyroid hormone and parathyroid hormone-related peptide. Science **254:** 1024–1025.
3. PARTRIDGE, N.C., J.J. JEFFREY, L.S. EHLICH, et al. 1987. Hormonal regulation of the production of collagenase and a collagenase inhibitor activity by rat osteogenic sarcoma cells. Endocrinology **120:** 1956–1962.
4. SELVAMURUGAN, N., M.R. PULUMATI, D.R. TYSON, et al. 2000. Parathyroid hormone regulation of the rat collagenase-3 promoter by protein kinase A-dependent transactivation of core binding factor a1. J. Biol. Chem. **275:** 5037–5042.
5. MUNDLOS, S., F. OTTO, C. MUNDLOS, et al. 1997. Mutations involving the transcription factor CBFA1 cause Cleidocranial Dysplasia. Cell **89:** 773–779.
6. OTTO, F., A.P. THORNEL, T. CROMPTON, et al. 1997. Cbfa1, a candidate gene for Cleidocranial Dysplasia Syndrome, is essential for osteoblast differentiation and bone development. Cell **89:** 765–771.
7. DUCY, P., R. ZHANG, V. GEOFFROY, et al. 1997. Osf2/Cbfa1: a transcriptional activator of osteoblast differentiation. Cell **89:** 747–754.
8. GRAY, S.G. & T.J. EKSTROM. 2001. The human histone deacetylase family. Exp. Cell Res. **262:** 75–83.
9. VERDIN, E., F. DEQUIEDT & H.G. KASLER. 2003. Class II histone deacetylases: versatile regulators. Trends Genet. **19:** 286–293.
10. ZAO, X., A. ITO, C.D. KANE, et al. 2001. The modular nature of histone deacetylase HDAC4 confers phosphorylation-dependent intracellular trafficking. J. Biol. Chem. **276:** 35042–35048.
11. GUARENTE, L. 2000. Sir2 links chromatin silencing, metabolism and aging. Genes Dev. **14:** 1021–1026.
12. LONGO, V.D. & B.K. KENNEDY. 2006. Sirtuins in aging and age-related disease. Cell **126:** 257–268.
13. GAO, L., M.A. CUETO, F. ASSELBERGS, et al. 2002. Cloning and functional characterization of HDAC11, a novel member of the human histone deacetylase family. J. Biol. Chem. **277:** 25748–25755.
14. GREGORETTI, I.V., Y.M. LEE & H.V. GOODSON. 2004. Molecular evolution of the histone deacetylase family: functional implications of phylogenetic analysis. J. Mol. Biol. **338:** 17–31.
15. SCHROEDER, T.M., R.A. KAHLER & J.J. WESTENDORF. 2004. Histone deacetylase 3 interacts with Runx2 to repress the osteocalcin promoter and regulate osteoblast differentiation. J. Biol. Chem. **279:** 41998–42007.
16. WESTENDORF, J.J., S.K. ZAIDI, J.E. CASCINO, et al. 2002. Runx2 (Cbfa1, AML-3) interacts with histone deacetylase 6 and represses the p21(CIP1/WAF1) promoter. Mol. Cell Biol. **22:** 7982–7992.

17. VEGA, R.B., K. MATSUDA, J. OH, *et al.* 2004. Histone deacetylase 4 controls chondrocyte hypertrophy during skeletogenesis. Cell **119:** 555–566.
18. ARNOLD, M.A., Y. KIM, M.P. CZUBRYT, *et al.* 2007. MEF2C transcription factor controls chondrocyte hypertrophy and bone development. Dev. Cell **12:** 377–389.
19. KANG, J.S., T. ALLISTON, R. DELSTON, *et al.* 2005. Repression of Runx2 function by TGF-beta through recruitment of class II histone deacetylases by Smad3. EMBO J. **24:** 2543–2555.
20. FISCHLE, W., F. DEQUIEDT, M.J. HENDZEL, *et al.* 2002. Enzymatic activity associated with class II HDACs is dependent on a multiprotein complex containing HDAC3 and SMRT/N-CoR. Mol. Cell **9:** 45–57.
21. ZHAO, X., T. STERNSDORF, T.A. BOLGER, *et al.* 2005. Regulation of MEF2 by histone deacetylase 4- and SIRT1 deacetylase-mediated lysine modifications. Mol. Cell Biol. **25:** 8456–8464.

Parathyroid Hormone Stimulation of Noncanonical Wnt Signaling in Bone

MARIKA K. BERGENSTOCK[a,b] AND NICOLA C. PARTRIDGE[a]

[a]*Department of Physiology and Biophysics, UMDNJ-Robert Wood Johnson Medical School and Graduate School of Biomedical Sciences, Piscataway, New Jersey, USA*

[b]*Department of Biomedical Engineering, Rutgers, The State University of New Jersey, Piscataway, New Jersey, USA*

ABSTRACT: In bone, parathyroid hormone (PTH) exerts either a catabolic or an anabolic effect depending on its method of administration. This paradoxical action has led to the use of PTH as an effective treatment for osteoporosis. The Wnt family of signaling proteins has a critical role in multiple events, which are necessary for proper animal development and survival, yet their exact method of action in bone remains elusive. We have uncovered a novel link between Wnt-4 and PTH. We think that, in bone, Wnt-4 signaling in response to PTH implicates cross-talk of multiple signaling pathways. This work hopes to further elucidate Wnt signaling in bone and provide greater understanding of PTH's anabolic effects in bone.

KEYWORDS: parathyroid hormone; bone; osteoblasts; Wnt signaling; β-catenin; Wnt-4

The microenvironment of bone is controlled systemically, through hormones such as parathyroid hormone (PTH) and locally, via paracrine and autocrine actions, by osteoblasts and osteoclasts through secreted growth factors and proteins. PTH is an 84-amino acid polypeptide hormone, which is secreted by the chief cells in the parathyroid gland.[1] PTH's primary targets are known to be the kidney and bone, and it is through these targets that PTH regulates calcium homeostasis and mediates bone remodeling. In addition to promoting calcium release from bone during times of low serum calcium, PTH effectively regulates calcium homeostasis in the kidney by stimulating renal 1-hydroxylase to accelerate the conversion of 25-hydroxyvitamin D to 1,25-dihydroxyvitamin D, which directly acts to stimulate intestinal calcium absorption.[2,3]

In this report, we will focus on PTH and the bone milieu. The delicate balance between bone formation and resorption in the skeleton results from the

Address for correspondence: Nicola C. Partridge, UMDNJ-RWJMS, Rm. 561 Research Tower, 675 Hoes Lane, Piscataway, NJ 08854. Voice: 732-235-4552; fax: 732-235-3977.
partrinc@umdnj.edu

Ann. N.Y. Acad. Sci. 1116: 354–359 (2007). © 2007 New York Academy of Sciences.
doi: 10.1196/annals.1402.047

dynamic relationship between osteoblasts and osteoclasts, respectively, and is known to be anabolically and catabolically stimulated by PTH. In bone, PTH can exert either a catabolic or an anabolic effect depending on its method of administration.[4,5] Continuous infusion of PTH results in severe hypercalcemia and a net decrease in trabecular bone volume. Daily intermittent PTH treatments cause an increase in bone formation surfaces accompanied by increased trabecular bone volume.[6,7] This dichotomy of PTH's action in the bone microenvironment is not clearly understood, and has led researchers to study PTH as an effective treatment for osteoporosis, the most common disease in the Western Hemisphere. Osteoporosis is characterized by reduced bone strength, low bone density, and altered macrogeometry and microscopic bone architecture, which leaves affected individuals, both men and women, with an increased risk for osteoporotic fractures.[8]

The Wnt family of signaling proteins has been shown to play a critical role in multiple events that are necessary for proper animal development and survival.[9] The *Wnt* gene family is known to regulate cell fate and cell–cell interactions of multipotential cells in many tissues; many Wnt molecules regulate and stimulate proliferation of hematopoietic progenitors and have been shown as integrated in Notch–Wnt signaling pathways for hematopoietic stem cell maintenance.[10] The Wnt proteins are cysteine-rich, disulfide-linked glycoproteins ranging from 39 to 46 kDa in size. They are evolutionarily conserved across different species and were originally isolated from mouse mammary tumor virus (MMTV)-induced mammary tumors.[11,12] Upon entry into the secretory pathway, Wnt proteins undergo removal of their N-terminal signal peptide and are further glycosylated. Additionally, Wnt molecules are posttranslationally lipid modified with a palmitate group addition through a thioester bond on the first cysteine (C77) residue, a modification that is essential for function and conserved through all Wnt proteins.[13,14]

As a family, Wnts are grouped into two functional classes, Wnt-1 or Wnt-5a as able or unable, respectively, to induce a secondary body axis in *Xenopus* embryos. Depending on their functional classification, Wnt proteins signal through the canonical and noncanonical pathways, which are composed of what were previously believed to be three very independent signal transduction pathways used to regulate the expression of different genes. The Wnt-1 class of proteins operates through the canonical Wnt pathway, which involves nuclear accumulation of β-catenin and subsequent activation of downstream transcription targets, such as Lef/Tcf. This pathway, commonly known as the Wnt/β-catenin pathway, is thought to be most active in osteoblast development since it effects cell proliferation and cell-fate determination.[15] The noncanonical Wnt pathway, which is employed by the Wnt-5a class of Wnts, is composed of two signaling pathways: (i) the activation of the PKC pathway (Wnt/Ca^{2+}) through cGMP proteins, through activation of phospholipase C and phosphodiesterase to increase intracellular concentrations of free calcium and (ii) the Wnt/PCP (planar cell polarity pathway), which effects cytoskeletal organization through JNK, a member of the MAP kinase family.[16,17] Our Wnt of interest, Wnt-4, is a noncanonical class Wnt molecule.

Bone and cartilage are the two main components of the skeleton, and are formed by intramembranous and endochondral ossification processes, respectively. The two key players in osteoblastogenesis and chondrogenesis are the osteoblast and chondrocyte. Both the osteoblast and chondrocyte differentiate from the mesenchymal stem cell lineage, which, in the bone environment, is located within the bone marrow. Bone formed via both intramembranous and endochondral ossification processes is remodeled by osteoclasts, which, in contrast to the osteoblast and chondrocyte, originate from the hematopoietic lineage.

Wnt signaling in chondrogenesis has been studied throughout chick limb development. During this process, the noncanonical *Wnt-4*, as well as *Wnt-5a* and *Wnt-5b* are found to be expressed in chondrogenic regions. More specifically, *Wnt-5b* is expressed in prehypertrophic chondrocytes, while *Wnt-4* is expressed in the joint regions.[18] Overexpression of *Wnt-4* accelerates chondrocyte differentiation, as well as the onset of bone collar formation, indicating *Wnt-4* may be involved in controlling osteoblast differentiation in the periosteum.[19] Conditional expression of *Wnt-4* in chondrogenic regions in mice causes dwarfism due to decreased proliferation and accelerated maturation of chondrocytes.[20] No further investigation of *Wnt-4* in bone has been conducted.

Wnt signaling has been intensely studied in kidney development and homeostasis. Wnt-4 has been implicated in kidney function.[21,22] Although Wnt-4 has traditionally been considered to be a member of the Wnt5a class, or noncanonical Wnt pathway, it has been shown to have canonical Wnt signaling functionality within kidney epithelial cells.[23]

Among many other developmental events, Wnts have been recently implicated in bone development. Due to the similar mesenchymal stem cell lineage of the chondrocyte and osteoblast, it is not surprising that Wnt signaling has been identified in osteoblasts and chondrocytes. Current data show that the Wnts affect gene expression and development in mesenchymal stem cells and osteoblastic cells.[24–26] Recent microarray data showed that PTH(1-34) regulates genes within the Wnt signaling family in osteoblastic cells as well as in *in vivo* studies conducted in rat bone.[27,28] These new data suggest that, depending on the maturity and differentiated state of the cell, there is significant cross-talk between the canonical and noncanonical Wnt pathways. Most importantly, new data show intermittent PTH treatment in Lrp5$^{-/-}$, and Lrp5$^{+/+}$ mice show equal enhancement of skeletal mass.[29] This evidence indicates that the anabolic effects due to PTH are independent of Lrp5, and thus canonical Wnt signaling.

Our study provides initial evidence for a link between PTH and *Wnt-4* expression in bone. Thus, this is the first analysis detailing PTH effects and pathways of action on *Wnt-4* expression in osteoblastic cells. This research proposes a role for the Wnt signaling protein, WNT-4, in the differentiation of osteoprogenitor cells as well as in the development of osteoblastic cells. Previously it had been shown that WNT-4 function was critical to the development of kidney tubules and that WNT-4 mediates its effects through the canonical Wnt signaling pathway in kidney epithelial cells. In addition, our laboratory has conducted an *in vivo* microarray analysis of intermittent and continuous PTH(1–34) injections and infusions, respectively, which showed that PTH regulates *Wnt-4* in bone. *In vitro* studies using developing osteoblastic cells confirm that PTH stimulation of *Wnt-4* is a universal response and not limited to certain cell types or certain stages of development. These results prompted us to examine Wnt-4 responsiveness and signaling action in bone cells. Our laboratory has demonstrated that PTH, *in vitro*, regulates most of its genes and response proteins through the protein kinase A (PKA) pathway.[30] To support this finding, our *in vivo* microarray data of intermittent injections of various PTH peptides, such as PTH(1–34), PTH(1–31), and PTH(3–34), show that most intracellular gene expression to PTH stimulation is in fact a result of the PKA pathway.[31] Thus, we have found that PTH stimulation of *Wnt-4* is primarily through the PKA pathway. As well, we have confirmed that this is a primary response to PTH. Preliminary data of WNT-4 treatment in osteoblast development indicate that it induces early expression of bone marker genes as well as stimulation of key canonical, Wnt/β-catenin pathway genes.

We support the hypothesis of cross-talk between Wnt signaling cascades in bone development. Our data suggest that osteoblast development is not solely dependent on the canonical Wnts or the Wnt/β-catenin signaling pathway, as previously believed. Ultimately, we think that Wnt-4 plays a crucial role in osteoblast differentiation to increase expression of bone marker genes as well as stimulate the differentiation of bone marrow stromal stem cells into osteoblast progenitors.

REFERENCES

1. STREWLER, G.J., P.H. STERN, J.W. JACOBS, *et al*. 1987. Parathyroid hormonelike protein from human renal carcinoma cells. Structural and functional homology with parathyroid hormone. J. Clin. Invest. **80:** 1803–1807.
2. DEMPSTER, D.W., F. COSMAN, M. PARISIEN, *et al*. 1993. Anabolic actions of parathyroid hormone on bone. Endocr. Rev. **14:** 690–709.
3. FELDMAN, D. 1999. Vitamin D, parathyroid hormone, and calcium: a complex regulatory network. Am. J. Med. **107:** 637–639.
4. DOBNIG, H. & R.T. TURNER. 1997. The effects of programmed administration of human parathyroid hormone fragment (1-34) on bone histomorphometry and serum chemistry in rats. Endocrinology **138:** 4607–4612.
5. ONYIA, J.E., L.M. HELVERING, L. GELBERT, *et al*. 2005. Molecular profile of catabolic versus anabolic treatment regimens of parathyroid hormone (PTH) in rat bone: an analysis by DNA microarray. J. Cell. Biochem. **95:** 403–418.
6. TAM, C.S., J.N. HEERSCHE, T.M. MURRAY & J.A. PARSONS. 1982. Parathyroid hormone stimulates the bone apposition rate independently of its resorptive action:

differential effects of intermittent and continuous administration. Endocrinology **110:** 506–512.

7. SCHILLER, P.C., G. D'IPPOLITO, B.A. ROOS & G.A. HOWARD. 1999. Anabolic or catabolic responses of MC3T3-E1 osteoblastic cells to parathyroid hormone depend on time and duration of treatment. J. Bone Miner. Res. **14:** 1504–1512.

8. COSMAN, F. 2005. The prevention and treatment of osteoporosis: a review. Med. Gen. Med. **7:** 73.

9. WESTENDORF, J.J., R.A. KAHLER & T.M. SCHROEDER. 2004. Wnt signaling in osteoblasts and bone diseases. Gene **341:** 19–39.

10. DUNCAN, A.W., F.M. RATTIS, L.N. DIMASCIO, et al. 2005. Integration of Notch and Wnt signaling in hematopoietic stem cell maintenance. Nat. Immunol. **6:** 314–322.

11. MILLER, J.R. 2002. The Wnts. Genome Biol. **3:**REVIEWS3001.

12. GAVIN, B.J., J.A. MCMAHON & A.P. MCMAHON. 1990. Expression of multiple novel *Wnt-1/int-1*-related genes during fetal and adult mouse development. Genes Dev. **4:** 2319–2332.

13. WILLERT, K., J.D. BROWN, E. DANENBERG, et al. 2004. Wnt proteins are lipid-modified and can act as stem cell growth factors. Nature **423:** 448–452.

14. NUSSE, R. 2003. Wnts and Hedgehogs: lipid-modified proteins and similarities in signaling mechanisms at the cell surface. Development **130:** 5297–5305.

15. BEHRENS, J., J.P. VON KRIES, M. KUHL, et al. 1996. Functional interaction of beta-catenin with the transcription factor LEF-1. Nature **382:** 638–642.

16. YAMANAKA, H., T. MORIGUCHI, N. MASUYAMA, et al. 2002. JNK functions in the non-canonical Wnt pathway to regulate convergent extension movements in vertebrates. EMBO Rep. **3:** 69–75.

17. HABAS, R., I.B. DAWID & X. HE. 2003. Coactivation of Rac and Rho by Wnt/Frizzled signaling is required for vertebrate gastrulation. Genes Dev. **17:** 295–309.

18. CHURCH, V., T. NOHNO, C. LINKER, et al. 2002. Wnt regulation of chondrocyte differentiation. J. Cell Sci. **115:** 4809–4818.

19. HARTMANN, C. & C.J. TABIN. 2000. Dual roles of Wnt signaling during chondrogenesis in the chicken limb. Development **127:** 3141–3159.

20. LEE, H.H. & R.R. BEHRINGER. 2007. Conditional expression of *Wnt4* during chondrogenesis leads to dwarfism in mice. PLoS ONE. **2:** e450.

21. KISPERT, A., S. VAINIO & A.P. MCMAHON. 1998. Wnt-4 is a mesenchymal signal for epithelial transformation of metanephric mesenchyme in the developing kidney. Development **125:** 4225–4234.

22. VAINIO, S.J. & M.S. UUSITALO. 2000. A road to kidney tubules via the Wnt pathway. Pediatr. Nephrol. **15:** 151–156.

23. LYONS, J.P., U.W. MUELLER, H. JI, et al. 2004. Wnt-4 activates the canonical beta-catenin-mediated Wnt pathway and binds Frizzled-6 CRD: functional implications of Wnt/beta-catenin activity in kidney epithelial cells. Exp. Cell Res. **298:** 369–387.

24. DE BOER, J. R. SIDDAPPA, C. GASPAR, et al. 2004. Wnt signaling inhibits osteogenic differentiation of human mesenchymal stem cells. Bone **34:** 818–826.

25. BOLAND, G.M., G. PERKINS, D.J. HALL & R.S. TUAN. 2004. Wnt 3a promotes proliferation and suppresses osteogenic differentiation of adult human mesenchymal stem cells. J. Cell. Biochem. **93:** 1210–1230.

26. KULKARNI, N.H., D.L. HALLADAY, R.R. MILES, et al. 2005. Effects of parathyroid hormone on Wnt signaling pathway in bone. J. Cell. Biochem. **95:** 1178–1190.

27. QIN, L., P. QIU, L. WANG, *et al.* 2003. Gene expression profiles and transcription factors involved in parathyroid hormone signaling in osteoblasts revealed by microarray and bioinformatics. J. Biol. Chem. **278:** 19723–19731.

28. ONYIA, J.E., L.M. HELVERING, L. GELBERT, *et al.* 2005. Molecular profile of catabolic versus anabolic treatment regimens of parathyroid hormone (PTH) in rat bone: an analysis by DNA microarray. J. Cell. Biochem. **95:** 403.

29. SAWAKAMI, K., A.G. ROBLING, M. AI, *et al.* 2006. The Wnt co-receptor LRP5 is essential for skeletal mechanotransduction but not for the anabolic bone response to parathyroid hormone treatment. J. Biol. Chem. **281:** 23698–23711.

30. SWARTHOUT, J.T., R.C. D'ALONZO, N. SELVAMURUGAN & N.J. PARTRIDGE. 2002. Parathyroid hormone-dependent signaling pathways regulating genes in bone cells. Gene **282:** 1–17.

31. PARTRIDGE, N.C., X. LI & L. QIN. 2006. Understanding parathyroid hormone action. Ann. N.Y. Acad. Sci. **1068:** 187–193.

T Cells: Unexpected Players in the Bone Loss Induced by Estrogen Deficiency and in Basal Bone Homeostasis

M. NEALE WEITZMANN AND ROBERTO PACIFICI

Division of Endocrinology, Metabolism, and Lipids, Department of Medicine, Emory University School of Medicine, Atlanta, Georgia, USA

ABSTRACT: The bone–immune interface has become a subject of intense interest in recent years. It has long been recognized that infection, inflammation, and autoimmune disorders are associated with systemic and local bone loss. Yet, it is only recently that T lymphocytes and their products have been recognized as key regulators of osteoclast formation, life span, and activity. Similarly, sex steroids and aging have been known to regulate the immune system and T cells for decades. In spite of the abundance of clinical and physiological clues, it is only in the last few years that investigators have linked immune cells to the etiology of postmenopausal and senile osteoporosis, as well as to the bone loss caused by a variety of endocrine conditions. As surprising is new evidence showing that in contrast to their bone destructive effects under certain pathological conditions, T cells are highly protective of basal bone homeostasis, through complex regulatory effects on osteoprotegerin (OPG) production by B cells, involving CD40 to CD40 Ligand (CD40L) costimulation. This article examines the experimental evidence suggesting that estrogen prevents bone loss by regulating T cell function, and that T cell costimulation with B cells is critical for OPG production and maintenance of basal bone homeostasis.

KEYWORDS: estrogen; T cells; B cells; bone loss; TNF; osteoclast

INTRODUCTION

Menopause is the most frequent cause of bone loss in humans. Both a decreased ovarian production of sex steroids and the resulting increase in follicle-stimulating hormone (FSH) production secondary to estrogen deficiency contribute to postmenopausal bone loss.[1,2]

Address for correspondence: Roberto Pacifici, M.D., Division of Endocrinology, Metabolism and Lipids, Emory University School of Medicine, 101 Woodruff Circle, Room 1309, Atlanta, GA 30322. Voice: 404-712-8420; fax: 404-727-1300.
roberto.pacifici@emory.edu

Ann. N.Y. Acad. Sci. 1116: 360–375 (2007). © 2007 New York Academy of Sciences.
doi: 10.1196/annals.1402.068

It is now accepted that the antiresorptive activity of estrogen is a result of multiple genomic and nongenomic effects on bone marrow (BM) and bone cells, which leads to decreased osteoclast (OC) formation, increased OC apoptosis, and decreased capacity of mature OCs to resorb bone.[3] It is also recognized that stimulation of bone resorption in response to estrogen deficiency is mainly due to cytokine-driven increased OC formation.[4,5] One of the cytokines responsible for the augmented osteoclastogenesis of estrogen deficiency is TNF.[6] The relevance of TNF in the mechanism by which estrogen causes bone loss has been demonstrated using multiple animal models. For example, ovariectomy fails to induce bone loss in TNF-/- mice,[7] transgenic mice insensitive to TNF due to the overexpression of a soluble TNF receptor,[8] and in mice treated with the TNF inhibitor TNF-binding protein.[9]

The presence of increased levels of TNF in the BM of ovariectomized animals and in the conditioned media of peripheral blood cells of postmenopausal women is well documented.[10–15] However, the cells responsible for this phenomenon had not been conclusively identified. A critical breakthrough in understanding the bone loss associated with estrogen deficiency was the identification of T cells as the critical source of enhanced TNF concentrations in ovariectomized mice.

T CELLS, A CRITICAL SOURCE OF TNF IN OVARIECTOMY-INDUCED BONE LOSS

Early studies revealed that estrogen does not regulate TNF expression in cultured stromal cells, some osteoblastic cell lines, and bone biopsies.[11,16,17] At the same time evidence began to emerge that estrogen deficiency increases TNF production by monocyte-enriched peripheral blood mononuclear cells, and unfractionated human and murine BM cells.[12,13] Based on these data the source of upregulated TNF production was ascribed to BM monocytes (BMMs). Surprisingly, some of these earlier observations showed that phytohemaglutinin stimulation was required to demonstrate increased TNF production from monocytes. Since phytohemaglutinin primarily promotes T cell activation, this requirement remained unexplained. Recent studies on highly purified cells have revealed that ovariectomy increases the production of TNF by T cells, but not by BMMs.[18] The earlier identification of TNF production by monocytes was due, in part, to the fact that adherent BM cells contain subsets of T cells ($\sim$ 10% of total cells) and that ovariectomy increases T cell content by approximately twofold. These findings in the mouse are concordant with studies in humans demonstrating that adherent mononuclear blood cells contain CD3$^+$ CD56$^+$ lymphocytes,[19] a TNF-producing subset of adherent T cells. In that study the number of CD3$^+$ CD56$^+$ T cells was decreased by estrogen treatment and inversely correlated with bone density. Thus, earlier findings

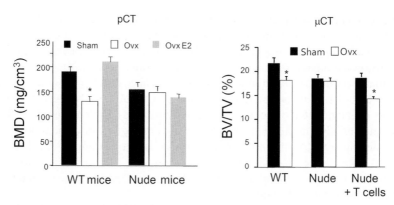

FIGURE 1. T cell-deficient nude mice are protected from bone loss associated with ovariectomy. (*Left*) Trabecular BMD was quantitated by peripheral quantitative CT (pQCT) in WT and nude mice 4 weeks after sham operation, ovariectomy, or ovariectomy with estrogen replacement. (*Right*) Trabecular BV/TV was quantitated by micro CT (μCT) in WT, nude, and nude mice reconstituted with T cells following sham operation, or ovariectomy. *$P < 0.005$.

in BM and adherent cell cultures are consistent with the stimulatory effect of estrogen deficiency on the T cell production of TNF observed in more recent studies and suggested that the induced increase in TNF levels was potentially due to T cell TNF production. Attesting to the relevance of T cells in estrogen deficiency-induced bone loss *in vivo*, athymic T cell-deficient nude mice are completely protected against the bone loss and the increase in bone turnover induced by ovariectomy while reconstitution of nude mice with T cells restores the capacity of estrogen deficiency to induced bone loss (FIG. 1).[7,18]

T CELLS AND THE BONE INTERFACE

The BM hosts fully functional mature T cells that exhibit several distinctive features. Both in humans and mice, T cells account for 3–8% of nucleated BM cells.[20] The percentage of activated T cells is much higher in the BM than in other secondary lymphoid organs and this feature is both cytokine (IL-7 and IL-15) and antigen driven.[21,22] As a result, the BM is the lymphoid organ with the highest percentage and number of proliferating T cells, apart from the thymus. One feature of activated T cells is that of expressing high levels of CD40L, a ligand for the costimulatory molecule CD40, which is expressed on antigen (Ag)-presenting cells (APCs) and stromal cells/osteoblasts.[23]

The BM has long been recognized as a primary lymphoid organ, but it is now clear that the BM plays a key role in the immune response by hosting and regulating adaptive immunity. The BM serves as a site for the initiation of naïve T cell responses[24,25] and as a reservoir of CD4 and CD8 memory T cells.[20] Donor T cells can be found in the BM a few hours after injection into recipient mice[26,27] but memory T cells home to BM in higher numbers than naïve T cells.[28] This selective T cell homing is more pronounced in recipient mice that possess a normal T cell repertoire.[28] The mechanism by which CD8 memory T cells are preferentially recruited in the BM has recently been described.[29] Since the entry into the BM of naïve T cells is limited by space availability and competition with other T cells,[28] senescent memory T cells accumulate in the BM with aging.[30] These T cells produce large amounts of TNF and exhibit increased reactivity to self-peptides and foreign Ag.[28] Furthermore, preliminary data demonstrate a close anatomical colocalization of T cells and OCs. These reasons may explain why the accumulation of aging lymphocytes correlates with an increased incidence of fractures.[30] In response to antigenic stimulation, memory CD4 and CD8 T cells of the BM produce effector cytokines.[28,29] BM T cells encounter Ag presented by DCs and BMMs that reside or have returned to the BM. Resident DCs and BMMs capture blood born Ag that circulates in the BM vessels, or Ag within the BM space.[31] BM T cells may move toward Ag-bearing DCs in the BM or return to the BM after having encountered Ag-loaded APCs outside the BM.

MECHANISMS OF ESTROGEN REGULATION OF T CELL TNF PRODUCTION

Ovariectomy upregulates T cell TNF production primarily by increasing the number of TNF-producing T cells.[7] This is the result of a complex pathway that involves multiple cytokines and immune cells.

Although the mechanism of T cell activation elicited by estrogen deficiency is similar to that triggered by infections, the intensity of the events that follows estrogen withdrawal is significantly less severe and this process should be envisioned as a partial increase in T cell autoreactivity to self-peptides resulting in a modest expansion in the pool of effector CD4$^+$ cells.

The relevance of this mechanism *in vivo* was established by utilizing DO11.10 mice, a strain in which all T cells recognize a single peptide epi-tope of chicken albumin (ovalbumin), which is not expressed in mice. In the absence of ovalbumin, APCs of DO11.10 mice are unable to induce T cell activation. Therefore, if APCs are a relevant target of estrogen, these mice should be protected from the increased T cell proliferation, the suppression of activation-induced T cell death, and the bone loss that follows ovx. As predicted ovx fails to increase the pool of T cells and to induce bone loss in these mice.[32] In addition, injection of ovalbumin, which permits the generation of the appropriate MHC-peptide antigen for these T cells, restores the capacity of ovx to expand the T cell pool by targeting proliferation and apoptosis and to induce bone loss. These data demonstrate that antigen presentation, specifically the generation of appropriate peptide–MHC complexes, is critical to the process by which ovx increases T cell proliferation and life span and leads to bone loss. Furthermore, the finding that T cells from ovx mice exhibit an increased response to ovalbumin demonstrates that ovx increases the reactivity of APCs to endogenous antigens rather than stimulating the production of a new antigen or modulating antigen levels.

The question thus arises as to the nature of the antigens. Estrogen deficiency is likely to increase T cell reactivity to a pool of self- and foreign antigens physiologically present in healthy animals and humans. This is consistent with the fact that T cell clones expressing T cell receptors (TCRs) directed against self-antigens not expressed in the thymus, survive negative selection during T cell maturation.[33] Such clones ("autoreactive" or "self-reactive" T cells) reside in peripheral lymphatic organs of adult individuals. In addition, foreign antigens of bacterial origin are physiologically absorbed in the gut. As these peptides come into contact with immune cells locally and systemically, they induce a low grade T cell activation.[34] Thus, a moderate immune response is constantly in place in healthy humans and rodents due to presentation by MHCII and MHCI molecules of both self- and foreign peptides to CD4$^+$ and CD8$^+$ T cells.[35] This autoreactive response is thought to be essential for immune cell survival and renewal.[36]

The effects of ovx on antigen presentation and the resulting changes in T cell activation, proliferation, and life span are explained by a stimulatory effect of ovx on the expression of the gene encoding Class II Transactivator (*CIITA*). The product of *CIITA* is a non-DNA binding factor induced by IFN-γ that functions as a transcriptional coactivator at the MHC II promoter.[37] Increased *CIITA* expression in macrophages derived from ovariectomized mice results from ovx-mediated increases in both T cell IFN-γ production and the responsiveness of *CIITA* to IFN-γ,[32] an inflammatory cytokine produced by helper T cell. The relevance of IFN-γ is shown by the failure of IFN-γR -/- and IFN-γ -/- mice to sustain bone loss in response to ovx.[32,38]

IFN-γ production by T cells is induced by either a cyclosporin-A-sensitive TCR-dependent mechanism, mediated by T cell activation, or by the cytokines IL-12 and IL-18 through activation of the MAP kinase p38. The increased production of IFN-γ by T cells from ovx mice is suppressed by *in vitro* treatment with the selective p38 inhibitor SB203580, but not by the activation inhibitor cyclosporin-A, indicating that increased IFN-γ production by CD4$^+$ cells in ovx mice is cytokine-driven. The expression of the IL-12 and IL-18 genes in BMMs is induced by NF-κB and AP-1, nuclear proteins whose transcriptional activity is directly repressed by estrogen.[39–41] Unstimulated BMMs, such as those from estrogen replete mice, are known to express low or undetectable levels of NF-κB and AP-1.[42] Accordingly, BM monocytes from estrogen-replete mice express minimal levels of IL-12 and IL-18 while those from ovx animals produce increased amounts of IL-12 and IL-18. Thus, one mechanism by which estrogen represses CIITA is by decreasing IFN-γ production via an inhibitory effect on the BM monocyte production of IL-12 and IL-18.

Another mechanism by which estrogen deficiency upregulates the production of IFN-γ is through TGF-β. Estrogen has a direct stimulatory effect on the production of this factor that is mediated through direct binding of estrogen/ER complex to an ERE region in the TGF-β promoter.[43]

TGF-β is recognized as a powerful repressor of T cell activation. Indeed, TGF-β exerts strong immunosuppressive effects by inhibiting the activation and the proliferation of T cells and their production of proinflammatory cytokines, including IFN-γ. Studies in a transgenic mouse that expresses a dominant negative form of the TGF-β receptor exclusively in T cells have allowed the significance of the repressive effects of this cytokine on T cell function in the bone loss associated with estrogen deficiency to be established.[44] This strain, known as CD4dnTGF-βRII, is severely osteopenic due to increased bone resorption. More importantly, mice with T cell-specific blockade of TGF-β signaling are completely resistant to the bone-sparing effects of estrogen.[44] This phenotype results from a failure of estrogen to repress IFN-γ production, which, in turn, leads to increased T cell activation and T cell TNF production. Gain of function experiments confirmed that elevation of the systemic levels of TGF-β prevents ovx-induced bone loss and bone turnover.[44]

A third mechanism by which estrogen regulates IFN-γ and TNF production is by repressing the production of IL-7. Levels of IL-7 are significantly elevated following ovx[45-47] and *in vivo* IL-7 blockade, using neutralizing antibodies, is effective in preventing ovx-induced bone destruction[45] by suppressing T cell expansion and TNF and IFN-γ production.[46] Furthermore, a recent study shows that liver-derived IGF-I is permissive for ovx-induced trabecular bone loss by modulation of the number of T cells and the expression of IL-7.[47] Indeed, the elevated BM levels of IL-7 contribute to the expansion of the T cell population in peripheral lymphoid organs through several mechanisms. First, IL-7 directly stimulates T cell proliferation by lowering tolerance to weak self-antigens. Second, IL-7 increases antigen presentation by upregulating the production of IFN-γ. Third, IL-7 and TGF-β inversely regulate each others production.[48,49]

The reduction in TGF-β signaling, characteristic of estrogen deficiency may serve to further stimulate IL-7 production, thus driving the cycle of osteoclastogenic cytokine production and bone wasting.

In estrogen deficiency IL-7 compounds bone loss by suppressing bone formation thus uncoupling bone formation from resorption. Recent studies have also identified elevated levels of IL-7 in patients suffering from multiple myeloma and in multiple myeloma-derived cell lines,[50] and have suggested a role for IL-7 in the enhanced bone resorption and suppressed bone formation associated with multiple myeloma. Increased IL-7 expression has also been implicated in the bone loss sustained by patients with rheumatoid arthritis.[51,52]

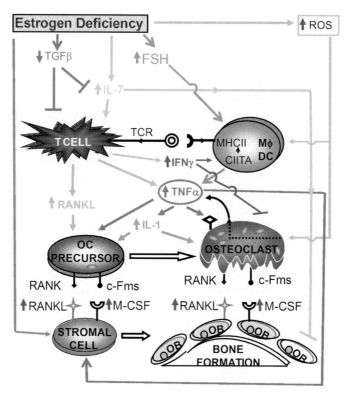

FIGURE 2. Schematic representation of the main mechanisms and feedback interactions by which estrogen deficiency leads to bone loss. The bone loss induced by estrogen deficiency is due to a complex interplay of hormones and cytokines that converge to disrupt the process of bone remodeling. Estrogen deficiency leads to a global increase in IL-7 production in target organs, such as bone, thymus, and spleen, in part through decreases in TGF-β, and increased IGF-1 production. This leads to a first wave of T cell activation. Activated T cells release IFN-γ, which increase Ag presentation by DCs and macrophages (Mφ) by upregulating MHCII expression through the transcription factor CIITA. Estrogen deficiency also amplifies T cell activation and osteoclastogenesis by downregulating antioxidant pathways leading to an upswing in ROS. The resulting increase in ROS stimulates Ag presentation and the production of TNF by mature OCs. The combined effect of IFN-γ and ROS markedly enhances Ag presentation, amplifying T cell activation and promoting release of the osteoclastogenic factors RANKL and TNF. TNF further stimulates SC and OB RANKL and M-CSF production, in part via IL-1 upregulation, driving up OC formation. TNF and IL-7 further exacerbate bone loss by blunting bone formation through direct repressive effects on OBs. Adapted from Reference 75.

In summary, a complex pathway links estrogen, the immune system, and the development of postmenopausal bone loss in experimental animals and is summarized diagrammatically in FIGURE 2.

T CELLS ARE CRITICAL REGULATORS OF BASAL BONE HOMEOSTASIS THROUGH REGULATION OF B CELL OPG PRODUCTION

T cells have the capacity to secrete a wide repertoire of cytokines, some pro-osteoclastogenic and some anti-osteoclastogenic. CD8[+] T cell have been reported to be anti-osteoclastogenic *in vitro*[53] while regulatory T cells inhibit OC differentiation,[54] and correlate negatively with T cell expression of RANKL in the bone resorptive lesions in periodontitis.[55] It has further been reported that depletion of CD4[+] and CD8[+] T lymphocyte subsets in mice *in vivo* enhances vitamin D3-stimulated OC formation *in vitro* by a mechanism involving decreased osteoprotegerin (OPG) production.[56] Activated T cells have also been suggested to inhibit osteoclastogenesis by diverting early OC precursors toward DC differentiation.[57] Indeed, T cells have the capacity to generate both osteoclastogenic cytokines, such as RANKL[58,59] and TNF, as well as anti-osteoclastogenic factors, such as IL-4[60–63] and IFN-γ.[64,65] It has also been suggested that the effects of activated T cells on osteoclastogenesis *in vitro* depends on the manner in which they are activated.[65] The net effect of T cells on OC formation may consequently represent the prevailing balance of anti- and pro-osteoclastogenic T cell cytokine secretion. However, it appears that during stimulated conditions, such as inflammation,[66] and during estrogen deficiency[7,18,32,45] pro-osteoclastogenic cytokines prevail.

Little is known about the anti-osteoclastogenic activities of T cells *in vivo*, however, we have previously reported that although BMD is normal or elevated in very young C57BL6 T cell-deficient nude mice,[18] BMD decreases significantly as the mice age[67] suggesting that T cells may play a protective role in postembryonic basal bone modeling. Recently, we performed a detailed characterization of the bone phenotype of nude mice, confirming an osteopenic phenotype by extensive DXA and micro-CT analyses. Histological analyses[67] and metabolic markers of bone turnover[68] revealed that this osteopenia was a consequence of increased osteoclastic bone resorption. As physiological bone turnover is regulated principally by the ratio of RANKL to OPG in the bone microenvironment we quantitated mRNA for RANKL and OPG in nude mice and wild-type controls. These data revealed a significant deficit in OPG mRNA in the BM of nude mice (FIG. 3A).[68] Surprisingly, this deficit in OPG production was largely accounted for by a significant reduction in OPG production by B cells (FIG. 3B).

B CELLS ARE A CRITICAL SOURCE OF OPG IN THE BONE MICROENVIRONMENT

Historically, an extensive *in vitro* data set involving the use of stromal and osteoblastic cells and cell lines[69,70] has led to the widely accepted precept that

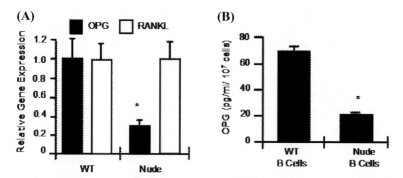

FIGURE 3. T cell-deficient nude mice display decreased total BM and B cell-specific OPG production. (**A**) Real time RT-PCR quantitation of OPG and RANKL mRNA expression in BM of WT and nude mice. Mean ± SD of three independent WT and nude mice. $^*P < 0.01$. (**B**) Decreased OPG production by purified B cells from nude mice relative to WT, quantitated in 48 h conditioned medium by ELISA. $n = 6$ mice per group. $^*P < 0.01$. This research was originally published in Blood. Li Y, Toraldo G, Li A, Yang X, Zhang H, Qian WP, Weitzmann MN; B cells and T cells are critical for the preservation of bone homeostasis and attainment of peak bone mass *in vivo*. Blood. 2007;109(9):3839–3848. © the American Society of Hematology.

the major source of BM OPG is the osteoblast and/or its immediate precursor, the BM stromal cell. In contrast to this view, we have recently reported that B cells are more likely the dominant producers of OPG in the bone microenvironment *in vivo*.[71] This conclusion was arrived at following an extensive series of investigations into the bone phenotype of B cell knockout (KO) (μMT/μMT) mice, instigated by our previous observations that B cells are inhibitory to osteoclastogenesis *in vitro*[72] and that nude mice have a deficit in OPG production, due to reduced B cell OPG production.[71] We found that B cell KO mice present at base line with an osteoporotic phenotype (FIG. 4A), a consequence of enhanced osteoclastic bone resorption.[68] Examination of the RANKL/OPG ratio in B cell KO BM identified a specific deficiency in OPG mRNA (FIG. 4B) and in protein expression. Reconstitution of young B cell KO mice with B cells by means of adoptive transfer, completely rescued mice from development of osteoporosis, by normalizing OPG production (FIG. 4A, B).[68]

COSTIMULATION BETWEEN T CELLS AND B CELLS THROUGH CD40 TO CD40L CROSS-TALK REGULATES B CELL OPG PRODUCTION

Interestingly, it has previously been reported that B cell OPG production by human tonsil-derived B cells could be significantly upregulated by the activation of CD40 signaling by an activating antibody.[73] CD40 is a

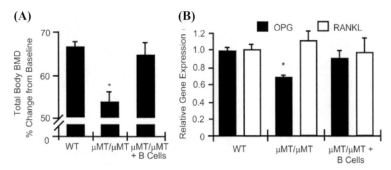

FIGURE 4. B cell KO mice display reduced BMD concurrent with reduced OPG expression. (**A**) B cell reconstitution of μMT/μMT B cell KO mice rescues the decline in BMD (*$P < 0.01$ vs. WT) and (**B**) OPG production (*$P < 0.05$ vs. WT). Mice were reconstituted at 4 weeks of age and BMD and RT-PCR studies preformed at 12 weeks of age. This research was originally published in Blood. Li Y, Toraldo G, Li A, Yang X, Zhang H, Qian WP, Weitzmann MN; B cells and T cells are critical for the preservation of bone homeostasis and attainment of peak bone mass *in vivo*. Blood. 2007;109(9):3839–3848. © the American Society of Hematology."

costimulatory molecule constitutively expressed by professional APCs, such as macrophages, DCs, and B cells, and partners with a receptor that is transiently upregulated on the surface of activated T cells. In agreement with the studies of Yun *et al.*, we found that mouse splenic B cells similarly produced upregulated concentrations of OPG in response to a recombinant soluble ligand to CD40 (sCD40L). In line with these data both CD40 and CD40L KO mice displayed an osteoporotic phenotype and a significant deficiency in BM OPG concentrations. This deficiency in total OPG further correlated with a B cell-specific deficiency in OPG production.[68]

While additional studies remain to be performed, the emerging data now suggest that the B lineage, rather than the osteoblast lineage, is likely the major source of OPG in the bone microenvironment and that T cell signaling to B cells, through the costimulatory molecules CD40L and CD40, play an important role in regulating basal OC formation and in regulating bone homeostasis. This model is summarized in (FIG. 5). Recent preliminary studies by O'Brien *et al.*, using an elegant transgenic mouse model in which osteoblasts and their immediate progenitors were specifically ablated further supports the contention that osteoblasts are not the dominant source of OPG in bone, as OPG transcript levels and OC number remained unaffected following osteoblast ablation.[74]

These findings may provide in part a novel explanation for the propensity for osteopenia and osteoporosis development in numerous pathological conditions in which altered immune function or immunodeficiency in B cells and/or T cells results. Such conditions include HIV infection, solid organ and BM

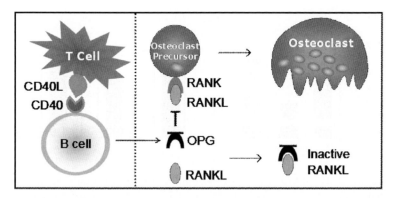

FIGURE 5. Model for the regulation of basal OC formation by T cell regulation of B cell OPG production through CD40 to CD40L costimulation. The differentiation of OC precursors into mature OCs is driven principally by binding of the key osteoclastogenic cytokine RANKL to its receptor RANK on the OC precursor. This process is opposed by the RANKL decoy receptors OPG, secreted by B cells. B cell OPG production is augmented by the interaction of B cells with T cells, through the costimulatory molecules CD40L on activated T cells with the CD40 receptor expressed constitutively on B cells.

transplantation, multiple myeloma in which normal B cells are significantly depleted, aging, and patients treated with immunosuppressive agents, such as glucocorticoids.

CONCLUSIONS

The role of the bone–immune interface in the regulation of not only inflammatory bone turnover, but also of models of postmenopausal osteoporosis, and in basal regulation of bone homeostasis, has only recently begun to be explored and exciting and unexpected new discoveries will likely continue to emerge in the future. Much of the data generated thus far is derived from animal models of pathological and physiological bone turnover. If these discoveries are ultimately translated into humans, this will open the door to exciting new potential therapeutic agents and strategies that target the bone–immune interface to ameliorate bone disease in numerous osteoporotic conditions. Furthermore, it may be possible to exploit the role played by the bone–immune interface in the maintenance of basal bone homeostasis to promote bone health and delay the onset of osteoporosis associated with menopause and aging.

REFERENCES

1. SUN, L., Y. PENG, A.C. SHARROW, *et al.* 2006. FSH directly regulates bone mass. Cell **125:** 247–260.

2. IQBAL, J., L. SUN, T.R. KUMAR, *et al.* 2006. Follicle-stimulating hormone stimulates TNF production from immune cells to enhance osteoblast and osteoclast formation. Proc. Natl. Acad. Sci. USA **103:** 14925–14930.

3. MANOLAGAS, S.C., S. KOUSTENI & R.L. JILKA. 2002. Sex steroids and bone. Recent Prog. Horm. Res. **57:** 385–409.

4. PFEILSCHIFTER, J., R. KODITZ, M. PFOHL & H. SCHATZ. 2002. Changes in proinflammatory cytokine activity after menopause. Endocr. Rev. **23:** 90–119.

5. HOFBAUER, L.C., S. KHOSLA, C.R. DUNSTAN, *et al.* 2000. The roles of osteoprotegerin and osteoprotegerin ligand in the paracrine regulation of bone resorption. J. Bone Miner. Res. **15:** 2–12.

6. SUDA, T., N. TAKAHASHI, N. UDAGAWA, *et al.* 1999. Modulation of osteoclast differentiation and function by the new members of the tumor necrosis factor receptor and ligand families. Endocr. Rev. **20:** 345–357.

7. ROGGIA, C., Y. GAO, S. CENCI, *et al.* 2001. Up-regulation of TNF-producing T cells in the bone marrow: a key mechanism by which estrogen deficiency induces bone loss *in vivo*. Proc. Natl. Acad. Sci. USA **98:** 13960–13965.

8. AMMANN, P., R. RIZZOLI, J.P. BONJOUR, *et al.* 1997. Transgenic mice expressing soluble tumor necrosis factor-receptor are protected against bone loss caused by estrogen deficiency. J. Clin. Invest. **99:** 1699–1703.

9. KIMBLE, R., S. BAIN & R. PACIFICI. 1997. The functional block of TNF but not of IL-6 prevents bone loss in ovariectomized mice. J. Bone Min. Res. **12:** 935–941.

10. PACIFICI, R., C. BROWN, E. PUSCHECK, *et al.* 1991. Effect of surgical menopause and estrogen replacement on cytokine release from human blood mononuclear cells. Proc. Natl. Acad. Sci. **88:** 5134–5138.

11. RALSTON, S.H. 1994. Analysis of gene expression in human bone biopsies by polymerase chain reaction: evidence for enhanced cytokine expression in postmenopausal osteoporosis. J. Bone Miner. Res. **9:** 883–890.

12. KITAZAWA, R., R.B. KIMBLE, J.L. VANNICE, *et al.* 1994. Interleukin-1 receptor antagonist and tumor necrosis factor binding protein decrease osteoclast formation and bone resorption in ovariectomized mice. J. Clin. Invest. **94:** 2397–2406.

13. KIMBLE, R.B., S. SRIVASTAVA, F.P. ROSS, *et al.* 1996. Estrogen deficiency increases the ability of stromal cells to support murine osteoclastogenesis via an interleukin-1and tumor necrosis factor- mediated stimulation of macrophage colony-stimulating factor production. J. Biol. Chem. **271:** 28890–28897.

14. RALSTON, S.H., S.J. GALLACHER, U. PATEL, *et al.* 1990. Use of bisphosphonates in hypercalcaemia due to malignancy [letter]. Lancet **335:** 737.

15. SHANKER, G., M. SORCI-THOMAS & M.R. ADAMS. 1994. Estrogen modulates the expression of tumor necrosis factor alpha mRNA in phorbol ester-stimulated human monocytic THP-1 cells. Lymphokine. Cytokine. Res. **13:** 377–382.

16. CHAUDHARY, L.R., T.C. SPELSBERG & B.L. RIGGS. 1992. Production of various cytokines by normal human osteoblast-like cells in response to interleukin-1β and tumor necrosis factor-α: lack of regulation by 17β-estradiol. Endocrinology **130:** 2528–2534.

17. ABRAHAMSEN, B., V. SHALHOUB, E.K. LARSON, *et al.* 2000. Cytokine RNA levels in transiliac bone biopsies from healthy early postmenopausal women. Bone **26:** 137–145.

18. CENCI, S., M.N. WEITZMANN, C. ROGGIA, *et al.* 2000. Estrogen deficiency induces bone loss by enhancing T-cell production of TNF-alpha. J. Clin. Invest. **106:** 1229–1237.

19. ABRAHAMSEN, B., K. BENDTZEN & H. BECK-NIELSEN. 1997. Cytokines and T-lymphocyte subsets in healthy post-menopausal women: estrogen retards bone loss without affecting the release of IL-1 or IL-1ra. Bone **20:** 251–258.

20. DI ROSA, F. & R. PABST. 2005. The bone marrow: a nest for migratory memory T cells. Trends Immunol. **26:** 360–366.

21. CLARK, P. & D.E. NORMANSELL. 1990. Phenotype analysis of lymphocyte subsets in normal human bone marrow. Am. J. Clin. Pathol. **94:** 632–636.

22. DI ROSA, F. & A. SANTONI. 2002. Bone marrow CD8 T cells are in a different activation state than those in lymphoid periphery. Eur. J. Immunol. **32:** 1873–1880.

23. AHUJA, S.S., S. ZHAO, T. BELLIDO, *et al.* 2003. CD40 ligand blocks apoptosis induced by tumor necrosis factor alpha, glucocorticoids, and etoposide in osteoblasts and the osteocyte-like cell line murine long bone osteocyte-Y4. Endocrinology **144:** 1761–1769.

24. FEUERER, M., P. BECKHOVE, N. GARBI, *et al.* 2003. Bone marrow as a priming site for T-cell responses to blood-borne antigen. Nat. Med. **9:** 1151–1157.

25. TRIPP, R.A., D.J. TOPHAM, S.R. WATSON & P.C. DOHERTY. 1997. Bone marrow can function as a lymphoid organ during a primary immune response under conditions of disrupted lymphocyte trafficking. J. Immunol. **158:** 3716–3720.

26. BERLIN-RUFENACH, C., F. OTTO, M. MATHIES, *et al.* 1999. Lymphocyte migration in lymphocyte function-associated antigen (LFA)-1-deficient mice. J. Exp. Med. **189:** 1467–1478.

27. KONI, P.A., S.K. JOSHI, U.A. TEMANN, *et al.* 2001. Conditional vascular cell adhesion molecule 1 deletion in mice: impaired lymphocyte migration to bone marrow. J. Exp. Med. **193:** 741–754.

28. DI ROSA, F. & A. SANTONI. 2003. Memory T-cell competition for bone marrow seeding. Immunology **108:** 296–304.

29. MAZO, I.B., M. HONCZARENKO, H. LEUNG, *et al.* 2005. Bone marrow is a major reservoir and site of recruitment for central memory CD8+ T cells. Immunity **22:** 259–270.

30. EFFROS, R.B. 2004. Replicative senescence of CD8 T cells: effect on human ageing. Exp. Gerontol. **39:** 517–524.

31. FEUERER, M., P. BECKHOVE, Y. MAHNKE, *et al.* 2004. Bone marrow microenvironment facilitating dendritic cell: CD4 T cell interactions and maintenance of CD4 memory. Int. J. Oncol. **25:** 867–876.

32. CENCI, S., G. TORALDO, M.N. WEITZMANN, *et al.* 2003. Estrogen deficiency induces bone loss by increasing T cell proliferation and lifespan through IFN-gamma-induced class II transactivator. Proc. Natl. Acad. Sci. USA **100:** 10405–10410.

33. ROBEY, E.A., F. RAMSDELL, J.W. GORDON, *et al.* 1992. A self-reactive T cell population that is not subject to negative selection. Int. Immunol. **4:** 969–974.

34. RAMMENSEE, H.G., K. FALK & O. ROTZSCHKE. 1993. Peptides naturally presented by MHC class I molecules. Annu. Rev. Immunol. **11:** 213–244.

35. GROSSMAN, Z. & W.E. PAUL. 2000. Self-tolerance: context dependent tuning of T cell antigen recognition. Semin. Immunol. **12:** 197–203; discussion 257–344.

36. TANCHOT, C., F.A. LEMONNIER, B. PERARNAU, *et al.* 1997. Differential requirements for survival and proliferation of CD8 naive or memory T cells. Science **276:** 2057–2062.

37. BOSS, J.M. & P.E. JENSEN. 2003. Transcriptional regulation of the MHC class II antigen presentation pathway. Curr. Opin. Immunol. **15:** 105–111.

38. GAO, Y., F. GRASSI, M.R. RYAN, *et al.* 2007. IFN-gamma stimulates osteoclast formation and bone loss *in vivo* via antigen-driven T cell activation. J. Clin. Invest. **117:** 122–132.

39. AN, J., R.C. RIBEIRO, P. WEBB, *et al.* 1999. Estradiol repression of tumor necrosis factor-alpha transcription requires estrogen receptor activation function-2 and is enhanced by coactivators. Proc. Natl. Acad. Sci. USA **96:** 15161–15166.

40. SHEVDE, N.K., A.C. BENDIXEN, K.M. DIENGER & J.W. PIKE. 2000. Estrogens suppress RANK ligand-induced osteoclast differentiation via a stromal cell independent mechanism involving c-Jun repression. Proc. Natl. Acad. Sci. USA **97:** 7829–7834.

41. GALIEN, R. & T. GARCIA. 1997. Estrogen receptor impairs interleukin-6 expression by preventing protein binding on the NF-kappaB site. Nucleic Acids Res. **25:** 2424–2429.

42. MUEGGE, K. & S. DURUM. 1990. Cytokines and transcription factors. [Review]. Cytokine **2:** 1–8.

43. YANG, N.N., M. VENUGOPALAN, S. HARDIKAR & A. GLASEBROOK. 1996. Identification of an estrogen response element activated by metabolites of 17β-estradiol and raloxifene. Science **273:** 1222–1225.

44. GAO, Y., W.P. QIAN, K. DARK, *et al.* 2004. Estrogen prevents bone loss through transforming growth factor beta signaling in T cells. Proc. Natl. Acad. Sci. USA **101:** 16618–16623.

45. WEITZMANN, M.N., C. ROGGIA, G. TORALDO, *et al.* 2002. Increased production of IL-7 uncouples bone formation from bone resorption during estrogen deficiency. J. Clin. Invest. **110:** 1643–1650.

46. RYAN, M.R., R. SHEPHERD, J.K. LEAVEY, *et al.* 2005. An IL-7-dependent rebound in thymic T cell output contributes to the bone loss induced by estrogen deficiency. Proc. Natl. Acad. Sci. USA **102:** 16735–16740.

47. LINDBERG, M.K., J. SVENSSON, K. VENKEN, *et al.* 2006. Liver-derived IGF-I is permissive for ovariectomy-induced trabecular bone loss. Bone **38:** 85–92.

48. HUANG, M., S. SHARMA, L.X. ZHU, *et al.* 2002. IL-7 inhibits fibroblast TGF-beta production and signaling in pulmonary fibrosis. J. Clin. Invest. **109:** 931–937.

49. DUBINETT, S.M., M. HUANG, S. DHANANI, *et al.* 1995. Down-regulation of murine fibrosarcoma transforming growth factor-beta 1 expression by interleukin 7. J. Natl. Cancer Inst. **87:** 593–597.

50. GIULIANI, N., S. COLLA, R. SALA, *et al.* 2002. Human myeloma cells stimulate the receptor activator of nuclear factor-kappa B ligand (RANKL) in T lymphocytes: a potential role in multiple myeloma bone disease. Blood **100:** 4615–4621.

51. VAN ROON, J.A., K.A. GLAUDEMANS, J.W. BIJLSMA & F.P. LAFEBER. 2003. Interleukin 7 stimulates tumour necrosis factor alpha and Th1 cytokine production in joints of patients with rheumatoid arthritis. Ann. Rheum. Dis. **62:** 113–119.

52. DE BENEDETTI, F., M. MASSA, P. PIGNATTI, *et al.* 1995. Elevated circulating interleukin-7 levels in patients with systemic juvenile rheumatoid arthritis. J. Rheumatol. **22:** 1581–1585.

53. JOHN, V., J.M. HOCK, L.L. SHORT, *et al.* 1996. A role for CD8+ T lymphocytes in osteoclast differentiation *in vitro*. Endocrinology **137:** 2457–2463.

54. KIM, Y.G., C.K. LEE, S.S. NAH, *et al.* 2007. Human CD4+CD25+ regulatory T cells inhibit the differentiation of osteoclasts from peripheral blood mononuclear cells. Biochem. Biophys. Res. Commun. **357:** 1046–1052.

55. ERNST, C.W., J.E. LEE, T. NAKANISHI, *et al.* 2007. Diminished forkhead box P3/CD25 double-positive T regulatory cells are associated with the increased

nuclear factor-kappaB ligand (RANKL+) T cells in bone resorption lesion of periodontal disease. Clin. Exp. Immunol. **148:** 271–280.

56. GRCEVIC, D., S.K. LEE, A. MARUSIC & J.A. LORENZO. 2000. Depletion of CD4 and CD8 T lymphocytes in mice *in vivo* enhances 1, 25- dihydroxyvitamin D(3)-stimulated osteoclast-like cell formation *in vitro* by a mechanism that is dependent on prostaglandin synthesis. J. Immunol. **165:** 4231–4238.

57. GRCEVIC, D., I.K. LUKIC, N. KOVACIC, *et al.* 2006. Activated T lymphocytes suppress osteoclastogenesis by diverting early monocyte/macrophage progenitor lineage commitment towards dendritic cell differentiation through down-regulation of receptor activator of nuclear factor-kappaB and c-Fos. Clin. Exp. Immunol. **146:** 146–158.

58. WEITZMANN, M.N., S. CENCI, L. RIFAS, *et al.* 2001. T cell activation induces human osteoclast formation via receptor activator of nuclear factor kappaB ligand-dependent and -independent mechanisms. J. Bone Miner. Res. **16:** 328–337.

59. HORWOOD, N.J., V. KARTSOGIANNIS, J.M. QUINN, *et al.* 1999. Activated T lymphocytes support osteoclast formation *in vitro*. Biochem. Biophys. Res. Commun. **265:** 144–150.

60. WEI, S., M.W. WANG, S.L. TEITELBAUM & F.P. ROSS. 2002. Interleukin-4 reversibly inhibits osteoclastogenesis via inhibition of NF-kappa B and mitogen-activated protein kinase signaling. J. Biol. Chem. **277:** 6622–6630.

61. RIANCHO, J.A., M.T. ZARRABEITIA, G.R. MUNDY, *et al.* 1993. Effects of interleukin-4 on the formation of macrophages and osteoclast- like cells. J. Bone Miner. Res. **8:** 1337–1344.

62. BIZZARRI, C., A. SHIOI, S.L. TEITELBAUM, *et al.* 1994. Interleukin-4 inhibits bone resorption and acutely increases cytosolic Ca2+ in murine osteoclasts. J. Biol. Chem. **269:** 13817–13824.

63. ABU-AMER, Y. 2001. IL-4 abrogates osteoclastogenesis through STAT6-dependent inhibition of NF-kappaB. J. Clin. Invest. **107:** 1375–1385.

64. TAKAYANAGI, H., K. OGASAWARA, S. HIDA, *et al.* 2000. T-cell-mediated regulation of osteoclastogenesis by signalling cross- talk between RANKL and IFN-gamma. Nature **408:** 600–605.

65. WYZGA, N., S. VARGHESE, S. WIKEL, *et al.* 2004. Effects of activated T cells on osteoclastogenesis depend on how they are activated. Bone **35:** 614–620.

66. KONG, Y.Y., U. FEIGE, I. SAROSI, *et al.* 1999. Activated T cells regulate bone loss and joint destruction in adjuvant arthritis through osteoprotegerin ligand. Nature **402:** 304–309.

67. TORALDO, G., C. ROGGIA, W.P. QIAN, *et al.* 2003. IL-7 induces bone loss *in vivo* by induction of receptor activator of nuclear factor kappa B ligand and tumor necrosis factor alpha from T cells. Proc. Natl. Acad. Sci. USA **100:** 125–130.

68. LI, Y., A. LI, X. YANG & M.N. WEITZMANN. 2007. Ovariectomy-induced bone loss occurs independently of B cells. J. Cell Biochem. **100:** 1370–1375.

69. YASUDA, H., N. SHIMA, N. NAKAGAWA, *et al.* 1998. Osteoclast differentiation factor is a ligand for osteoprotegerin/osteoclastogenesis-inhibitory factor and is identical to TRANCE/RANKL. Proc. Natl. Acad. Sci. USA **95:** 3597–3602.

70. UDAGAWA, N., N. TAKAHASHI, H. YASUDA, *et al.* 2000. Osteoprotegerin produced by osteoblasts is an important regulator in osteoclast development and function. Endocrinology **141:** 3478–3484.

71. LI, Y., G. TORALDO, A. LI, *et al.* 2007. B cells and T cells are critical for the preservation of bone homeostasis and attainment of peak bone mass *in vivo*. Blood **109:** 3839–3848.

72. WEITZMANN, M.N., S. CENCI, J. HAUG, *et al.* 2000. B lymphocytes inhibit human osteoclastogenesis by secretion of TGFbeta. J. Cell Biochem. **78:** 318–324.
73. YUN, T.J., P.M. CHAUDHARY, G.L. SHU, *et al.* 1998. OPG/FDCR-1, a TNF receptor family member, is expressed in lymphoid cells and is up-regulated by ligating CD40. J. Immunol. **161:** 6113–6121.
74. O'BRIEN, C., Q. FU, L. MOMMSEN, *et al.* 2006. Osteoblasts are not the source of RANKL and OPG in bone but are required for maintenance of osteoclast function. J. Bone Miner. Res. **22:** abstract.
75. WEITZMANN, M.N. & R. PACIFICI. 2006. Estrogen deficiency and bone loss: an inflammatory tale. J. Clin. Invest. **116:** 1186–1194.

Proresorptive Actions of FSH and Bone Loss

MONE ZAIDI,[a] HARRY C. BLAIR,[b] JAMEEL IQBAL,[a] LING LING ZHU,[a] T. RAJENDRA KUMAR,[c] ALBERTA ZALLONE,[d] AND LI SUN[a]

[a] Mount Sinai Bone Program, The Mount Sinai School of Medicine, New York, New York, USA

[b] Departments of Pathology and Cell Biology, University of Pittsburgh, Pittsburgh, Pennsylvania, USA

[c] Molecular and Integrative Physiology, University of Kansas Medical Center, Kansas City, Missouri, USA

[d] University of Bari, Italy

ABSTRACT: We review studies that propose follicle-stimulating hormone (FSH) as a physiologic stimulator of osteoclastic bone resorption. We hypothesize that, in addition to low estrogen, a rising FSH contributes to the increased bone resorption and bone loss in hypergonadism. This is of particular relevance to the perimenopausal transition, wherein profound bone loss is accompanied by trabecular perforations in the face of high FSH and normal estrogen levels. Potential therapeutic implications include the development of antagonists to both circulating FSH and its osteoclastic receptor.

KEYWORDS: osteoclast; osteoporosis; hypogonadism

Since Fuller Albright's discovery of hypogonadal bone loss, estrogen deficiency has been considered its sole culprit.[1] Evidence for more than two decades is overwhelming in that reduced estrogen causes bone loss, and that estrogen replacement prevents this loss.[2] This concept holds well across species. In contrast to *in vivo* and clinical data, *in vitro* studies explaining how estrogen prevents bone loss are far and few. There is evidence for weak, but direct inhibitory effects of estrogen on osteoclast precursors, which are exerted by interfering with the phosphorylation of the enzyme *janus* N-terminal kinase downstream of receptor activator for NF-κB-ligand (RANK-L).[3,4] By this mechanism, estrogen inhibits the osteoclastogenesis, which increases dramatically upon hormone withdrawal. Estrogen also uses multiple pathways to

Address for correspondence: Sun Li, M.D., Ph.D., The Mount Sinai School of Medicine, Division of Endocrinology, Box 1055, One Gustave L. Levy Place, Atran Building AB4-02, New York, NY 10029. Voice: 212-241-3054; fax: 212-534-4820.
Li.sun@mssm.edu

Ann. N.Y. Acad. Sci. 1116: 376–382 (2007). © 2007 New York Academy of Sciences.
doi: 10.1196/annals.1402.056

alter the secretion of the critical osteoclastogenic cytokine (RANK-L) and its decoy receptor osteoprotegerin (OPG) from osteoblasts,[5] thereby secondarily affecting osteoclast formation. Effects on osteoclastogenesis are also exerted via T lymphocytes, wherein estrogen has been shown in elegant studies to regulate the production of the cytokine and tumor necrosis factor-α (TNF-α), and also to modulate the production of nitric oxide.[6–8] Similarly, an anabolic effect of estrogen is well established in animal models.[9] We neither dispute the anabolic actions of estrogen, nor do we doubt its antiresorptive function.

However, sex steroid withdrawal in mammals, either acutely through gonadal ablation or gradually as in the natural menopause, is invariably associated with a sharp rise in the serum levels of the pituitary hormone follicle-stimulating hormone (FSH). A primary function of FSH, *hitherto* considered to be its only role, is to stimulate ovarian folliculogenesis and estrogen production. This results in the positive feedback regulation of FSH secretion from the pituitary in response to failing ovaries. Thus, in women undergoing menopause, serum FSH levels rise almost exponentially, even before serum estrogen levels have declined (FIG. 1). It has also been gleaned recently that during the "menopausal transition," despite no reductions in circulating estrogen, and in the presence of elevated FSH levels, there is a substantial increase in the rate of bone loss, most prominently during the late perimenopause.[10,19] This bone loss is associated with significant trabecular perforations,[11] which in themselves cause a dramatic reduction in bone strength and a predisposition to first fractures. The fact that circulating FSH is high during this period of rapid bone loss promoted us to investigate whether FSH directly affects the skeleton.

We reported that independently of the species studies, mice or human, FSH stimulated osteoclast formation from bone marrow precursors in the presence of RANK-L.[12] This finding has recently been confirmed by others.[13] In addition, we showed that FSH stimulated the resorption of bone by mature human osteoclasts.[12] The osteoclastogenic and proresorptive actions of FSH, we found, were exerted through a $G_{i2\alpha}$-coupled FSH receptor identified on both human and mouse osteoclasts and osteoclast precursors using RT-PCR, Western blotting, immune staining, and flow cytometry. We also found that FSH enhanced the phosphorylation of downstream RANK-L-sensitive kinases, namely Erk and Akt, as well as Iκ-Bα, an inhibitor for NF-κB; all of these pathways are known to transduce the proresorptive effects of RANK-L.[12] Thus, the attenuation of such phosphorylation by appropriately selected chemical inhibitors was further shown to reduce osteoclast formation, as was the use of $G_{i2\alpha}^{-/-}$ cells.[12] Together, the studies established a link between FSHR activation, G_{i2a} coupling, and kinase phosphorylation in mediating the effects of FSH.

The strength of our *in vitro* observations suggests that elevated FSH levels could potentially contribute to the enhanced osteoclastogenesis and resorption that characterize hypogonadal states. However, to provide genetic evidence for an effect of FSH on the skeleton that was exerted independently of estrogen,

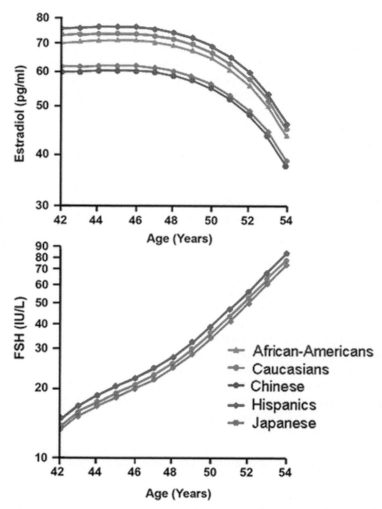

FIGURE 1. FSH and estradiol levels across the menopausal transition. Modified and redrawn from Randolph et al.[10]

we used mice lacking the β subunit of FSH or the FSHR.[12] Genetic evidence from mice lacking FSHβ was compelling in that haploinsufficiency (50% reduction) of circulating FSH in FSHβ heterozygote mice resulted in a high bone mass, despite these mice being eugonadal.[12] Importantly, we found that the high bone mass arose from reduced bone resorption, as was evident on histomorphometry and *ex vivo* cultures of bone marrow cells.[12] The enhanced bone mass and osteoclast defect in eugonadal, but FSH haploinsufficient mice, clearly separated the skeletal actions of FSH and estrogen.

Hypogonadal FSHβ$^{-/-}$ mice also failed to lose bone despite severe hypogonadism, as did FSHR$^{-/-}$ mice in our hands.[12] Nonetheless, both homozygous strains failed to show an increase in resorption parameters *in vivo* and *ex vivo.*[12] It has been shown recently that FSHR$^{-/-}$ mice do show an age-dependent loss of bone mass, which is further reduced upon androgenic withdrawal.[14] However, all of these data demonstrate substantial inhibitory effects of estrogen and androgen depletion on bone formation, with a remarkably small ~4–5% increase in bone resorption.[14]

If one examines bone resorption quantitated *in vivo* as TRAP-positive surfaces in genetic models of hypogonadism (FIG. 2), it is clear that significant elevations in resorption do *not* occur in the absence of a high FSH level. Thus, aromatase$^{-/-}$ mice, a model for chronic hypogonadism, display a doubling of resorption surfaces in the presence of high FSH, while ERα$^{-/-}$β$^{-/-}$ or GnRH^{mut} *hpg* mice, in which estrogen deficiency is as severe, but is not accompanied by high FSH, do not show the expected increases in bone resorption.[15,16] Some of these mice do lose bone, but this loss is due to reduced bone formation rather than an elevated bone resorption.[13,16] Estrogen therefore appears to have a primary and profound anabolic action in these mouse models. With that said, we are by no means excluding an antiresorptive effect of estrogen, the withdrawal of which in humans treated with GnRH agonists, causes high-turnover bone loss despite reduced FSH levels.[17] Acute estrogen withdrawal may cause resorption increases and skeletal loss *per se,* even in the absence of a high FSH in humans.

Our *in vitro* and *in vivo* evidence for distinct effects of estrogen and FSH on bone, in which FSH is proresorptive, and estrogen is both anabolic and antiresorptive, creates a paradigm shift in our understanding of hypogonadal, and more importantly, peri- and postmenopausal bone loss. We attribute bone loss in the early years of the menopause not only to reduced estrogen, but also to elevated FSH levels. This, we propose, assumes particular relevance during the late perimenopause when estrogen levels are normal, despite which bone loss is most profound.

Several new clinical observations attest to a role of FSH in bone loss across the perimenopausal transition. The Study of Women Across nations (SWAN), a cross-sectional and longitudinal study of more than 2,300 perimenopausal women between 42 and 52 years, showed that not only was there a strong correlation between basal FSH and bone remodeling markers,[18] but a change in FSH levels over 4 years could predict a change in bone mass [19] (FIG. 3). It is indeed difficult to establish causality from SWAN, as high FSH is a more sensitive marker for hypogonadism than reduced estrogen; hence this association may simply be correlative. However, a step toward establishing causality is a smaller study, which shows that, with similar estrogen levels, amenorrheic women having a mean serum FSH of 35 IU/L had considerably greater bone loss than women whose mean serum FSH level was 8 U/L.[20] Thus, the mouse and human studies together make a compelling argument for

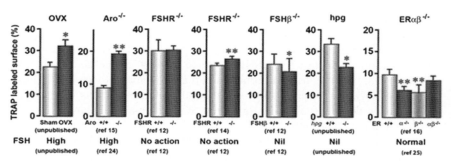

FIGURE 2. Histomorphometric evaluation of bone resorption (expressed as tartrate-resistant acid phosphatase-labeled surfaces,%) in various genetic mouse models, including mice lacking aromatase (Aro$^{-/-}$), follicle-stimulating hormone receptor (FSHR$^{-/-}$), FSHβ$^{-/-}$, gonadotropin-releasing hormone (*hpg*), and estrogen receptors α and β (ERαβ$^{-/-}$), displaying hypogonadism, as well as the effect of acute hypogonadism due to ovariectomy (OVX). These data have been adopted from the references, as listed. *P*-values *$P < 0.05$; **$P < 0.01$.

further investigations on a role for FSH in causing hypogonadal bone loss, traditionally attributable solely to low estrogen.

In considering this topic, it deserves mention that the menopause also affects estrogen-related signals including activin and inhibin that are involved in regulation of FSH production, for which receptors are also present on bone cells including osteoblasts and osteoblast precursors.[21] There is evidence that both hormones can modulate osteoclast and osteoblast formation *in vitro*.[21] More recent evidence suggests that both follistatin and activin A regulate osteoblastic bone formation.[22] Thus, in considering the effects of FSH an important *caveat* is that these effects may be modified, positively or negatively, by additional hormones. The skeletal actions of these hormones, traditionally regarded to be important only in the pituitary–endocrine axis, reflect that there is a much broader functional profile than has previously been appreciated.

Finally, it is likely that the mechanism of action of FSH on the skeleton is more complex that we currently envisage. Of particular note is that FSH does not affect bone formation due to absent FSHRs on mature osteoblasts.[12] The role of FSHRs on human mesenchymal cells, however, remains unknown.[12] Furthermore, while we have documented direct effects of FSH on osteoclast signaling pathways,[12] we also find that FSH stimulates the production of TNF-α from bone marrow precursor macrophages.[23] We have not studied FSH-induced increases in TNF-α from T cells, and it is quite possible that T cells express FSHRs. If this is the case, FSH may be the primary trigger through which estrogen deficiency increases TNF-α production from T cells to contribute to the bone loss in hypogonadism.

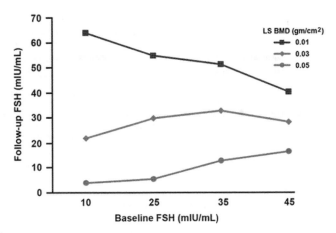

FIGURE 3. Changes in serum FSH predict changes in bone mass over a 4-year period in the SWAN study. Modified and redrawn from Sowers *et al.*[19]

ACKNOWLEDGMENTS

We gratefully acknowledge support from the National Institutes of Health. J.I. won the Anthony Means Award from the Endocrine Society and S.L. won the Award for Outstanding Research in the Pathophysiology of Osteoporosis from the American Society for Bone and Mineral Research.

REFERENCES

1. ALBRIGHT, F., P.H. SMITH & A.M. RICHARDSON. 1941. Post menopausal osteoporosis. JAMA **1167:** 2465–2474.
2. RIGGS, B.L., S. KHOSLA, L.J. MELTON III. 2002. Sex steroids and the construction and conservation of the adult skeleton. Endoc. Rev. **23:** 279–302.
3. SHEVDE, N.K., A.C. BENDIXEN, K.M. DIENGER & J.W. PIKE. 2000. Estrogens suppress RANK ligand-induced osteoclast differentiation via a stromal cell independent mechanism involving c-jun repression. Proc. Natl. Acad. Sci. **97:** 7829–7834.
4. SRIVASTAVA, S., G. TORALDO, M.N. WEITZMANN, *et al.* 2002. Estrogen decreases osteoclast formation by downregulating receptor activator for NF-kappa B ligand induced JNK activation. J. Biol. Chem. **276:** 8836–8840.
5. SRIVASTAVA, S., M.N. WEITZMANN, R.B. KIMBLE, *et al.* 1998. Estrogen blocks M-CSF gene expression and osteoclast formation by regulating phosphorylation of Egr-1 and its interaction with Sp-1. J. Clin. Invest. **102:** 1850–1859.
6. ROGGIA, C., Y. GAO, S. CENCI, *et al.* 2001. Up-regulation of TNF producing T cells in the bone marrow: a key mechanism by which estrogen deficiency induces bone loss *in vivo*. Proc. Natl. Acad. Sci. **98:** 13960–13965.
7. GARCÍA PALACIOS, V., L.J. ROBINSON, C.W. BORYSENKO, *et al.* 2005. Negative regulation of RANKL induced osteoclastic differentiation in RAW264.7 cells by estrogen and phytoestrogens. J. Biol. Chem. **280:** 13720–13727.

8. ARMOUR, K.E., K.J. ARMOUR, M.E. GALLAGHER, *et al.* 2001. Defective bone formation and anabolic response to exogenous estrogen in mice with targeted disruption of endothelial nitrix oxide synthase. Endocrinology **142:** 760–766.

9. JAGGER, C.J., J.W. CHOW & T.J. CHAMBERS. 1996. Estrogen suppresses activation but enhances formation phase of osteogenic response to mechanical stimulation in rat bone. J. Clin. Invest. **98:** 2351–2357.

10. RANDOLPH, J.F., M. SOWERS, I.V. BONDARENKO, *et al.* 2004. Change in estradiol and follicle-stimulating hormone across the early menopausal transition: effects of ethnicity and age. J. Clin. Endocrinol. Metabol. **89:** 1555–1561.

11. AKHTER, M.P., J.M. LAPPE, K.M. DAVIES & R.R. RECKER. 2007. Transmenopausal changes in the trabecular bone structure. Bone **41:** 111–116.

12. SUN, L., Y. PENG, A.C. SHARROW, *et al.* 2006. FSH directly regulates bone mass. Cell **125:** 247–260.

13. WU, Y., J. TORCHIA, W. YAO, *et al.* 2007. Bone microenvironment specific roles of ITAM adapter signaling during bone remodeling induced by acute estrogen deficiency. PLoS ONE **2:** e586.

14. GAO, J., R. TIWARI-PANDEY, R. SAMADFAM, *et al.* 2007. Altered ovarian function affects skeletal homeostasis independent of the action of follicle stimulating hormone. Endocrinology **148:** 2613–2621.

15. MIYAURA, C., K. TODA, M. INADA, *et al.* 2001. Sex- and age-related response to aromatase deficiency in bone. Biochem. Biophys. Res. Commun. **280:** 1062–1068.

16. SIMS, N.A., S. DUPONT, A. KRUST *et al.* 2002. Deletion of estrogen receptors reveals a regulatory role for estrogen receptor-beta in bone remodeling in females but not in males. Bone **30:** 18–25.

17. DAWSON-HUGHES, B. 2001. Bone loss accompanying medical therapies. N. Engl. J. Med. **345:** 989–992.

18. SOWERS, M.R., G.A. GREENDALE, J. BONDARENKO, *et al.* 2003. Endogenous hormones and bone turnover markers in pre- and perimenopausal women: SWAN. Ost. Internat. **14:** 191–197.

19. SOWERS, M.R., M. JANNAUSCH, D. MCCONNELL, *et al.* 2006. Hormone predictors of bone mineral density changes during the menopausal transition. J. Clin. Endocrinol. Metabol. **91:** 1261–1267.

20. DEVLETA, B., B. ADEM & S. SENEDA. 2004. Hypergonadotropic amenorrhea and bone density: new approach to an old problem. J. Bone Miner. Res. **22:** 360–364.

21. GADDY-KURTEN, D., J.K. COKER, E. ABE, *et al.* 2002. Inhibin suppresses and activin stimulates osteoblastogenesis and osteoclastogenesis in murine bone marrow cultures. Endocrinology **143:** 74–83.

22. EIJKEN, M., S. SWAGEMAKERS, M. KOEDAM, *et al.* 2007. Activin A-follistatin system: potent regulator of human extracellular matrix mineralization. FASEB J. Epub. April 20.

23. IQBAL, J., L. SUN, T.R. KUMAR, *et al.* 2006. Follicle stimulating hormone stimulates TNF production from immune cells to enhance osteoblast and osteoclast formation. Proc. Natl. Acad. Sci. **103:** 14925–14930.

24. BRITT, K.L., A.E. DRUMMOND, M. DYSON, *et al.* 2001. The ovarian phenotype of the aromatase knockout (ArKO) mouse. J. Steroid Biochem. Mol. Biol. **79:** 181–185.

25. COUSE, J.F., M.M. YATES, V.R. WALKER, *et al.* 2003. Characterization of the hypothalamic-pituitary-gonadal axis in estrogen receptor (ER) null mice reveals hypergonadism and endocrine sex reversal in females lacking ERalpha but not ERbeta. Mol. Endocrinol. **17:** 1039–1053.

Bone Loss in Thyroid Disease

Role of Low TSH and High Thyroid Hormone

ETSUKO ABE, LI SUN, JEFFREY MECHANICK, JAMEEL IQBAL,
KOSJ YAMOAH, RAMKUMARIE BALIRAM, ARIO ARABI, BALJIT S.
MOONGA, TERRY F. DAVIES, AND MONE ZAIDI

Mount Sinai Bone Program and The Thyroid Research Unit at the James Peter's Veterans Affairs Medical Center, Mount Sinai School of Medicine, New York New York, USA

ABSTRACT: More than 10% of postmenopausal women in the United States receive thyroid hormone replacement therapy and up to 20% of these women are over-replaced inducing subclinical hyperthyroidism. Because hyperthyroidism and post menopausal osteoporosis overlap in women of advancing age, it is urgent to understand the effect of thyroid hormone excess on bone. We can now provide results that not thyroid hormones but also TSH itself has an equally important role to play in bone remodeling.

KEYWORDS: thyroid receptor; osteoporosis; bone remodeling

INTRODUCTION

Bone loss in thyroid dysfunction was first described by von Recklinghausen. Thyrotoxicosis and thyroid hormone therapy are both associated with significant bone loss.[1] Elevated bone turnover and decreased bone mass result when thyroid hormone levels are high and thyroid-stimulating hormone (TSH) levels suppressed. The changes in bone mass are impressive even in patients with subclinical hyperthyroidism, where thyroid hormones are within the normal range, but TSH is low or undetectable.[2–5] Recent evidence from genetically manipulated mice shows that, in addition to thyroid hormone receptors (TRs), TSH receptors (TSHRs) also play a role in bone remodeling.[6–8] Here, we review this genetic evidence and surmise on the relative contributions of a high thyroid hormone and low TSH in the pathophysiology of human thyrotoxic bone disease.

Address for correspondence: Etsuko Abe, Ph.D., P.O. Box 1055, Mount Sinai School of Medicine, New York, NY 10029. Voice: 212-241-8735; fax: 212-241-4218.
Etsuko.abe@mssm.edu

Ann. N.Y. Acad. Sci. 1116: 383–391 (2007). © 2007 New York Academy of Sciences.
doi: 10.1196/annals.1402.062

THYROID HORMONE EFFECTS ON BONE

A decade ago, the prevailing paradigm was that alterations in bone metabolism in hyperthyroid states arose directly and solely from thyroid hormones (T3/T4) excess.[1] This notion was supported by numerous longitudinal studies demonstrating associations between the thyroid state and bone status (TABLE 1). In laboratory studies, thyroid hormones were shown to activate bone resorption directly through nuclear TRs α and β.[9,10] Further studies showed that thyroid hormones increased osteoclastic resorptive activity indirectly through the osteoblast.[11,12] Various inflammatory cytokines and hormones have been implicated in causing osteoclastic activation, including IL-6, PGE_2, PTH, and 1,25-dihydroxyvitamin D_3.[13]

In bone, TRα is expressed at higher levels than TRβ and has thus been thought to be functionally predominant.[6,14] Detail analysis of TR knockouts has not only highlighted functional differences between the α and β isoforms *vis-à-vis* their effects on bone cells, but also yielded unexpected bone phenotypes in adult mice (TABLE 2). For example, biochemically euthyroid $TR\alpha^{0/0}$ mice, with all α isoforms deleted, display delayed chondrocyte maturation, including growth and ossification defects in young mice, but show osteosclerosis in adults.[15] In contrast, $TR\beta^{-/-}$ mice, which have elevated thyroid hormone and TSH levels, exhibit accelerated chondrocyte maturation, but induce osteoporosis in adults.[15] The same group recently generated functionally null mice for both receptors, $TR\alpha1^{+/m}\beta^{+/-}$ and $TR\alpha1^{+/m}\beta^{-/-}$; their phenotypes were similar to those observed in $TR^{0/0}$ or $TR\beta^{-/-}$ mice.[16] Thus, $TR\alpha1^{+/m}\beta^{+/-}$ and $TR\alpha1^{+/m}\beta^{-/-}$ mice exhibited osteosclerosis and osteoporosis, respectively. These genetic data indicate that TRs play a critical role in bone maturation *via* both α and β isoforms. However, it is unclear whether the adult bone phenotype results from bone maturation defects or represents a direct effect of TRs on the adult skeleton.

TABLE 1. Various thyroid disorders and modalities of thyroid hormone therapies associated with bone loss and fracture risk

Scenario	Markers	Clinical correlates	Refs.
Primary hypothyroidism		↑ fracture risk	24,44,45
Subclinical hyperthyroidism		↑ fracture risk	44
TSH suppression therapy			
Benign disease		↑ fracture risk	44
Thyroid cancer	↑ cross-links	ns to ↓ spine or hip	26
		BMD up to 5–9%	46–48
Thyroid hormone replacement	↑ cross-links	ns to ↓ spine BMD by 5%	49–51
Normal TSH		ns to ↓ hip BMD by <1 to 7%	49–52
		No effect on fracture risk	24,44

Only longitudinal studies included: ns = not statistically different; Refs. = references.

TABLE 2. Bone phenotype of thyroid hormone receptor knockout mice

Mice	Receptor isotypes	Survival	TSH	TH†	Bone Phenotype Juvenile	Bone Phenotype Adult	Refs.#
TRα−/−	(α1−/−, α2−/−, Δα1+/+, Δα2+/+)	Die	↓	↓	Delayed maturation		53
TRα1−/−	(α1−/−, α2+/+, Δα1−/−, Δα2+/+)	Survive (bradycardia)	↑	↑	Normal		54
TRα0/0	(α−/−, α2−/−, Δα1−/−, Δα2−/−)	Survive (hypothermia)	↑	↑	Delayed maturation	Osteosclerosis	15,55,56
TRβ−/−	(β1−/−, β2−/−)	Survive (deaf)	↗	↗	Advanced maturation	Osteoporosis	15,55,56,57,58
TRβ2−/−	(β1+/+, β2−/−)	Survive	↗	↗	Normal		59,60
TRα1−/−β−/−	(α1−/−, α2+/+, Δα1−/−, Δα2+/+, β−/−)	Survive (small)	↓	↓	Delayed maturation	Normal/low BMD	61,62
TRα−/−β−/−	(α1−/−, α2−/−, Δα1+/+, Δα2+/+, β−/−)	Die	↓	↓	Impaired growth plate		57
TRα1+/m*β+/−	(α1+/m, α2−/−, Δα1+/m, Δα2+/+, β+/−)	Survive	↑	↑	Delayed maturation	Osteosclerosis	16
TRα1+/m*β−/−	(α1+/m, α2−/−, Δα1+/m, Δα2+/+, β−/−)	Survive	↗	↗	Advanced maturation	Osteoporosis	16
TRα0/0β−/−	(α1−/−, α2−/−, Δα1−/−, Δα2−/−, β−/−)	Die (hypothermia)	↓	↓	Delayed maturation		55

m* = dominant negative mutation; TH† = thyroid hormone; Refs.# = References.

Indeed, the cellular and molecular basis of the bone maturation by thyroid hormone is clear. Thyroid hormones promote chondrocyte differentiation by modifying the set-point of the Indian hedgehog (Ihh)–bone morphogenic protein (BMP)–parathyroid hormone-related protein (PTHrP) long feedback loop.[17] They also induce cyclin-dependent kinase inhibitors to regulate the G1-S cell cycle checkpoint[18] and, in addition, increase alkaline phosphatase and proangiogenic factor collagenase 3.[19,20] In this manner, thyroid hormones can promote cartilage anabolism during growth and development. Loss of TRs thus affects predominantly growth and maturation.

EFFECTS OF TSH ON BONE

Despite evidence linking thyroid hormones to skeletal morphogenesis and growth, clinically, the adverse effects of thyrotoxicosis are due to excess bone resorption.[21–29] Specifically, patients with subclinical hyperthyroidism, that is, those having normal thyroid hormone levels and low TSH levels, display excessive resorption and bone loss.[2–5] These points of incongruence prompted us to challenge the notion that thyroid hormone was the sole hormone responsible for altering bone metabolism in thyrotoxic states.

In 2003, we discovered that TSH (thyrotropin) can directly suppress both components of bone remodeling, bone formation and bone resorption.[7,8] Two key lines of genetic evidence provided proof that TSH directly regulates bone remodeling.[8] First, mice heterozygous for a deletion of the TSHR gene were osteopenic, despite normal thyroid hormone levels. Second, repletion of homozygous hypothyroid TSHR$^{-/-}$ mice with thyroid hormone extracts failed to correct the osteopenia. These genetic observations suggested that TSHRs had dominant effects on bone remodeling that were exerted independently of thyroid hormone levels. On further characterizing the *in vivo* bone phenotype of these animals, it was apparent that they had high-turnover osteoporosis with focal osteosclerosis, which is typically associated with uncoupling between bone formation and resorption.[8]

Ex vivo studies indicated that TSH affects osteoclast and osteoblast function and formation. We found that TSH decreased osteoblast differentiation and collagen synthesis.[8] RANKL-induced osteoclastogenesis was also inhibited by recombinant TSH at two signal transduction steps involving inhibitor IκBα and JNK.[8] This, in turn, led to depressed nuclear levels of the proosteoclastogenic transcription factors c-*jun* and p65. The depressed c-*jun* levels impaired tartrate-resistant acidic phosphatase (TRAP) and cathepsin K gene transcription, which are critical to osteoclastic bone resorption.[30–33]

Further experiments defined TNF-α as the critical cytokine mediating the downstream effects of TSH.[34,35] Specifically, TSH inhibited cytokine-induced TNF-α production in the bone marrow microenvironment. Crossing the TSHR$^{-/-}$ animals to TNF-α$^{-/-}$ animals showed that the effect of TSHR inactivation on bone could be reversed, thus providing *in vivo* evidence for a role of TNF-α in the osteopenia associated with TSH receptor deficiency.[34]

More recently, studies by Sampath and colleagues have demonstrated that TSH suppresses bone remodeling in rats, and by doing so, restores the bone loss induced upon ovariectomy.[36] Our own data in mice support the premise that recombinant human TSH can prevent ovariectomy-induced bone loss (unpublished). Similarly, a single injection of human TSH given to postmenopausal women reduced markers of bone turnover to premenopausal levels within 2 days: this was followed by full recovery.[37] It is notable that, where measured, serum thyroid hormone levels did not change, again confirming a direct effect of TSH on bone remodeling *in vivo*.

CLINICAL RELEVANCE

In line with the discovery that TSH can directly regulate bone metabolism in a thyroid hormone-independent manner, a 2004 review of population-based studies by Murphy and Williams concluded that endogenous and exogenous TSH suppression, but not thyroid hormone therapy *per se*, was associated with an increased fracture risk.[1] This adverse effect did not necessarily correlate with decreased bone mineral density (BMD). A high fracture risk in the face of a normal bone density is not surprising and is usually an accompaniment of high dose glucocorticoid therapy, immunosuppression, and acute immobilization.[38] Consistent with this, two cross-sectional studies, which did not report fracture risk, showed no adverse effect of thyroid hormone suppression therapy on biochemical or densitometric parameters of skeletal integrity.[39,40]

In an impressive epidemiological study, Bauer and colleagues used serum TSH levels to predict the risk of fracture in hyperthyroid postmenopausal women.[41] Most impressive was the increase in vertebral and hip fracture risk by 4.5- and 3.3-fold, respectively, when serum TSH was <0.1 mIU/mL. No correlations between thyroid hormone levels and fracture risk were observed. Rather than being a causal relationship, this correlation could arise because TSH suppression is more sensitive index of thyroid hormone excess than thyroid hormone levels *per se*. Subjects with TSH receptor mutations reveal subclinical hypothyroidism with slightly high levels of TSH.[42] Thus, in the light of genetic and pharmacological evidence for direct effects of TSH on bone, it is highly likely that a low TSH contributes to hyperthyroid bone loss.

Because of the strong correlation between low TSH levels and a high fracture risk, which appears to be dissociable from long-term decrements in BMD,[43] we suggest maintaining TSH levels during replacement therapy to above 1 mU/mL, unless there is a clinical rationale for TSH suppression as in thyroid cancer patients. In these patients, admittedly without clinical evidence of efficacy, we propose the empiric use of an oral bisphosphonate to prevent the high-turnover osteoporosis and associated fracture risk, which appears to be highly correlated to a TSH level of <0.1 mU/mL.

ACKNOWLEDGMENTS

The studies on TSH and bone were supported by the National Institutes of Health and the Department of Veterans Affairs (to E.A., M.Z., and T.F.D.), as well as the Medical Science Training Program (J.I.).

REFERENCES

1. MURPHY, E. & G.R. WILLIAMS. 2004. The thyroid and the skeleton. Clin. Endocr. **61:** 285–298.
2. DE MENIS, E *et al.* 1992. Bone turnover in overt and subclinical hyperthyroidism due to autonomous thyroid adenoma. Horm. Res. **37:** 217–220.
3. KISAKOL, G. *et al.* 2003. Bone and calcium metabolism in subclinical autoimmune hyperthyroidism and hypothyroidism. Endocr. J. **50:** 657–661.
4. FOLDES, J. *et al.* 1993. Bone mineral density in patients with endogenous subclinical hyperthyroidism: is this thyroid status a risk factor for osteoporosis? Clin. Endocr. **39:** 521–527.
5. GURLEK, A. & O. GEDIK. 1999. Effect of endogenous subclinical hyperthyroidism on bone metabolism and bone mineral density in premenopausal women. Thyroid **9:** 539–543.
6. BASSETT, J.H. & G.R. WILLIAMS. 2003. The molecular actions of thyroid hormone in bone. Trends Endocrinol. Metab. **14:** 356–353.
7. MARIANS, R.C. *et al.* 2002. Defining thyrotropin-dependent and -independent steps of thyroid hormone synthesis by using thyrotropin receptor-null mice. Proc. Natl. Acad. Sci. USA **99:** 15776–15781.
8. ABE, E. *et al.* 2003. TSH is a negative regulator of skeletal remodeling. Cell **115:** 151–162.
9. MUNDY, G.R. *et al.* 1976. Direct stimulation of bone resorption by thyroid hormones. J. Clin. Invest. **58:** 529–534.
10. KANATANI, M. *et al.* 2004. Thyroid hormone stimulates osteoclast differentiation by a mechanism independent of RANKL-RANK interaction. J. Cell. Physiol. **201:** 17–25.
11. ALLAIN, T.J. *et al.* 1992. Tri-iodothyronine stimulates rat osteoclastic bone resorption by an indirect effect. J. Endocrinol. **133:** 327–331.
12. BRITTO, J.M. *et al.* 1994. Osteoblasts mediate thyroid hormone stimulation of osteoclastic bone resorption. Endocrinology **134:** 169–176.
13. KIM, C.H. *et al.* 1999. Thyroid hormone stimulates basal and interleukin (IL)-1-induced IL-6 production in human bone marrow stromal cells: a possible mediator of thyroid hormone-induced bone loss. J. Endocrinol. **160:** 97–102.
14. ABU, E.O. *et al.* 2000. The localization of thyroid hormone receptor mRNAs in human bone. Thyroid **10:** 287–293.
15. BASSETT, J.H. *et al.* 2007. Thyroid hormone excess rather than thyrotropin deficiency induces osteoporosis in hyperthyroidism. Mol. Endocrinol. **21:** 1095–1107.

16. BASSETT, J.H. *et al.* 2007. Thyroid status during skeletal development determines adult bone structure and mineralization. Mol. Endocrinol. May 8; [Epub ahead of print].

17. STEVENS, D.A. *et al.* 2000. Thyroid hormones regulate hypertrophic chondrocyte differentiation and expression of parathyroid hormone-related peptide and its receptor during endochondral bone formation. J. Bone Miner. Res. **15:** 2431–2442.

18. BALLOCK, R.T. *et al.* 2000. Expression of cyclin-dependent kinase inhibitors in epiphyseal chondrocytes induced to terminally differentiate with thyroid hormone. Endocrinology **141:** 4552–4557.

19. ISHIKAWA, Y. *et al.* 1998. Thyroid hormone inhibits growth and stimulates terminal differentiation of epiphyseal growth plate chondrocytes. J. Bone Miner. Res. **13:** 1398–1411.

20. HIMENO, M. *et al.* 2002. Impaired vascular invasion of Cbfa1-deficient cartilage engrafted in the spleen. J. Bone Miner. Res. **17:** 1297–1305.

21. TOH, S.H. *et al.* 1985. Effect of hyperthyroidism and its treatment on bone mineral content. Arch. Intern. Med. **145:** 883–886.

22. FRANKLYN, J. *et al.* 1994. Bone mineral density in thyroxine treated females with or without a previous history of thyrotoxicosis. Clin. Endocrinol. (Oxf.) **41:** 425–432.

23. DIAMOND, T. *et al.* 1994. Thyrotoxic bone disease in women: a potentially reversible disorder. Ann. Intern. Med. **120:** 8–11.

24. CUMMINGS, S.R. *et al.* 1995. Risk factors for hip fracture in white women. Study of Osteoporotic Fractures Research Group. N. Engl. J. Med. **332:** 767–773.

25. GRANT, D.J. *et al.* 1995. Is previous hyperthyroidism still a risk factor for osteoporosis in post-menopausal women? Clin. Endocrinol. (Oxf.) **43:** 339–345.

26. JODAR, E. *et al.* 1997. Antiresorptive therapy in hyperthyroid patients: longitudinal changes in bone and mineral metabolism. J. Clin. Endocrinol. Metab. **82:** 1989–1994.

27. VESTERGAARD, P. *et al.* 2000. Fracture risk in patients treated for hyperthyroidism. Thyroid **10:** 341–348.

28. SERRACLARA, A. *et al.* 2001. Bone mass after long-term euthyroidism in former hyperthyroid women treated with (131)I influence of menopausal status. J. Clin. Densitom. **4:** 249–255.

29. VESTERGAARD, P. & L. MOSEKILDE. 2003. Hyperthyroidism, bone mineral, and fracture risk-a meta-analysis. Thyroid **13:** 585–593.

30. IKEDA, F. *et al.* 2004. Critical roles of c-Jun signaling in regulation of NFAT family and RANKL-regulated osteoclast differentiation. J. Clin. Invest. **114:** 475–484.

31. SHEVDE, N.K. *et al.* 2000. Estrogens suppress RANK ligand-induced osteoclast differentiation via a stromal cell independent mechanism involving c-Jun repression. Proc. Natl. Acad. Sci. USA **97:** 7829–7834.

32. SRIVASTAVA, S. *et al.* 2001. Estrogen decreases osteoclast formation by down-regulating receptor activator of NF-kappa B ligand (RANKL)-induced JNK activation. J. Biol. Chem. **276:** 8836–8840.

33. KAWAIDA, R. *et al.* 2003. Jun dimerization protein 2 (JDP2), a member of the AP-1 family of transcription factor, mediates osteoclast differentiation induced by RANKL. J. Exp. Med. **197:** 1029–1035.

34. HASE, H. *et al.* 2006. TNFalpha mediates the skeletal effects of thyroid-stimulating hormone. Proc. Natl. Acad. Sci. USA **103:** 12849–12854.

35. SIMSEK, G. *et al.* 2003. Osteoporotic cytokines and bone metabolism on rats with induced hyperthyroidism; changes as a result of reversal to euthyroidism. Chin. J. Physiol. **46:** 181–186.
36. SAMPATH, T.K. *et al.* 2007. Thyroid-stimulating hormone restores bone volume, microarchitecture, and strength in aged ovariectomized rats. J. Bone Miner. Res. **22:** 849–859.
37. MAZZIOTTI, G. *et al.* 2005. Recombinant human TSH modulates in vivo C-telopeptides of type-1 collagen and bone alkaline phosphatase, but not osteoprotegerin production in postmenopausal women monitored for differentiated thyroid carcinoma. J. Bone Miner. Res. **20:** 480–486.
38. EPSTEIN, S. *et al.* 2003. Disorders associated with acute rapid and severe bone loss. J. Bone Miner. Res. **18:** 2083–2094.
39. HEIJCKMANN, A.C. *et al.* 2005. Hip bone mineral density, bone turnover and risk of fracture in patients on long-term suppressive L-thyroxine therapy for differentiated thyroid carcinoma. Eur. J. Endocrinol. **153:** 23–29.
40. REVERTER, J. L *et al.* 2005. Lack of deleterious effect on bone mineral density of long-term thyroxine suppressive therapy for differentiated thyroid carcinoma. Endocr. Relat. Cancer **12:** 973–981.
41. BAUER, D.C. *et al.* 2001. Risk for fracture in women with low serum levels of thyroid-stimulating hormone. Ann. Intern. Med. **134:** 561–568.
42. ALBERTI, L. *et al.* 2002. Germline mutations of TSH receptor gene as cause of nonautoimmune subclinical hypothyroidism. J. Clin. Endocrinol. Metab. **87:** 2549–2555.
43. KIM, D.J. *et al.* 2006. Low normal TSH levels are associated with low bone mineral density in healthy postmenopausal women. Clin. Endocrinol. (Oxf.) **64:** 86–90.
44. BAUER, D.C. *et al.* 2001. Study of Osteoporotic Fractures Research Group. Risk for fracture in women with low serum levels of thyroid-stimulating hormone. Ann. Intern. Med. **134:** 561–568.
45. SEELEY, D.G. *et al.* 1996. Predictors of ankle and foot fractures in older women. The Study of Osteoporotic Fractures Research Group. J. Bone Miner. Res. **11:** 1347–1355.
46. KUNG, A.W. & S.S. YEUNG. 1996. Prevention of bone loss induced by thyroxine suppressive therapy in postmenopausal women: the effect of calcium and calcitonin. J. Clin. Endocrinol. Metab. **81:** 1232–1236.
47. MCDERMOTT, M.T. *et al.* 1995. A longitudinal assessment of bone loss in women with levothyroxine-suppressed benign thyroid disease and thyroid cancer. Calcif. Tissue Int. **56:** 521–525.
48. PIOLI, G. *et al.* 1992. Longitudinal study of bone loss after thyroidectomy and suppressive thyroxine therapy in premenopausal women. Acta. Endocrinol. (Copenh.) **126:** 238–242.
49. RIBOT, C. *et al.* 1990. Bone mineral density and thyroid hormone therapy. Clin. Endocrinol. (Oxf.) **33:** 143–153.
50. GUO, C.Y. *et al.* 1997. Longitudinal changes of bone mineral density and bone turnover in postmenopausal women on thyroxine. Clin. Endocrinol. (Oxf.) **46:** 301–307.
51. ROSS, D.S. 1993. Bone density is not reduced during the short-term administration of levothyroxine to postmenopausal women with subclinical hypothyroidism: a randomized, prospective study. Am. J. Med. **95:** 385–388.

52. GARTON, M *et al.* 1994. Bone mineral density and metabolism in premenopausal women taking L-thyroxine replacement therapy. Clin. Endocrinol. (Oxf.) **41:** 747–755.

53. FRAICHARD, A. *et al.* 1997. The T3R alpha gene encoding a thyroid hormone receptor is essential for post-natal development and thyroid hormone production. EMBO J. **16:** 4412–4420.

54. WIKSTROM, *et al.* 1998. Abnormal heart rate and body temperature in mice lacking thyroid hormone receptor alpha 1. EMBO J. **17:** 455–461.

55. GAUTHIER, K. *et al.* 2001. Genetic analysis reveals different functions for the products of the thyroid hormone receptor alpha locus. Mol. Cell. Biol. **21:** 4748–4760.

56. MACCHIA, P.E. *et al.* 2001. Increased sensitivity to thyroid hormone in mice with complete deficiency of thyroid hormone receptor alpha. Proc. Natl. Acad. Sci. USA **98:** 349–354.

57. GAUTHIER, K. *et al.* 1999. Different functions for the thyroid hormone receptors TRalpha and TRbeta in the control of thyroid hormone production and post-natal development. EMBO J. **18:** 623–631.

58. FORREST, D. *et al.* 1996. Recessive resistance to thyroid hormone in mice lacking thyroid hormone receptor beta: evidence for tissue-specific modulation of receptor function. EMBO J. **15:** 3006–3015.

59. ABEL, E.D. *et al.* 1999. Divergent roles for thyroid hormone receptor beta isoforms in the endocrine axis and auditory system. J. Clin. Invest. **104:** 291–300.

60. NG, *et al.* 2001. A thyroid hormone receptor that is required for the development of green cone photoreceptors. Nat. Genet. **27:** 94–98.

61. KINDBLOM, J.M. *et al.* 2005. Increased adipogenesis in bone marrow but decreased bone mineral density in mice devoid of thyroid hormone receptors. Bone **36:** 607–616.

62. GOTHE, S. *et al.* 1999. Mice devoid of all known thyroid hormone receptors are viable but exhibit disorders of the pituitary-thyroid axis, growth, and bone maturation. Genes Dev. **13:** 1329–1341.

Hematopoietic Stem Cell Trafficking

Regulated Adhesion and Attraction to Bone Marrow Microenvironment

SIMÓN MÉNDEZ-FERRER AND PAUL S. FRENETTE

Departments of Medicine and Gene and Cell Medicine, Immunology Institute, and Black Family Stem Cell Institute, Mount Sinai School of Medicine, New York, New York, USA

ABSTRACT: Hematopoiesis takes place preferentially within bone cavities, suggesting that bone-derived factors contribute to blood formation. Hematopoietic stem and progenitor cells (HSPCs) can be mobilized from the bone marrow parenchyma to the circulation by various agonists whose common downstream action leads to alteration in the expression or function of the chemokine CXCL12 and adhesion molecules mediating migration. Granulocyte colony-stimulating factor (G-CSF), the most prevalent drug used to mobilize HSPCs, dramatically suppresses osteoblast function. Recent studies suggest that G-CSF-mediated suppression requires signals from the sympathetic nervous system (SNS). This review summarizes emerging concepts thought to contribute to stem cell migration.

KEYWORDS: hematopoietic stem and progenitor cells; adhesion; homing; mobilization; sympathetic nervous system; stem cell niche

INTRODUCTION

Hematopoietic stem and progenitor cells (HSPCs) reside in specific niches that regulate their survival, proliferation, self-renewal, or differentiation in the bone marrow (BM). The concept of microenvironments supporting the self-renewal of stem cells and differentiation toward specific lineages was introduced at least 35 years ago.[1] Schofield coined the term "niche" to describe specific areas where stem cells can self-renew.[2] Searching the location of stem cells in the BM, initial studies have found a higher concentration of colony-forming units in spleen (CFU-S) near the bone surface than in the central longitudinal axis of the femoral cavity.[3] Further studies then revealed that fluorescently labeled Lin[neg] cells tend to localize predominantly in the endosteum 15 h after transplantation.[4] Confocal microscopy imaging with

Address for correspondence: Paul Frenette, M.D., Department of Medicine, Mount Sinai School of Medicine, One Gustave L Levy Place, Box 1079, New York, NY, 10029. Voice: +1-212-659-9693; fax: +1-212-849-2574.

paul.frenette@mssm.edu or simon.mendez-ferrer@mssm.edu

Ann. N.Y. Acad. Sci. 1116: 392–413 (2007). © 2007 New York Academy of Sciences.
doi: 10.1196/annals.1402.086

lineage staining and BrdU retention studies has revealed that stem cells associate closely with spindle-shaped N-cadherin-expressing osteoblasts that line the endosteal bone,[5] and that the stem cell pool size expands in mice genetically engineered to exhibit enhanced osteoblastic activity.[5,6] Ablation of osteoblasts using a thymidine kinase suicide approach has revealed a dramatic loss in BM cellularity upon ganciclovir administration, suggesting again a role for the osteoblast in the maintenance of hematopoiesis.[7] However, further studies suggest that other areas in the BM may contain stem cells. For example, the SLAM marker combination ($CD150^+CD48^-$) has localized stem cells in association with sinusoids.[8] Whether these niches contain different stem cell subsets is unclear. The presence of stem cells near blood vessels may explain why circulating stem cells can be elicited so rapidly (within minutes) by certain agonists, although this has not yet been demonstrated.

Circulating HSPC activity was inferred many years ago from experiments in which lethally irradiated rats could regain normal peripheral blood counts and BM cellularity when in parabiosis with unirradiated rats.[9,10] In normal individuals, the continuous trafficking of HSPCs between the BM and blood compartments likely fills empty or damaged niches and contributes to the maintenance of normal hematopoiesis.[11] The ability to enhance these physiological mechanisms and enforce the egress of HSPCs from the BM into the blood (termed "mobilization") has proven invaluable for harvesting stem cells for clinical BM transplantation procedures. This process can be triggered by various, structurally distinct, agonists at variable kinetics.[12,13] Among these, granulocyte colony-stimulating factor (G-CSF) has emerged as the most powerful and commonly used drug in the clinic and has served as the prototype for gaining mechanistic insights into the phenomenon. Several excellent reviews have summarized recent gain of knowledge in hematopoietic stem cell niches.[14-18] Here we will review niche components that are thought to participate in hematopoietic stem cell trafficking.

CELLULAR RECEPTORS AND MATRIX COMPONENTS CONTRIBUTING TO HSPC MIGRATION

Selectins and Their Ligands

Selectins are membrane-bound C-type lectins that bind to cell-surface glycosylated ligands.[19,20] The three members of the selectin family of adhesion molecules mediate interactions between endothelial cells and HSPCs. P-selectin (encoded by the *SELP* gene) and E-selectin (*SELE*) are expressed on endothelial cells, whereas L-selectin (*SELL*) is expressed on HSPCs. Several functional selectin receptors have been described, including P-selectin glycoprotein ligand-1 (PSGL-1/*SELPLG*), which can bind all three selectins[21-24]; CD34[25] and the podocalyxin-like protein 1 (PCLP1),[26] which can interact

TABLE 1. Potential pathways contributing to HSPCs adhesion in BM microenvironment

HSPC receptor	BM ligand	Function	Reference
CXCR4	CXCL12/SDF-1	mobilization	118,124,127,128
PSGL-1/CD162	E-,P-,L-Sel/CD62E,P,L	homing	38,39,45
$\alpha 4\beta 1$/VLA-4	VCAM-1/CD106, fibronectin	homing, mobilization	50,51,116
$\alpha 5\beta 1$/VLA-5	fibronectin	homing	59,60
$\alpha 4\beta 7$/LPAM-1	MAdCAM-1	homing	48
$\alpha 6\beta 1$, $\alpha 6\beta 4$	Laminin-8,10	homing	64
CD44/Pgp-1/ ECMRIII	HA	homing, mobilization	50,79,81
N-cadherin	N-cadherin	lodgement?	5,85
c-kit	kit ligand/SCF	lodgement, mobilization	90–92,121
Tie-2	Ang-1	lodgement	98
OPN	integrins, CD44, fibronectin, collagen, Ca^{2+}	lodgement	105
CaR	Ca^{2+}	lodgement	108

with L-selectin; ESL-1[27,28] and CD44,[29] which bind to E-selectin (TABLE 1). L-selectin expression is higher on the more primitive population of mobilized CD34[+] cells,[30] and the degree of hematopoietic recovery following chemotherapy correlates with the expression of L-selectin on CD34[+] human HSPCs,[31] although L-selectin does not appear to contribute to HSPC interactions in BM microvessels in mice.[32] Selectin counter-receptor activity is borne by glycoproteins that undergo specific post-translational modifications, typically O- or N-linked oligosaccharides capped with the sialyl Lewis × moiety, an $\alpha 2,3$-sialylated, $\alpha 1,3$-fucosylated tetrasaccharide synthesized by specific glycosyltransferases.[33]

Loss-of-function studies have revealed a critical role of selectins in hematopoiesis and HSPCs trafficking. Mice deficient in endothelial selectins exhibit severe leukocytosis, elevated cytokine levels, and abnormal hematopoiesis.[34] The administration of anti-P-selectin antibody to E-selectin-deficient mice produces changes similar to those present in mice lacking both selectins from birth, suggesting that increased circulating HSPCs likely arise from selectin function blockade and not from the secondary abnormalities related to compromised leukocyte trafficking.[34,35] However, the exact mechanism responsible for the leukocytosis remains unclear. One study has suggested that the leukocytosis in adhesion molecule-deficient mice results from the disruption of an IL-23/IL-17-dependent feedback loop that controls the number of leukocytes that have emigrated to the extravascular space.[36] On the other hand, other studies suggest that the loss of myeloid homeostasis may result from a deficit in suppressive signaling mediated by selectins in the BM microenvironment.[24,37] Selectins and their ligands clearly participate in the initial interactions of progenitors with BM endothelial cells, the first step of "homing" following transplantation. Indeed, homing of HSPCs to the BM is impaired in animals deficient in endothelial selectins,[38] and human adult CD34[+] cell rolling on BM endothelium and retention in the BM compartment are drastically reduced in endothelial selectin-deficient NOD/SCID mice.[39] Treatment with fucoidan, a sulfated glycan able to inhibit P- and L-selectins *in vivo*, can induce a rapid HSPCs mobilization response[35,40] (see below). A critical role for fucosylation in the synthesis of selectin ligands was revealed using mice deficient in leukocyte fucosyltransferases, which show a phenotype similar to mice null for endothelial selectins.[41,42] Indeed, a large subset (30%) of circulating CD34[+] cells from cord blood, enriched in CD34[+] CD38[low/−] stem cells, expresses a nonfunctional form of PSGL-1,[39] due to reduced fucosylation.[43,44] In fact, enforced fucosylation corrects the abnormality in nonbinding cells and is sufficient to generate selectin ligands that enhance the initial interactions with microvessels,[43] although whether enforced fucosylation enhances homing[44] or not[43] is unclear. E-selectin ligands cooperate with $\alpha 4$ integrins rather than P-selectin ligands for HSPCs homing, underscoring a major difference between the adhesion mechanisms of mature myeloid cells and immature hematopoietic cells.[45] It is interesting that E-selectin, unlike P-selectin or PECAM-1, has been reported to be expressed in discrete microdomains that co-localize with the expression of the chemokine CXCL12 on the endothelium of the BM in areas of HSPC entry.[46]

Integrins

Integrins are receptors that interact with extracellular matrix proteins and constitute a family exclusively expressed in all metazoan phyla. Integrins are $\alpha\beta$ heterodimers whose polypeptide chains, largely exposed to the extracellular space, connect to the intracellular milieu through short cytoplasmic domains. Integrins and their ligands play various roles in development, leukocyte trafficking, immune response, hemostasis and cancer (see Ref. 47 for a review and TABLE 1). Two $\alpha 4$ integrins are expressed on adult HSPCs: $\alpha 4 \beta 1$ integrin (also called very late antigen-4, or VLA-4), and $\alpha 4 \beta 7$ (named lymphocyte-Peyer patch adhesion molecule-1 or LPAM-1 too).[48] Both integrins bind to VCAM-1/CD106[49–51] and fibronectin.[52] In addition, $\alpha 4 \beta 7$ can interact with the mucosal addressin cell adhesion molecule-1 (MAdCAM-1).[48] The αIIb integrin (CD41), also expressed in fetal murine HSPCs,[53] identifies the onset of primitive and definitive hematopoiesis in the murine embryo, and persists on some stem and progenitor cell populations in the fetal liver and adult marrow.[54] Recent studies suggest that the $\beta 3$ integrin subunit is also expressed in quiescent HSCs.[55] The use of chimeric mice generated with $\beta 1$ integrin-deficient embryonic stem cells has revealed that HSCs lacking β1-integrins can form and differentiate but cannot colonize the fetal liver.[56] Functional blockade using antibodies[38,57] or conditional deletion of α4 integrins[58] has shown reduced homing capability of HSPCs. Homing into the BM is further compromised in mice with additional deletion of E-selectin, indicating that E-selectin ligands cooperate with α4 integrins for HSPC homing.[45] The α5β1 integrin (VLA-5) is also expressed in murine and human HSPCs, where its presence is required for engraftment in NOD/SCID mice.[59,60] The β2 integrin leukocyte function antigen-1 (LFA-1, αLβ2) can also synergize with α4β1 integrins controlling HSPC adhesion.[61] In addition, functional blockade using antibodies against LFA-1 and macrophage antigen-1 (Mac-1, αMβ2) can enhance G-CSF-induced mobilization,[62] in agreement with a role for Mac-1 in the retention of progenitors in the BM.[63] The laminin receptors, α6 integrins (α6 heterodimerizes with the β1 or β4 chains) also collaborate with α4 integrins in HSPC homing.[64] Integrin activity can be switched on by inside-out signaling, allowing the integrin to bind to extracellular ligands that, in turn, can transduce intracellular signals.[47,65] A number of hematopoietic growth factors secreted in the stem cell niche, such as G-CSF, granulocyte-macrophage colony-stimulating factor (GM-CSF), kit ligand,[66,67] Flk-2/Flt-3/CD135 ligand,[68] hepatocyte growth factor (HFG),[69] thrombopoietin,[70] and chemokines, like CXCL12,[60,71,72] are able to activate integrins. These studies suggest that integrins, through their multiple cellular and extracellular binding partners and elicited signals, have versatile functions in HSPC migration into, within, and out of the marrow microenvironment.

CD44

CD44, also known as Pgp-1, ECMRIII, or Hermes antigen, is a polymorphic integral membrane glycoprotein that binds to several extracellular matrix components, such as hyaluronic acid,[73] fibronectin, collagen,[74] and also transmembrane receptors like E-selectin.[29,75] Different studies have suggested that CD44 may play a role in hematopoiesis, since the addition of function-blocking antibodies prevented myelopoiesis and lymphopoiesis in murine and human long-term BM cultures.[74,76,77] CD44-mediated adhesion of human CD34$^+$ cells to hyaluronic acid can be modulated by various activating molecules, including phorbol 12-myristate 13-acetate (PMA) and cytokines, such as kit ligand, interleukin-3 (IL-3) and GM-CSF,[78] or chemokines, like CXCL12.[79] Studies using mice with abolished expression of all known isoforms of CD44 by targeting exons encoding the invariant N terminus region of the molecule, revealed defective G-CSF-induced mobilization in CD44-deficient animals.[80] The use of function-blocking antibodies against CD44 also decreased homing[79] and induced mobilization of HSPCs.[50] Thus, combining CD44 blockade with G-CSF administration may improve the reconstitutive capacity of mobilized HSPCs.[81] However, other studies have revealed normal hematopoiesis and homing in CD44-deficient mice.[82] The discrepancies in published data on the role of CD44 may originate from methodological differences and from the various glycoforms of CD44 that may provide distinct ligand specificities.

N-Cadherin

The family of cadherins comprises cell-surface glycoproteins responsible for Ca^{2+}-dependent homophilic cell–cell adhesion. Among members of this family, the vascular endothelial cadherin (VE-cadherin) is expressed in fetal HSCs[83] but only N-cadherin is expressed in both osteoblasts and quiescent HSCs.[5] It has been suggested that N-cadherin may participate in bone formation since targeted expression of a dominant negative truncated form of N-cadherin delays acquisition of peak bone mass in mice and retards osteoblast differentiation.[84] However, the role of N-cadherin in HSC biology has been challenged in recent studies showing that N-cadherin-expressing cells in the BM do not contain stem cell activity.[85]

c-Kit

The protooncogene c-kit belongs to the class III family of receptor tyrosine kinases. After binding kit ligand/stem cell factor (SCF), it activates an intracellular signaling pathway that is critical for proliferation, migration, and differentiation of stem cells belonging to the melanogenic, gametogenic, and hematopoietic lineages (see Ref. 86 for a review). SCF regulates adhesion of HSPCs through its chemotactic and chemokinetic activity,[87] and by promoting adhesion via activation of the integrins $\alpha 4\beta 1$ and $\alpha 5\beta 1$.[66,88] The infusion of SCF has been shown to induce or enhance HSPC mobilization in preclinical and clinical studies (reviewed in Ref. 89). The administration of the soluble portion of c-kit in mice has also been shown to enhance G-CSF-induced mobilization.[90] Remarkably, c-kit seems to be important for mobilization, since mice carrying inactivating mutations of this receptor mobilize poorly.[91,92]

Other Factors Linking HSPCs and Bone: Tie-2, Osteopontin, Calcium-Sensing Receptor

Tie-2, a receptor tyrosine kinase containing Ig-like loops and epidermal growth factor domains, is a receptor tyrosine kinase expressed on endothelial cells and HSCs.[93,94] Although Tie-2-deficient cells migrate normally from the fetal liver to the BM during ontogeny, they fail to be maintained in the adult BM microenvironment.[95] Angiopoietin-1 (Ang-1), an angiogenic factor that functions as a soluble ligand for Tie-2,[96] promotes the adhesion of Tie-2$^+$ HSCs to BM fibronectin through integrins,[97] and the quiescent state in Tie-2$^+$ HSCs bound to Ang-1$^+$ bone-lining osteoblasts.[98]

Osteopontin (OPN), a multidomain phosphorylated glycoprotein, is produced by osteoblasts and osteoclasts in the BM.[99] OPN inhibits hydroxyapatite formation.[100,101] Although bone development proceeds normally in $Opn^{-/-}$ mice,[102] the reduction in trabecular bone volume after ovariectomy is decreased by sixfold in $Opn^{-/-}$ animals, suggesting that OPN may contribute to postmenopausal osteoporosis in women.[103] Further studies showed that OPN is directly required for bone resorption activated by PTH-RANKL axis via increasing the number of osteoclasts in the microenvironment of bone.[104] OPN was shown to contribute to HSC quiescence and trans-marrow migration toward the endosteal region, as demonstrated by the markedly aberrant distribution and enhanced proliferation of HSCs in Opn$^{-/-}$ mice after transplantation.[105,106] OPN can bind a number of integrins, CD44, extracellular matrix proteins (such as fibronectin and collagen) and calcium (recently reviewed in Ref. 107). The calcium-sensing receptor ($Casr$) was proposed to mediate the attachment of HSCs to the endosteal niche since $Casr^{-/-}$ mice exhibited aberrant distribution of HSCs in the endosteum due to defective adhesion to the extracellular matrix protein collagen I.[108] The attraction of HSCs to areas of free Ca^{2+} suggests that osteoblast-associated HSCs may form a functional unit with osteoclasts in areas of bone resorption, although this idea remains speculative at present.

MECHANISMS INVOLVED IN G-CSF-INDUCED MOBILIZATION

G-CSF, the most commonly used agent in the clinical arena, can elicit robust mobilization in 5–10 days. Owing to its widespread use, mobilization by G-CSF has served as the prototype to obtain mechanistic insights into this phenomenon. Mice deficient in the G-CSF receptor ($Csf3r^{-/-}$) have reduced number of HSPCs in the BM and are unresponsive to G-CSF stimulation.[109,110] However, $Csf3r^{-/-}$ HSPCs can be mobilized by G-CSF in chimeric mice that harbor mixtures of $Csf3r^{+/+}$ and $Csf3r^{-/-}$ hematopoietic cells, suggesting the contribution of "*trans*-acting" signals.[110] Further studies (outlined below) have suggested that the release of soluble proteases and norepinephrine in the BM microenvironment may provide such signals.

Proteolytic Cleavage of Niche Components

G-CSF has been reported to induce neutrophils proliferation, activation, and degranulation,[111,112] with the subsequent release of serine proteases, such as elastase and cathepsin G, in the BM environment.[112,113] In addition, natural inhibitors of serine proteases are downregulated in the BM following G-CSF administration.[114] The neutrophil serine proteases can cleave VCAM-1 *in vitro*[113,115] and seem to be the only proteases that cleave it *in vivo*.[116] However, VCAM-1 cleavage is not required for mobilization since VCAM-1 is not cleaved from stromal cells of the BM of mice lacking both elastase and cathepsin G, but these mice mobilize HSPCs normally. Elastase and cathepsin G have also been reported to cleave murine and human c-kit,[117] CXCL12, and CXCR4.[118]

Other proteolytic enzymes may contribute to HSPC mobilization. Matrix metalloproteinase-9 (MMP-9) is released rapidly after IL-8 administration in monkeys, and the subsequent HSPC mobilization can be prevented by pretreatment with an inhibitory anti-gelatinase B antibody, suggesting that MMP-9 is a mediator of IL-8-induced mobilization of HSPCs.[119] However, unlike the presence of neutrophils, the expression of MMP-9 is dispensable for IL-8-induced mobilization in mice.[120] Studies with mice show that MMP-9 can release soluble kit ligand and either be critical[121] or dispensable[122,123] for G-CSF-induced mobilization. CXCL12 and human CXCR4 can also be cleaved by MMP-9.[118,124] Another proteolytic enzyme, the membrane-bound extracellular peptidase CD26 (DPPIV/dipeptidylpeptidase IV) is also able to cleave the functional N terminus of CXCL12 and participate in G-CSF-induced mobilization.[125,126] Thus, these data suggest that protease cleavage of CXCL12 and other retention signals are involved in mobilization. However, the role of many of these proteases has been challenged by the normal mobilization found in mice lacking virtually all serine protease activity (dipeptidylpeptidase-I-deficient), even when combined with a broad inhibitor of metalloproteinases.[116] Therefore, functional redundancy with other proteases and/or other mechanisms must be involved.

CXCL12-CXCR4 Chemotaxis

During mobilization, the levels of CXCL12 sharply drop in the BM[118,124,127] and bone.[128] Interestingly, transgenic mice carrying truncated forms of *Csf3r* display variable responses to G-CSF, but the decrease in BM CXCL12 protein expression closely correlates with the degree of HSPC mobilization.[127] The data currently available suggest that at least two different mechanisms disrupt the CXCL12-CXCR4 chemotactic axis during HSC egress. First, CXCL12 reductions coincide with the induction of proteolytic activity in the BM. As mentioned above, a number of proteases, including neutrophil elastase, cathepsin G, MMP-9, and CD26 are capable to cleave CXCL12[118,124,125,129,130] and CXCR4.[118,131] Inhibitors of the neutrophil serine proteases significantly affect CXCL12 reductions and subsequent mobilization following G-CSF administration, suggesting that proteolytic cleavage of CXCL12 mediates at least in part HSPC mobilization.[123,124] On the other hand, G-CSF treatment

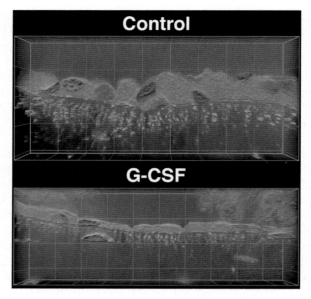

FIGURE 1. Osteoblasts are suppressed after G-CSF treatment. Volume rendering from image stacks of bone-lining osteoblasts from wild-type mice injected with recombinant human G-CSF or vehicle (control). Osteoblasts cell bodies and projections into the bone matrix are dramatically reduced after G-CSF treatment[128] (Grid, 10 μM).

decreases BM CXCL12 mRNA and this closely mirrors the fall in CXCL12 protein.[127,128] Moreover, CXCL12 levels decrease during mobilization in animals deficient in both neutrophil serine proteases and MMP-9, suggesting that CXCL12 regulation mainly occurs at the mRNA level.[127] Consistent with this last possibility, G-CSF induces a profound suppression of osteoblasts, a major producer of CXCL12, as shown by the marked reduction of osteocalcin,[127] Runx2, and α1(I)collagen,[128] three indicators of osteoblast activity. Osteoblast suppression is also manifested by flattened cell bodies and shorter projections into the bone matrix[128] (FIG. 1, see below). However, osteoblasts do not express the G-CSF receptor, suggesting that a different target cell mediates this effect.

Role of the Sympathetic Nervous System (SNS)

A role for the SNS in regulating osteoblast proliferation, function, and bone formation has been suggested by numerous clinical and experimental observations. Patients with neurological disorders were reported many years ago to exhibit localized bone fragility and osteopenia,[132] excessive callus formation,[133] and altered fracture healing.[134] Therapeutic sympathectomy was formerly used to decrease the discrepancies in limb length in children affected with poliomyelitis, underscoring the possibility of a dual effect, positive and negative, of bone innervation on bone formation.[135] Reflex sympathetic dystrophy, a disease characterized by local hyperadrenergic activity, is accompanied by focal osteopenia, and the administration of β-blockers is indeed one of the most effective treatments for this disease.[136] The role of the SNS on bone formation has only recently been elucidated using genetic models.[137] These elegant studies show that the hormone leptin decreases bone mass through SNS-derived signals originating in the ventromedial hypothalamic nuclei.[138,139] Further studies have shown that sympathetic signaling also favors bone resorption by promoting the differentiation of osteoclasts.[140] Moreover, recent data indicate that the SNS regulates osteoblast proliferation through the core genes of the molecular clock *Per1* and *Cry*, suggesting that bone remodeling is subjected to circadian regulation.[141]

A rich network of myelinated and nonmyelinated fibers innervates the bone and reaches stromal cells in the BM.[142] Immunolabeling studies have revealed a close association between catecholamine-, glutamate-, or vasoactive intestinal peptide-containing nerve fibers and osteoblasts or osteoclasts in the endosteum.[139,143,144] In addition, the expression of neurotransmitters, neuropeptides, and their receptors has been reported in hematopoietic and stromal cells in BM.[145–147] Neurotrophins, such as the nerve growth factor, are also expressed by stromal cells and synergize with SCF supporting HSPCs.[148] Other studies have suggested that α_1-adrenergic agonists can inhibit the growth of granulocyte macrophage colony-forming units (CFU-GM) and may protect mice from chemotherapy-induced BM toxicity.[149,150] Because only a very thin layer of sinusoidal endothelium separates nerve terminals from stromal cells, and abundant gap junctions connect BM reticular cells with nerve fibers, a "neuroreticular complex" has been proposed to have a role in BM homeostasis.[151]

While investigating the role of selectins in HSPC mobilization, we and others have found that a classic selectin inhibitor, the sulfated fucose polymer fucoidan, increases circulating HSPCs in a manner that does not require selectin binding.[35,40] Based on these data, we hypothesized that analogous sulfated glycans expressed in the BM microenvironment might regulate HSPC trafficking. The sulfated glycolipid sulfatide was a good candidate because its presence has been reported in leukocytes[152] and it shares biological activities with fucoidan.[153–157] The synthesis of sulfatide and its nonsulfated form, galactosylceramide (GalCer), is initiated by the addition of UDP-galactose to ceramide in a reaction mediated by UDP-galactose:ceramide galactosyltransferase (Cgt), an enzyme highly expressed in oligodendrocytes and Schwann cells.[158] The products of Cgt are major components of the myelin sheaths that allow the propagation of saltatory conduction. As a consequence, $Cgt^{-/-}$ mice display impaired nerve transmission because of the reduced insulative capacity of myelin sheaths and postnatal lethality due to severe tremor and ataxia.[159,160]

Administration of G-CSF to $Cgt^{-/-}$ and normal littermate controls has revealed a profound defect in mobilization in $Cgt^{-/-}$ mice. Investigations into the mechanisms have revealed that $Cgt^{-/-}$ mice exhibit a severe defect in lymphopoiesis (both B and T) owing to deficits in stromal cells that specifically support postnatal BM lymphoid progenitors.[161] However, the defect in mobilization is not due to the lymphopenia or to the absence of sulfatide, but caused by an altered neural influence on BM stroma.[128]

Interestingly, bone-lining osteoblasts in steady-state $Cgt^{-/-}$ mice exhibit a markedly flattened appearance with much shorter, less complex projections anchored into the underlying bone matrix. In addition, osteoblast function is suppressed in $Cgt^{-/-}$ mice, as measured by serum osteocalcin levels or $Runx2$ and $\alpha1(I)collagen$ expression in the BM. The facts that osteoblast suppression also occurs following the administration of G-CSF to wild-type mice (FIG. 1), and that osteoblasts do not express the $Csf3r$ or Cgt genes, indicate that the G-CSF-induced osteoblast suppression may depend on neural inputs. Indeed, we have recently shown that SNS signals play an important role in G-CSF-induced mobilization. The evidence supporting this claim includes: (i) impaired mobilization in mice chemically sympathectomized with 6-hydroxydopamine (6OHDA); (ii) reduced mobilization in mice injected with a β-adrenergic antagonist; (iii) reduced mobilization in mice deficient in dopamine β-hydroxylase (Dbh), an enzyme required for the synthesis of norepinephrine; (iv) the fact that the mobilization defect present in $Dbh^{-/-}$ mice is partially rescued by the administration of a β_2-adrenergic agonist, which also enhances mobilization in wild-type mice.[128] All together, these data suggest that the autonomic nervous system regulates HSPC attraction in the BM. It is worth pointing out, however, that the degree of inhibition varies among these loss-of-function approaches. For example, the effect of 6OHDA is more profound than the administration of a β-adrenergic antagonist, suggesting the contribution of other adrenergic or nonadrenergic signals. Whether β-adrenergic signaling is sufficient to alter $Cxcl12$ expression is currently unclear but a β_2-adrenergic agonist cannot elicit circulating HSPCs without costimulation with G-CSF. Osteoclasts, which participate in HSPC egress following stress,[162] represent a prime cellular candidate for additional G-CSF-mediated signals, although other studies have revealed a paradoxical increase of circulating HSPCs when mice were administered pamidronate, a potent inhibitor of bone resorption, together with G-CSF.[163]

Using a knockin mouse model expressing green fluorescent protein (GFP) under the $Cxcl12$ promoter, recent studies have confirmed the presence of a large subset of HSCs adjacent to CXCL12-abundant reticular (CAR) cells, which are largely near BM vessels.[164] It is intriguing that most SNS innervation of the BM is associated with blood vessels and numerous gap junctions have been noted by electron microscopy between nerve fibers and reticular cells,[151] suggesting that stromal cells are neurally regulated. Whether the SNS can

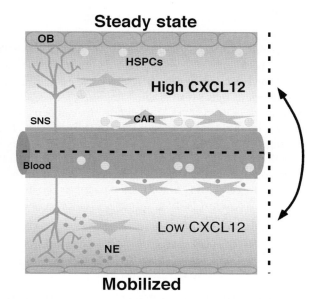

FIGURE 2. Model of regulation of hematopoietic stem and progenitor cell (HSPC) trafficking by the sympathetic nervous system (SNS). Under steady-state conditions, high levels of CXCL12 in the BM and bone maintain HSPCs attached to osteoblasts (OB) and CXCL12-abundant reticular (CAR) cells in the endosteal and vascular niches, respectively. G-CSF promotes both proteolytic cleavage and activation of the SNS. Norepinephrine (NE) release by the sympathetic nerve terminals and other unknown signals suppress osteoblasts and reduce CXCL12 content in both bone and BM, triggering HSPC egress to the blood-stream.

directly regulate Cxcl12 expression in stromal cells and affect HSPC retention in the vascular niche is currently unknown.

FUTURE PERSPECTIVES

It has become clear that the mechanisms mediating HSPC migration out (mobilization) of the BM (Fig. 2) are not simply the mirror image of those mediating the migration into (homing) the BM. Although much progress has been made in our understanding of HSPC trafficking, much more remains to be elucidated about the mechanisms controlling the migration and distribution of HSPCs within BM niches. Little is known in particular about the stromal components forming the specialized niches that promote lineage differenti-ation. Burgeoning evidence of stromal cell specification in the BM exists. Endothelial cells, for example, appear to play key roles in megakaryopoiesis since mature megakaryocytes are almost exclusively localized near thin-walled sinusoids.[165] The combined deficits in stromal elements supporting lymphoid progenitors and osteoblast function in the BM of $Cgt^{-/-}$ mice,[128,161] suggest that osteoblasts may support the commitment of HSCs toward the earliest lymphoid progenitor. This notion is further suggested by the *in vivo* ablation of osteoblasts, which severely reduces the B lymphocyte content in the BM[7] and other studies confirming these results and showing that osteoblasts support lymphoid commitment *in vitro*.[166] IL-7-secreting stromal cells may constitute a separate niche for lymphocyte committed to the B lineage.[167] The presence of two putative stem cell niches, vascular and osteoblastic, raises interesting unanswered questions about differences in repopulating ability, cell cycling, and susceptibility to be mobilized. Finally, the contribution of the autonomic nervous system in enforced HSPC mobilization suggests the intriguing possibility that this central keeper of body homeostasis may regulate physiological HSPC trafficking during steady state.

ACKNOWLEDGMENTS

We thank Dr. Andres Hidalgo for comments on the article. We are grateful to the National Institutes of Health (R01 Grants DK056638, HL69438, HG003950, AI069402) and the Department of Defense (Idea Development Award PC060271) for their support. P.S.F. is an Established Investigator of the American Heart Association.

REFERENCES

1. TRENTIN, J.J. *et al.* 1968. Factors controlling stem cell differentiation and proliferation: the hemopoietic inductive microenvironment. *In* The Proliferation and Spread of Neoplastic Cells. E. Frei, III, Ed.: 713–731. Williams & Wilkins. Baltimore, MD.
2. SCHOFIELD, R. 1978. The relationship between the spleen colony-forming cell and the haemopoietic stem cell. Blood Cells **4:** 7–25.
3. LORD, B.I., N.G. TESTA & J.H. HENDRY. 1975. The relative spatial distributions of CFUs and CFUc in the normal mouse femur. Blood **46:** 65–72.
4. NILSSON, S.K., H.M. JOHNSTON & J.A. COVERDALE. 2001. Spatial localization of transplanted hemopoietic stem cells: inferences for the localization of stem cell niches. Blood **97:** 2293–2299.
5. ZHANG, J. *et al.* 2003. Identification of the haematopoietic stem cell niche and control of the niche size. Nature **425:** 836–841.
6. CALVI, L.M. *et al.* 2003. Osteoblastic cells regulate the haematopoietic stem cell niche. Nature **425:** 841–846.
7. VISNJIC, D. *et al.* 2004. Hematopoiesis is severely altered in mice with an induced osteoblast deficiency. Blood **103:** 3258–3264.
8. KIEL, M.J. *et al.* 2005. SLAM family receptors distinguish hematopoietic stem and progenitor cells and reveal endothelial niches for stem cells. Cell **121:** 1109–1121.

9. BRECHER, G. & E.P. CRONKITE. 1951. Post-radiation parabiosis and survival in rats. Proc. Soc. Exp. Biol. Med. **77:** 292–294.

10. WARREN, S., R.N. CHUTE & E.M. FARRINGTON. 1960. Protection of the hematopoietic system by parabiosis. Lab. Invest. **9:** 191–198.

11. WRIGHT, D.E. *et al.* 2001. Physiological migration of hematopoietic stem and progenitor cells. Science **294:** 1933–1936.

12. LAPIDOT, T. & I. PETIT. 2002. Current understanding of stem cell mobilization: the roles of chemokines, proteolytic enzymes, adhesion molecules, cytokines, and stromal cells. Exp. Hematol. **30:** 973–981.

13. NERVI, B., D.C. LINK & J.F. DIPERSIO. 2006. Cytokines and hematopoietic stem cell mobilization. J. Cell Biochem. **99:** 690–705.

14. SUDA, T., F. ARAI & A. HIRAO. 2005. Hematopoietic stem cells and their niche. Trends Immunol. **26:** 426–433.

15. WILSON, A. & A. TRUMPP. 2006. Bone-marrow haematopoietic-stem-cell niches. Nat. Rev. Immunol. **6:** 93–106.

16. SCADDEN, D.T. 2006. The stem-cell niche as an entity of action. Nature **441:** 1075–1079.

17. MOORE, K.A. & I.R. LEMISCHKA. 2006. Stem cells and their niches. Science **311:** 1880–1885.

18. LI, Z. & L. LI. 2006. Understanding hematopoietic stem-cell microenvironments. Trends Biochem. Sci. **31:** 589–595.

19. LEY, K. 2003. The role of selectins in inflammation and disease. Trends Mol. Med. **9:** 263–268.

20. MCEVER, R.P. 2004. Interactions of selectins with PSGL-1 and other ligands. Ernst. Schering Res. Found Workshop **44:** 137–147.

21. ZANNETTINO, A.C. *et al.* 1995. Primitive human hematopoietic progenitors adhere to P-selectin (CD62P). Blood **85:** 3466–3477.

22. SPERTINI, O. *et al.* 1996. P-selectin glycoprotein ligand 1 is a ligand for L-selectin on neutrophils, monocytes, and CD34 +hematopoietic progenitor cells. J. Cell Biol. **135:** 523–531.

23. SNAPP, K.R. *et al.* 1997. P-selectin glycoprotein ligand-1 is essential for adhesion to P-selectin but not E-selectin in stably transfected hematopoietic cell lines. Blood **89:** 896–901.

24. LEVESQUE, J.P. *et al.* 1999. PSGL-1-mediated adhesion of human hematopoietic progenitors to P-selectin results in suppression of hematopoiesis. Immunity **11:** 369–378.

25. BAUMHETER, S. *et al.* 1993. Binding of L-selectin to the vascular sialomucin CD34. Science **262:** 436–438.

26. SASSETTI, C. *et al.* 1998. Identification of podocalyxin-like protein as a high endothelial venule ligand for L-selectin: parallels to CD34. J. Exp. Med. **187:** 1965–1975.

27. HIDALGO, A. *et al.* 2007. Complete identification of E-selectin ligands on neutrophils reveals distinct functions of PSGL-1, ESL-1, and CD44. Immunity **26:** 477–489.

28. STEEGMAIER, M. *et al.* 1995. The E-selectin-ligand ESL-1 is a variant of a receptor for fibroblast growth factor. Nature **373:** 615–620.

29. DIMITROFF, C.J. *et al.* 2001. CD44 is a major E-selectin ligand on human hematopoietic progenitor cells. J. Cell Biol. **153:** 1277–1286.

30. MOHLE, R. *et al.* 1995. Differential expression of L-selectin, VLA-4, and LFA-1 on CD34+ progenitor cells from bone marrow and peripheral blood during G-CSF-enhanced recovery. Exp. Hematol. **23:** 1535–1542.

31. DERCKSEN, M.W. *et al*. 1995. Expression of adhesion molecules on CD34+ cells: CD34+ L-selectin+ cells predict a rapid platelet recovery after peripheral blood stem cell transplantation. Blood **85:** 3313–3319.

32. MAZO, I.B. *et al*. 1998. Hematopoietic progenitor cell rolling in bone marrow microvessels: parallel contributions by endothelial selectins and vascular cell adhesion molecule 1. J. Exp. Med. **188:** 465–474.

33. LOWE, J.B. 2002. Glycosylation in the control of selectin counter-receptor structure and function. Immunol. Rev. **186:** 19–36.

34. FRENETTE, P.S. *et al*. 1996. Susceptibility to infection and altered hematopoiesis in mice deficient in both P- and E-selectins. Cell **84:** 563–574.

35. FRENETTE, P.S. & L. WEISS. 2000. Sulfated glycans induce rapid hematopoietic progenitor cell mobilization: evidence for selectin-dependent and independent mechanisms. Blood **96:** 2460–2468.

36. STARK, M.A. *et al*. 2005. Phagocytosis of apoptotic neutrophils regulates granulopoiesis via IL-23 and IL-17. Immunity **22:** 285–294.

37. ETO, T. *et al*. 2005. Contrasting effects of P-selectin and E-selectin on the differentiation of murine hematopoietic progenitor cells. Exp. Hematol. **33:** 232–242.

38. FRENETTE, P.S. *et al*. 1998. Endothelial selectins and vascular cell adhesion molecule-1 promote hematopoietic progenitor homing to bone marrow. Proc. Natl. Acad. Sci. USA **95:** 14423–14428.

39. HIDALGO, A., L.A. WEISS & P.S. FRENETTE. 2002. Functional selectin ligands mediating human CD34(+) cell interactions with bone marrow endothelium are enhanced postnatally. J. Clin. Invest. **110:** 559–569.

40. SWEENEY, E.A. *et al*. 2000. Mobilization of stem/progenitor cells by sulfated polysaccharides does not require selectin presence. Proc. Natl. Acad. Sci. USA **97:** 6544–6549.

41. MALY, P. *et al*. 1996. The alpha(1,3)fucosyltransferase Fuc-TVII controls leukocyte trafficking through an essential role in L-, E-, and P-selectin ligand biosynthesis. Cell **86:** 643–653.

42. HOMEISTER, J.W. *et al*. 2001. The alpha(1,3)fucosyltransferases FucT-IV and FucT-VII exert collaborative control over selectin-dependent leukocyte recruitment and lymphocyte homing. Immunity **15:** 115–126.

43. HIDALGO, A. & P.S. FRENETTE. 2005. Enforced fucosylation of neonatal CD34+ cells generates selectin ligands that enhance the initial interactions with microvessels but not homing to bone marrow. Blood **105:** 567–575.

44. XIA, L. *et al*. 2004. Surface fucosylation of human cord blood cells augments binding to P-selectin and E-selectin and enhances engraftment in bone marrow. Blood **104:** 3091–3096.

45. KATAYAMA, Y. *et al*. 2003. PSGL-1 participates in E-selectin-mediated progenitor homing to bone marrow: evidence for cooperation between E-selectin ligands and alpha4 integrin. Blood **102:** 2060–2067.

46. SIPKINS, D.A. *et al*. 2005. In vivo imaging of specialized bone marrow endothelial microdomains for tumour engraftment. Nature **435:** 969–973.

47. HYNES, R.O. 2002. Integrins: bidirectional, allosteric signaling machines. Cell **110:** 673–687.

48. KATAYAMA, Y. *et al*. 2004. Integrin alpha4beta7 and its counterreceptor MAdCAM-1 contribute to hematopoietic progenitor recruitment into bone marrow following transplantation. Blood **104:** 2020–2026.

49. ARROYO, A.G. *et al*. 2000. In vivo roles of integrins during leukocyte development and traffic: insights from the analysis of mice chimeric for alpha 5, alpha v, and alpha 4 integrins. J. Immunol. **165:** 4667–4675.

50. VERMEULEN, M. *et al*. 1998. Role of adhesion molecules in the homing and mobilization of murine hematopoietic stem and progenitor cells. Blood **92:** 894–900.

51. ZANJANI, E.D. *et al*. 1999. Homing of human cells in the fetal sheep model: modulation by antibodies activating or inhibiting very late activation antigen-4-dependent function. Blood **94:** 2515–2522.

52. RUEGG, C. *et al*. 1992. Role of integrin alpha 4 beta 7/alpha 4 beta P in lymphocyte adherence to fibronectin and VCAM-1 and in homotypic cell clustering. J. Cell Biol. **117:** 179–189.

53. EMAMBOKUS, N.R. & J. FRAMPTON. 2003. The glycoprotein IIb molecule is expressed on early murine hematopoietic progenitors and regulates their numbers in sites of hematopoiesis. Immunity **19:** 33–45.

54. FERKOWICZ, M.J. *et al*. 2003. CD41 expression defines the onset of primitive and definitive hematopoiesis in the murine embryo. Development **130:** 4393–4403.

55. UMEMOTO, T. *et al*. 2006. Expression of Integrin beta3 is correlated to the properties of quiescent hemopoietic stem cells possessing the side population phenotype. J. Immunol. **177:** 7733–7739.

56. HIRSCH, E. *et al*. 1996. Impaired migration but not differentiation of haematopoietic stem cells in the absence of beta1 integrins. Nature **380:** 171–175.

57. PAPAYANNOPOULOU, T. *et al*. 1995. The VLA4/VCAM-1 adhesion pathway defines contrasting mechanisms of lodgement of transplanted murine hemopoietic progenitors between bone marrow and spleen. Proc. Natl. Acad. Sci. USA **92:** 9647–9651.

58. SCOTT, L.M., G.V. PRIESTLEY & T. PAPAYANNOPOULOU. 2003. Deletion of alpha4 integrins from adult hematopoietic cells reveals roles in homeostasis, regeneration, and homing. Mol. Cell Biol. **23:** 9349–9360.

59. VAN DER LOO, J.C. *et al*. 1998. VLA-5 is expressed by mouse and human long-term repopulating hematopoietic cells and mediates adhesion to extracellular matrix protein fibronectin. J. Clin. Invest. **102:** 1051–1061.

60. PELED, A. *et al*. 2000. The chemokine SDF-1 activates the integrins LFA-1, VLA-4, and VLA-5 on immature human CD34(+) cells: role in transendothelial/stromal migration and engraftment of NOD/SCID mice. Blood **95:** 3289–3296.

61. PAPAYANNOPOULOU, T. *et al*. 2001. Synergistic mobilization of hemopoietic progenitor cells using concurrent beta1 and beta2 integrin blockade or beta2-deficient mice. Blood **97:** 1282–1288.

62. VELDERS, G.A. *et al*. 2002. Enhancement of G-CSF-induced stem cell mobilization by antibodies against the beta 2 integrins LFA-1 and Mac-1. Blood **100:** 327–333.

63. HIDALGO, A. *et al*. 2004. The integrin alphaMbeta2 anchors hematopoietic progenitors in the bone marrow during enforced mobilization. Blood **104:** 993–1001.

64. QIAN, H. *et al*. 2006. Contribution of alpha6 integrins to hematopoietic stem and progenitor cell homing to bone marrow and collaboration with alpha4 integrins. Blood **107:** 3503–3510.

65. GINSBERG, M.H., A. PARTRIDGE & S.J. SHATTIL. 2005. Integrin regulation. Curr. Opin. Cell Biol. **17:** 509–516.

66. LEVESQUE, J.P. *et al*. 1995. Cytokines increase human hemopoietic cell adhesiveness by activation of very late antigen (VLA)-4 and VLA-5 integrins. J. Exp. Med. **181:** 1805–1815.

67. LEVESQUE, J.P., D.N. HAYLOCK & P.J. SIMMONS. 1996. Cytokine regulation of proliferation and cell adhesion are correlated events in human CD34+ hemopoietic progenitors. Blood **88:** 1168–1176.
68. SOLANILLA, A. *et al.* 2003. Flt3-ligand induces adhesion of haematopoietic progenitor cells via a very late antigen (VLA)-4- and VLA-5-dependent mechanism. Br. J. Haematol. **120:** 782–786.
69. WEIMAR, I.S. *et al.* 1998. Hepatocyte growth factor/scatter factor (HGF/SF) is produced by human bone marrow stromal cells and promotes proliferation, adhesion and survival of human hematopoietic progenitor cells (CD34+). Exp. Hematol. **26:** 885–894.
70. CUI, L. *et al.* 1997. Thrombopoietin promotes adhesion of primitive human hemopoietic cells to fibronectin and vascular cell adhesion molecule-1: role of activation of very late antigen (VLA)-4 and VLA-5. J. Immunol. **159:** 1961–1969.
71. HIDALGO, A. *et al.* 2001. Chemokine stromal cell-derived factor-1alpha modulates VLA-4 integrin-dependent adhesion to fibronectin and VCAM-1 on bone marrow hematopoietic progenitor cells. Exp. Hematol. **29:** 345–355.
72. WRIGHT, N. *et al.* 2002. The chemokine stromal cell-derived factor-1 alpha modulates alpha 4 beta 7 integrin-mediated lymphocyte adhesion to mucosal addressin cell adhesion molecule-1 and fibronectin. J. Immunol. **168:** 5268–5277.
73. ARUFFO, A. *et al.* 1990. CD44 is the principal cell surface receptor for hyaluronate. Cell **61:** 1303–1313.
74. VERFAILLIE, C.M. *et al.* 1994. Adhesion of committed human hematopoietic progenitors to synthetic peptides from the C-terminal heparin-binding domain of fibronectin: cooperation between the integrin alpha 4 beta 1 and the CD44 adhesion receptor. Blood **84:** 1802–1811.
75. KATAYAMA, Y. *et al.* 2005. CD44 is a physiological E-selectin ligand on neutrophils. J. Exp. Med. **201:** 1183–1189.
76. MIYAKE, K. *et al.* 1990. Monoclonal antibodies to Pgp-1/CD44 block lymphohemopoiesis in long-term bone marrow cultures. J. Exp. Med. **171:** 477–488.
77. GUNJI, Y. *et al.* 1992. Expression and function of adhesion molecules on human hematopoietic stem cells: CD34+ LFA-1- cells are more primitive than CD34+ LFA-1+ cells. Blood **80:** 429–436.
78. LEGRAS, S. *et al.* 1997. CD44-mediated adhesiveness of human hematopoietic progenitors to hyaluronan is modulated by cytokines. Blood **89:** 1905–1914.
79. AVIGDOR, A. *et al.* 2004. CD44 and hyaluronic acid cooperate with SDF-1 in the trafficking of human CD34+ stem/progenitor cells to bone marrow. Blood **103:** 2981–2989.
80. SCHMITS, R. *et al.* 1997. CD44 regulates hematopoietic progenitor distribution, granuloma formation, and tumorigenicity. Blood **90:** 2217–2233.
81. CHRIST, O. *et al.* 2001. Combining G-CSF with a blockade of adhesion strongly improves the reconstitutive capacity of mobilized hematopoietic progenitor cells. Exp. Hematol. **29:** 380–390.
82. OOSTENDORP, R.A., S. GHAFFARI & C.J. EAVES. 2000. Kinetics of in vivo homing and recruitment into cycle of hematopoietic cells are organ-specific but CD44-independent. Bone Marrow Transplant. **26:** 559–566.
83. KIM, I., O.H. YILMAZ & S.J. MORRISON. 2005. CD144 (VE-cadherin) is transiently expressed by fetal liver hematopoietic stem cells. Blood **106:** 903–905.

84. LAI, C.F. *et al.* 2006. Accentuated ovariectomy-induced bone loss and altered osteogenesis in heterozygous N-cadherin null mice. J. Bone Miner. Res. **21:** 1897–1906.

85. KIEL, M.J., G.L. RADICE & S.E. MORRISON. 2007. Lack of evidence that hematopoietic stem cells depend on N-cadherin-mediated adhesion to osteoblasts for their maintenance. Cell Stem Cell **1:** 204–217.

86. SHARMA, S. *et al.* 2006. Stem cell c-KIT and HOXB4 genes: critical roles and mechanisms in self-renewal, proliferation, and differentiation. Stem Cells Dev. **15:** 755–778.

87. OKUMURA, N. *et al.* 1996. Chemotactic and chemokinetic activities of stem cell factor on murine hematopoietic progenitor cells. Blood **87:** 4100–4108.

88. KINASHI, T. & T.A. SPRINGER. 1994. Steel factor and c-kit regulate cell-matrix adhesion. Blood **83:** 1033–1038.

89. MOHLE, R. & L. KANZ. 2007. Hematopoietic growth factors for hematopoietic stem cell mobilization and expansion. Semin. Hematol. **44:** 193–202.

90. NAKAMURA, Y. *et al.* 2004. Soluble c-kit receptor mobilizes hematopoietic stem cells to peripheral blood in mice. Exp. Hematol. **32:** 390–396.

91. CYNSHI, O. *et al.* 1991. Reduced response to granulocyte colony-stimulating factor in W/Wv and Sl/Sld mice. Leukemia **5:** 75–77.

92. PAPAYANNOPOULOU, T., G.V. PRIESTLEY & B. NAKAMOTO. 1998. Anti-VLA4/VCAM-1-induced mobilization requires cooperative signaling through the kit/mkit ligand pathway. Blood **91:** 2231–2239.

93. IWAMA, A. *et al.* 1993. Molecular cloning and characterization of mouse TIE and TEK receptor tyrosine kinase genes and their expression in hematopoietic stem cells. Biochem. Biophys. Res. Commun. **195:** 301–309.

94. HSU, H.C. *et al.* 2000. Hematopoietic stem cells express Tie-2 receptor in the murine fetal liver. Blood **96:** 3757–3762.

95. PURI, M.C. & A. BERNSTEIN. 2003. Requirement for the TIE family of receptor tyrosine kinases in adult but not fetal hematopoiesis. Proc. Natl. Acad. Sci. USA **100:** 12753–12758.

96. DAVIS, S. *et al.* 1996. Isolation of angiopoietin-1, a ligand for the TIE2 receptor, by secretion-trap expression cloning. Cell **87:** 1161–1169.

97. TAKAKURA, N. *et al.* 1998. Critical role of the TIE2 endothelial cell receptor in the development of definitive hematopoiesis. Immunity **9:** 677–686.

98. ARAI, F. *et al.* 2004. Tie2/angiopoietin-1 signaling regulates hematopoietic stem cell quiescence in the bone marrow niche. Cell **118:** 149–161.

99. MAZZALI, M. *et al.* 2002. Osteopontin—a molecule for all seasons. QJM **95:** 3–13.

100. HUNTER, G.K., C.L. KYLE & H.A. GOLDBERG. 1994. Modulation of crystal formation by bone phosphoproteins: structural specificity of the osteopontin-mediated inhibition of hydroxyapatite formation. Biochem. J. **300**(Pt 3): 723–728.

101. HUNTER, G.K. *et al.* 1996. Nucleation and inhibition of hydroxyapatite formation by mineralized tissue proteins. Biochem J. **317**(Pt 1): 59–64.

102. RITTLING, S.R. *et al.* 1998. Mice lacking osteopontin show normal development and bone structure but display altered osteoclast formation in vitro. J. Bone Miner. Res. **13:** 1101–1111.

103. YOSHITAKE, H. *et al.* 1999. Osteopontin-deficient mice are resistant to ovariectomy-induced bone resorption. Proc. Natl. Acad. Sci. USA **96:** 8156–8160.

104. IHARA, H. *et al.* 2001. Parathyroid hormone-induced bone resorption does not occur in the absence of osteopontin. J. Biol. Chem. **276:** 13065–13071.

105. NILSSON, S.K. *et al.* 2005. Osteopontin, a key component of the hematopoietic stem cell niche and regulator of primitive hematopoietic progenitor cells. Blood **106:** 1232–1239.
106. STIER, S. *et al.* 2005. Osteopontin is a hematopoietic stem cell niche component that negatively regulates stem cell pool size. J. Exp. Med. **201:** 1781–1791.
107. HAYLOCK, D.N. & S.K. NILSSON. 2006. Osteopontin: a bridge between bone and blood. Br. J. Haematol. **134:** 467–474.
108. ADAMS, G.B. *et al.* 2006. Stem cell engraftment at the endosteal niche is specified by the calcium-sensing receptor. Nature **439:** 599–603.
109. LIU, F. *et al.* 1996. Impaired production and increased apoptosis of neutrophils in granulocyte colony-stimulating factor receptor-deficient mice. Immunity **5:** 491–501.
110. LIU, F., J. POURSINE-LAURENT & D.C. LINK. 2000. Expression of the G-CSF receptor on hematopoietic progenitor cells is not required for their mobilization by G-CSF. Blood **95:** 3025–3031.
111. DE HAAS, M *et al.* 1994. Granulocyte colony-stimulating factor administration to healthy volunteers: analysis of the immediate activating effects on circulating neutrophils. Blood **84:** 3885–3894.
112. FALANGA, A. *et al.* 1999. Neutrophil activation and hemostatic changes in healthy donors receiving granulocyte colony-stimulating factor. Blood **93:** 2506–2514.
113. LEVESQUE, J.P. *et al.* 2001. Vascular cell adhesion molecule-1 (CD106) is cleaved by neutrophil proteases in the bone marrow following hematopoietic progenitor cell mobilization by granulocyte colony-stimulating factor. Blood **98:** 1289–1297.
114. WINKLER, I.G. *et al.* 2005. Serine protease inhibitors serpina1 and serpina3 are down-regulated in bone marrow during hematopoietic progenitor mobilization. J. Exp. Med. **201:** 1077–1088.
115. LEVESQUE, J.P. *et al.* 2002. Mobilization by either cyclophosphamide or granulocyte colony-stimulating factor transforms the bone marrow into a highly proteolytic environment. Exp. Hematol. **30:** 440–449.
116. LEVESQUE, J.P. *et al.* 2004. Characterization of hematopoietic progenitor mobilization in protease-deficient mice. Blood **104:** 65–72.
117. LEVESQUE, J.P. *et al.* 2003. Granulocyte colony-stimulating factor induces the release in the bone marrow of proteases that cleave c-KIT receptor (CD117) from the surface of hematopoietic progenitor cells. Exp. Hematol. **31:** 109–117.
118. LEVESQUE, J.P. *et al.* 2003. Disruption of the CXCR4/CXCL12 chemotactic interaction during hematopoietic stem cell mobilization induced by GCSF or cyclophosphamide. J. Clin. Invest. **111:** 187–196.
119. PRUIJT, J.F. *et al.* 1999. Prevention of interleukin-8-induced mobilization of hematopoietic progenitor cells in rhesus monkeys by inhibitory antibodies against the metalloproteinase gelatinase B (MMP-9). Proc. Natl. Acad. Sci. USA **96:** 10863–10868.
120. PRUIJT, J.F. *et al.* 2002. Neutrophils are indispensable for hematopoietic stem cell mobilization induced by interleukin-8 in mice. Proc. Natl. Acad. Sci. USA **99:** 6228–6233.
121. HEISSIG, B. *et al.* 2002. Recruitment of stem and progenitor cells from the bone marrow niche requires MMP-9 mediated release of kit-ligand. Cell **109:** 625–637.

122. PAPAYANNOPOULOU, T. *et al*. 2003. The role of G-protein signaling in hematopoi-
     etic stem/progenitor cell mobilization. Blood **101:** 4739–4747.
123. PELUS, L.M. *et al*. 2004. Neutrophil-derived MMP-9 mediates synergistic mo-
     bilization of hematopoietic stem and progenitor cells by the combination of
     G-CSF and the chemokines GRObeta/CXCL2 and GRObetaT/CXCL2delta4.
     Blood **103:** 110–119.
124. PETIT, I. *et al*. 2002. G-CSF induces stem cell mobilization by decreasing bone
     marrow SDF-1 and up-regulating CXCR4. Nat. Immunol. **3:** 687–694.
125. CHRISTOPHERSON, K.W., II, S. COOPER & H.E. BROXMEYER. 2003. Cell surface
     peptidase CD26/DPPIV mediates G-CSF mobilization of mouse progenitor
     cells. Blood **101:** 4680–4686.
126. CHRISTOPHERSON, K.W. *et al*. 2003. CD26 is essential for normal G-CSF-induced
     progenitor cell mobilization as determined by CD26-/- mice. Exp. Hematol. **31:**
     1126–1134.
127. SEMERAD, C.L. *et al*. 2005. G-CSF potently inhibits osteoblast activity and
     CXCL12 mRNA expression in the bone marrow. Blood **106:** 3020–3027.
128. KATAYAMA, Y. *et al*. 2006. Signals from the sympathetic nervous system regulate
     hematopoietic stem cell egress from bone marrow. Cell **124:** 407–421.
129. MCQUIBBAN, G.A. *et al*. 2001. Matrix metalloproteinase activity inactivates the
     CXC chemokine stromal cell-derived factor-1. J. Biol. Chem. **276:** 43503–
     43508.
130. DELGADO, M.B. *et al*. 2001. Rapid inactivation of stromal cell-derived factor-1
     by cathepsin G associated with lymphocytes. Eur. J. Immunol. **31:** 699–707.
131. VALENZUELA-FERNANDEZ, A. *et al*. 2002. Leukocyte elastase negatively regulates
     Stromal cell-derived factor-1 (SDF-1)/CXCR4 binding and functions by amino-
     terminal processing of SDF-1 and CXCR4. J. Biol. Chem. **277:** 15677–15689.
132. GILLESPIE, J.A. 1954. The nature of bone changes associated with nerve injuries
     and disease. J. Bone Joint Surg. **36B:** 464–473.
133. HARDY, A.G. & J.W. DICKSON. 1963. Pathological ossification in traumatic para-
     plegia. J. Bone Joint Surg. **45B:** 76–87.
134. FREEHAFER, A.A. & W.A. MAST. 1965. Lower extremity fractures in patients with
     spinal-chord injury. J. Bone Joint Surg. **47A:** 683–694.
135. BARR, J.S., A.J. STINCHFIELD & J.A. REIDY. 1950. Sympathetic ganglionectomy
     and limb length in poliomyelitis. J. Bone Joint Surg. Am. **32:** 793–802.
136. SCHWARTZMAN, R.J. & T.L. MCLELLAN. 1987. Reflex sympathetic dystrophy. A
     review. Arch. Neurol. **44:** 555–561.
137. CHIEN, K.R. & G. KARSENTY. 2005. Longevity and lineages: toward the integrative
     biology of degenerative diseases in heart, muscle, and bone. Cell **120:** 533–544.
138. DUCY, P. *et al*. 2000. Leptin inhibits bone formation through a hypothalamic relay:
     a central control of bone mass. Cell **100:** 197–207.
139. TAKEDA, S. *et al*. 2002. Leptin regulates bone formation via the sympathetic
     nervous system. Cell **111:** 305–317.
140. ELEFTERIOU, F. *et al*. 2005. Leptin regulation of bone resorption by the sympathetic
     nervous system and CART. Nature **434:** 514–520.
141. FU, L. *et al*. 2005. The molecular clock mediates leptin-regulated bone formation.
     Cell **122:** 803–815.
142. CALVO, W. 1968. The innervation of the bone marrow in laboratory animals. Am.
     J. Anat. **123:** 315–328.
143. HOHMANN, E.L. *et al*. 1986. Innervation of periosteum and bone by sympathetic
     vasoactive intestinal peptide-containing nerve fibers. Science **232:** 868–871.

144. SERRE, C.M. *et al.* 1999. Evidence for a dense and intimate innervation of the bone tissue, including glutamate-containing fibers. Bone **25:** 623–629.

145. FELDMAN, R.D., G.W. HUNNINGHAKE & W.L. MCARDLE. 1987. Beta-adrenergic-receptor-mediated suppression of interleukin 2 receptors in human lymphocytes. J. Immunol. **139:** 3355–3359.

146. SPENGLER, R.N. *et al.* 1990. Stimulation of alpha-adrenergic receptor augments the production of macrophage-derived tumor necrosis factor. J. Immunol. **145:** 1430–1434.

147. RAMESHWAR, P. & P. GASCON. 1996. Induction of negative hematopoietic regulators by neurokinin-A in bone marrow stroma. Blood **88:** 98–106.

148. AUFFRAY, I. *et al.* 1996. Nerve growth factor is involved in the supportive effect by bone marrow–derived stromal cells of the factor-dependent human cell line UT-7. Blood **88:** 1608–1618.

149. MAESTRONI, G.J. & A. CONTI. 1994. Modulation of hematopoiesis via alpha 1-adrenergic receptors on bone marrow cells. Exp. Hematol. **22:** 313–320.

150. MAESTRONI, G.J., M. TOGNI & V. COVACCI. 1997. Norepinephrine protects mice from acute lethal doses of carboplatin. Exp. Hematol. **25:** 491–494.

151. YAMAZAKI, K. & T.D. ALLEN. 1990. Ultrastructural morphometric study of efferent nerve terminals on murine bone marrow stromal cells, and the recognition of a novel anatomical unit: the "neuro-reticular complex." Am. J. Anat. **187:** 261–276.

152. ARUFFO, A. *et al.* 1991. CD62/P-selection recognition of myeloid and tumor cell sulfatides. Cell **67:** 35–44.

153. BRENNER, B. *et al.* 1996. L-selectin activates the Ras pathway via the tyrosine kinase p56lck. Proc. Natl. Acad. Sci. USA **93:** 15376–15381.

154. IMAI, Y. *et al.* 1990. Direct demonstration of the lectin activity of gp90MEL, a lymphocyte homing receptor. J. Cell Biol. **111:** 1225–1232.

155. ROBERTS, D.D. *et al.* 1986. Comparison of the specificities of laminin, thrombospondin, and von Willebrand factor for binding to sulfated glycolipids. J. Biol. Chem. **261:** 6872–6877.

156. SKINNER, M.P. *et al.* 1991. GMP-140 binding to neutrophils is inhibited by sulfated glycans. J. Biol. Chem. **266:** 5371–5374.

157. WADDELL, T.K. *et al.* 1995. Signaling functions of L-selectin. Enhancement of tyrosine phosphorylation and activation of MAP kinase. J. Biol. Chem. **270:** 15403–15411.

158. SPRONG, H. *et al.* 1998. UDP-galactose:ceramide galactosyltransferase is a class I integral membrane protein of the endoplasmic reticulum. J. Biol. Chem. **273:** 25880–25888.

159. BOSIO, A., E. BINCZEK & W. STOFFEL. 1996. Functional breakdown of the lipid bilayer of the myelin membrane in central and peripheral nervous system by disrupted galactocerebroside synthesis. Proc. Natl. Acad. Sci. USA **93:** 13280–13285.

160. COETZEE, T. *et al.* 1996. Myelination in the absence of galactocerebroside and sulfatide: normal structure with abnormal function and regional instability. Cell **86:** 209–219.

161. KATAYAMA, Y. & P.S. FRENETTE. 2003. Galactocerebrosides are required postnatally for stromal-dependent bone marrow lymphopoiesis. Immunity **18:** 789–800.

162. KOLLET, O. *et al.* 2006. Osteoclasts degrade endosteal components and promote mobilization of hematopoietic progenitor cells. Nat. Med. **12:** 657–664.

163. TAKAMATSU, Y. *et al.* 1998. Osteoclast-mediated bone resorption is stimulated during short-term administration of granulocyte colony-stimulating factor but is not responsible for hematopoietic progenitor cell mobilization. Blood **92:** 3465–3473.
164. SUGIYAMA, T. *et al.* 2006. Maintenance of the hematopoietic stem cell pool by CXCL12-CXCR4 chemokine signaling in bone marrow stromal cell niches. Immunity **25:** 977–988.
165. KOPP, H.G. *et al.* 2005. The bone marrow vascular niche: home of HSC differentiation and mobilization. Physiology (Beth.) **20:** 349–356.
166. ZHU, J. *et al.* 2007. Osteoblasts support B-lymphocyte commitment and differentiation from hematopoietic stem cells. Blood **109:** 3706–3712.
167. TOKOYODA, K. *et al.* 2004. Cellular niches controlling B lymphocyte behavior within bone marrow during development. Immunity **20:** 707–718.

Regulation of Skeletal Remodeling by the Endocannabinoid System

ITAI A. BAB

Bone Laboratory, The Hebrew University of Jerusalem, Jerusalem, Israel

ABSTRACT: Since the discovery of the endocannabinoid system, its presence and involvement have been reported in a handful of biological systems. Recently, the skeleton has been identified as a major endocannabinoid target through both the neuronal CB1 and predominantly peripheral CB2 cannabinoid receptors. CB1 is present in sympathetic nerve terminals in bone, whereas CB2 is expressed in osteoblasts and osteoclasts, the respective bone-forming and -resorbing cells. Furthermore, the skeleton appears as the main system physiologically regulated by CB2. CB2-deficient mice show a markedly accelerated age-related bone loss and the CB2 locus in women is associated with low bone density and osteoporotic fractures. Since activation of CB2 attenuates experimentally induced bone loss by inhibiting bone resorption and stimulating bone formation, and because synthetic cannabinoids are stable and orally available, a therapy based on synthetic CB2 agonists is a promising novel target for antiosteoporotic drug development.

KEYWORDS: adrenergic signaling; anandamide; bone formation; bone mass; bone remodeling; bone resorption; cannabinoid receptors; CB1; CB2; endocannabinoid system; norepinephrine; osteoblast; osteoclast; osteoporosis; 2-arachidonoylglycerol

INTRODUCTION

In vertebrates, bone structure presents substantial changes throughout life displaying three temporal phases: (i) rapid skeletal growth and accrual of peak bone mass; (ii) steady state during which bone mass remains constant; (iii) bone loss.[1] These changes result from a continuous destruction/formation process referred to as bone remodeling. Imbalances in bone remodeling lead to bone mass accrual (positive imbalance) or bone loss (negative imbalance).[2] The remodeling process occurs at the same time in multiple foci that in humans encompass approximately 5% of trabecular, endosteal, and osteonal

Address for correspondence: Prof. Itai Bab, Bone Laboratory, The Hebrew University of Jerusalem, P.O.B. 12272, Jerusalem 91120, Israel. Voice: 972-675-8572; fax: 972-675-7623.
babi@cc.huji.ac.il

Ann. N.Y. Acad. Sci. 1116: 414–422 (2007). © 2007 New York Academy of Sciences.
doi: 10.1196/annals.1402.014

surfaces.[3] The remodeling cycle in individual foci consists initially of a relatively rapid (i.e., a few weeks) resorption of preexisting bone by a bone-specific, bone marrow hematopoietic cell type, the osteoclast, derived from the monocyte/macrophage lineage.[4] It is then followed by a slower (i.e., a few months) step of *de novo* bone formation by another bone-specific cell type, the osteoblast,[3] which belongs to the stromal cell lineage of bone marrow.[5] Different foci present different phases of the cycle; hence, the net effect on bone mass reflects the overall balance between bone destruction and formation. The physiologic importance of balanced bone remodeling is best illustrated in osteoporosis, the most frequent degenerative disease in developed countries, which results from a net increase in bone resorption that leads to bone loss and increased fracture risk, mainly in females but also in males.

The synchronized occurrence of multiple remodeling sites has long been viewed as suggestive of a complex, local, autocrine/paracrine as well as endocrine regulation.[6] Indeed, experiments in knockout and transgenic mice have demonstrated paracrine regulation of osteoclast differentiation and activity by factors, such as receptor activator of NF-κB ligand (RANKL), osteoprotegerin (OPG), macrophage colony-stimulating factor (M-CSF), and interleukin 6 (IL-6), which are derived from neighboring stromal cells, including osteoblasts and osteoblast precursors.[7–12] The most convincing evidence for local osteoblast regulation is by bone morphogenetic proteins (BMPs).[13] Systemically, ablation of gonadal hormones in females and males has been repeatedly demonstrated to favor bone loss in humans, rats, and mice.[14–16] In addition, parathyroid hormone (PTH),[17,18] calcitonin,[19] insulin-like growth factor I (IGF-I),[20] and the osteogenic growth peptide (OGP),[21] have been shown to regulate bone formation. More recently, it has been demonstrated that bone remodeling is also subject to a potent central control consisting of hypothalamic leptin and neuropeptide Y signaling,[22,23] as well as downstream noradrenergic signaling via osteoblastic β2 adrenergic receptors.[24] Moreover, it has been lately suggested that imbalances in bone remodeling, previously attributed to excessive thyroid activity and estrogen depletion, actually result from the direct action of the pituitary-derived thyroid-stimulating hormone (TSH) and follicular-stimulating hormone (FSH) on bone cells.[25,26]

The discovery of the endocannabinoid system followed the identification and cloning of receptors for Δ^9-tetrahydrocannabinol (Δ^9-THC), the major psychoactive component of marijuana and hashish.[27] The actions of Δ^9-THC are mediated mainly through binding to and activation of CB1 and CB2.[28] Both are seven-transmembrane domain, G protein-coupled receptors. CB1 contains 472 (human) or 473 (rat and mouse) amino acids. CB2 consists of 360 amino acids. CB1 and CB2 share 44% overall identity (68% identity for the transmembrane domains). CB1 is the most abundantly expressed receptor in the central nervous system. It is also present in peripheral neurons and the gonads and to a lesser extent in several other peripheral tissues. Until recently it was perceived that in health, CB2 is expressed almost exclusively in the immune system.[29] It has also been reported in liver cirrhosis, arteriosclerotic plaques, and brain inflammation.[30,31] That CB1 and CB2 are not functionally identical is demonstrated by the presence of cannabinoid agonists and antagonists with distinct binding specificities to either receptor.[32,33] Relatively little is known about the intracellular signaling events triggered by CB activation, particularly those initiated by CB2. In general, both receptors are coupled directly to the G(i/o) subclass of G proteins and inhibit stimulated overall adenylyl cyclase activity. Still, several adenylyl cyclase subtypes may be enhanced by CB activation,[34] probably depending on the cell type and specific cannabinoid ligands. In addition, the CBs have been shown to induce the activation of p42/44 mitogen-activated protein kinase (MAPK),[35–37] p38 MAPK,[38] c-Jun N-terminal kinase,[38,39] AP-1,[40] the neural form of focal adhesion kinase,[41] protein kinase B,[42] and Ca^{2+} transients.[43]

The identification and cloning of CB1 and CB2 were soon followed by the discovery of their main endogenous ligands, N-arachidonoylethanolamine (AEA or anandamide) and 2-arachidonoylglycerol (2-AG).[44,45] Anandamide is present in a variety of tissues, such as the brain (including the hypothalamus), kidney, liver, spleen, testis, uterus, and blood in picomol/g tissue concentrations. The highest levels were found in the central nervous system. The low tissue levels of anandamide have been attributed to low substrate (arachidonic acid esterified at the 1-position) levels for this pathway,[46] and/or the short anandamide half-life *in vivo* ($t1/2 < 5$ min).[47] The main anandamide degrading enzyme is fatty acid amide hydrolase (FAAH), a membrane-associated serine hydrolase enriched in the brain and liver.[48] In general, the tissue distribution of 2-AG is similar to that of anandamide. However, its concentration in the various tissues is 300–1000 higher (ng/g range). Unlike anandamide, which is regarded as a CB1 partial agonist, 2-AG is considered a full agonist of both CB1 and CB2 receptors. Production of 2-AG has been demonstrated in the central nervous system. Even larger amounts are produced in platelets and macrophages, especially in response to stimulation by inflammatory agents, such as lipopolysaccharide.[49,50] It has been shown that 2-AG is generated from arachidonic acid-enriched membrane phospholipids, such as inositol phospholipids, through the combined actions of phospholipase C and diacylglycerol lipases (DAGLα and DAGLβ).[51,52] The most plausible mechanism of 2AG degradation is that 2AG is metabolized by a monoacylglycerol lipase, like other monoacylglycerols.[53]

A couple of striking findings led us to study the involvement of the endocannabinoid system in the regulation of skeletal remodeling. One is that, as in the case of bone formation and bone mass, the central production of at least one major endocannabinoid, 2-AG, is subject to negative regulation by leptin.[54] The other observation is that traumatic head injury stimulates both bone formation and central 2-AG production.[55–57]

EXPRESSION OF CANNABINOID RECEPTORS IN BONE

To assess the feasibility of a skeletal endocannabinoid system, we initially explored the expression of cannabinoid receptors in bone cells. Undifferentiated osteoblast progenitors, such as mouse bone marrow-derived stromal cells and MC3T3 E1 preosteoblasts,[58,59] exhibit very low levels, if any, of CB1. CB2 mRNA levels in these cells are also very low.[60,61] When these cells are grown for 2 to 4 weeks in "osteogenic medium,"[62] CB1 mRNA remains at the same levels. However, CB2 mRNA expression increases progressively in parallel to the expression of osteoblastic marker genes, such as tissue nonspecific alkaline phosphatase (*TNSALP*),[63] parathyroid hormone receptor 1 (*PTHRc1*),[64] and particularly the osteoblastic master regulatory gene, *RUNX2*.[65] CB1 is expressed at low levels in monocytic cells undergoing osteoclastogenesis induced by RANK ligand and M-CSF.[66] By contrast, CB2 mRNA transcripts in these cells are present in high abundance.[60,61,67] *In vivo*, we identified CB2 protein in trabecular osteoblasts and their decedents, the osteocytes,[65] as well as in osteoclasts.[61] CB1 protein is highly abundant in skeletal sympathetic nerve terminals.[68]

CANNABINOID LIGANDS REGULATE BONE CELL DIFFERENTIATION AND ACTIVITY

CB2 activation has different effects in early preosteoblasts and in more mature osteoblastic cells. In bone marrow derived, partially differentiated osteoblasts, with limited CB2 expression, the specific CB2 agonist HU-308,[33] but not the specific CB1 agonist noladin ether,[69] triggers a G_i protein-mediated mitogenic effect, leading to a dose-response expansion of the preosteoblastic pool.[60] *Ex vivo* osteoblastic colony (CFU-Ob) formation by bone marrow stromal *cb2*[−/−] cells is markedly diminished, whereas CFU-Ob formation by wild-type cells is enhanced by CB2-specific agonists.[61,67] In mature osteoblastic cells, represented by the MC3T3 E1 cell line, these ligands stimulate osteoblast differentiated functions, such as alkaline phosphatase activity and matrix mineralization.[60,61] Thus, CB2 signaling has multiple regulatory osteogenic anabolic functions along the osteoblast differentiation pathway.

In bone marrow-derived osteoclastogenic cultures and in the RAW 264.7 osteoclast-like cell line, CB2 activation inhibits osteoclast differentiation by restraining mitogenesis at the monocytic stage, prior to incubation with RANKL. It also suppresses osteoclast formation by repressing RANKL expression in osteoblasts.[61]

CANNABINOID RECEPTOR SIGNALING
REGULATES BONE MASS *IN VIVO*

We used cannabinoid receptor null mice to assess the physiologic role of CB1 and CB2 in the control of bone mass. The skeletal phenotype of CB1 depends on the mouse strain and/or the methods of gene ablation. In one CB1-deficient line, backcrossed to CD1 mice, the N-terminal 233 codons of *cb1* were ablated.[70] The skeletal phenotype of these mice shows a gender disparity. Females have normal trabecular bone with a slight cortical expansion, whereas male CD1[CB1−/−] animals exhibit a high bone mass phenotype.[68] Sexually mature CD1[CB1−/−] mice of either gender display normal bone formation and resorption parameters, suggesting that the male phenotype is acquired early in life. In the second line, backcrossed to C57Bl/6J mice (C57[CB1−/−]), almost the entire protein-encoding sequence was removed.[72] Both male and female C57[CB1−/−] have a low bone mass phenotype accompanied by increased osteoclast counts and decreased bone formation rate.[68] It appears that CB1 controls osteoblast function by negatively regulating norepinephrine (NE) release from sympathetic nerve terminals in the immediate proximity of these cells. NE suppresses osteoblast function by binding to osteoblastic β2-adrenergic receptors,[24] which is alleviated by activation of sympathetic CB1.[71]

CB2-deficient mice have a skeletal phenotype that is gender independent. Both male and female *cb2*[−/−] accrue a normal peak trabecular bone mass, but later display a markedly enhanced age-related bone loss; their trabecular bone volume density at 1 year of age is approximately half compared to wild-type controls.[61] Reminiscent of human postmenopausal osteoporosis,[73] the *cb2*[−/−] mice have a high bone turnover characterized by increases in both bone resorption and formation, which are at a net negative balance.[61] Because healthy CB2 null mice are otherwise normal, it appears that the main physiologic role of CB2 is in maintaining bone remodeling at balance.

ATTENUATION OF BONE LOSS BY CB2-SPECIFIC AGONIST

All the known cannabinoid psychoactivity is mediated by CB1 and is not at all shared by CB2. Hence, CB2-specific ligands could provide an opportunity to augment bone mass while avoiding the adverse psychological effects typical of cannabinoids. Indeed, HU-308, a specific CB2 agonist, which is nonpsychoactive,[33] attenuates bone loss induced by estrogen depletion in ovariectomized (OVX) animals using either preventive,[61] or rescue (unpublished data) protocols, commencing immediately after OVX or 6 weeks later (to allow for bone loss to occur), respectively. The attenuation results from both inhibition of bone resorption and stimulation of bone formation.[60] Hence, CB2 agonists may become an orally available, combined antiresorptive and anabolic therapy for osteoporosis.

ASSOCIATION OF CB2 WITH OSTEOPOROSIS IN HUMANS

The genetic factors involved in the pathogenesis of human osteoporosis are largely unknown. Thus, because of the skeletal phenotypes of the CB1 and CB2 mice we analyzed the human *CNR1* (encoding CB1) and *CNR2* (encoding CB1) genes in a systematic genetic association study in a sample of French postmenopausal osteoporosis patients and matched female controls. We found a significant association of single polymorphisms ($P = 0.0014$) and haplotypes ($P = 0.0001$) encompassing the *CNR2* gene on human chromosome 1p36, whereas no convincing association was found for *CNR1*.[74] Similar findings were also reported in a Japanese sample demonstrating that *CNR2* is a susceptibility locus for reduced bone mass in women.[75]

These genetic studies in humans and mice demonstrate a role for CB2 in the etiology of osteoporosis and provide a novel target for the diagnosis and treatment of this severe and common disease.

REFERENCES

1. SEGEV, E. *et al.* 2006. Age- and gender-related variations in bone growth and structure: a μCT analysis in mice. J. Bone Min. Res. **21:** S47.
2. KARSENTY, G. 2001. Leptin controls bone formation through a hypothalamic relay. Recent Prog. Horm. Res. **56:** 401–415.
3. PARFITT, A.M. 1982. The coupling of bone formation to bone resorption: a critical analysis of the concept and of its relevance to the pathogenesis of osteoporosis. Metab. Bone Dis. Relat. Res. **4:** 1–6.
4. ROODMAN, G.D. 1999. Cell biology of the osteoclast. Exp. Hematol. **27:** 1229–1241.
5. BAB, I. *et al.* 1986. Kinetics and differentiation of marrow stromal cells in diffusion chambers *in vivo*. J. Cell Sci. **84:** 139–151.
6. MANOLAGAS, S.C. 2000. Birth and death of bone cells: basic regulatory mechanisms and implications for the pathogenesis and treatment of osteoporosis. Endocr. Rev. **21:** 115–137.
7. POLI, V. *et al.* 1994. Interleukin-6 deficient mice are protected from bone loss caused by estrogen depletion. EMBO J. **13:** 1189–1196.
8. SUDA, T. *et al.* 2001. The molecular basis of osteoclast differentiation and activation. Novartis Found Symp. **232:** 235–247; discussion 247–250.
9. SIMONET, W.S. *et al.* 1997. Osteoprotegerin: a novel secreted protein involved in the regulation of bone density. Cell **89:** 309–319.
10. LACEY, D.L. *et al.* 1998. Osteoprotegerin ligand is a cytokine that regulates osteoclast differentiation and activation. Cell **93:** 165–176.
11. BUCAY, N. *et al.* 1998. Osteoprotegerin-deficient mice develop early onset osteoporosis and arterial calcification. Genes Dev. **12:** 1260–1268.
12. KONG, Y.Y. *et al.* 1999. OPGL is a key regulator of osteoclastogenesis, lymphocyte development and lymph-node organogenesis. Nature **397:** 315–323.
13. YOSHIDA, Y. *et al.* 2000. Negative regulation of BMP/Smad signaling by Tob in osteoblasts. Cell **103:** 1085–1097.

14. MOST, W. *et al.* 1997. Ovariectomy and orchidectomy induce a transient increase in the osteoclastogenic potential of bone marrow cells in the mouse. Bone **20:** 27–30.
15. GABET, Y. *et al.* 2005. Intermittently administered parathyroid hormone 1–34 reverses bone loss and structural impairment in orchiectomized adult rats. Osteoporos. Int. **16:** 1436–1443.
16. ALEXANDER, J.M. *et al.* 2001. Human parathyroid hormone 1–34 reverses bone loss in ovariectomized mice. J. Bone Miner. Res. **16:** 1665–1673.
17. POTTS, J.T. & H. JUPPNER 1998. Parathyroid Hormone and Parathyroid Hormone-Related Peptidein Calcium Homeostasis, Bone Metabolism, and Bone Development: The Proteins, Their Genes, and Receptors. Academic Press. San Diego, CA.
18. GUNTHER, T. *et al.* 2000. Genetic ablation of parathyroid glands reveals another source of parathyroid hormone. Nature **406:** 199–203.
19. NICHOLSON, G.C. *et al.* 1986. Abundant calcitonin receptors in isolated rat osteoclasts. Biochemical and autoradiographic characterization. J. Clin. Invest. **78:** 355–360.
20. YAKAR, S. *et al.* 2002. Circulating levels of IGF-1 directly regulate bone growth and density. J. Clin. Invest. **110:** 771–781.
21. BAB, I. & M. CHOREV. 2002. Osteogenic growth peptide: from concept to drug design. Biopolymers **66:** 33–48.
22. DUCY, P. *et al.* 2000. Leptin inhibits bone formation through a hypothalamic relay: a central control of bone mass. Cell **100:** 197–207.
23. BALDOCK, P.A. *et al.* 2002. Hypothalamic Y2 receptors regulate bone formation. J. Clin. Invest. **109:** 915–921.
24. TAKEDA, S. *et al.* 2002. Leptin regulates bone formation via the sympathetic nervous system. Cell **111:** 305–317.
25. ABE, E. *et al.* 2003. TSH is a negative regulator of skeletal remodeling. Cell **115:** 151–162.
26. SUN, L. *et al.* 2006. FSH directly regulates bone mass. Cell **125:** 247–260.
27. GAONI Y. & R. MECHOULAM. 1964. Isolation, structure and partial synthesis of an active constituent of Hashish. J. Am. Chem. Soc. **86:** 1646.
28. HOWLETT, A.C. 2002. The cannabinoid receptors. Prostaglandins Other Lipid Mediat. **68–69:** 619–631.
29. SUGIURA, T. *et al.* 2002. Biosynthesis and degradation of anandamide and 2-arachidonoylglycerol and their possible physiological significance. Prostaglandins Leukot. Essent. Fatty Acids **66:** 173–192.
30. JULIEN, B. *et al.* 2005. Antifibrogenic role of the cannabinoid receptor CB2 in the liver. Gastroenterology **128:** 742–755.
31. STEFFENS, S. *et al.* 2005. Low dose oral cannabinoid therapy reduces progression of atherosclerosis in mice. Nature **434:** 782–786.
32. SHIRE, D. *et al.* 1999. Cannabinoid receptor interactions with the antagonists SR 141716A and SR 144528. Life Sci. **65:** 627–635.
33. HANUS, L. *et al.* 1999. HU-308: a specific agonist for CB(2), a peripheral cannabinoid receptor. Proc. Natl. Acad. Sci. USA **96:** 14228–14233.
34. RHEE, M.H. *et al.* 1998. Cannabinoid receptor activation differentially regulates the various adenylyl cyclase isozymes. J. Neurochem. **71:** 1525–1534.
35. WARTMANN, M. *et al.* 1995. The MAP kinase signal transduction pathway is activated by the endogenous cannabinoid anandamide. FEBS Lett. **359:** 133–136.

36. ABBOTT, N.J. 2000. Inflammatory mediators and modulation of blood-brain barrier permeability. Cell Mol. Neurobiol. **20:** 131–147.
37. MELCK, D. *et al.* 1999. Involvement of the cAMP/protein kinase A pathway and of mitogen-activated protein kinase in the anti-proliferative effects of anandamide in human breast cancer cells. FEBS Lett. **463:** 235–240.
38. DERKINDEREN, P. *et al.* 2001. Cannabinoids activate p38 mitogen-activated protein kinases through CB1 receptors in hippocampus. J. Neurochem. **77:** 957–960.
39. RUEDA, D. *et al.* 2000. The CB(1) cannabinoid receptor is coupled to the activation of c-Jun N-terminal kinase. Mol. Pharmacol. **58:** 814–820.
40. LIU, J. *et al.* 2000. Functional CB1 cannabinoid receptors in human vascular endothelial cells. Biochem. J. **346**(Pt 3): 835–840.
41. DERKINDEREN, P. *et al.* 1996. Regulation of a neuronal form of focal adhesion kinase by anandamide. Science **273:** 1719–1722.
42. GOMEZ DEL PULGAR, T., G. VELASCO & M. GUZMAN. 2000. The CB1 cannabinoid receptor is coupled to the activation of protein kinase B/Akt. Biochem. J. **347:** 369–373.
43. MOMBOULI, J.V. *et al.* 1999. Anandamide-induced mobilization of cytosolic Ca2+ in endothelial cells. Br. J. Pharmacol. **126:** 1593–1600.
44. DEVANE, W.A. *et al.* 1992. Isolation and structure of a brain constituent that binds to the cannabinoid receptor. Science **258:** 1946–1949.
45. MECHOULAM, R. *et al.* 1995. Identification of an endogenous 2-monoglyceride, present in canine gut, that binds to cannabinoid receptors. Biochem. Pharmacol. **50:** 83–90.
46. HANSEN, H.S. *et al.* 2000. N-Acylethanolamines and precursor phospholipids—relation to cell injury. Chem. Phys. Lipids **108:** 135–150.
47. WILLOUGHBY, K.A. *et al.* 1997. The biodisposition and metabolism of anandamide in mice. J. Pharmacol. Exp. Ther. **282:** 243–247.
48. CRAVATT, B.F. *et al.* 2001. Supersensitivity to anandamide and enhanced endogenous cannabinoid signaling in mice lacking fatty acid amide hydrolase. Proc. Natl. Acad. Sci. USA **98:** 9371–9376.
49. VARGA, K. *et al.* 1998. Platelet- and macrophage-derived endogenous cannabinoids are involved in endotoxin-induced hypotension. FASEB J. **12:** 1035–1044.
50. DI MARZO, V. *et al.* 1999. Biosynthesis and inactivation of the endocannabinoid 2-arachidonoylglycerol in circulating and tumoral macrophages. Eur. J. Biochem. **264:** 258–267.
51. STELLA, N., P. SCHWEITZER & D. PIOMELLI 1997. A second endogenous cannabinoid that modulates long-term potentiation. Nature **388:** 773–778.
52. BISOGNO, T. *et al.* 2003. Cloning of the first sn1-DAG lipases points to the spatial and temporal regulation of endocannabinoid signaling in the brain. J. Cell Biol. **163:** 463–468.
53. KONRAD, R.J., C.D. MAJOR & B.A. WOLF. 1994. Diacylglycerol hydrolysis to arachidonic acid is necessary for insulin secretion from isolated pancreatic islets: sequential actions of diacylglycerol and monoacylglycerol lipases. Biochemistry **33:** 13284–13294.
54. DI MARZO, V. *et al.* 2001. Leptin-regulated endocannabinoids are involved in maintaining food intake. Nature **410:** 822–825.
55. ORZEL, J.A. & T.G. RUDD. 1985. Heterotopic bone formation: clinical, laboratory, and imaging correlation. J. Nucl. Med. **26:** 125–132.

56. WILDBURGER, R. *et al.* 1998. Post-traumatic hormonal disturbances: prolactin as a link between head injury and enhanced osteogenesis. J. Endocrinol. Invest. **21:** 78–86.
57. PANIKASHVILI, D. *et al.* 2001. An endogenous cannabinoid (2-AG) is neuroprotective after brain injury. Nature **413:** 527–531.
58. SUDO, H. *et al.* 1983. *In vitro* differentiation and calcification in a new clonal osteogenic cell line derived from newborn mouse calvaria. J. Cell Biol. **96:** 191–198.
59. JORGENSEN, N.R. *et al.* 2004. Dexamethasone, BMP-2, and 1,25-dihydroxyvitamin D enhance a more differentiated osteoblast phenotype: validation of an *in vitro* model for human bone marrow-derived primary osteoblasts. Steroids **69:** 219–226.
60. BAB, I. 2005. The skeleton: stone bones and stoned heads? *In* Cannabinoids as Therapeutics. Milestones in Drug Therapy Series. R. Mechoulam, Ed.: 201–207. Birkhäuser. Basel.
61. OFEK, O. *et al.* 2006. Peripheral cannabinoid receptor, CB2, regulates bone mass. Proc. Natl. Acad. Sci. USA **103:** 696–701.
62. BELLOWS, C.G. *et al.* 1986. Mineralized bone nodules formed *in vitro* from enzymatically released rat calvaria cell populations. Calcif. Tissue Int. **38:** 143–154.
63. ZHOU, H. *et al.* 1994. *In situ* hybridization to show sequential expression of osteoblast gene markers during bone formation *in vivo.* J. Bone Miner. Res. **9:** 1489–1499.
64. ZHANG, R.W. *et al.* 1995. Expression of selected osteogenic markers in the fibroblast-like cells of rat marrow stroma. Calcif. Tissue Int. **56:** 283–291.
65. LIAN, J.B. *et al.* 2004. Regulatory controls for osteoblast growth and differentiation: role of Runx/Cbfa/AML factors. Crit. Rev. Eukaryot. Gene Expr. **14:** 1–41.
66. ZOU, W. *et al.* 2002. CpG oligonucleotides: novel regulators of osteoclast differentiation. FASEB J. **16:** 274–282.
67. SCUTT, A. & E.M. WILLIAMSON. 2007. Cannabinoids stimulate fibroblastic colony formation by bone marrow cells indirectly via CB2 receptors. Calcif. Tissue Int. **80:** 50–59.
68. TAM, J. *et al.* 2006. Involvement of neuronal cannabinoid receptor CB1 in regulation of bone mass and bone remodeling. Mol. Pharmacol. **70:** 786–792.
69. HANUS, L. *et al.* 2001. 2-arachidonyl glyceryl ether, an endogenous agonist of the cannabinoid CB1 receptor. Proc. Natl. Acad. Sci. USA **98:** 3662–3665.
70. LEDENT, C. *et al.* 1999. Unresponsiveness to cannabinoids and reduced addictive effects of opiates in CB1 receptor knockout mice. Science **283:** 401–404.
71. TAM, J. *et al.* 2007. The Cannabinoid CB1 receptor regulates bone formation by modulating adrenergic signaling. FASEB J. In press.
72. ZIMMER, A. *et al.* 1999. Increased mortality, hypoactivity, and hypoalgesia in cannabinoid CB1 receptor knockout mice. Proc. Natl. Acad. Sci. USA **96:** 5780–5785.
73. BROWN, J.P. *et al.* 1984. Serum bone Gla-protein: a specific marker for bone formation in postmenopausal osteoporosis. Lancet **1:** 1091–1093.
74. KARSAK, M. *et al.* 2005. Cannabinoid receptor type 2 gene is associated with human osteoporosis. Hum. Mol. Genet. **14:** 3389–3396.
75. YAMADA, Y., F. ANDO & H. SHIMOKATA. 2007. Association of candidate gene polymorphisms with bone mineral density in community-dwelling Japanese women and men. Int. J. Mol. Med. **19:** 791–801.

Androgens Promote Preosteoblast Differentiation via Activation of the Canonical Wnt Signaling Pathway

XIN-HUA LIU, ALEXANDER KIRSCHENBAUM,* SHEN YAO, AND ALICE C. LEVINE

Department of Medicine and Department of Urology, Mount Sinai School of Medicine, New York, New York, USA*

ABSTRACT: Although androgens stimulate bone formation the precise events underlying these effects have not been elucidated. Wnt signaling plays a central role in osteoblast development and bone formation. We demonstrated that dihydrotestosterone (DHT) significantly stimulates MC3T3 preosteoblast differentiation with no effect on cell growth. This effect of DHT was accompanied by increased Wnt signaling in the same cells. Moreover, the stimulatory effects of DHT on preosteoblast differentiation were inhibited by overexpression of soluble frizzed-related protein (sFRP), a naturally occurring Wnt antagonist. These results suggest that androgens promote preosteoblastic differentiation via effects on the canonical Wnt signaling pathway.

KEYWORDS: androgens; preosteoblasts; Wnt signaling

INTRODUCTION

Androgens play a critical role in the development and maintenance of the skeleton in both sexes. Androgen receptors (AR) are detected in a variety of bone cells, including osteoblasts, osteocytes, and bone marrow stromal cells.[1] Studies using nonaromatizable androgens in ovariectomized rats demonstrate direct actions of androgens on the preservation of bone mass.[2] These reports suggest that androgens, in contrast to estrogens, have little effect on the inhibition of bone resorption but, rather, directly stimulate bone formation. However, the precise mechanisms underlying the observed anabolic effects of androgens on bone cells are unclear. Moreover, several *in vitro* studies have yielded conflicting and contradictory data. Androgens have been reported to enhance both the proliferation and differentiation of primary cultures of osteoblast-like cells.[3] However, that same report demonstrated a direct inhibitory effect of

Address for correspondence: Xin-Hua Liu, Department of Medicine, Box 1055, Mount Sinai School of Medicine, New York, NY 10029. Voice: 212-241-4130; fax: 212-241-4218.
Liu.Xinhua@mssm.edu

Ann. N.Y. Acad. Sci. 1116: 423–431 (2007). © 2007 New York Academy of Sciences.
doi: 10.1196/annals.1402.017

DHT on a human fetal osteoblast cell line *in vitro*.[3] Davey and Morris also reported that androgens can inhibit osteoblast development in the absence of estrogen and in states of low bone turnover.[4] The reported discordant effects of androgens on bone cells warranted further investigation.

The canonical Wnt signaling pathway plays a central role in bone mesenchymal cell lineage commitment, promoting osteoblastic differentiation and stimulating bone formation. While various bone-related cells, including osteoblasts, express androgen receptor (AR) protein, the cross talk between the AR and Wnt signaling pathways in bone cells has not previously been investigated.

This study demonstrates that although the nonaromatizable androgen, dihydrotestosterone (DHT), has no effect on preosteoblastic cell growth, androgens significantly promote preosteoblast differentiation. This effect of DHT is accompanied by activation of canonical Wnt signaling, as evidenced by increases in Wnt-dependent transcriptional reporter activity, enhanced GSK3β^{S-9} phosphorylation, accumulation of nuclear β-catenin and increased nuclear Runx2 expression. Moreover, this effect of DHT can be prevented in the same cell line by forced overexpression of soluble frizzed-related protein (sFRP)-1, a natural inhibitor of the Wnt signaling pathway. These results suggest that androgen-promoted preosteoblast differentiation is mediated, at least in part, by activation of canonical Wnt signaling.

MATERIALS AND METHODS

Cell Culture and Reagents

The MC3T3-E1 cell line was obtained from the American Type Tissue Collection (ATCC). DHT was purchased from Sigma (St Louis, MO, USA). Enhanced Luciferase Assay kit was obtained from BD Pharmingen (San Diego, CA, USA). Antibodies against AR were purchased from Santa Cruz BioTech, Inc. (Santa Cruz, CA, USA). Antibodies against GSK3β, phosphor-GSK3β^{ser-9} and phosphor Akt$^{ser-473}$ were obtained from Cell Signaling Tech. (Beverly, MA, USA). β-catenin antibodies were obtained from BD Pharmingen Inc. (San Diego, CA, USA). Runx2 antibodies were purchased from MBL International Co. (Woburn, MA, USA).

Alkaline Phosphatase Activity (ALP) Assay

ALP activity was measured in cell lysates and by staining of cultured cells using an ALP assay kit and ALP staining kit, respectively (Sigma), according to the manufacturer's instructions.

Protein Isolation, Immunoblotting, and Immunofluorescence Assay

Total protein isolation and immunoblotting were performed as described previously.[5] Proteins from the cytosolic and nuclear fractions were isolated using a commercial kit purchased from PIERCE (Rockford, IL, USA), according to the manufacturer's instructions. Immunofluorescence assay was performed as previously reported.[6]

Transient Transfection and Luciferase Reporter Assay

The generation of the TCF luciferase reporter construct pGL3-OT and control vector were described previously.[6] TransLucent AR Reporter Vector and control plasmid were purchased from Panomics Inc. (Redwood City, CA, USA). The methodology for luciferase reporter assay and transient transfection were carried out as previously described.[6]

RESULTS

Effect of DHT on AR Expression and AR Luciferase Activity in MC3T3 Preosteoblasts

We initially examined AR protein expression in MC3T3 and NIH3T3 cells (used as a negative control). As shown in FIGURE 1A, MC3T3 cells expressed relatively low basal levels of AR protein and DHT increased AR expression after 24 h. This AR protein induction was significant ($P < 0.01$) with a mean 3.2-fold increase after DHT addition (three separate assays). In contrast, NIH3T3 fibroblast cells expressed neither basal nor inducible AR protein. AR transcriptional activity was next assessed. As shown in FIGURE 1B, treatment of MC3T3 cells with DHT enhanced AR luciferase activity in a time-dependent manner.

DHT-Induced Tcf Transcriptional Activity in MC3T3 Preosteoblasts

We tested the effects of DHT on activation of the Tcf transcription factor using a Tcf-dependent luciferase transcriptional reporter. As shown in FIGURE 2, DHT addition resulted in a significant enhancement of Tcf-luciferase activity in MC3T3 cells. This induction was both time- and dose-dependent.

DHT Promoted Akt and GSK3β Phosphorylation

The activation of GSK3, a critical component in the Wnt pathway, is regulated by Akt activity.[7] Activated Akt by phosphorylation at the site of serine[473]

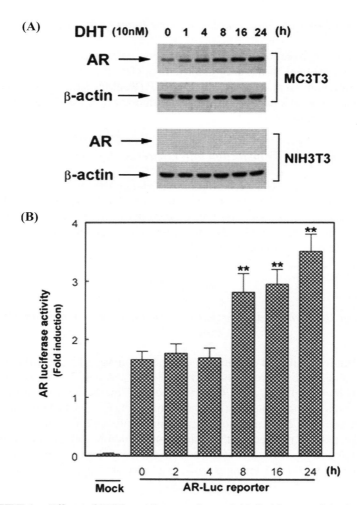

FIGURE 1. Effects of DHT on AR expression and AR luciferase activity in MC3T3 preosteoblast cells. (**A**) DHT increases AR protein expression. MC3T3 cell lysates were prepared and subjected to Western blotting. (**B**) DHT induces AR transcriptional activity. Data are the mean ± SE from three determinations.* $P < 0.05$, and** $P < 0.01$ versus vehicle control.

promotes GSK3α^{ser-21} and GSK3β^{ser-9} phosphorylation, thereby inhibiting both GSK3β and GSK3α activity.[8] We next examined the effects of DHT on Akt and GSK3α/β protein phosphorylation. Although MC3T3 cells expressed low basal levels of phosphorylated Akt and GSK3α/β protein, DHT significantly enhanced Akt^{ser473} and GSK3β phosphorylation, particularly at serine 9, in MC3T3 cells (FIG. 3A, B). There was no effect of DHT on endogenous

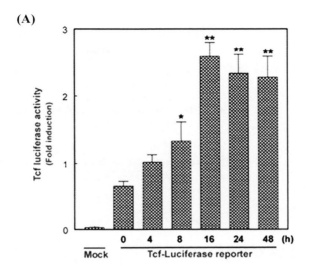

(A)

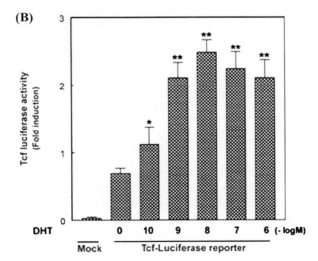

(B)

FIGURE 2. DHT-induced Tcf transcriptional activity in MC3T3 preosteoblast cells. (A) Time course of DHT (10^{-8} M) induction. (B) Dose-response of DHT effects. Results represent the mean $\pm$ SE from three determinations.* $P < 0.05$, and** $P < 0.01$ versus vehicle control.

Akt and GSK3 protein levels in these cells. These data demonstrate an association between DHT-induced Akt^{ser473} phosphorylation and GSK3β activity, which may be an important component of canonical Wnt signaling activation in preosteoblasts.

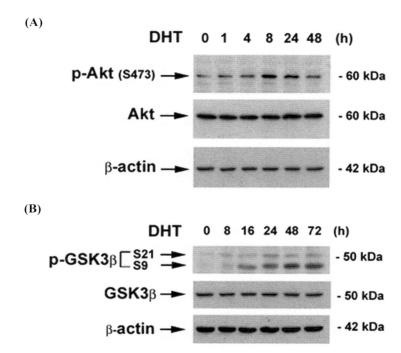

FIGURE 3. DHT promotes both Akt and GSK3β phosphorylation in MC3T3 pre-osteoblast cells. MC3T3 cell lysates were prepared after treatment with 10^{-8} M DHT, and subjected to Western blotting. Data shown are representative of three separate experiments.

Effect of DHT on the Nuclear Levels of β-Catenin and Runx2

Activation of the canonical Wnt signaling pathway results in accumulation of β-catenin in the nucleus, heterodimer formation with Tcf, and activation of Tcf-dependent transcription. Thus, we examined the effects of DHT on the subcellular localization of β-catenin and nuclear Runx2 levels (a direct target gene of the Wnt signaling in bone-related cell lines)[9] in MC3T3 preosteoblast cells. As shown in FIGURE 4A, DHT significantly increased both β-catenin and Runx2 nuclear protein levels. The DHT effect on β-catenin nuclear localization was confirmed in MC3T3 cells by immunofluorescence assay (FIG. 4B).

Inhibition of Wnt Signaling Abrogated DHT Effects on MC3T3 Preosteoblast Cells

To determine whether DHT-induced Wnt signaling activation is critical to its biological activity in preosteoblasts, we established an MC3T3 subline that stably overexpresses sFRP, a naturally occurring Wnt inhibitor (sFRP-subline). The effects of DHT on cell proliferation and differentiation were

(A)

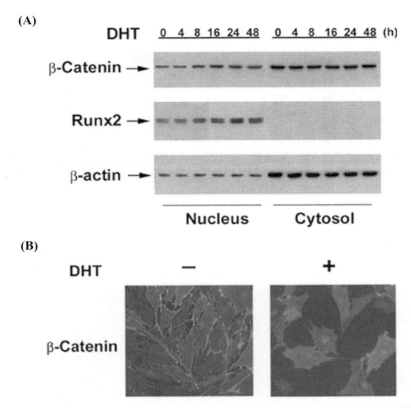

FIGURE 4. Effect of DHT on the nuclear levels of β-catenin and Runx2 in MC3T3 preosteoblast cells. **(A)** Western blot. The cytosolic and nuclear proteins were isolated from MC3T3 cells, and subjected to Western blotting. **(B)** Immunofluorescence assay. MC3T3 cells were treated with either vehicle or DHT (10^{-8}M) for 24 h and subjected to immunofluorescence stain. Data shown are representative of three separate experiments.

examined and compared between the two MC3T3 cell lines. As shown in FIGURE 5A, DHT did not influence parental MC3T3 cell proliferation, and had a moderate inhibitory effect on the growth of the sFRP-subline. However, DHT significantly stimulated parental MC3T3 differentiation as determined by ALP assay and ALP staining (FIG. 5B, C). In contrast, the MC3T3-sFRP cells exhibited lower basal ALP expression and this expression was not increased by DHT addition.

SUMMARY

Our data indicate that the potent, nonaromatizable androgen dihydro-13testosterone (DHT) promotes the differentiation of the MC3T3 preosteoblast

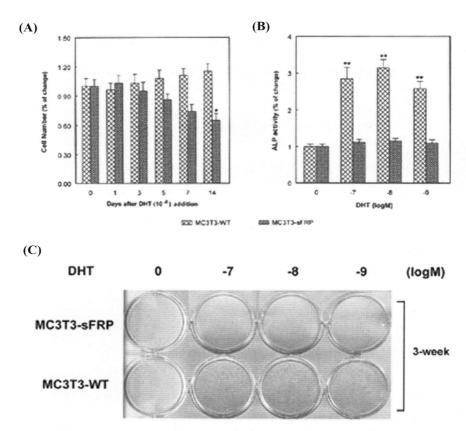

FIGURE 5. Inhibition of Wnt signaling mitigated DHT effects on MC3T3 pre-osteoblast cell differentiation. (**A**) Cells were treated with DHT for 7days and cell numbers were counted using a hemacytometer. (**B**) Effect of DHT on alkaline phosphatase (ALP) activity. MC3T3 cells were treated with 10^{-8} M DHT for 21 days, and ALP activity was assayed in cell lysates. Data shown (A&B) are means $\pm$ SE of three independent assays. (**C**) Both cell lines were exposed to varying levels of DHT for 3 weeks and stained for ALP activity.

cell line, as determined by alkaline phosphatase expression. In addition, DHT activated canonical Wnt signaling in these same cells, and this effect appears to be initiated by activation of Akt followed by phosphorylation and inactivation of GSK-3β. Finally, forced overexpression of the Wnt inhibitor sFRP blocked the effect of DHT on alkaline phosphatase expression. These results indicate that androgens promote preosteoblast cell differentiation and that this effect is mediated, at least in part, by activation of the canonical Wnt signaling pathway.

REFERENCES

1. HOFBAUER, L.C. & S. KHOSLA. 1999. Androgen effects on bone metabolism: recent progress and controversies. Eur. J. Endocrin. **140:** 271–286.
2. COXAM, V., B.M. BOWMAN, C.M. ROTH, *et al.* 1996. Effects of DHT alone and combined with estrogen on bone mineral density, bone growth, and formation rates in ovariectomized rats. Bone **19:** 107–114.
3. KASPERK, CH., J.E. WERGEDAL, J.R. FARLEY, *et al.* 1989. Androgens directly stimulate proliferation of bone cells *in vitro*. Endocrinology **124:** 1576–1578.
4. DAVEY, R.A. & H.A. MORRIS. 2004. Effects of dihydrotestosterone on osteoblast gene expression in osteopenic ovariectomized rats. Endocrine Res. **30:** 361–368.
5. LIU, X.H., A. KIRSCHENBAUM, A.C. LEVINE, *et al.* 2005. Cross-talk between the IL-6 and PGE2 signaling systems results in enhancement of osteoclastogenesis through effects on the OPG/RANKL/RANK system. Endocrinology **146:** 1991–1998.
6. LIU, X.H., A. KIRSCHENBAUM, A.C. LEVINE, *et al.* 2007. Androgen-induced Wnt signaling in pre-osteoblasts promotes the growth of MDA-PCa-2b human prostate cancer cells. Cancer Res. **67:** 5747–5753.
7. CROSS, D.A., D.R. ALESSI, P. COHEN, *et al.* 1995. Inhibition of glycogen synthase kinase-3 by insulin mediated by protein kinase B. Nature **378:** 785–789.
8. WESTENDORF, J., R.A. KAHLER & T.M. SCHROEDER. 2004. Wnt signaling in osteoblasts and bone diseases. Gene **341:** 19–39.
9. GAUR, T, C.J. LENGNER, H. HOVHANNISYAN, *et al.* 2005. Canonical Wnt signaling promotes osteogenesis by directly stimulating Runx2 gene expression. J. Biol. Chem. **280:** 33132–33140.

Tumor Necrosis Factor Family Receptors Regulating Bone Turnover

New Observations in Osteoblastic and Osteoclastic Cell Lines

LISA J. ROBINSON,[a] CHRISTOPHER W. BORYSENKO,[b] AND HARRY C. BLAIR[a]

[a]Departments of Pathology, and of Cell Biology and Physiology, University of Pittsburgh and Veterans' Affairs Health System, Pittsburgh, Pennsylvania, USA

[b]Interdisciplinary Laboratory, Carnegie Mellon University, Pittsburgh, Pennsylvania, USA

ABSTRACT: While the tumor necrosis factor (TNF) family members RANKL and TNF-α are critical regulators of osteoclast formation, functions of other TNFs in bone are poorly understood. Here we consider the roles in regulating bone turnover of TNF receptors (TNF-R) also expressed by osteoblasts and osteoblast precursors. TNF receptors in osteoblasts and preosteoblasts include TNFR1 (p55), DR3 (TNFR25), DR5 (TRAIL-R2) and Fas, and possibly FN14 and DR4 (TRAIL-R1). Osteoblasts also produce soluble TNF receptors, DcR2, osteoprotegerin, and sDR3; these bind the TNFs TRAIL, RANKL, TL1A, and Apo3L and block ligand effects on cell surface receptors. Activation of DR3 regulates osteoblast maturation and may control the decision to exit the pool of differentiation-competent preosteoblasts. A major natural ligand for DR3, TL1A, is produced by vascular cells adjacent to differentiating osteoblasts and possibly by Fcγ-stimulated osteoclast precursors. The activity of DR3 is regulated by osteoblast production of its soluble DR3 splice variant. Activation of TNFR1 or DR5 by TNF-α or TRAIL may regulate osteoblast connectivity, which is important to bone turnover. When there is a source for Fas ligand, such as an inflammatory infiltrate, activation of Fas may lead to apoptosis of any bone cell. TNF receptors are thus implicated in multiple aspects of bone turnover.

KEYWORDS: apo3; death receptor-3; death receptor-5; rheumatoid arthritis; osteoporosis; LARD; TRAMP; TWEAK

Address for correspondence: Harry C. Blair, 705 Scaife Hall, University of Pittsburgh, Pittsburgh, PA 15261. Voice: 412-383-9616; fax: 412-647-8567.
    hcblair@imap.pitt.edu

Ann. N.Y. Acad. Sci. 1116: 432–443 (2007). © 2007 New York Academy of Sciences.
doi: 10.1196/annals.1402.025

Tumor necrosis factor (TNF) family receptors are critical to the differentiation and survival of many cell types, including hematopoietic precursors and cells of the immune system.[1,2] Expression of TNF receptors by osteoblasts and their precursors have been defined by transcriptional profiling studies, and there are many studies examining the expression and function of individual receptors (reviewed in Ref. 3). However, the function of these receptors in osteoblasts and related cells is, for the most part, incompletely characterized.

Interest in the TNFs and their receptors in bone led to important findings regarding the role of RANKL, along with cosignaling pathways, in osteoclast differentiation.[4] There is also a long-standing interest in TNF-α because of its association with osteoporosis and rheumatoid diseases. TNF-α is an accelerator of osteoclast formation and in humans, in some circumstances, may permit limited osteoclast formation even in the absence of RANKL.[5] Although RANKL and TNF-α are of key clinical importance in bone losing states and in arthritis, the role of these signals in bone catabolism is a subject of extensive reviews elsewhere and will not be considered further here.

On the other hand, bone degradation is balanced by bone formation, and the presence of other TNF receptors and ligands in bone, including in osteoblasts, argues that the process of bone formation, distal to the bone morphogenetic protein (BMP), parathyroid hormone (PTH), and Wnt-related signals regulating self-replication in stem cell compartments (reviewed elsewhere in this volume by Blair, Li, and Kohansky[6]), is also likely to be regulated by TNFs. Since TNF-R1 and DR5 are expressed on osteoblasts while osteoclasts are also known to respond to their ligands, it is possible that both bone-forming and bone-degrading cells are coregulated by TNF-α. However, other osteoblast TNF-Rs, such as DR3 (TNFSFR12/TRAMP/LARD) and possibly FN14, appear to be unique to mesenchymal stem cell-derived components. Of these, DR3 is of particular interest because it is alternatively spliced to produce soluble decoy receptor and cell membrane forms, and there are precedents in lymphoid cells for the expression of membrane DR3 versus soluble DR3 functioning as a developmental switch. Further, it is implicated by linkage analysis with bone and joint diseases (discussed in Ref.7). Thus, following a brief discussion of emerging principles regarding the function of TNF-regulated systems in complex organs, we will focus discussion on DR3 biology. We will discuss TRAIL, TNF-α, and Fas and finally focus on the potential importance of FN14, a TNF receptor lacking the death domain, which some report to be expressed by osteoblasts. To keep discussion to a manageable length, however, we will not discuss in detail TNF signals or receptors, such as the TNF LIGHT,[8] that are reported to be important in bone turnover but for which the ligand–receptor pairs are not well defined. The interested reader will be referred to the references at these points.

IMPORTANCE OF LOCAL EXPRESSION OF
MEMBRANE-BOUND TNF LIGANDS, AND CAVEATS
REGARDING THE FUNCTIONAL SIGNIFICANCE
OF TNF INTERACTIONS

Consideration of TNF-Rs and their ligands is complicated by their degeneracy. The TNF-Rs occur in groups with 3–5 extracellular cysteine-rich globular domains that interact with ligand(s). Typically ligands form trimers and cause the association of receptor subunits upon binding.[9] Destabilization of cysteine-rich domains is a common cause of defective signaling, which may be associated with malignancy and diseases, such as rheumatoid arthritis.[7] The intracellular portion of the receptors also include common domains that are linked to specific differentiation and death pathways. Many of the receptors contain TRAF-activating (differentiation) domains that are linked to further pathways including jun-kinase activation. Some of the TNF receptors also contain death domains that, when present, may activate caspases leading to cell death.

TNF receptor signaling is complicated not only by parallel activation of common intracellular signal pathways by multiple receptors but also by nonspecificity of ligands ("ligand degeneracy"). The subject was recently reviewed with some key interactions highlighted.[9] High-affinity interactions of ligands with multiple receptors usually occur with receptors showing greater homology, but low-affinity interactions are probably quite common and may be responsible for quirky experimental cross-reactions between TNF ligands *in vitro*, where pharmacological quantities of ligands are often added to cells.

There are many well-characterized examples of receptors with multiple ligands. One of the first receptors found to show high-affinity binding of multiple ligands was TNF-R1, which can be activated by both lymphotoxin and TNF-α. Lymphotoxin and TNF-α share about 40% amino acid sequence identity and show great similarity in structure with conservation of key amino acids.[10] An example relevant to bone turnover is that of TRAIL and RANKL, both of which bind the soluble TNF-R family decoy receptor, osteoprotegerin.[11]

Additional interactions of this type that may be important in bone include that of Apo3L, a high-affinity ligand for the small, nondeath receptor, TNF-R family member FN14,[12,13] while Apo3L also binds to DR3, albeit probably at lower affinity.[14,15] DR3 is an important TNF family receptor that functions either as a transmembrane receptor with differentiation and death-regulating domains, or as a "decoy" receptor in a secreted form that lacks the transmembrane domain. A higher affinity ligand for DR3 is TL1A,[15] which is likely to be its major natural ligand under many circumstances.

There have been heated arguments as to which ligand(s) of TNF-Rs are the "real" ligand(s). Most of these arguments are difficult to justify. While some receptor–ligand interactions detected *in vitro* may be biologically unimportant, this is not necessarily established, one way or the other, by studies using soluble recombinant proteins showing differences in affinity on the order of 2- to 10-fold. *In vivo*, most interactions of TNFs with their receptors do not involve binding of a ligand free in solution: almost all naturally occurring TNFs are transmembrane proteins that signal locally. Usually, TNFs released from the cell surface are rapidly scavenged, and physiologically significant interactions are thus typically those in which ligand and receptor co-localize. If co-localization of a putative receptor and ligand pair is confirmed, knockout cells or RNA interference to eliminate ligand or receptor can be useful in determining whether or not the interaction is biologically meaningful.

TNF-α may be an important exception to the principle that the membrane-bound ligands are the primary active form since significant quantities of TNF-α circulate, although TNF-α is also synthesized as a membrane-bound molecule. In addition, it is not clear which, if any, of the scavenger receptors have adequate affinity for TNF-α to clear it efficiently from circulation, although this is an important unresolved point.

There are also a few TNFs, such as TL1A, that are synthesized as both membrane[16] and soluble proteins (often called vascular endothelial growth inhibitor [VEGI] in this context).[9] The purpose of producing both soluble and membrane-bound TL1A may be an important question in bone biology because of the increasing evidence of a key role for its receptor, DR3, in the regulation of bone.

THE ROLE OF DR3 IN OSTEOBLAST
MATURATION AND SURVIVAL

Nontransformed human osteoblasts and a transformed cell line, MG63, express transmembrane and soluble forms of DR3. DR3 regulates osteoblast differentiation and, in narrowly defined contexts, may induce apoptosis.[7,15] In addition, it was recently found that TNF-α upregulates the DR3 ligand VEGI (an isoform of TL1A) in bone cells.[17] Thus, DR3 signaling may be amplified as a downstream action of increased TNF-α activity, which occurs in rheumatoid arthritis and postmenopausal osteoporosis.

In addition, genetic studies demonstrate that DR3 functions in skeletal regulation. Single nucleotide polymorphisms in the extracellular ligand-binding domain of DR3 are linked to rheumatoid arthritis.[18,19] These polymorphisms are related to single amino acid substitutions that destabilize the third cysteine-rich ligand-binding domain of DR3, and occur in positions homologous to ligand-destabilizing mutations in other TNF family receptors.[7] Further, sequencing of the cDNA for full-length transmembrane DR3 from human MG63 preosteoblast-like osteosarcoma cells disclosed a novel mutation in the intracellular domain proximal to the death domain, suggesting that DR3 may contribute to dysregulation in some bone cancers.[15]

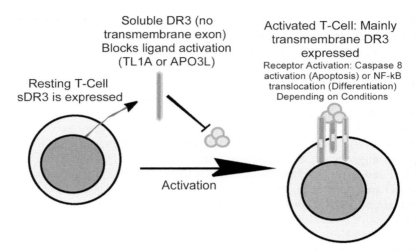

FIGURE 1. DR3 is expressed in a soluble form in resting T cells and as a transmembrane protein in activated T cells. Resting T cells express soluble DR3, which binds DR3 ligand(s) to prevent ligand engagement with the transmembrane form. Thus, it is a decoy receptor similar to OPG. Activated T cells, however, express transmembrane DR3 but not the soluble form, permitting ligand activation of the receptor. In cell lines, activation of transmembrane DR3 leads to either apoptosis or activation of transcription factors, such as NF-κB. The death receptor Fas also is made in soluble and transmembrane forms, but its transmembrane form mediates only death signals.

Other cells expressing DR3 include endothelial cells and lymphocytes. Indeed, the mechanisms of action of DR3 are best understood from the extensive studies of its function in lymphocytes in the context of immune cell differentiation. In T cells, DR3 regulates survival and proliferation.[20,21] In quiescent T cells, a soluble form of DR3 is secreted and is the main form of DR3 expressed. It protects the cells from activation of membrane-bound DR3. After activation of T cells, DR3 transcripts are alternatively spliced to produce mainly transmembrane DR3. DR3 signaling stimulates antigen-stimulated T cells to survive, and regulation of activation of the cells is also mediated by DR3 ligands.[12] DR3 activity can be reproduced *in vitro* by oligomerization of DR3 using anti-DR3 antibodies.

As with most TNF family receptors, DR3 activation can alternatively lead to apoptosis, and transmembrane forms of DR3 are important in negative clonal selection of pre-T cells. Effects of DR3 on cell differentiation and survival are associated with NF-κB activation (FIG. 1). Overexpression of DR3 can activate caspase 8, leading to apoptosis, or may mediate NF-κB activation with enhanced cell survival and cytokine responses.[12,14]

The Apo3L (TNF-like weak apoptosis inducer, or TWEAK) was the first activating ligand for DR3 identified.[14] It is highly similar in structure to TNF-α, but no direct interaction of Apo3L with TNF-α receptors is known. Apo3L may, however, indirectly activate TNF-R1 by binding soluble receptors, common to APO3L and TNF-α, that would otherwise limit TNF-α activity.[22] Apo3L is also a strong ligand for FN14,[12,13] (discussed below), and there is evidence that additional receptors activated by Apo3L occur on osteoclasts.[23] RANK, however, is effectively ruled out as a receptor for Apo3L.

An additional close TNF-α homolog, TL1A, and its splice variants, including the soluble VEGI, also activate DR3.[12,24] However, in the main, a naturally occurring form of TLIA is a type II transmembrane protein that acts mostly through direct cell–cell interactions. Thus, its function is probably mainly as a local cell–cell stimulus. TL1A and its isoforms are induced by TNF-α in bone,[17] and are also well-characterized endothelial products that are strong candidates for the regulation of the paravascular preosteoblastic component in bone.

Our studies of human osteoblasts and osteosarcoma cells[15] showed that these cells produce both signaling and soluble forms of DR3. Mainly at low cell densities, DR3 activation can cause osteoblast apoptosis; media containing soluble DR3 prevent this. However, the most interesting findings were that at high cell density, DR3 activation does not cause apoptosis but inhibits differentiation and promotes quiescence of osteoblast precursors. The effect was not coupled with any marked increase in cell numbers, so it is unlikely that the DR3 effect in inhibiting differentiation is directly linked to proliferation. This system may function together with more fundamental regulatory signals mediated by PTH, insulin-like growth factor, or frizzled family seven transmembrane-domain receptors to regulate the number or rate of cells entering the secretory phase of bone production. Interestingly, activation of the Fcγ receptor in human monocytic cells induces expression of the membrane form of TL1A,[16] and Fcγ receptor activation is a required costimulus, with RANK and Fms (the CSF-1 receptor), for osteoclast formation.[4] This raises the possibility of TL1A as a coregulator coupling bone formation and degradation preventing osteoblast differentiation at sites where osteoclast activity is also induced.

THE ROLE OF TRAIL AND TNF-α IN OSTEOCLAST AND OSTEOBLAST REGULATION: RESISTANCE TO APOPTOSIS BUT POSSIBLE ROLE REGULATION IN CELL–CELL INTERACTIONS

While TNF-α promotes osteoclast formation,[5] it has been reported that TRAIL can interfere with RANKL signaling during osteoclast formation.[25] It is also possible that TRAIL participates in cell death decisions in osteoclasts.[26] However, a physiological role for TRAIL is at this point uncertain; as discussed above, TRAIL signals almost entirely locally, through cell–cell interactions, and in bone there are many highly expressed TRAIL scavenger proteins to bind free TRAIL in bone.[3]

On the other hand, the role of the receptors TNF-R1 (p55) and TRAIL-R2 (DR5) in osteoblasts is unclear. Rodent osteoblasts can undergo apoptosis in response to TNF-α, but only when cells have been sensitized, for example, by serum starvation.[27,28] Generally, human osteoblasts are resistant to TNF-α- or (unlike osteoclasts) TRAIL-mediated apoptosis, but some human osteoblast-like cells may undergo apoptosis with TNF-α under specific conditions, such as after treatment with sublethal doses of cytotoxic drugs or after serum starvation.[29,30]

Normal human osteoblasts do not produce TNF-α,[3] and TNF-α effects in bone probably require cells in the monocyte lineage (which also may produce osteoclasts) or inflammatory cells. Circulating soluble TNF-α may be the functionally significant form; perhaps as much as 50% of TNF-α biological activity is mediated by the soluble factor, whereas concentrations of other TNF family ligands are mainly vanishingly small with the transmembrane forms, active locally, the sole relevant product. In contrast, TRAIL is produced in surprisingly large quantities by osteoblasts; however, it appears to occur mainly as a high-molecular weight form that may be inactive,[3] an observation that will require further analysis to determine its full meaning.

Osteoblast apoptosis is an important cause of focal bone loss that precedes collapse in high-dose glucocorticoid therapy,[31] In studies of TNF-α and TRAIL effects on both nontransformed and transformed human osteoblasts, we found that osteoblasts were resistant to apoptosis induced by TNF-α or TRAIL.[3] As with DR3, some inhibition of receptor effects may actually reflect sequestration of the ligand by soluble decoy receptors. There are several soluble TNF receptors that can bind TRAIL. However, it is clear that under most circumstances the cells simply do not die when exposed to these TNF family ligands. Instead, TRAIL appeared to affect cell–cell connectivity.[3] This may be similar to the reduction in fibroblast connexin43 protein quantity and cell connectivity when exposed to TNF-α.[32] These findings raise the prospect that late osteoblast phenotype characteristics, including connectivity and connexon structure, may be regulated by TNF-α or TRAIL, although further study will be required to confirm this hypothesis.

OTHER RECEPTORS ACTIVE IN OSTEOBLASTS INCLUDE FAS AND APO3L RECEPTORS

FN14 is a TNF family receptor expressed on many epithelial cells, such as hepatocytes, and endothelial cells.[33] It is activated by Apo3L (TWEAK) and possibly by other unknown ligands. There are two reports of FN14 inhibiting rodent osteoblast maturation *in vitro*, with additional very interesting associations, including observations such as the association of elevated levels of

Apo3L with arthritis and the increase in RANKL expression by osteoblasts after Apo3L exposure.[34,35] These studies used Apo3L to induce osteoblast production of RANTES and RANKL and showed cell staining with an antibody to FN14 on flow analysis, but did not show the antibody was specific for FN14 only, or isolate FN14 mRNA.[34] The potential clinical significance of FN14 is clear from the finding that antibody to Apo3L protected against arthritis progression in a murine model, although this finding, or interference of soluble FN14 in Apo3L binding, do not necessarily indicate that FN14 is mediating the cellular response.

In our hands, in studies using transformed and nontransformed human cells, Apo3L had minor effects on osteoblast maturation and osteoblast apoptosis in low density cultures.[15] There were possible effects of Apo3L on NF-κB nuclear localization that were consistent with the results from murine work on FN14 responses. Additionally, when human osteosarcoma cells or osteoblasts were grown at low density, Apo3L increased osteoblast apoptosis slightly.[15] However, we could not detect FN14 protein by Western blots of nontransformed or transformed human osteoblasts. With reverse transcriptase-polymerase chain reaction (RT-PCR) we obtained only very low levels of an amplification product of appropriate size, and results of real-time PCR using published probes were consistent with only very low levels, if any, of FN14 mRNA.[15] On the other hand, FN14 expression may vary with cell maturation or other conditions, and very small quantities of receptor may have activity. Further, some inconsistencies between DR3 activation using anti-DR3 and the effects of Apo3L were observed. These could be due to the involvement of multiple receptors— DR3, FN14, and possibly others— in the signaling response, or may reflect variable biological activity of recombinant soluble Apo3L due to its synthesis in nonmammalian cells.

Some studies of DR3 showed effects on osteoblast maturation[15] cross-linking of DR3 using antibodies in transformed and nontransformed cells, and solutions containing soluble DR3 blocked activation, so it is unlikely that the DR3 effect is an artifact. In contrast, studies using Apo3L could also reflect activation of DR3 as well as FN14. Thus, it will be important to study FN14 and DR3 further. It is most likely that activation of either DR3 or FN14 can retard osteoblast differentiation: related effects for two TNF receptors on the same cell have many precedents, including interactions between TNF-α and RANKL in osteoclasts. Indeed, redundancy or partial overlap of TNF receptor activity and of TNF ligands with multiple receptors is more the rule than the exception. It will only be clear whether FN14, DR3, or both are of physiological importance in osteoblast differentiation *in vitro* when further characterization is done.

It will be useful in further studies to determine the effects of Apo3L in the absence of DR3, and of TL1A in the absence of FN14, as part of this additional characterization. In this regard, osteoclast precursors, on which FN14 was undetectable, responded to Apo3L,[23] while we found no DR3 at all in osteoclasts,[15] thus indicating that Apo3L (TWEAK) may have additional receptors beyond DR3 and FN14. Homology studies[9] suggest that BCMA, TAC1, or BAFF might be candidates. This is not yet supported by experimental work, but TAC1 is proposed as a myeloma-produced osteoclast survival factor.[36]

Another major TNF family receptor expressed in osteoblasts and pre-osteoblastic cells is the death receptor Fas.[3] Activation of Fas with pentameric anti-Fas led to near-quantitative osteoblast apoptosis. However, a natural source for Fas-ligand in bone is not known. Fas-ligand was not detected as an osteoblast product. These findings raise the possibility that osteoblasts are killed by Fas-mediated apoptosis during inflammatory remodeling, such as in postfracture reorganization where Fas-L may be brought into contact with osteoblasts by inflammatory cells, and suggest that the source of Fas-ligand is normally external to bone.

SUMMARY AND CONCLUSIONS

The complex interactions and biological roles of TNF ligand and receptors in bone are still in many ways incompletely understood. It is clear that TNF-α itself is important as an "accelerator" of bone turnover, both enhancing osteoclast formation and activity, and possibly facilitating osteoblastic participation in bone degradation (TABLE 1). As with many cell types, the pure death receptor Fas was also expressed on most bone cells, but presumably this functions only in widespread tissue death as the ligand is not produced normally by bone cells. In the mid 1990s, the important discovery that RANKL is important in osteoclast differentiation added a new dimension to the role of TNFs in bone. Indeed, this was an example of a T cell-related differentiation factor adapted for use in regulating bone cell differentiation. Initially seen as an isolated observation, it is now clear that the osteoblastic component of bone participates in related regulation through DR3, which is also a T cell differentiation controlling agent. Further complexity includes the involvement of TRAIL, a TNF-α-like factor that is produced by osteoblasts and which has receptors on both osteoblasts and osteoclasts. TRAIL interferes with osteoclast differentiation *in vitro* and may mediate short-term reductions in osteoblast–osteoblast communication, but these, if active *in vivo*, are likely to be controlled by a finely regulated spatial and temporal restriction of TRAIL expression. In addition, there are vascular-related TNFs, including TL1A, which may participate in maintaining the undifferentiated state in osteoblast precursors. There may also be osteoblast maturation-promoting TNF receptors, such as FN14, but the physiological role for this and which cells make it and its ligand in meaningful quantities, is unknown. Thus, the TNF receptor system in bone is integrated into regulation of differentiation and activity of both bone-forming and -degrading cells. Its activity is not only vital for osteoclast formation, but

TABLE 1. Major TNF receptors regulating differentiation or activity in bone-forming or -degrading cells

Cell	TNF-R	Ligand(s) / source(s)	Actions	References
Osteoblasts or osteoblast precursors	TNF-R1	TNF-α / macrophage osteoclast(?) lymphocyte	Cell–cell communication (?)	[32]
	DR5	TRAIL / Osteoblast, other bone cells(?)	Cell–cell attachment	[3]
	DR3	TL1A, Apo3L(?) / Endothelial cell monocyte-family cells	Inhibition of differentiation / conditional apoptosis	[15–16]
	FN14	Apo3L/ (?)	Osteoblast differentiation	[15, 34–35]
	Fas	FasL/ inflammatory cells (?)	Osteoblast apoptosis	[3]
Osteoclasts and precursor cells	RANK	RANKL / MSC preosteoblast, osteoblast	Osteoclast differentiation	[4]
	TRAIL-R1, R2	TRAIL / osteoblast (?)	Inhibition of differentiation apoptosis (?)	[25–26]
	TNF-R1, R2	TNF-α / inflammatory cells and lymphocytes	Accelerated differentiation and increased activity	[5]
	Fas	FasL / inflammatory cells (?)	Osteoclast apoptosis (?)	[3]

Entries marked (?) are incompletely characterized or hypothetical. This list is not comprehensive; additional TNFs and receptors have been associated with bone disease (e.g., [8, 23, 36]). Vascular endothelial cells, which are spatially related in bone to differentiation of osteoblasts, may be significant sources of TNFs, probably including TL1A, although these cells are otherwise beyond the scope of this review.

may also regulate the relative rate and localization of osteoblast differentiation for bone production. The complex interactions between multiple TNF receptors may also be important in balancing the anabolic and catabolic activities of bone cells.

ACKNOWLEDGMENTS

This work was supported in part by the National Institutes of Health (USA), National Institute for Arthritis and Musculoskeletal Diseases, and by the Department of Veterans' Affairs (USA).

REFERENCES

1. WAJANT, H. 2003. Death receptors. Essays Biochem. **39:** 53–71.
2. BLAIR, H.C., M. ZAIDI & P.H. SCHLESINGER. 2002. Mechanisms balancing skeletal matrix synthesis and degradation. Biochem. J. **364:** 329–341.
3. BU, R. *et al.* 2003. Expression of TNF-family proteins and receptors in human osteoblasts. Bone **33:** 760–770.
4. BLAIR, H.C., L.J. ROBINSON & M. ZAIDI. 2005. Osteoclast signalling pathways. Biochem. Biophys. Res. Comm. **328:** 728–738.
5. BLAIR, H.C. & N.A. ATHANASOU. 2004. Recent advances in osteoclast biology and pathological bone resorption. Histol. Histopathol. **19:** 189–199.
6. BLAIR, H.C., S. LI & R.A. KOHANSKY. 2007. Balanced regulation of proliferation, growth, differentiation, and degradation in skeletal cells. Ann. N.Y. Acad. Sci. In press.
7. BORYSENKO, C.W., W.F. FUREY & H.C. BLAIR. 2005. Comparative modeling of TN-FRSF25 (DR3) predicts receptor destabilization by a mutation linked to rheumatoid arthritis. Biochem. Biophys. Res. Commun. **328:** 794–799.
8. EDWARDS, J.R. *et al.* 2006. LIGHT (TNFSF14), a novel mediator of bone resorption, is elevated in rheumatoid arthritis. Arthritis Rheum. **54:** 1451–1462.
9. BODMER, J.L., P. SCHNEIDER & J. TSCHOPP. 2002. The molecular architecture of the TNF superfamily. Trends Biochem. Sci. **27:** 19–26.
10. ECK, M.J. & S.R. SPRANG. 1989. The structure of tumor necrosis factor-alpha at 2.6 resolution. Implications for receptor binding. J. Biol. Chem. **264:** 17595–17605.
11. EMERY, J.G. *et al.* 1998. Osteoprotegerin is a receptor for the cytotoxic ligand TRAIL. J. Biol. Chem. **273:** 14363–14367.
12. MIGONE, T.S. *et al.* 2002. TL1A is a TNF-like ligand for DR3 and TR6/DcR3 and functions as a T cell costimulator. Immunity **16:** 479–492.
13. NAKAYAMA, M. *et al.* 2003. Fibroblast growth factor-inducible 14 mediates multiple pathways of TWEAK-induced cell death. J. Immunol. **170:** 341–348.
14. MARSTERS, S.A. *et al.* 1996. Apo-3, a new member of the tumor necrosis factor receptor family, contains a death domain and activates apoptosis and NF-kB. Curr. Biol. **6:** 1669–1676.
15. BORYSENKO, C.W. *et al.* 2006. Death receptor-3 mediates apoptosis in human osteoblasts under narrowly regulated conditions. J. Cell Physiol. **209:** 1021–1028.
16. PREHN, J.L. *et al.* 2007. The T cell costimulator TL1A is induced by Fcγ R signaling in human monocytes and dendritic cells. J. Immunol. **178:** 4033–4038.

17. LEHMANN, W. *et al.* 2005. TNFα coordinately regulates the expression of specific matrix metalloproteinases and angiogenic factors during fracture healing. Bone **36:** 300–310.

18. CORNELIS, F. *et al.* 1998. New susceptibility locus for rheumatoid arthritis suggested by a genome-wide linkage study. Proc. Natl. Acad. Sci. USA **95:** 10746–10750.

19. SHIOZAWA, S. *et al.* 2002. The molecular genetics of rheumatoid arthritis disease gene. Nippon Rinsho **60:** 2269–2275.

20. CHINNAIYAN, A.M. *et al.* 1996. Signal transduction by DR3, a death domain-containing receptor related to TNFR-1 and CD95. Science **274:** 990–992.

21. WANG, E.C. *et al.* 2001. DR3 regulates negative selection during thymocyte development. Mol. Cell Biol. **21:** 3451–3461.

22. SCHNEIDER, P. *et al.* 1999. TWEAK can induce cell death via endogenous TNF and TNF receptor 1. Eur. J. Immunol. **29:** 1785–1792.

23. POLEK, T.C. *et al.* 2003. TWEAK mediates signal transduction and differentiation of RAW264.7 cells in the absence of Fn14/TweakR. Evidence for a second TWEAK receptor. J. Biol. Chem. **278:** 32317–32323.

24. KAPTEIN, A. *et al.* 2000. Studies on the interaction between TWEAK and the death receptor WSL-1/TRAMP (DR3). FEBS Lett. **485:** 135–141.

25. ZAULI, G. *et al.* 2004. TNF-related apoptosis-inducing ligand (TRAIL) blocks osteoclastic differentiation induced by RANKL plus M-CSF. Blood **104:** 2044–2050.

26. ROUX, S. *et al.* 2005. Death receptors, Fas and TRAIL receptors, are involved in human osteoclast apoptosis. Biochem. Biophys. Res. Commun. **333:** 42–50.

27. HILL, P.A., A. TUMBER & M.C. MEIKLE. 1997. Multiple extracellular signals promote osteoblast survival and apoptosis. Endocrinology **138:** 3849–3858.

28. JILKA, R.L. *et al.* 1998. Osteoblast programmed cell death (apoptosis): modulation by growth factors and cytokines. J. Bone Miner. Res. **13:** 793–802.

29. EVODOKIOU, A. *et al.* 2002. Chemotherapeutic agents sensitize osteogenic sarcoma cells, but not normal human bone cells, to apo2L/Trail induced apoptosis. Int. J. Cancer **99:** 491–504.

30. WELSH, J. 1997. 1,25(OH)$_2$D3 protects MG-63 osteoblasts from TNFa and ceramide induced apoptosis. *In* Vitamin D. Chemistry, Biology, and Clinical Applications of the Steroid Hormone. A.W. Norman, R. Bouillon, M. Thomaset, Eds.: 405–406. U. California, Riverside.

31. EBERHARDT, A.W., A. YEAGER-JONES & H.C. BLAIR. 2001. Regional trabecular bone matrix degeneration and osteocyte death in femora of glucocorticoid treated rabbits. Endocrinology **142:** 1333–1340.

32. HAO, J.L. *et al.* 2005. Inhibition of gap junction-mediated intercellular communication by TNF-a in cultured human corneal fibroblasts. Invest. Ophthalmol. Vis. Sci. **46:** 1195–1200.

33. JAKUBOWSKI, A. *et al.* 2005. TWEAK induces liver progenitor cell proliferation. J. Clin. Invest. **115:** 2330–2340.

34. ANDO, T. *et al.* 2006. TWEAK/Fn14 interaction regulates RANTES production, BMP-2-induced differentiation, and RANKL expression in mouse osteoblastic MC3T3-E1 cells. Arthritis Res. Ther. **8:** R146.

35. PERPER, S.J. *et al.* 2006. TWEAK is a novel arthritogenic mediator. J. Immunol. **177:** 2610–2620.

36. ABE, M. *et al.* 2006. BAFF and APRIL as osteoclast-derived survival factors for myeloma cells: a rationale for TACI-Fc treatment in patients with multiple myeloma. Leukemia **20:** 1313–1315.

Bone Loss after Temporarily Induced Muscle Paralysis by Botox Is Not Fully Recovered After 12 Weeks

SUSAN K. GRIMSTON,[a] MATTHEW J. SILVA,[b]
AND ROBERTO CIVITELLI[a]

[a]Division of Bone and Mineral Disease, Department of Internal Medicine,
Washington University School of Medicine, St. Louis, Missouri, USA

[b]Orthopaedic Research Laboratory, Department of Orthopaedic Surgery,
Washington University School of Medicine, St. Louis, Missouri, USA

ABSTRACT: To study the effect of unloading followed by reloading on hindlimb bone mineral content (BMC), we used botulinum toxin A (Botox). We studied the timing and degree of recovery upon restoration of muscle function. We also tested to see if reaction to Botox injection occurred as a function of the degree of expression of connexin43 (Cx43). Sixteen mice were divided by gender and genotype; wild-type equivalent ($Gja1^{+/flox}$) and heterozygous ($Gja1^{+/-}$) mice were injected with a single dose of 2U/100 g Botox i.m. in the quadriceps, hamstrings, and posterior calf muscle groups (day 0). Regional BMC was monitored for 12 weeks in the injected and contralateral control limb. Significant bone loss was observed in the injected limb by week 2 in the $Gja1^{+/flox}$ mice, and by week 3–4 in the $Gja1^{+/-}$ mice. By week 12, BMC in the Botox-treated limb of both male and female wild-type and heterozygous mice was still at least 14% less than that of the noninjected limb. Cortical thickness and BV/TV were lower in the Botox femur compared to the noninjected femur. The heterozygous mice tended to show a slower response to Botox injection as reflected in a slower loss of BMC. Our results show that the rapid and profound bone loss following temporary muscle paralysis is not fully recovered upon restoration of muscle function within 12 weeks. These results underscore the significance of normal muscle function in the maintenance of bone mass.

KEYWORDS: connexin43; skeletal unloading; skeletal reloading; Botox

Address for correspondence: Susan K. Grimston, Ph.D., Division of Bone and Mineral Diseases, Washington University in St. Louis, 4940 Parkview, St. Louis, MO 63110. Voice: 314-454-8431; fax: 314-454-5047.
SGrimsto@im.wustl.edu

Ann. N.Y. Acad. Sci. 1116: 444–460 (2007). © 2007 New York Academy of Sciences.
doi: 10.1196/annals.1402.009

INTRODUCTION

Osteoporosis is a chronic disease affecting millions of Americans and with a high financial burden. Decreased physical activity and skeletal unloading may contribute to osteoporotic bone loss. Prolonged immobilization consequent to long periods of bed rest, decreased muscle function caused by neurological conditions and injuries, as well as weightlessness of spaceflight result in a very rapid and severe bone loss attended by uncoupling between bone formation and resorption, with the latter prevailing.[1,2] Although weightlessness increases osteoclast number and bone resorption in short-term spaceflight,[3,4] for longer periods of weightlessness bone turnover decreases.[5] Therefore, removal of physical loading decreases osteoblast function, whereas mechanical stimulation in general has been shown to have an anabolic effect on bone cell dynamics, bone mass, and quality.[6]

Prolonged bed rest and hindlimb suspension have been used to study the mechanisms of bone loss associated with disuse osteoporosis in humans and small animals, respectively. However, these methods do not allow distinguishing effects of muscle activity from gravity, that is, body weight. The positive effect of muscle contraction on bone mass has been widely acknowledged[7,8] and profound loss of both muscle and bone mass occurs with decreased muscle activity, such as spinal cord injury.[9] Therefore, an animal model of isolated muscle loss would be useful in studying the potential role of muscle dysfunction in the initiation and progression of disuse osteoporosis.

Botulinum toxin type A (Botox) is one of seven serologically distinct neuromuscular blocking agents produced by the bacterium Clostridium botulinum.[10] Acetylcholine release at the presynaptic terminal is inhibited when Botox cleaves a 25-kDa synaptosome-associated protein.[10,11] Botox is highly specific for motor nerve terminals and has a high capacity for diffusion through muscle upon injection. These factors make Botox an ideal agent for induction of muscle weakness and paralysis. Importantly, Botox is not lethal for motorneurons, but recovery from exposure is dose-dependent, with little recovery observed following higher doses. Recovery occurs by neurogenesis within 10 days of moderate exposure to Botox, with the formation of axonal sprouts from the original axon terminal and new motor endplates.[12] In the rabbit longissimus dorsi, muscle atrophy is seen within 2 weeks of toxin injection, continues for about 4 weeks, then reverses.[13] More recently, Botox has been used to paralyze the musculature of the lower limb in C57BL/6 mice, resulting in a significant decrease in trabecular bone mass in the distal femoral epiphysis and proximal tibial metaphysis of the injected limb.[14] In this study mice were sacrificed 21 days after Botox injection and results indicated significant bone loss despite resumption of normal activity.[14] Other investigators have used Botox to study muscle function in human and animal studies,[15–18] supporting the use of Botox as a paralyzing agent.

Growing evidence suggests that cell-to-cell communication via gap junctions is involved in sensing and responding to mechanical factors.[19–22] Gap junctions consist of intercellular channels connecting adjacent cells and allowing the passage of small molecules and ions between cells.[23] Each cell contributes half of the intercellular unit (hemichannel), which is composed of six subunits, connexins, arranged in a hexagonal array. The most abundant connexins in bone are connexin43 (Cx43) and connexin45 (Cx45). *In vitro* studies have shown that the Cx43 protein is critical for bone cell response to a number of stimuli and pharmacologic agents. In particular, inhibition of gap junction communication and/or interference with Cx43 gene (*Gja1*) expression hinders osteoblast responses to fluid flow[19,23] and to mechanically induced calcium waves.[20] A previous study using Botox suggested that increased bone resorption was the mechanism of bone loss.[14] In mice haploinsufficient for *Gja1* we would not expect to see differences in bone loss but may expect to find a slower recovery from Botox-induced bone loss due to compromised Cx43 function and abundance in osteoblasts.

The objective of this study was to determine the effect of unloading followed by reloading on hindlimb bone mass in normal and *Gja1* haploinsufficient mice. We used Botox to induce muscle paralysis,[14] and we studied the timing and degree of recovery upon restoration of muscle function after Botox injection. We found that bone loss was only partially reversible following muscle recovery and that full restoration of bone mass was not achieved in the 12-week period of observation, irrespective of *Gja1* status. We also found that mice haploinsufficient for *Gja1* had a slower skeletal response to Botox injection compared to their wild-type equivalent littermates.

MATERIAL AND METHODS

Animal Models and Experimental Design

Sixteen skeletally mature, 4-month-old mice in a C57BL/6 background were studied with the approval of the Animal Studies Committee at Washington University in St. Louis. We used mouse strains harboring mutations of the Gjal specifically, $Gja1^{+/flox}$ (wild-type equivalent) and $Gja1^{+/-}$ (heterozygous) mice for this pilot project. We are interested in the bone response to unloading in Gja1-deficient mice ($ColCre;Gja1^{-/flox}$) and decided that if we found differences between wild-type and heterozygous mice in this pilot study, embarking on a larger study using the conditional Gja1 knockout mice would be reasonable. Details on genotyping and features of these mice are given elsewhere.[24] Animals were caged in groups, fed standard rodent chow, and given fresh water *ad libitum*. They were housed in a 12-h light versus dark cycle at a controlled temperature of 25°C. All animals were allowed normal cage activity. Four groups of animals were generated, divided by genotype and gender, and four animals per group were included in the study.

Mice were injected intramuscularly with Botox (20 μL; Allergan Inc., Irvine CA, USA) at the dose of 2.0 Units per 100 g body weight (2.5 U/100 μL volume) in the quadriceps, hamstrings, and posterior calf muscles of the right hind limb. The Botox dose used was based on results of a previous study.[14] Before the injection, mice were sedated using an anesthetic cocktail (100 mg/kg ketamine and 10 mg/kg xylazine i.p.[24]). The anterior compartment of the thigh containing the quadriceps musculature was manually palpated and visualized into three compartments, the posterior compartment of the thigh containing the hamstrings was manually palpated and visualized into two compartments and the posterior calf muscles palpated manually. In contrast to the previous study[14] we chose to inject the hamstring muscle group to ensure paralysis of the limb. One-sixth of the Botox volume (20 μL) was injected into each compartment to increase toxin diffusion.[11] We did not use saline controls in this study based on data indicating few changes in the noninjected limb due to Botox.[14] Data did suggest some systemic effects of Botox but the local bone losses induced through muscle paralysis far exceeded any changes noted in the contralateral limb.[14] The contralateral limb was therefore used as an internal control in this study.

Overall Health and Mobility

Daily monitoring of overall health and mobility was conducted during the first week post injection. Mice were weighed every other day for the first 2 weeks and then every week thereafter. Overall health was subjectively determined based on grooming, activity level, and general appearance. In all cases food pellets were placed on the floor of the cage post injection to facilitate feeding. If overall health was compromised, mice were isolated and food pellets coated with Nutri-Cal (EVSCO; Buena NJ, USA) were placed on the floor of the cage until health was restored. Overall mobility was assessed based upon the degree to which the mice used the Botox-injected limb during normal cage activity and was assessed in a similar manner to that reported previously.[14]

Bone Mineral Content

Hindlimb bone mineral content (BMC) was monitored weekly for 12 weeks by dual energy X-ray absorptiometry (DXA; PIXImus Lunar-GE). BMC of the right and left hind limbs were isolated using the region of interest feature of the software. Briefly the femur was isolated from the hip joint, and the tibia and calcaneus isolated from the foot of the mouse. Thus the region-specific BMC included the femur, tibia, fibula, patella, and calcaneous thereby including both trabecular and cortical bone. In this case the animals were positioned with the femur at a 45° angle to the spinal column and the tibia. Calibration of the DXA was performed daily with a standard phantom as per the manufacturer's instructions. The precision of whole body bone mineral density (BMD) assessed by the root mean square method is 1.34% (coefficient of variation).[24,25] All hindlimb analyses were conducted by a single investigator (SKG).

Bone Microstructure

Mice were euthanized 12 weeks post injection under anesthesia by cervical dislocation. Right and left femurs were excised and cleaned of muscle and connective tissue. Micro-CT scans were taken of each bone sample (μCT 40, Scanco Medical AG, Basserdorf, Switzerland), using a modification of a previously described procedure.[26] Bones were stabilized in 1.5% agarose gel and 16 μm voxel resolution images of the distal femoral epiphysis and midshaft of the femoral diaphysis were scanned. Specifically, a 1 mm thick middiaphyseal region and a 0.5 mm thick section encompassing the distal femoral epiphysis were analyzed. The epiphysis was measured distal to the growth plate as determined by assessment of serial slices. At the femoral epiphysis user controlled subroutines were used to isolate the cortical bone from the trabecular bone. Trabecular parameters measured were the tissue volume (TV mm^3), bone volume (BV mm^3), trabecular number (Tb.N.#/mm), trabecular thickness (Tb.Th mm), and trabecular spacing (Tb.Sp.mm). Bone volume/tissue volume (BV/TV) was determined. Mean cortical thickness was determined by measuring cortical width at four different locations around the middiaphysis and taking the average value. Total tissue area (Tt.Ar mm^2) and marrow area (Ma.Ar mm^2) were also measured and cortical area (Ct.Ar mm^2) calculated (Tt.Ar.–Ma.Ar.).

Statistical Analyses

Multifactorial analysis of variance (ANOVA) was used to assess differences due to time, genotype, and treatment on BMC. *Post hoc* Bonferroni's multiple comparison procedure was used to determine which values were significantly different. Paired *t*-tests were used to compare means for μCT data between injected and noninjected limbs. Data were managed and analyzed using Excel 2000 (Microsoft Corp., Redmond, WA, USA) and Statgraphic Plus 3.0 (Manugistic Inc., Rockville, MD, USA) with the level of significance for comparison set at $P < 0.05$. Trends were noted at $P < 0.1$. All data are expressed as the mean $\pm$ SD (unless otherwise noted).

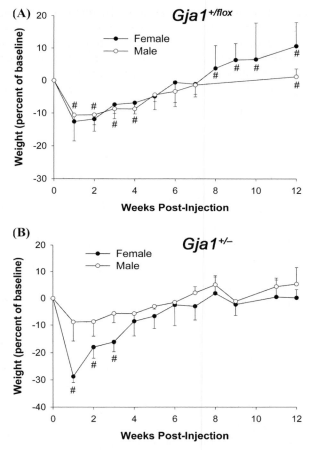

FIGURE 1. Body weight expressed as a percentage of change from baseline (mean ± SD.). All groups lost weight within the first week. This was significant for the male groups and particularly evident in the female heterozygous group ($P < 0.05$). Weight was restored throughout the course of the experiment. (**A**) Wild-type. (**B**) Heterozygous. $P < 0.05$ versus baseline.

RESULTS

Overall Health and Mobility

Maximum limb dysfunction occurred by days 2–3 after Botox injection in both genotype groups without gender differences. General health was significantly affected in these mice, particularly the $Gja1^{+/-}$ females as evidenced by the dramatic and profound loss of body weight (FIG. 1A, B) and reduction in overall cage activity. Cage activity was gradually restored as mice regained full use of the injected limb. Restoration of body weight was facilitated by the coating of pellets with Nutri-Cal (EVSCO; Bueno, NJ, USA). Two weeks after the injection, mice were using the injected limb for balance as they moved about the cage. By 3–4 weeks post injection, full activity was restored as mice were observed to ambulate and climb using both hind limbs equally. Male and female wild-type mice showed similar patterns of body mass loss (FIG. 1A). Nonetheless, body mass was restored by 12 weeks post injection regardless of genotype or gender.

BMC

There was a rapid and profound bone loss in the injected limb following Botox injection in all genotypes and genders. The difference with the non-injected limb was significant by week 2 in the wild-type equivalent mice, but only at weeks 3–4 in the heterozygous groups (FIG. 2A–D). After 3–4 weeks, bone loss stopped and BMC stabilized at a new lower level relative to the noninjected limb, with minimal bone mass recovery by week 12. In all groups there was a rapid decrease of BMC of the noninjected leg over the first 4 weeks but the decrease only reached statistical significance in the female $Gja1^{+/-}$ mice at week 4 (FIG. 2A–D). This bone loss was quickly restored, however, until there were no differences compared to baseline values.

To control for the potential systemic effects of the Botox injection and the potential impact of reduced overall cage activity after Botox injection we calculated the ratio of the Botox leg BMC relative to the noninjected leg BMC. There was a trend for the heterozygous mice to react slower to the Botox injection than the wild-type equivalent mice (FIG. 3A, B). However, after 4 weeks, bone loss reached a similar level in both genotypes—about 20–25% below that of baseline (FIG. 2A–D). Twelve weeks after the injection there remained approximately 14% discrepancy between the Botox and noninjected limb in wild-type and 17% discrepancy in the heterozygous mice compared to baseline (FIG. 3A, B).

Bone Microstructure

Microstructural parameters assessed by μCT at 12 weeks after Botox injection revealed a substantial reduction of trabecular bone volume in the injected limb compared to the noninjected limb in all groups, reaching statistical significance in the two groups of male mice ($P < 0.05$) and showing a similar trend in the female heterozygous mice ($P = 0.07$) (FIG. 4A, B). The differences in BV/TV were due to differences in Bone Volume since Tissue Volume did not show significant differences (TABLE 1). In the Botox femur of male

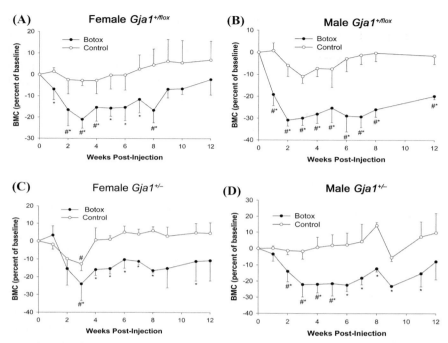

FIGURE 2. Hind limb bone mineral content (BMC) of the Botox limb and the Control limb. Significant loss of BMC was observed for the wild-type groups by week 2 post injection compared to Control. In the heterozygous mice BMC loss was significant by week 3–4. Mean ± SD. (**A**) Female wild-type; (**B**) Male wild-type; (**C**) Female heterozygous, and (**D**) Male Heterozygous. # Significantly less than baseline; * Significantly less than control ($P < 0.05$).

$Gja1^{+/-}$ mice trabecular thickness was significantly less than in the noninjected limb, while trabecular spacing was significantly larger (TABLE 1). In the female heterozygous mice there was a trend for trabecular number to be lower and trabecular spacing higher in the injected limb.

Cortical thickness was lower in the Botox-injected limb compared to the noninjected limb in all groups at week 12 post injection (FIG. 5A, B). This difference was statistically significant in the $Gja1^{+/\text{flox}}$ female mice and the $Gja1^{+/-}$ mice. Marrow area was greater in Botox femurs of all groups compared to control. This was statistically significant in the two groups of females and approaching significance ($P = 0.06$) in the male wild-type equivalent group (TABLE 1). Cortical area was significantly smaller in the Botox limb compared to Control limb in all groups excluding the male $Gja1^{+/-}$ mice (TABLE 1).

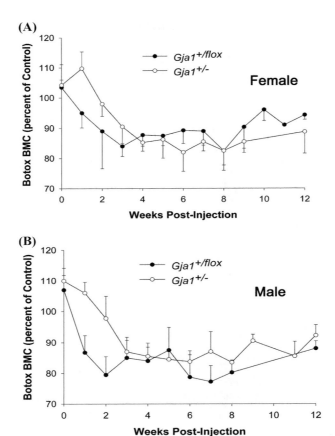

FIGURE 3. Comparison of wild-type equivalent versus heterozygous mice for BMC of the Botox limb expressed as a percentage of the noninjected limb BMC. There were no significant differences between groups although a trend was observed for the heterozygous to react more slowly than the wild-type mice to Botox injection. (**A**) Female comparison; (**B**) Male comparison. Mean ± SD.

DISCUSSION

In this study we used a single dose of Botox injected into the musculature around the knee joint of mice to induce significant paralysis of the injected limb for a period of 3–4 weeks. Mice did use the injected limb for balance as they moved around the cage so the effects of body weight on bone were still present. Therefore, the rapid bone loss in the injected limb was most likely the result of loss of muscle function induced by Botox. We took serial measures of BMC over a period of 12 weeks thereby providing some insight into the course of BMC recovery with restoration of muscle function. A previous study used

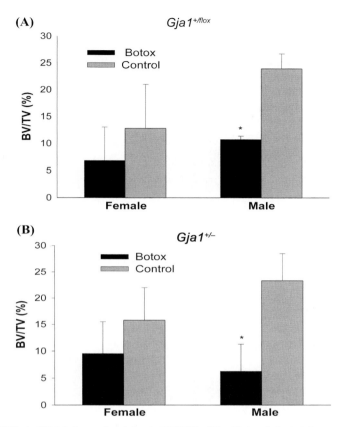

FIGURE 4. Distal femoral epiphysis BV/TV of the Botox-injected femur versus the control femur at 12 weeks post injection. In all groups BV/TV was lower in the Botox femur than the noninjected femur, reaching statistical significance for the male wild-type and heterozygous mice. Mean ± SD. (**A**) Wild-type; (**B**) Heterozygous. * Significantly less than control $P < 0.05$.

Botox to transiently paralyze the right hind limb of female C57BL/6 mice.[14] Mice were monitored over a 21-day period and found to regain use of the right hind limb within that time. Nevertheless, Botox treatment significantly reduced muscle mass compared to saline controls as measured 21 days post injection.[14] Bone loss was still significant when compared with control animals after 21 days.[14] In our study we provided an average of 7–8 weeks recovery from muscle paralysis and found that the bone loss measured was only partially reversible, as there continued to be differences between the Botox-injected leg and the noninjected leg 12 weeks post injection.

Interestingly, *Gja1* haploinsufficient mice lost bone mass at a slower rate than did wild-type equivalent mice. The differences measured were

TABLE 1. Femoral microstructural parameters at end of study

	Female $Gja1^{+/fox}$		Female $Gja1^{+/-}$		Male $Gja1^{+/flox}$		Male $Gja1^{+/-}$	
	Botox	Control	Botox	Control	Botox	Control	Botox	Control
Femoral Epiphysis								
TV (mm^3)	0.64 ± 0.20	1.01 ± 0.33	1.29 ± 0.47	1.36 ± 0.53	1.10 ± 0.50	1.08 ± 0.31	1.12 ± 0.15	1.35 ± 0.20
BV (mm^3)	0.24 ± 0.35	0.26 ± 0.27	0.10 ± 0.06	0.23 ± 0.16	0.12 ± 0.06	0.26 ± 0.08	0.08 ± 0.07	0.31 ± 0.06
Tb.N. (#/mm)	17.7 ± 6.2	16.9 ± 6.8	7.0 ± 2.2	7.4 ± 2.2	9.9 ± 2.4	9.9 ± 2.4	8.3 ± 1.4	8.2 ± 1.3
Tb.Th.(mm)	0.04 ± 0.01	0.05 ± 0.01	0.06 ± 0.01	0.07 ± 0.01	0.06 ± 0.01	0.07 ± 0.01	0.05 ± 0.01*	0.08 ± 0.01
Tb.Sp.(mm)	0.09 ± 0.02	0.09 ± 0.02	0.18 ± 0.06	0.16 ± 0.04	0.14 ± 0.03	0.13 ± 0.02	0.15 ± 0.02*	0.13 ± 0.02
Femoral Diaphysis								
M.Ar (mm^2)	0.97 ± 0.08*	0.87 ± 0.11	1.09 ± 0.06*	0.90 ± 0.08	1.54 ± 0.19	1.40 ± 0.09	1.43 ± 0.47	1.35 ± 0.37
Ct.Ar. (mm^2)	0.80 ± 0.03*	0.90 ± 0.03	0.78 ± 0.07*	0.91 ± 0.05	0.99 ± 0.07*	1.18 ± 0.04	0.69 ± 0.47	0.92 ± 0.45
Tt.Ar (mm^2)	1.77 ± 0.04	1.76 ± 0.08	1.87 ± 0.05	1.81 ± 0.11	2.54 ± 0.20	2.58 ± 0.09	2.13 ± 0.05	2.28 ± 0.04

*$P < 0.05$ versus Control (t-test for paired samples).

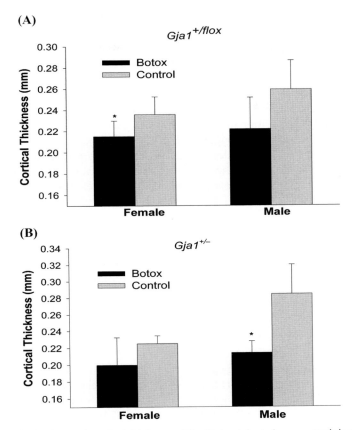

FIGURE 5. Femoral cortical thickness of the Botox injected versus noninjected femur at week 12 post injection. In all groups cortical thickness was reduced in the Botox femur compared to the noninjected femur, reaching statistical significance for the female wild-type and male heterozygous mice. Mean ± SD. (**A**) Wild-type; (**B**) Heterozygous. *Significantly less than control $P < 0.05$.

approaching significance with $0.05 < P < 0.1$. This would be compatible with the idea that *Gja1* haploinsufficiency, and attendant decrease of Cx43 protein abundance,[27] reduces the effect of muscle unloading on bone remodeling. Attenuation of biologic effects has been reported after interference with *Gja1* expression or function *in vitro*,[28,29] and we have found that the anabolic action of parathyroid hormone is reduced *in vivo* in mice with osteoblast-specific deletion of *Gja1*.[25] Interestingly, the differences in BV/TV between the Botox and noninjected limbs were greater in the *Gja1*[+/−] mice than in the *Gja1*[+/flox] mice. However, the DXA results suggest no differences in the recovery of the haploinsufficient mice compared to the wild-type. The apparent discrepancy in the degree of osteopenia and the course of its recovery between DXA and μCT is not uncommon and reflects both a lower sensitivity of DXA and different skeletal sites specifically measured.

Our results are consistent with a similar previous study using Botox to induce muscle paralysis, in which rapid trabecular bone loss was also detected by μCT after Botox injection.[14] In our study, we allowed approximately 7–8 weeks of reloading, but BMC, cortical thickness, trabecular bone volume, and trabecular thickness were still lower in the Botox-injected femur compared with the noninjected femur in animals of both genotypes and genders. Such microstructural changes are suggestive of increased bone resorption during the paralysis phase. Indeed, decreased cortical area and increased marrow area remaining at the end of the study, with no changes in total tissue cross-sectional area, imply an increased endocortical bone resorption. Accelerated bone resorption shortly after loss of muscle function may also explain the loss of trabecular bone, as previously suggested.[14] This type of bone loss is similar to that observed in humans exposed to acute disuse, such as spinal cord injury or bed rest.[2,9,30]

Remobilization studies have also used the hind limb suspension model. However, this method does not eliminate the effect of muscular contraction on bone during the immobilization phase since the mice can fully extend and flex the hind limb at the hip and knee while suspended. Nevertheless, substantial bone loss occurred after hind limb suspension, and recovery followed after reloading.[31] Bone loss using this model arises as a consequence of both decreased bone formation as well as increased resorption.[31–33]

Bone loss consequent to bed rest can be restored at a rate equal to or greater than that at which the bone was lost.[34,35] On the contrary, bone lost as a consequence of trauma in adult humans or cast immobilization in adult dogs, rats, and monkeys is not fully restored after reloading and restoration rate takes much longer than the original bone loss.[4,31,34–37] In our study, loss of weight and the concurrent moderate bone loss in the noninjected limb suggested that Botox injection at this dosage represented a traumatic event for these mice. This trauma could be minimized by using a lower dosage of Botox.

The finding of bone mass recovery taking longer than the time period required for bone mass loss in animals is consistent with findings in humans. After prolonged spaceflight bone density was still lower than preflight levels even 1 year after return to earth.[38] Our data support the notion that more time is required to restore bone than to lose bone in adult mice subjected to temporary and traumatic muscle paralysis, since both DXA and μCT data indicate incomplete recovery at 12 weeks.

Between weeks 3–4 and weeks 9–10 BMC appears to stabilize, suggesting that bone formation and bone resorption reach a new steady state reflecting an adaptation to a new strain environment. It is possible that although muscle function was restored in our study, muscle mass and strength remained impaired. Indeed, in the study by Warner et al.,[14] wet muscle mass was significantly reduced after 21 days despite significant improvement in the lameness inventory used. Decreased muscle strength would be expected to reduce the muscle contractile effects on bone. This proposed adaptation to a new strain environment in our study is supported by previous immobilization studies in which a new steady state of bone remodeling has been found histologically, subsequent to a period of immobilization.[32,37] This pattern of bone loss followed by a new lower steady state of bone mass may be attributed theoretically to a reduction in the minimum effective strain threshold for remodeling, as proposed by Frost and others.[39–41] The gradual reapplication of muscle function in the injected limb could be expected to at least reinstate the equilibrium between resorption and formation and halt the precipitous bone loss seen in the first 4 weeks, as suggested by our data. As muscle strength is fully regained, we may expect to see enhanced formation relative to resorption with some recovery of bone mass. Clearly, such a scenario would require longer than the 12 weeks allowed in this study. The notion that reapplication of muscle function might increase bone formation relative to resorption is supported by studies in humans. A 78% increase in serum osteoclacin with no changes in urinary calcium and hydroxyproline was detected after muscle stimulation in spinal cord injury patients, with a reduction in the amount of paralysis-induced bone loss.[42]

We expected to see a delayed recovery of bone mass in the $Gja1^{+/-}$ mice compared to the $Gja1^{+//flox}$ mice based on their compromised gap junction intercellular communication (GJIC) due to reduction in Cx43. We anticipated that reduction of Cx43 would hinder the osteoblast response to increasing bone strain as a function of muscle reinnervation. However, recovery rates for both genotypes were similar, with little recovery evident. A period of study longer than the 12 weeks of this study would be required to study differences in recovery rates as a function of $Gja1$ gene expression.

The major limitation to this study was the absence of saline control mice for each genotype and gender. With these controls it would have been possible to delineate specific Botox effects from those of, for example, aging. However, mice were brought into the study at 4 months of age so growth effects would have been minimized. In addition, body weight was rapidly recovered post injection until it was at or above baseline, suggesting no long-term health consequences of the Botox injection. For this pilot study we determined that the contralateral limb could be used as an effective control based on previous data.[14]

In summary, we successfully adapted a model for induced muscle paralysis using a Botox injection,[14] which resulted in profound loss of bone that was not fully restored 12 weeks post injection. Our results underscore the significance of normal muscle function in the maintenance of BMC. The use of Botox provides a convenient model to test the effect of muscle mechanics on bone, providing a period of skeletal reloading after unloading in mice.

ACKNOWLEDGMENTS

The authors would like to thank Dr. Stavros Thomopoulus and Rosalina Das for their help with the Botox administration. Supported by NIH grant R01 AR041255 and supplement (RC). Presented in part at the 28th annual meeting of the American Society for Bone and Mineral Research, Philadelphia, PA, September 15–19, 2006.

REFERENCES

1. DONALDSON, C.L., S.B. HULLEY, J.M. VOGEL, et al. 1970. Effects of prolonged bed rest on bone mineral. Metabolism **19**: 1071–1084.
2. LEBLANC, A.D., V.S. SCHNEIDER, H.J. EVANS, et al. 1990. Bone mineral loss and recovery after 17 weeks of bed rest. J. Bone Miner. Res. **5**: 843–850.
3. VICO, L, D. CHAPPARD, C. ALEXANDRE, et al. 1987. Effects of weightlessness on bone mass and osteoclast number in pregnant rats after a five-day spaceflight (COSMS 1514). Bone **8**: 95–103.
4. WRONSKI, T.J., E.R. MOREY-HOLTON, A.C. DOTY, et al. 1987. Histomorphometric analysis of rat skeleton following spaceflight. Am. J. Physiol. **252**: R252–R255.
5. JEE, W.S.S., T.J. WRONSKI, T.J. MOREY & D.B. KIMMEL. 1983. Effects of spaceflight on trabecular bone in rats. Am. J. Physiol. **244**: R310–R314.
6. BIKLE, D.D., B.P. HALLORAN & E.R. MOREY-HOLTON. 1997. Space flight and the skeleton: lessons for the earthbound. Endocrinologist **7**: 10–22.
7. ADAMS, D.J., A.A. SPIRT, T.D. BROWN, et al. 1997. Testing the daily stress stimulus theory of bone adaptation with natural and experimentally controlled strain histories. J. Biomech. **30**: 671–678.
8. FRITTON, S.P., K.J. MCLEOD & C.T. RUBIN. 2000. Quantifying the strain history of bone: spatial uniformity and self-similarity of low-magnitude strains. J. Biomech. **33**: 317–325.
9. KIRATLI, B.J., A.E. SMITH, T. NAUENBERG, et al. 2000. Bone mineral and geometric changes through the femur with immobilization due to spinal cord injury. J. Rehabil. Res. Dev. **37**: 225–233.
10. BRIN, M.F. 1997. Botulinum toxin: chemistry, pharmacology, toxicity, and immunology. Muscle Nerve Suppl. **6**: S146–S168.
11. PEARCE, L.B., E.R. FIRST & G.E. BORODIC. 1995. Botulinum toxin: death versus localized denervation. J. R. Soc. Med. **88**: 239–240.
12. HUANG, W, J.A. FOSTER & A.S. ROGACHEFSKY. 2000. Pharmacology of botulinum toxin. J. Am. Acad. Dermatol. **43**: 249–259.
13. BORODIC, G.E., R. FERRANTE, L.B. PEARCE & K. SMITH. 1994. Histologic assessment of dose-related diffusion and muscle fiber response after therapeutic botulinum A toxin injections. Mov. Disord. **9**: 31–39.
14. WARNER, S.E., D.A. SANFORD, B.A. BECKER, et al. 2006. Botox induced muscle paralysis rapidly degrades bone. Bone **38**: 257–264.
15. HARRIS, C.P., K. ALDERSON, J. NEBEKER, et al. 1991. Histologic features of human orbicularis oculi treated with botulinum A toxin. Arch. Ophthalmol. **109**: 393–395.
16. LONGINO, D, C. FRANK, T.R. LEONARD, et al. 2005. Proposed model of botulinum toxin-induced muscle weakness in the rabbit. J. Orthop. Res. **23**: 1411–1418.

17. LONGINO, D, T.A. BUTTERFIELD & W. HERZOG. 2005. Frequency and length-dependent effects of Botulinum toxin-induced muscle weakness. J. Biomech. **38:** 609–613.

18. SPENCER, R.F. & K.W. MCNEER. 1987. Botulinum toxin paralysis of adult monkey extraocular muscle. Structural alterations in orbital, singly innervated muscle fibers. Arch. Ophthalmol. **105:** 1703–1711.

19. CHERIAN, P.P., A.J. SILLER-JACKSON, S. GU, *et al.* 2005. Mechanical strain opens connexin 43 hemichannels in osteocytes: a novel mechanism for the release of prostaglandin. Mol. Biol. Cell **16:** 3100–3106.

20. JØRGENSEN, N.R., Z. HENRIKSEN, C. BROT, *et al.* 2000. Human osteoblastic cells propagate intercellular calcium signals by two different mechanisms. J. Bone Miner. Res. **15:** 1024–1032.

21. ROBINSON, J.A., M CHATTERJEE-KISHORE, P.J. YAWORSKY, *et al.* 2006. Wnt/beta-catenin signaling is a normal physiological response to mechanical loading in bone. J. Biol. Chem. **281:** 31720–31728.

22. SAUNDERS, M.M., J. YOU, J.E. TROSKO, *et al.* 2001. Gap junctions and fluid flow response in MC3T3-E1 cells. Am. J. Physiol. Cell Physiol. **281:** C1917–C1925.

23. GOODENOUGH, D.A., J.A. GOLIGER & D.L. PAUL. 1996. Connexins, connexons, and intercellular communication. Annu. Rev. Biochem. **65:** 475–502.

24. CASTRO, C.H., J.P. STAINS, S. SHEIKH, *et al.* 2003. Development of mice with osteoblast-specific connexin43 gene deletion. Cell Commun. Adhes. **10:** 445–450.

25. CHUNG, D.J., C.H. CASTRO, M. WATKINS, *et al.* 2006. Low peak bone mass and attenuated anabolic response to parathyroid hormone in mice with an osteoblast-specific deletion of connexin43. J. Cell Sci. **119:** 4187–4198.

26. ARMAMENTO-VILLAREAL, R., S. SHEIKH, A. NAWAZ, *et al.* 2005. A new selective estrogen receptor modulator, CHF 4227.01, preserves bone mass and microarchitecture in ovariectomized rats. J. Bone Miner. Res. **20:** 2178–2188.

27. LECANDA, F., P.M. WARLOW, S. SHEIKH, *et al.* 2000. Connexin43 deficiency causes delayed ossification, craniofacial abnormalities, and osteoblast dysfunction. J. Cell Biol. **151:** 931–944.

28. DONAHUE, H.J., K.J. MCLEOD, C.T. RUBIN, *et al.* 1995. Cell-to-cell communication in osteoblastic networks: cell line-dependent hormonal regulation of gap junction function. J. Bone Miner. Res. **10:** 881–889.

29. SCHILLER, P.C., G. D'IPPOLITO, R. BRAMBILLA, *et al.* 2001. Inhibition of gap-junctional communication induces the trans-differentiation of osteoblasts to an adipocytic phenotype in vitro. J. Biol. Chem. **276:** 14133–14138.

30. CHAPPARD, D., P. MINAIRE, C. PRIVAT, *et al.* 1995. Effects of tiludronate on bone loss in paraplegic patients. J. Bone Miner. Res. **10:** 112–118.

31. SESSIONS, N.D., B.P. HALLORAN, D.D. BIKLE, *et al.* 1989. Bone response to normal weight bearing after a period of skeletal unloading. Am. J. Physiol. **257:** E606–E610.

32. JEE, W.S. & Y. MA. 1999. Animal models of immobilization osteopenia. Morphologie **83:** 25–34.

33. WEINREB, M., G.A. RODAN & D.D. THOMPSON. 1989. Osteopenia in the immobilized rat hind limb is associated with increased bone resorption and decreased bone formation. Bone **10:** 187–194.

34. WESTLIN, N.E. 1974. Loss of bone mineral after Colles' fracture. Clin. Orthop. Relat. Res. 194–199.

35. JAWORSKI, Z.F. & H.K. UHTHOFF. 1986. Reversibility of nontraumatic disuse osteoporosis during its active phase. Bone **7:** 431–439.
36. YOUNG, D.R., W.J. NIKLOWITZ & C.R. STEELE. 1983. Tibial changes in experimental disuse osteoporosis in the monkey. Calcif. Tissue Int. **35:** 304–308.
37. UHTHOFF, H.K. & Z.F. JAWORSKI. 1978. Bone loss in response to long-term immobilisation. J. Bone Joint Surg. Br. **60-B:** 420–429.
38. LANG, T.F., A.D. LEBLANC, H.J. EVANS & Y. LU. 2006. Adaptation of the proximal femur to skeletal reloading after long-duration spaceflight. J. Bone Miner. Res. **21:** 1224–1230.
39. FROST, H.M. 1996. Perspectives: a proposed general model of the "mechanostat" (suggestions from a new skeletal-biologic paradigm). Anat. Rec. **244:** 139–147.
40. DONAHUE, H.J. 1998. Gap junctional intercellular communication in bone: a cellular basis for the mechanostat set point [editorial]. Calcif. Tissue Int. **62:** 85–88.
41. GRIMSTON, S.K. 1993. An application of mechanostat theory to research design: a theoretical model. Med. Sci. Sports Exerc. **25:** 1293–1297.
42. BLOOMFIELD, S.A. & W.J. JACKSON. 1996. Bone mass and endocrine adaptations to training in spinal cord injured individuals. Bone **19:** 61–68.

Genetics of Hypercalciuric Nephrolithiasis

Renal Stone Disease

MICHAEL J. STECHMAN, NELLIE Y. LOH, AND RAJESH V. THAKKER

Academic Endocrine Unit, Nuffield Department of Clinical Medicine, Oxford Centre for Diabetes, Endocrinology and Metabolism, Churchill Hospital, Headington, Oxford, United Kingdom

ABSTRACT: Renal stone disease (nephrolithiasis) affects 5% of adults and is often associated with hypercalciuria. Hypercalciuric nephrolithiasis is a familial disorder in more than 35% of patients, and may occur as a monogenic disorder, or as a polygenic trait involving 3 to 5 susceptibility loci in man and rat, respectively. Studies of monogenic forms of hypercalciuric nephrolithiasis in man, for example, Bartter syndrome, Dent's disease, autosomal dominant hypocalcemic hypercalciuria (ADHH), hypercalciuric nephrolithiasis with hypophosphatemia, and familial hypomagnesemia with hypercalciuria have helped to identify a number of transporters, channels, and receptors that are involved in regulating the renal tubular reabsorption of calcium. Thus, Bartter syndrome, an autosomal recessive disease, is caused by mutations of the bumetanide-sensitive Na-K-Cl (NKCC2) cotransporter, the renal outer-medullary potassium channel (ROMK), the voltage-gated chloride channel, CLC-Kb, or in its β subunit, Barttin. Dent's disease, an X-linked disorder characterized by low molecular weight proteinuria, hypercalciuria, and nephrolithiasis, is due to mutations of the chloride/proton antiporter, CLC-5; ADHH is associated with activating mutations of the calcium-sensing receptor, which is a G protein-coupled receptor; hypophosphatemic hypercalciuric nephrolithiasis associated with rickets is due to mutations in the type 2c sodium–phosphate cotransporter (NPT2c); and familial hypomagnesemia with hypercalciuria is due to mutations of paracellin-1, which is a member of the claudin family of membrane proteins that form the intercellular tight junction barrier in a variety of epithelia. These studies have provided valuable insights into the renal tubular pathways that regulate calcium reabsorption and predispose to kidney stones and bone disease.

KEYWORDS: idiopathic hypercalciuria; nephrolithiasis; calcium; magnesium; phosphate; inheritance; hereditary; renal tubular disorders; animal models

Address for correspondence: Prof. R.V. Thakker, M.D., F.R.C.P., F.R.C.Path, F.Med.Sci., Academic Endocrine Unit, Nuffield Department of Clinical Medicine, Oxford Centre for Diabetes, Endocrinology and Metabolism (OCDEM), Churchill Hospital, Headington, Oxford, OX3 7LJ, UK. Voice: +441865-857501; fax: +441865-857502.
rajesh.thakker@ndm.ox.ac.uk

Ann. N.Y. Acad. Sci. 1116: 461–484 (2007). © 2007 New York Academy of Sciences.
doi: 10.1196/annals.1402.030

INTRODUCTION

Nephrolithiasis (kidney stone disease), which is a common disorder that affects approximately 8% of the population by the seventh decade, is usually associated with a metabolic abnormality that may include hypercalciuria, hyperphosphaturia, hyperoxaluria, hypocitraturia, hyperuricosuria, cystinuria, a low urinary volume, and a defect of urinary acidification.[1,2] The etiology of these metabolic abnormalities and of renal stones is multifactorial and involves interactions between environmental and genetic determinants.[3–6] The environmental determinants, which include dietary intake of salt, protein, calcium and other nutrients, fluid intake, urinary tract infections, socioeconomic status of the individual, lifestyle, and climate have been comprehensively reviewed elsewhere,[7,8] and this review will focus on the progress made on the genetic determinants (TABLE 1) and in particular those associated with hypercalciuric nephrolithiasis.

HYPERCALCIURIC NEPHROLITHIASIS

Hypercalciuria, which is defined as a urinary calcium excretion in excess of 0.1 mmol/kg/24 h or 4 mg/kg/24 h,[9,10] is the most common metabolic abnormality associated with nephrolithiasis and is found in approximately 60% of patients with renal stones.[1,11] Hypercalciuria may occur either as an isolated trait, or in association with other metabolic abnormalities, and also as part of a renal tubular disorder. Between 35% and 65% of patients with hypercalciuric nephrolithiasis may have a family history of the disorder[12,13] and the inheritance may occur as that of a polygenic quantitative trait, or of a monogenic trait with either autosomal dominant, autosomal recessive, or X-linked recessive modes of transmission (TABLE 1). These hypercalciuric nephrolithiasis disorders and their associated underlying molecular genetic mechanisms that may be associated with defects of intestinal calcium absorption, bone calcium resorption, and renal calcium reabsorption will be reviewed.

GENETIC EPIDEMIOLOGY

The greatest risk factor for nephrolithiasis, after controlling for known dietary determinants, is having an affected family member[13] and it has been known since the 1890s that patients with renal stones were more likely than nonrenal stone formers, to have other family members affected with nephrolithiasis.[14] Thus, between 35% and 65% of renal stone formers will have relatives with nephrolithiasis, whereas only 5–20% of nonrenal stone formers will have relatives with nephrolithiasis.[6,12,13] The first-degree relative risk (λ_R) among recurrent stone formers has been estimated to be in the range of 2 to 16.[13,15,16] The wide range of these estimates is largely due to differences in the study designs and the methods used to ascertain the occurrence of renal stones in relatives. Moreover, genetic contributions to nephrolithiasis and hypercalciuria have been confirmed by two studies that have investigated the occurrence of these conditions in monozygotic (MZ) and dizygotic (DZ) twins (TABLE 2).

MZ twins are genetically identical and share 100% of their genes, whereas DZ twins share only 50% of their genes and are hence like other siblings. Both types of twins usually share the same environment during their childhood. Thus, a substantial discrepancy between the disease concordance rates in MZ twins and DZ twins, with a higher concordance of the disease in MZ twins, indicates a high likelihood for genetic factors causing the disease. A complete 100% concordance in disease between MZ twins would mean that only genetic factors were of importance, and the degree of divergence in concordance from 100% indicates the influence of environmental factors. Two twin studies examining the heritability (h^2) of kidney stones and urinary calcium excretion (TABLE 2) have reported heritability values of 56% and 52%, respectively.[17,18] In one study, 3,391 male–male twins were studied for the occurrence of kidney stones, and the concordance rates in MZ and DZ twins were 32% and 17%, respectively.[17] In the other study 1,068 female–female twins were investigated for urinary calcium excretion and the concordance rates in MZ and DZ twins were 52% and 35%, respectively.[18] Both of these studies support the notion that there is a strong genetic contribution to renal stone disease and in the regulation of renal calcium excretion. However, the responsible genes remain to be identified, as linkage studies using affected sibling pairs or small families in genome-wide searches have to date not been undertaken. This may be partly due to the lack of availability of appropriate patients and their relatives, and also because of difficulties in correctly ascertaining the phenotype, which may require radiological investigations and 24-h urine collections. Such studies are likely to be highly worthwhile, as has been illustrated by the identification of polygenic loci causing other complex diseases, such as diabetes mellitus, obesity, hypertension, and the metabolic syndrome.[19–22] Indeed, such quantitative trait loci (QTLs) contributing to hypercalciuric nephrolithiasis have been identified by investigations of a genetic hypercalciuric stone-forming (GHS) rat model.[23]

IDENTIFICATION OF FIVE QTLS
IN GHS RAT MODEL

The GHS rat resembles human idiopathic hypercalciuria (IH), as the rats have excessive intestinal calcium absorption, increased bone resorption, and impaired renal calcium reabsorption.[24] The GHS rat model was established by selective breeding of Sprague–Dawley rats that had the highest urinary

TABLE 1. Genetic defects associated with monogenic forms of nephrolithiasis

Renal stone disease[a]	Mode of inheritance[b]	Gene[c]	Human chromosomal location (mouse)	Reference[d]
Associated with hypercalciuria				
IH nephrolithiasis				
	A-d	SAC	1q23.3-q24	36
	A-d	VDR	12q12-q14	31
	A-d	?	9q33.2-q34.2	32
ADHH	A-d	CASR	3q21.1	40
Hypercalcemia with hypercalciuria	A-d	CASR	3q21.1	38
Bartter syndromes				
Type I	A-r	SLC12A1/NKCC2	15q15-q21.1	106
Type II	A-r	KCNJ1/ROMK	11q24	107
Type III	A-r	CLCNKB	1q36	108
Type IV	A-r	BSND	1q31	109
Type V	A-d	CASR	3q21.1	48
Type VI	X-r	CLCN5	Xp11.22	50
Dent's disease	X-r	CLCN5	Xp11.22	52
Lowe syndrome	X-r	OCRL1	Xq25	58
HHRH	A-r	NPT2c/SLC34A3	9q34	69
Nephrolithiasis, osteoporosis and hypophosphatemia	A-d	NPT2a/SLC34A1	5q35	64
Familial hypomagnesemia with hypercalciuria and nephrocalcinosis	A-r	PCLN1/CLDN16	3q28	73
Familial hypomagnesemia with hypercalciuria and nephrocalcinosis with ocular abnormalities	A-r	CLDN19	1p34.2	77
dRTA	A-d	SLC4A1/kAE1	17q21.31	83
dRTA with sensorineural deafness	A-r	ATP6B1/ATP6V1B1	2p13	85
dRTA with preserved hearing	A-r	ATP6N1B/ATP6V0A4	7q34	88

Continued

TABLE 1. *Continued*

Renal stone disease[a]	Mode of inheritance[b]	Gene[c]	Human chromosomal location (mouse)	Reference[d]
Calcium oxalate nephrolithiasis[e]	A-r	*Slc26a6*	3q21.3 (9F2)	94
Hypercalciuric calcium phosphate nephrolithiasis[e]	A-r	*Cav1*	7q31.1 (6A2)	91
Nephrocalcinosis, hypophosphatemia, and secondary ossification[e]	A-r	*Npt2a/Slc34a1*	5q35 (13B2)	66
Absorptive hypercalciuria with reduced bone mass[e]	A-r	*Trpv5*	7q34 (6B2)	101
Not associated with hypercalciuria				
Primary hyperoxaluria type 1	A-r	AGXT	2q37.3	110
Primary hyperoxaluria type 2	A-r	GRHPR	9p13.2	110
APRT deficiency	A-r	APRT	16q24.3	111
Cystinuria type A	A-r	SLC3A1	2p16.3	112
Cystinuria type B	A-r	SLC7A9	19q13.1	112
Wilson's disease	A-r	ATP7B	13q14.3	113
Aprt/osteopontin null mouse[e]	A-r	*Aprt/Spp1*	16q24.3(8E2)/4q22.1(5E4)	114
Osteopontin null mouse[e]	A-r	*Spp1*	4q22.1(5E4)	98

[a]IH = idiopathic hypercalciuria; ADHH = autosomal dominant hypocalcemia with hypercalciuria; HHRH = hereditary hypophosphataemic rickets with hypercalciuria; dRTA = distal renal tubular acidosis; APRT = adenine phosphoribosyltransferase.

[b]A-d = autosomal dominant; A-r = autosomal recessive; X-r = X-linked recessive.

[c]*SAC* = human soluble adenylyl cyclase; *VDR* = vitamin D receptor; *SLC12A1* = solute carrier family 12, member 1; *NKCC2* = sodium-potassium-chloride cotransporter 2; *ROMK* = renal outer-medullary potassium channel; *CLCNKB* = chloride channel Kb; *BSND* = Barttin; *CLCN5* = chloride channel 5; *OCRL1* = oculocerebrorenal syndrome of Lowe; *NPT2c/a* = sodium-phosphate cotransporter type 2c/a; *SLC34A1/3* = solute carrier family 34, member 1/3; *PCLN1* = paracellin; *CLDN16/19* = claudin 16/19; kAE1 = kidney anion exchanger 1; *ATP6B1* = ATPase, H⁺ transporting (vacuolar proton pump), V1 subunit B1; *ATP6N1B* = ATPase, H⁺ transporting, lysosomal V0 subunit a4; *Cav1* = caveolin 1; *AGXT* = alanine glyoxylate aminotransferase; *GRHPR* = glyoxalate reductase/hydroxypyruvate reductase; *APRT* = adenine phosphoribosyl transferase; *SLC3A1* = solute carrier family 3, member 1; *SLC7A9* = solute carrier family 7, member 9; *ATP7B* = ATPase, Cu⁺⁺ transporting, beta polypeptide; *Spp1* = secreted phosphoprotein 1 (osteopontin); *Trpv5* = transient receptor potential cation channel; subfamily V, member 5.

[d]Numbers refer to citation in the reference section.

[e]Mouse models for nephrolithiasis, for which a human disease equivalent has not yet been identified, have been included and are shown in italics.

TABLE 2. Heritability rates of renal stones and urinary calcium excretion, determined from twin studies

Registry	VET[a], USA	St. Thomas'[a], UK
Twin pairs (n)	3,391	1,068
Sex	Male–Male	Female–Female
Phenotype	Kidney stones	Urinary Ca
Concordance (%)		
MZ	32	52
DZ	17	35
Heritability (h^2)%	56	52
Reference[b]	17	18

[a]VET = Vietnam Era Twin Study; St Thomas' Hospital Twin Registry.
[b]Number refers to citation in the reference section.
h^2 = additive genetic variance/total phenotypic variance.

calcium excretion, to produce the next generation. Subsequent selection and inbreeding of the most hypercalciuric offspring, and repeating this for each generation established the GHS rat model, such that by the 30th generation the GHS rats had a urinary calcium excretion that was 8 to 10 times greater than that of control Sprague–Dawley rats.[24] Moreover, the GHS rats developed kidney stones, which were not found in the Sprague–Dawley control rats. To identify the QTLs, the GHS rats were bred with normocalciuric Wistar Kyoto (WKY) rats and the progeny (F2) were phenotyped for urinary calcium excretion and used for a whole-genome scan. Linkage was observed between hypercalciuria and loci on chromosomes 1, 4, 7, 10, and 14.[23] Further breeding between the GHS and WKY rats to yield congenic rats with the chromosome 1 locus on a WKY background, has revealed that rats that are homozygous for the GHS alleles remain hypercalciuric.[25] The gene associated with this rat chromosome 1 QTL, and the other 4 QTLs remains to be identified. However, investigation of the vitamin D receptor (VDR) and the calcium-sensing receptor (CaSR), which are two key receptors involved in the pathways of calcium homeostasis, has revealed interesting alterations. The GHS rats have elevated levels of VDR in the intestinal mucosa, bone, and renal cortex, and this is analogous to the increased levels of VDR that are found in circulating monocytes from human stone-formers.[26] Investigations have established that the increased levels of VDR in the GHS rats are due to a prolonged half-life of the VDR.[27] In addition, the renal expression of the CaSR mRNA and protein is also raised in the GHS rat.[28] The VDR and CaSR are not located in the region of the 5 QTLs linked to hypercalciuria in the GHS rat model, and the genetic alterations underlying these QTLs and their signaling pathways that lead to the increased levels of the VDR and CaSR and hypercalciuria remain to be elucidated.

MONOGENIC FORMS OF HYPERCALCIURIC NEPHROLITHIASIS

Idiopathic Hypercalciuria (IH)

Families with IH and recurrent calcium oxalate stones usually reveal an autosomal dominant mode of inheritance.[29] Studies of such families have established linkage between hypercalciuric nephrolithiasis and loci on: chromosome 1q23.3-q24, which contains the human soluble adenylyl cyclase (*SAC*) gene;[30] chromosome 12q12–q14, which contains the *VDR* gene;[31] and chromosome 9q33.2–q34.2, from which an appropriate candidate gene remains to be identified.[32]

Absorptive Hypercalciuria Locus on 1q23.3-q24

Absorptive hypercalciuria (AH), which is characterized by increased intestinal calcium uptake, normocalcemia, normal circulating parathyroid hormone (PTH) concentrations, and low bone mineral density,[33] may occur as an autosomal dominant trait.[34] Linkage studies in three families with AH, mapped the locus to chromosome 1q23.3-q24[30] and to a region that contained a gene encoding a soluble adenylyl cyclase (*SAC*), which is a divalent cation and bicarbonate sensor.[35] The SAC protein, which does not respond to the heterotrimeric G protein regulators, exists freely in cytosolic and membrane-associated forms and its cyclase catalytic activity helps to facilitate the generation of cAMP in the vicinity of its targets. Mutational analysis of *SAC* in patients with AH, revealed six sequence variations and four of these were shown to be associated with a significantly increased relative risk for AH.[36] The manner in which these four sequence variations of SAC result in hypercalciuria and nephrolithiasis remains to be elucidated.

VDR and IH

Increased levels of VDR are found in circulating monocytes from patients with nephrolithiasis,[26] and studies in a cohort of large French Canadian families have revealed an association between IH nephrolithiasis and polymorphic loci from chromosome 12q12-q14, a region that contains the *VDR* gene.[31] However, DNA sequence analysis of the *VDR* gene did not identify any VDR mutations, but did reveal conservative substitutions within the coding region. The role of these VDR polymorphisms in the etiology of IH nephrolithiasis remains to be explained.

Chromosome 9q33.2-q34.2 Locus for Autosomal Dominant Nephrolithiasis

A form of autosomal dominant nephrolithiasis (NPL1) has been reported in a Spanish kindred that originates from La Gomera in the Canary Islands, and resides in Tenerife.[32] The renal stones were reported to consist of calcium oxalate, and serum and urine analyses did not reveal any significant abnormalities, although some affected members had mild hypercalciuria and some had hypomagnesemia. Linkage analysis mapped the NPL1 locus to chromosome 9q33.2-q34.2. This region contains approximately 170 genes, and to date the gene causing NPL1 has not been identified.[32]

CALCIUM-SENSING RECEPTOR AND HYPERCALCIURIC DISORDERS

The human calcium-sensing receptor (CaSR) is a 1,078 amino acid cell surface protein, which is predominantly expressed in the parathyroids and kidney, and is a member of the family of G protein-coupled receptors. The CaSR allows regulation of PTH secretion and renal tubular calcium reabsorption in response to alterations in extracellular calcium concentrations. The human CaSR gene is located on chromosome 3q21.1 and loss-of-function CaSR mutations have been reported in the hypercalcemic disorders of familial benign (hypocalciuric) hypercalcemia (FBHH), neonatal severe primary hyperparathyroidism (NSHPT), and familial isolated hyperparathyroidism (FIHP). However, gain-of-function CaSR mutations result in autosomal dominant hypocalcemia with hypercalciuria (ADHH) and Bartter syndrome type V.

Familial Isolated Primary Hyperparathyroidism Due to CaSR Mutations

Hereditary disorders associated with hypercalcemia include familial isolated primary hyperparathyroidism (FIHP), familial benign hypocalciuric hypercalcemia (FBHH), multiple endocrine neoplasia type 1 (MEN1), MEN type 2 (MEN2), and the hyperparathyroidism jaw–tumor (HPT–JT) syndrome. The hypercalcemia of FIHP, MEN1, MEN2, and HPT–JT (TABLE 3) is associated with hypercalciuria and sometimes kidney stones, whereas that of FBHH is associated with a low urinary calcium–creatinine clearance ratio (<0.01). In FBHH, which is an autosomal dominant disorder, the hypercalcemia is usually mild to moderate, that is, within 10% of the upper limit of normal, although some patients do have more severe hypercalcemia.[37] Other biochemical features include mild hypermagnesemia and normal or mildly elevated serum PTH concentrations. FBHH is due to inactivating mutations of the CaSR, which is

TABLE 3. Hereditary diseases associated with hypercalcaemia and hypercalciuria

Disorder[a]	Clinical features	Gene product	Chromosomal location of the gene
FIHP	Familial isolated parathyroid tumors	MENIN	11q13
		PARAFIBROMIN	1q31.2
		CASR	3q21.1
MEN1	Parathyroid hyperplasia and/or tumors associated with pituitary and pancreatico-duodenal neuro-endocrine tumours	MENIN	11q13
MEN2a	Parathyroid tumors with medullary thyroid cancer and pheochromacytoma	RET	10q11.2
HPT-JT	Parathyroid tumors with ossifying fibromas of the jaw	PARAFIBROMIN	1q31.2

[a]FIHP = familial isolated hyperparathyroidism; MEN = multiple endocrine neoplasia; HPT-JT = hyperparathyroidism-jaw tumor syndrome.

located on chromosome 3q21.1, whereas the genes causing MEN1, MEN2, and HPT–JT are located on chromosomes 11q13, 10q11.2, and 1q31.2, respectively. Some patients with FIHP have mutations of the MEN1, HPT–JT, and CaSR genes, although the majority of FIHP patients do not have such mutations and the genes involved need to be characterized. Five CaSR mutations (Thr100Ile, Lys336deletion, Leu650Pro, Val689Met, and Phe881Leu) have been reported in FIHP.[38,39] Functional characterization of the mutant Phe881Leu CaSR only, has been undertaken in HEK293 cells and this demonstrated that the Phe881Leu mutation resulted in a loss of function.[38] Thus, although the majority of loss-of-function CaSR mutations will lead to FBHH, some may result in hypercalcemic adenoma formation.

ADHH

Patients with ADHH usually have mild hypocalcemia, which is generally asymptomatic, but may in some patients be associated with carpopedal spasm and seizures.[40] The serum phosphate concentrations in patients with ADHH, are either elevated or in the upper-normal range and the serum magnesium concentrations are either low or in the low-normal range.[40] These biochemical features of hypocalcemia, hyperphosphatemia, and hypomagnesemia, are consistent with hypoparathyroidism and pseudohypoparathyroidism. However, these patients have serum PTH concentrations that are in the low-normal range.[40–44] Thus, they are not hypoparathyroid, which would be associated with undetectable serum PTH concentrations, or pseudohypoparathyroid, which would be associated with elevated serum PTH concentrations. These patients were therefore classified as having autosomal dominant hypocalcemia (ADH),[41] and the association of hypercalciuria with this condition leads to it being referred to as ADHH.[40] Treatment with active metabolites of vitamin D to correct the hypocalcemia, has been reported to result in marked hypercalciuria, nephrocalcinosis, nephrolithiasis, and renal impairment, which was partially reversible after cessation of the vitamin D treatment.[40] Thus, it is important to identify and avoid vitamin D treatment in such ADHH patients and their families whose hypocalcemia is due to a gain-of-function CaSR mutation and not hypoparathyroidism.[40] More than 40 different CaSR mutations have been identified in ADHH patients, and more than 50% of these are in the extracellular domain.[40–45] Almost every ADHH family has its own unique missense heterozygous CaSR mutation.[45] Expression studies of the ADHH associated CaSR mutations have demonstrated a gain-of-function, whereby there is a leftward shift in the dose-response curve, such that the extracellular calcium concentration needed to produce a half-maximal (EC_{50}) increase in the total intracellular calcium ions (or inositol trisphosphate, IP_3), is significantly lower than that required for the wild-type receptor.[40,41]

Bartter Syndrome

Bartter syndrome is a heterogeneous group of disorders of electrolyte homeostasis characterized by hypokalemic alkalosis, renal salt wasting that may lead to hypotension, hyper-reninemic hyperaldosteronism, increased urinary prostaglandin excretion, and hypercalciuria with nephrocalcinosis.[46,47] Mutations of several ion transporters and channels have been associated with Bartter syndrome, and 6 types (TABLE 4) are now recognized.[47] Thus, type I is due to mutations involving the bumetanide-sensitive sodium–potassium–chloride cotransporter (NKCC2 or SLC12A2); type II is due to mutations of the renal outer-medullary potassium channel (ROMK); type III is due to mutations of the voltage-gated chloride channel (CLC-Kb); type IV is due to mutations of Barttin, which is a β subunit that is required for trafficking of CLC-Kb and CLC-Ka, and this form is also associated with deafness as Barttin, CLC-Ka, and CLC-Kb are also expressed in the marginal cells of the scala media of the inner ear that secretes potassium ion-rich endolymph; and type V is due to activating mutations of the CaSR. Patients with Bartter syndrome type V have the classical features of the syndrome, that is, hypokalemic metabolic alkalosis, hyper-reninemia, and hyperaldosteronism.[48,49] In addition, they develop hypocalcemia, which may be symptomatic and leads to carpopedal spasm, and an elevated fractional excretion of calcium, that may be associated with nephrocalcinosis.[48,49] Such patients have been reported to have heterozygous gain-of-function CaSR mutations, and *in vitro* functional expression of these mutations not only revealed a leftward shift in the dose-response curve for the receptor, but also showed them to have a much lower

TABLE 4. Types of Bartter syndrome

Bartter type	Gene/channel/transporter	Autosomal dominant/recessive	Gain/loss of function
I	SLC12A2 / NKCC2	Recessive	Loss
II	ROMK	Recessive	Loss
III[a]	CLC-Kb	Recessive	Loss
IV[b]	Barttin	Recessive	Loss
V[c]	CaSR	Dominant	Gain
VI	CLC-5	X-linked recessive	Loss

[a]Patients with type III have not been reported to have nephrocalcinosis, despite the occurrence of hypercalciuria in 65% of them.
[b]Patients with type IV have deafness.
[c]Patients with type V have hypocalcemia.

EC_{50} than that found in patients with ADHH.[47–49] This suggests that the additional features that occur in Bartter syndrome type V, when compared to ADHH, are due to severe gain-of-function mutations of the CaSR.[47] Bartter syndrome type VI has been reported in one child from Turkey[50] and was associated with a *CLCN5* mutation; mutations in this gene are usually observed in Dent's disease (see below).

DENT'S DISEASE

Dent's disease is an X-linked recessive renal tubular disorder characterized by a low molecular weight proteinuria, hypercalciuria, nephrocalcinosis, nephrolithiasis, and eventual renal failure.[51] Dent's disease is also associated with the other multiple proximal tubular defects of the renal Fanconi syndrome, which include aminoaciduria, phosphaturia, glycosuria, kaliuresis, uricosuria, and impaired urinary acidification.[51] With the exception of rickets, which occurs in a minority of patients, there appear to be no extrarenal manifestations in Dent's disease.[51] The gene causing Dent's disease, *CLCN5*, encodes the chloride/proton antiporter, CLC-5.[52] CLC family members, which are usually voltage-gated chloride channels, have important diverse functions that include the control of membrane excitability, transepithelial transport, and regulation of cell volume.[53] CLC-5, which is predominantly expressed in the kidney and in particular the proximal tubule, thick ascending limb of Henle, and the alpha intercalated cells of the collecting duct, has been reported to be critical for acidification in the endosomes that participate in solute reabsorption and membrane recycling in the proximal tubule.[54,55] CLC-5 is also known to alter membrane trafficking via the receptor-mediated endocytic pathway that involves megalin and cubulin.[56] CLC-5 mutations associated with Dent's disease impair chloride flow and likely lead to impaired acidification of the endosomal lumen, and thereby also disrupt trafficking of endosomes back to the apical surface.[56]

This will result in impairment of solute reabsorption by the renal tubule and in the defects observed in Dent's disease.[57] Mice that are deficient for CLC-5 develop the phenotypic abnormalities associated with Dent's disease.[57] Mutations of the gene encoding an inositol polyphosphate 5-phosphatase result in Lowe syndrome (see below) and also Dent's disease.[58]

OCULOCEREBRORENAL SYNDROME OF LOWE (OCRL)

OCRL is an X-linked recessive disorder that is characterized by congenital cataracts, mental retardation, muscular hypotonia, rickets, and defective proximal tubular reabsorption of bicarbonate, phosphate, and amino acids. Some patients may also develop hypercalciuria and renal calculi.[59] The disease is nearly always confined to males, who develop renal dysfunction in the first year of life, have delayed bone age and reduced height, and may die in childhood. Female carriers who have normal neurological and renal function, can be identified in 80% of cases by micropunctate cortical lens opacities. The Lowe syndrome gene, *OCRL1*, is located on Xq25 and encodes a member of the type II family of inositol polyphosphate 5-phosphatases.[60] These enzymes hydrolyze the 5-phosphate of inositol 1, 4, 5-trisphosphate and of inositol 1, 3, 4, 5-tetrakisphosphate, phosphatidylinositol 4, 5-bisphosphate, and phosphatidylinositol 3, 4, 5-trisphosphate, thereby presumably inactivating them as second messengers in the phosphatidylinositol signaling pathway.[61] The preferred substrate of OCRL1 is phosphatidylinositol 4, 5-bisphosphate, and this lipid accumulates in renal proximal tubular cells from patients with Lowe syndrome.[61] OCRL1 has been localized to lysosomes in renal proximal tubular cells and to the trans-golgi network in fibroblasts. This localization is consistent with the role for OCRL1 in lysosomal enzyme trafficking from the trans-golgi network to lysosomes, and the activities of several lysosomal hydrolases are found to be elevated in plasma from affected patients.[62] OCRL1 has also been shown to interact with clathrin and indeed colocalizes with clathrin on endosomal membranes that contain transferrin and mannose 6-phosphate receptors.[63] Mannose 6-phosphate receptor-bound lysosomal enzymes are recruited by appendage subunits and golgi-localized binding proteins into clathrin-coated vesicles that transport them from the trans-golgi network to endosomes.[63] Thus, it seems likely the *OCRL1* mutations in Lowe syndrome patients result in OCRL1 protein deficiency, which leads to disruptions in lysosomal trafficking and endosomal sorting. This abnormality is similar to that observed in Dent's disease, and it is of interest to note that some patients with the latter disease, who had no demonstrable CLC-5 mutations, were found instead to have *OCRL1* mutations.[58] The absence of cataracts in patients with Dent's disease due to *OCRL1* mutations was the major phenotypic difference when compared to patients with Lowe syndrome.[58] The molecular and cellular basis of these phenotypic differences still remains to be elucidated.

HEREDITARY HYPOPHOSPHATEMIC RICKETS
WITH HYPERCALCIURIA (HHRH)

Two different heterozygous mutations (Ala48Phe and Val147Met) in NPT2a (also referred to as *SLC34A1*), the gene encoding a sodium-dependent phosphate transporter, have been reported in patients with urolithiasis or osteoporosis and persistent idiopathic hypophosphatemia due to decreased renal tubular phosphate reabsorption.[64] When expressed in *Xenopus laevis* oocytes, the mutant NPT2a showed impaired function. However, these *in vitro* findings were not confirmed in another study using oocytes and OK cells, raising the concern that the identified NPT2a mutation could not explain the findings in the described patients.[65] However, homozygous ablation of *Npt2a* in mice (*Npt2a-/-*) results in increased urinary phosphate excretion, hypophosphatemia, an appropriate elevation in the serum levels of 1,25-dihydroxyvitamin D, hypercalcemia, decreased serum parathyroid hormone levels, increased serum alkaline phosphatase activity, and hypercalciuria (see below).[66] Some of these biochemical features are observed in patients with HHRH, but there are important differences.[67] Thus, HHRH patients develop rickets, short stature, with an increased renal phosphate clearance, hypercalciuria but have normal serum calcium levels, an increased gastrointestinal absorption of calcium and phosphate due to an elevated serum concentration of 1, 25-dihydroxyvitamin D, suppressed parathyroid function, and normal urinary cyclic AMP excretion.[67] However, HHRH patients do not have NPT2a mutations[68] and studies have demonstrated that HHRH patients harbor homozygous or compound heterozygous mutations of *SLC34A3*, the gene encoding the sodium–phosphate cotransporter NPT2c.[69,70] These findings indicate that NPT2c has a more important role in phosphate homeostasis than previously thought.

FAMILIAL HYPOMAGNESEMIA WITH HYPERCALCIURIA
AND NEPHROCALCINOSIS DUE TO PARACELLIN-1
(CLAUDIN 16) MUTATIONS

Familial hypomagnesemia with hypercalciuria and nephrocalcinosis (FHHNC) is an autosomal recessive renal tubular disorder that is frequently associated with progressive kidney failure.[71] FHHNC often presents in childhood with seizures, or tetany due to hypocalcemia and hypomagnesemia. Other recurrent clinical manifestations include urinary tract infections, polyuria, polydipsia, and failure to thrive. Investigations reveal hypomagnesemia, hypocalcemia, hyperuricemia, hypermagnesuria, hypercalciuria, incomplete distal renal tubular acidosis, hypocitraturia, and renal calcification.[72] Treatment consists of high-dose enteral magnesium to restore normomagnesemia. Children with FHHNC who receive such treatment early develop normally. Linkage studies in 12 FHHNC kindreds localized the disease locus to chromosome 3q, and positional cloning studies identified mutations in the gene encoding Paracellin-1 (PCLN-1), which is also referred to as claudin 16 (*CLDN16*).[72] FHHNC patients were either homozygotes or compound heterozygotes for PCLN-1 mutations, consistent with the autosomal recessive inheritance of the disorder.[73] The PCLN-1 mutations consisted of premature termination codons, splice-site mutations, and missense mutations.[72–74] The PCLN-1 protein, which consists of 305 amino acids, has sequence and structural similarity to the members of the claudin family, and is therefore also referred to as CLDN16.[72] Claudins are membrane-bound proteins that form the intercellular tight junction barrier in a variety of epithelia.[75] Claudins have four transmembrane domains and intracellular amino- and carboxy-termini. The two luminal loops mediate cell–cell adhesion via homo- and heterotypic interactions with claudins on a neighboring cell. In addition, claudins form paracellular ion channels, which facilitate renal tubular paracellular transport of solutes.[75] CLDN16 is exclusively expressed in the thick ascending limb of Henle's loop, where it forms the paracellular channels that are driven by an electrochemical gradient and allow reabsorption of calcium and magnesium.[76] Hence, loss of function of CLDN16 that would arise from FHHNC mutations would result in urinary calcium and magnesium loss and lead to hypocalcemia and hypomagnesemia, respectively. A CLDN16 missense mutation (Thr233Arg) has also been identified in two families with self-limiting childhood hypercalciuria.[74] The hypercalciuria decreased with age and was not associated with progressive renal failure. The Thr233Arg mutation resulted in inactivation of a PDZ-domain binding motif and this disrupted the association with the tight junction scaffolding protein, ZO-1,[74] with accumulation of the mutant CLDN16 protein in lysosomes and no localization to the tight junctions. Thus, CLDN16 mutations may result in different abnormalities of renal tubular cell function and hence lead to differences in the clinical phenotype. A form of FHHNC with severe ocular involvement reported in 1 Swiss and 8 Spanish/Hispanic families was recently mapped to chromosome 1p34.2.[77] This region contains *CLDN19*, the gene that encodes claudin 19, a tight-junction protein expressed in kidney and eye. A Gly20Asp mutation located in the first transmembrane domain of CLDN19 was identified in all but one of the Spanish/Hispanic families and a Gln57Glu mutation in the first extracellular loop of CLDN19 was found in the Swiss family. In addition, a Leu90Pro mutation in CLDN19 was identified in a consanguineous family of Turkish origin with FHHNC and severe ocular involvement.[77]

DISTAL RENAL TUBULAR ACIDOSIS

In distal renal tubular acidosis (dRTA) the tubular secretion of hydrogenions in the distal nephron is impaired, and this results in a metabolic acidosis that is often associated with hypokalemia due to renal potassium wasting, hypercalciuria with nephrocalcinosis, and metabolic bone disease. Distal RTA may be familial with autosomal dominant or recessive inheritance. One form of autosomal dominant dRTA is due to mutations of the erythrocyte anion exchanger (Band 3, AE1). Two autosomal recessive forms of dRTA are caused by mutations of subunits of the H^+-ATPase (proton) pump. Thus, dRTA associated with sensorineural deafness is associated with mutations of the gene encoding the B1 subunit of the apical H^+-ATPase pump (referred to as ATP6B1 or ATP6V1B1) while dRTA without deafness is caused by mutations of the gene encoding a different subunit, ATP6N1B (also referred to as ATP6V0A4), which is an isoform of ATP6N1A that is the 116-kDa noncatalytic accessory subunit of the proton pump.

Autosomal Dominant dRTA due to Erythrocyte Anion Exchanger (Band 3, AE1) Mutations

The family of anion exchangers (AEs) is widely distributed and involved in the regulation of transcellular transport of acid and base across epithelial cells, cell volume, and intracellular pH.[78] For example, AE1, which is a major glycoprotein of the erythrocyte membrane, mediates exchange of chloride and bicarbonate.[79] AE1 is also found in the basolateral membrane of the α-intercalated cells of renal collecting ducts that are involved in acid secretion.[80] Patients with autosomal dominant dRTA, the majority of whom had hypercalciuria, renal stones, and nephrocalcinosis, and a few of them who had erythrocytosis, were found to have AE1 mutations.[81] These AE1 mutations resulted in several functional abnormalities that included reductions in chloride transport, and trafficking defects that lead to a cellular retention of AE1 or mis-targeting of AE1 to the apical membrane.[82,83] AE1 mutations may also be associated with autosomal recessive dRTA in Southeast Asian kindreds that have ovalocytosis.[84]

Autosomal Recessive Distal Renal Tubular Acidosis due to Proton Pump (H^+-ATPase) Mutations

Proton pumps are ubiquitously expressed, and one such multiunit H^+-ATPase is found in abundance on the apical (luminal) surface of the α-intercalated cells of the cortical collecting duct, which regulates urinary acidification. Failure of vectorial proton transport by these α-intercalated cells results in an inability of urinary acidification and in disorders of dRTA. The molecular basis of two types of autosomal recessive dRTA due to proton pump abnormalities has been characterized. The gene causing one type of autosomal recessive dRTA that was associated with sensorineural hearing loss was mapped to chromosome 2p13, which contained the ATP6B1 gene that encodes the B1 subunit of the apical proton pump (H^+ATPase).[85] Mutations, which would likely result in a functional loss of ATP6B1, were identified in more than 30% of families with autosomal recessive dRTA that occurred with deafness in >85% of families.[85] The association of dRTA and deafness is consistent with the renal and cochlear expression of ATP6B1.[86,87] ATP6B1 plays a critical role in regulating the pH of the inner ear endolymph and dysfunction of this would lead to an alkaline microenvironment in the inner ear, which has been proposed to impair hair cell function and result in progressive deafness.[85,87] The gene causing autosomal recessive dRTA with normal hearing was localized to chromosome 7q33-q34, which contained the ATP6N1B gene that encodes the noncatalytic accessory subunit of the proton pump of the α-intercalated cells of the collecting duct.[88] ATP6N1B mutations, which are predicted to result in a functional loss, were identified in >85% of kindreds with autosomal recessive dRTA associated with normal hearing, and this is consistent with the expression of ATP6N1B in the kidney and not other organs. Approximately 15% of families with autosomal recessive dRTA were not found to have mutations in ATP6B1 or ATP6N1B mutations and this indicates mutations in other genes are likely to be involved in the etiology of autosomal recessive dRTA.[89]

MOUSE MODELS FOR NEPHROLITHIASIS FOR WHICH A HUMAN DISEASE EQUIVALENT HAS NOT BEEN IDENTIFIED

Caveolin 1 (Cav-1) Knockout Mice Display Hypercalciuric Nephrolithiasis

Cav-1 is an important structural protein of caveolae, which are small plasma membrane invaginations that are involved in potocytosis (i.e., small molecule transport across cell membranes), transcytosis, and signal transduction.[90] Male mice that are deleted for Cav-1 ($Cav1^{-/-}$) have been reported to develop calcium phosphate bladder calculi in association with hypercalciuria.[91] $Cav1^{-/-}$ female mice did not develop urinary tract stones or hypercalciuria,[91] and the basis of this gender difference remains to be elucidated. Cav-1 is predominantly expressed on the basolateral side of the distal convoluted tubular cells and colocalizes with calcium pumps.[91,92] This is consistent with the role of caveolae as major determinants of calcium transport and calcium-mediated signaling.[93]

Mice Lacking Slc26a6 Anion Transporter Develop Calcium Oxalate Stones

The SLC26 family of conserved anion exchangers comprises at least 10 members that are expressed in a tissue-specific manner. Slc26a6 is expressed in the mouse pancreas, duodenum, and kidney, and among its transport activities, Slc26a6 mediates chloride–oxalate exchange. Mice deleted for Slc26a6 (*Slc26a6* $^{-/-}$) develop a high incidence of calcium oxalate stones in the urinary tract.[94] This was associated with hyperoxaluria and an elevated plasma oxalate concentration, that was attenuated by a reduction in dietary oxalate. *In vitro* studies revealed that *Slc26a6* $^{-/-}$ mice had defective intestinal oxalate secretion. Indeed, *in vivo* studies confirmed that *Slc26a6* $^{-/-}$ male mice had reduced fecal oxalate concentrations. Thus, Slc26a6 plays a critical role in oxalate homeostasis, and its lack in *Slc26a6*$^{-/-}$ male mice results in a reduction of the enteric secretion of oxalate such that the net amount of absorbed oxalate is increased, which leads to an elevation of plasma oxalate with a resultant hyperoxaluria that is associated with urolithasis.[94] To date, mapping studies have not implicated the *SLC26A6* locus on human chromosome 3p21.3 as a cause of calcium oxalate urolithiasis, but it remains an interesting possibility that polymorphisms or mutations of this gene may play a role in the disorder.

Mice Deficient for Npt2a Develop Hypophosphatemia, Hypercalciuria, and Skeletal Abnormalities

The renal proximal tubular specific, brush border membrane sodium–phosphate cotransporter, NPT2a, is important for phosphate reabsorption. Indeed, homozygous ablation of NPT2a in mice (NPT2a$^{-/-}$) results in urinary phosphate loss and hypophosphatemia. As a result of the hypophosphatemia, NPT2a$^{-/-}$ mice have an inappropriate elevation in circulating 1,25-dihydroxyvitamin D concentrations, which results in hypercalcemia, and hypercalciuria that is associated with nephrocalcinosis.[66,95] NPT2a$^{-/-}$ mice also have decreased serum PTH concentrations and an increased serum alkaline phosphatase activity that is not associated with rickets or osteomalacia. Instead, NPT2a$^{-/-}$ mice have poorly developed trabecular bone and retarded secondary ossification that improves with age. This phenotype in NPT2a$^{-/-}$ mice differs from that found in human HHRH patients (see above) and it has been shown that HHRH is due to mutations of the NPT2c gene.[69,70]

Calcium Oxalate Crystal Formation in Osteopontin Knockout Mice

Urine is a supersaturated solution that contains solubilizing proteins to maintain the solutes in solution.[96] Proteins, such as thrombin, uromodulin, nephrocalcin, bikunin, and osteopontin, have been reported to fulfill this role.[97] For example, osteopontin (secreted phosphoprotein 1, Spp1) null mice (*Spp1*$^{-/-}$), although having normal renal function and histologically normal kidneys, showed abnormal responses to injurious stimuli. Thus, *Spp1*$^{-/-}$ mice, but not wild-type mice, developed extensive intratubular deposits of calcium oxalate when administered drinking water containing 1% ethylene glycol for 1 month.[98] Interestingly, wild-type $^{+/+}$ mice, when administered 1% ethylene glycol in the drinking water revealed significant upregulation of osteopontin in their kidneys, thereby supporting a protective role for osteopontin in the pathogenesis of renal stones.[98]

Absorptive Hypercalciuria and Reduced Bone Thickness in Mice Deleted For TRPV5

The superfamily of transient receptor potential (TRP) channel proteins, which form six transmembrane cation-permeable channels, consists of six subfamilies on the basis of their amino acid sequence homology and these are referred to as TRPC, TRPV, TRPM, TRPA, TRPP, and TRPML. Many TRP channels are expressed in the kidney and are involved in the etiologies of renal disorders.[99] For example, TRPC6, TRPM6, and TRPP2 are associated with hereditary focal segmental glomerulosclerosis, hypomagnesemia with secondary hypocalcemia, and polycystic kidney disease, respectively.[99] Moreover, TRPV5, which is an epithelial calcium channel located on the apical surface of distal renal epithelial cells and that facilitates calcium absorption,[100] has been implicated in the etiology of hypercalciuria in mice. Thus, $Trpv^{-/-}$ mice have severe hypercalciuria, raised circulating concentrations of 1,25-dihydroxyvitamin D, enhanced absorption of intestinal calcium, and reduced trabecular bone thickness.[101] Trpv5 function is contingent upon hydrolysis of sugar residues sited on its extracellular domain by the β-glucuronidase enzyme klotho,[102] which is predominantly expressed in the kidney. Indeed, homozygous mutations of the klotho gene in mice, produce a syndrome of accelerated aging including reduced life span, skin atrophy, infertility, bone abnormalities, vascular calcification, and hypercalcemia.[103] Circulating 1,25-dihydroxyvitamin D positively regulates the expression of both TRPV5 and klotho,[104,105] however, the exact mechanisms remain to be elucidated.

ACKNOWLEDGMENTS

This study was funded by grants from the Kidney Research UK (MJS, RVT), The European Union, EuReGene FP6 (NYL, RVT), The Medical Research Council (RVT), and The Wellcome Trust (NYL, RVT).

REFERENCES

1. FRICK, K.K. & D.A. BUSHINSKY. 2003. Molecular mechanisms of primary hypercalciuria. J. Am. Soc. Nephrol. **14:** 1082–1095.
2. SCHEINMAN, S.J. 1999. Nephrolithiasis. Semin. Nephrol. **19:** 381–388.

3. ROBERTSON, W.G. *et al.* 1975. Seasonal variations in the composition of urine in relation to calcium stone-formation. Clin. Sci. Mol. Med. **49:** 597–602.
4. MOE, O.W. & O. BONNY. 2005. Genetic hypercalciuria. J. Am. Soc. Nephrol. **16:** 729–745.
5. PARRY, E.S. & I.S. LISTER. 1975. Sunlight and hypercalciuria. Lancet **1:** 1063–1065.
6. CURHAN, G.C. *et al.* 1997. Family history and risk of kidney stones. J. Am. Soc. Nephrol. **8:** 1568–1573.
7. CURHAN, G.C. *et al.* 1993. A prospective study of dietary calcium and other nutrients and the risk of symptomatic kidney stones. N. Engl. J. Med. **328:** 833–838.
8. SERIO, A. & A. FRAIOLI. 1999. Epidemiology of nephrolithiasis. Nephron **81**(Suppl 1): 26–30.
9. AUDRAN, M. & E. LEGRAND. 2000. Hypercalciuria. Joint Bone Spine **67:** 509–515.
10. CURHAN, G.C. *et al.* 2001. Twenty-four-hour urine chemistries and the risk of kidney stones among women and men. Kidney Int. **59:** 2290–2298.
11. PAK, C.Y. 1997. Nephrolithiasis. Curr. Ther. Endocrinol. Metab. **6:** 572–576.
12. POLITO, C. *et al.* 2000. Idiopathic hypercalciuria and hyperuricosuria: family prevalence of nephrolithiasis. Pediatr. Nephrol. **14:** 1102–1104.
13. RESNICK, M., D.B. PRIDGEN & H.O. GOODMAN. 1968. Genetic predisposition to formation of calcium oxalate renal calculi. N. Engl. J. Med. **278:** 1313–1318.
14. CLUBBE, W.H. 1894. Family disposition to urinary concretions. Lancet **1:** 823.
15. MCGEOWN, M.G. 1960. Heredity in renal stone disease. Clin. Sci. **19:** 465–471.
16. TRINCHIERI, A. *et al.* 1988. Familial aggregation of renal calcium stone disease. J. Urol. **139:** 478–481.
17. GOLDFARB, D.S. *et al.* 2005. A twin study of genetic and dietary influences on nephrolithiasis: a report from the Vietnam Era Twin (VET) Registry. Kidney Int. **67:** 1053–1061.
18. HUNTER, D.J. *et al.* 2002. Genetic contribution to renal function and electrolyte balance: a twin study. Clin. Sci. (Lond). **103:** 259–265.
19. GAUGUIER, D. 2006. Diabetes quantitative trait locus research: from physiology to genetics and back. Diabetologia **49:** 431–433.
20. KOTCHEN, T.A. *et al.* 2002. Identification of hypertension-related QTLs in African American sib pairs. Hypertension **40:** 634–639.
21. SHMULEWITZ, D. *et al.* 2006. Linkage analysis of quantitative traits for obesity, diabetes, hypertension, and dyslipidemia on the island of Kosrae, Federated States of Micronesia. Proc. Natl. Acad. Sci. USA **103:** 3502–3509.
22. FRAYLING, T.M. *et al.* 2007. A common variant in the FTO gene is associated with body mass index and predisposes to childhood and adult obesity. Science **316:** 889–894.
23. HOOPES, R.R., Jr. *et al.* 2003. Quantitative trait loci for hypercalciuria in a rat model of kidney stone disease. J. Am. Soc. Nephrol. **14:** 1844–1850.
24. BUSHINSKY, D.A. 1999. Genetic hypercalciuric stone-forming rats. Curr. Opin. Nephrol. Hypertens. **8:** 479–488.
25. HOOPES, R.R., Jr. *et al.* 2006. Isolation and confirmation of a calcium excretion quantitative trait locus on chromosome 1 in genetic hypercalciuric stone-forming congenic rats. J. Am. Soc. Nephrol. **17:** 1292–1304.
26. FAVUS, M.J. *et al.* 2004. Peripheral blood monocyte vitamin D receptor levels are elevated in patients with idiopathic hypercalciuria. J. Clin. Endocrinol. Metab. **89:** 4937–4943.

27. KARNAUSKAS, A.J. *et al.* 2005. Mechanism and function of high vitamin D receptor levels in genetic hypercalciuric stone-forming rats. J. Bone Miner. Res. **20:** 447–454.

28. YAO, J.J. *et al.* 2005. Regulation of renal calcium receptor gene expression by 1,25-dihydroxyvitamin D3 in genetic hypercalciuric stone-forming rats. J. Am. Soc. Nephrol. **16:** 1300–1308.

29. COE, F.L., J.H. PARKS & E.S. MOORE. 1979. Familial idiopathic hypercalciuria. N. Engl. J. Med. **300:** 337–340.

30. REED, B.Y. *et al.* 1999. Mapping a gene defect in absorptive hypercalciuria to chromosome 1q23.3-q24. J. Clin. Endocrinol. Metab. **84:** 3907–3913.

31. SCOTT, P. *et al.* 1999. Suggestive evidence for a susceptibility gene near the vitamin D receptor locus in idiopathic calcium stone formation. J. Am. Soc. Nephrol. **10:** 1007–1013.

32. WOLF, M.T. *et al.* 2005. Mapping a new suggestive gene locus for autosomal dominant nephrolithiasis to chromosome 9q33.2-q34.2 by total genome search for linkage. Nephrol. Dial. Transplant. **20:** 909–914.

33. PAK, C.Y. *et al.* 1975. A simple test for the diagnosis of absorptive, resorptive and renal hypercalciurias. N. Engl. J. Med. **292:** 497–500.

34. PAK, C.Y. *et al.* 1981. Familial absorptive hypercalciuria in a large kindred. J. Urol. **126:** 717–719.

35. GENG, W. *et al.* 2005. Cloning and characterization of the human soluble adenylyl cyclase. Am. J. Physiol. Cell Physiol. **288:** C1305–C1316.

36. REED, B.Y. *et al.* 2002. Identification and characterization of a gene with base substitutions associated with the absorptive hypercalciuria phenotype and low spinal bone density. J. Clin. Endocrinol. Metab. **87:** 1476–1485.

37. HEATH, H., III. 1989. Familial benign (hypocalciuric) hypercalcemia. A troublesome mimic of mild primary hyperparathyroidism. Endocrinol. Metab. Clin. North Am. **18:** 723–740.

38. CARLING, T. *et al.* 2000. Familial hypercalcemia and hypercalciuria caused by a novel mutation in the cytoplasmic tail of the calcium receptor. J. Clin. Endocrinol. Metab. **85:** 2042–2047.

39. WARNER, J. *et al.* 2004. Genetic testing in familial isolated hyperparathyroidism: unexpected results and their implications. J. Med. Genet. **41:** 155–160.

40. PEARCE, S.H. *et al.* 1996. A familial syndrome of hypocalcemia with hypercalciuria due to mutations in the calcium-sensing receptor. N. Engl. J. Med. **335:** 1115–1122.

41. POLLAK, M.R. *et al.* 1994. Autosomal dominant hypocalcaemia caused by a Ca(2+)-sensing receptor gene mutation. Nat. Genet. **8:** 303–307.

42. FINEGOLD, D.N. *et al.* 1994. Preliminary localization of a gene for autosomal dominant hypoparathyroidism to chromosome 3q13. Pediatr. Res. **36:** 414–417.

43. PERRY, Y. *et al.* 1994. A missense mutation in the Ca-sensing receptor causes familial autosomal dominant hypoparathyroidism. Am. J. Hum. Genet. **55:** A17.

44. BARON, J. *et al.* 1996. Mutations in the Ca(2+)-sensing receptor gene cause autosomal dominant and sporadic hypoparathyroidism. Hum. Mol. Genet. **5:** 601–606.

45. HENDY, G.N. *et al.* 2000. Mutations of the calcium-sensing receptor (CASR) in familial hypocalciuric hypercalcemia, neonatal severe hyperparathyroidism, and autosomal dominant hypocalcemia. Hum. Mutat. **16:** 281–296.

46. THAKKER, R.V. 2000. Molecular pathology of renal chloride channels in Dent's disease and Bartter's syndrome. Exp. Nephrol. **8:** 351–360.
47. HEBERT, S.C. 2003. Bartter syndrome. Curr. Opin. Nephrol. Hypertens. **12:** 527–532.
48. WATANABE, S. *et al.* 2002. Association between activating mutations of calcium-sensing receptor and Bartter's syndrome. Lancet **360:** 692–694.
49. VARGAS-POUSSOU, R. *et al.* 2002. Functional characterization of a calcium-sensing receptor mutation in severe autosomal dominant hypocalcemia with a Bartter-like syndrome. J. Am. Soc. Nephrol. **13:** 2259–2266.
50. BESBAS, N. *et al.* 2005. CLCN5 mutation (R347X) associated with hypokalaemic metabolic alkalosis in a Turkish child: an unusual presentation of Dent's disease. Nephrol. Dial. Transplant. **20:** 1476–1479.
51. WRONG, O.M., A.G. NORDEN & T.G. FEEST. 1994. Dent's disease; a familial proximal renal tubular syndrome with low-molecular-weight proteinuria, hypercalciuria, nephrocalcinosis, metabolic bone disease, progressive renal failure and a marked male predominance. QJM **87:** 473–493.
52. LLOYD, S.E. *et al.* 1996. A common molecular basis for three inherited kidney stone diseases. Nature **379:** 445–449.
53. JENTSCH, T.J., I. NEAGOE & O. SCHEEL. 2005. CLC chloride channels and transporters. Curr. Opin. Neurobiol. **15:** 319–325.
54. GUNTHER, W. *et al.* 1998. ClC-5, the chloride channel mutated in Dent's disease, colocalizes with the proton pump in endocytotically active kidney cells. Proc. Natl. Acad. Sci. USA **95:** 8075–8080.
55. DEVUYST, O. *et al.* 1999. Intra-renal and subcellular distribution of the human chloride channel, CLC-5, reveals a pathophysiological basis for Dent's disease. Hum. Mol. Genet. **8:** 247–257.
56. PIWON, N. *et al.* 2000. ClC-5 Cl—channel disruption impairs endocytosis in a mouse model for Dent's disease. Nature **408:** 369–373.
57. GUNTHER, W., N. PIWON & T.J. JENTSCH. 2003. The ClC-5 chloride channel knockout mouse—an animal model for Dent's disease. Pflugers Arch. **445:** 456–462.
58. HOOPES, R.R., Jr. *et al.* 2005. Dent disease with mutations in OCRL1. Am. J. Hum. Genet. **76:** 260–267.
59. SLIMAN, G.A. *et al.* 1995. Hypercalciuria and nephrocalcinosis in the oculocerebrorenal syndrome. J. Urol. **153:** 1244–1246.
60. LEAHEY, A.M., L.R. CHARNAS & R.L. NUSSBAUM. 1993. Nonsense mutations in the OCRL-1 gene in patients with the oculocerebrorenal syndrome of Lowe. Hum. Mol. Genet. **2:** 461–463.
61. ZHANG, X. *et al.* 1998. Cell lines from kidney proximal tubules of a patient with Lowe syndrome lack OCRL inositol polyphosphate 5-phosphatase and accumulate phosphatidylinositol 4,5-bisphosphate. J. Biol. Chem. **273:** 1574–1582.
62. UNGEWICKELL, A.J. & P.W. MAJERUS. 1999. Increased levels of plasma lysosomal enzymes in patients with Lowe syndrome. Proc. Natl. Acad. Sci. USA **96:** 13342–13344.
63. LOWE, M. 2005. Structure and function of the Lowe syndrome protein OCRL1. Traffic **6:** 711–719.
64. PRIE, D. *et al.* 2002. Nephrolithiasis and osteoporosis associated with hypophosphatemia caused by mutations in the type 2a sodium-phosphate cotransporter. N. Engl. J. Med. **347:** 983–991.

65. VIRKKI, L.V. *et al.* 2003. Functional characterization of two naturally occurring mutations in the human sodium-phosphate cotransporter type IIa. J. Bone Miner. Res. **18:** 2135–2141.

66. BECK, L. *et al.* 1998. Targeted inactivation of Npt2 in mice leads to severe renal phosphate wasting, hypercalciuria, and skeletal abnormalities. Proc. Natl. Acad. Sci. USA **95:** 5372–5377.

67. TIEDER, M. *et al.* 1985. Hereditary hypophosphatemic rickets with hypercalciuria. N. Engl. J. Med. **312:** 611–617.

68. JONES, A. *et al.* 2001. Hereditary hypophosphatemic rickets with hypercalciuria is not caused by mutations in the Na/Pi cotransporter NPT2 gene. J. Am. Soc. Nephrol. **12:** 507–514.

69. BERGWITZ, C. *et al.* 2006. SLC34A3 mutations in patients with hereditary hypophosphatemic rickets with hypercalciuria predict a key role for the sodium-phosphate cotransporter NaPi-IIc in maintaining phosphate homeostasis. Am. J. Hum. Genet. **78:** 179–192.

70. LORENZ-DEPIEREUX, B. *et al.* 2006. Hereditary hypophosphatemic rickets with hypercalciuria is caused by mutations in the sodium-phosphate cotransporter gene SLC34A3. Am. J. Hum. Genet. **78:** 193–201.

71. PAUNIER, L. *et al.* 1968. Primary hypomagnesemia with secondary hypocalcemia in an infant. Pediatrics **41:** 385–402.

72. SIMON, D.B. *et al.* 1999. Paracellin-1, a renal tight junction protein required for paracellular Mg2+ resorption. Science **285:** 103–106.

73. WEBER, S. *et al.* 2001. Novel paracellin-1 mutations in 25 families with familial hypomagnesemia with hypercalciuria and nephrocalcinosis. J. Am. Soc. Nephrol. **12:** 1872–1881.

74. MULLER, D. *et al.* 2003. A novel claudin 16 mutation associated with childhood hypercalciuria abolishes binding to ZO-1 and results in lysosomal mistargeting. Am. J. Hum. Genet. **73:** 1293–1301.

75. COLEGIO, O.R. *et al.* 2002. Claudins create charge-selective channels in the paracellular pathway between epithelial cells. Am. J. Physiol. Cell Physiol. **283:** C142–C147.

76. KONRAD, M., K.P. SCHLINGMANN & T. GUDERMANN. 2004. Insights into the molecular nature of magnesium homeostasis. Am. J. Physiol. Renal Physiol. **286:** F599–F605.

77. KONRAD, M. *et al.* 2006. Mutations in the tight-junction gene claudin 19 (CLDN19) are associated with renal magnesium wasting, renal failure, and severe ocular involvement. Am. J. Hum. Genet. **79:** 949–957.

78. KOPITO, R.R. 1990. Molecular biology of the anion exchanger gene family. Int. Rev. Cytol. **123:** 177–199.

79. TANNER, M.J. 1993. Molecular and cellular biology of the erythrocyte anion exchanger (AE1). Semin. Hematol. **30:** 34–57.

80. WAGNER, S. *et al.* 1987. Immunochemical characterization of a band 3-like anion exchanger in collecting duct of human kidney. Am. J. Physiol. **253:** F213–F221.

81. BRUCE, L.J. *et al.* 1997. Familial distal renal tubular acidosis is associated with mutations in the red cell anion exchanger (Band 3, AE1) gene. J. Clin. Invest. **100:** 1693–1707.

82. BRUCE, L.J. *et al.* 2000. Band 3 mutations, renal tubular acidosis and South-East Asian ovalocytosis in Malaysia and Papua New Guinea: loss of up to 95% band 3 transport in red cells. Biochem. J. **350**(Pt 1): 41–51.

83. TOYE, A.M. 2005. Defective kidney anion-exchanger 1 (AE1, Band 3) trafficking in dominant distal renal tubular acidosis (dRTA). Biochem. Soc. Symp. **72:** 47–63.

84. WRONG, O. *et al.* 2002. Band 3 mutations, distal renal tubular acidosis, and Southeast Asian ovalocytosis. Kidney Int. **62:** 10–19.

85. KARET, F.E. *et al.* 1999. Mutations in the gene encoding B1 subunit of H+-ATPase cause renal tubular acidosis with sensorineural deafness. Nat. Genet. **21:** 84–90.

86. NELSON, R.D. *et al.* 1992. Selectively amplified expression of an isoform of the vacuolar H(+)-ATPase 56-kilodalton subunit in renal intercalated cells. Proc. Natl. Acad. Sci. USA **89:** 3541–3545.

87. STANKOVIC, K.M. *et al.* 1997. Localization of pH regulating proteins H+ATPase and Cl-/HCO3- exchanger in the guinea pig inner ear. Hear Res. **114:** 21–34.

88. SMITH, A.N. *et al.* 2000. Mutations in ATP6N1B, encoding a new kidney vacuolar proton pump 116-kD subunit, cause recessive distal renal tubular acidosis with preserved hearing. Nat. Genet. **26:** 71–75.

89. STOVER, E.H. *et al.* 2002. Novel ATP6V1B1 and ATP6V0A4 mutations in autosomal recessive distal renal tubular acidosis with new evidence for hearing loss. J. Med. Genet. **39:** 796–803.

90. PARTON, R.G. 1996. Caveolae and caveolins. Curr. Opin. Cell Biol. **8:** 542–548.

91. CAO, G. *et al.* 2003. Disruption of the caveolin-1 gene impairs renal calcium reabsorption and leads to hypercalciuria and urolithiasis. Am. J. Pathol. **162:** 1241–1248.

92. FUJIMOTO, T. 1993. Calcium pump of the plasma membrane is localized in caveolae. J. Cell Biol. **120:** 1147–1157.

93. SHAUL, P.W. & R.G. ANDERSON. 1998. Role of plasmalemmal caveolae in signal transduction. Am. J. Physiol. **275:** L843–L851.

94. JIANG, Z. *et al.* 2006. Calcium oxalate urolithiasis in mice lacking anion transporter Slc26a6. Nat. Genet. **38:** 474–478.

95. CHAU, H. *et al.* 2003. Renal calcification in mice homozygous for the disrupted type IIa Na/Pi cotransporter gene Npt2. J. Bone Miner. Res. **18:** 644–657.

96. SAYER, J.A., G. CARR & N.L. SIMMONS. 2004. Nephrocalcinosis: molecular insights into calcium precipitation within the kidney. Clin. Sci (Lond). **106:** 549–561.

97. WATTS, R.W. 2005. Idiopathic urinary stone disease: possible polygenic aetiological factors. QJM **98:** 241–246.

98. WESSON, J.A. *et al.* 2003. Osteopontin is a critical inhibitor of calcium oxalate crystal formation and retention in renal tubules. J. Am. Soc. Nephrol. **14:** 139–147.

99. HSU, Y.J., J.G. HOENDEROP & R.J. BINDELS. 2007. TRP channels in kidney disease. Biochim. Biophys. Acta. doi: 10.1016/j.bbadis.2007.02.001.

100. HOENDEROP, J.G. *et al.* 2001. Function and expression of the epithelial Ca(2+) channel family: comparison of mammalian ECaC1 and 2. J. Physiol. **537:** 747–761.

101. HOENDEROP, J.G. *et al.* 2003. Renal Ca2+ wasting, hyperabsorption, and reduced bone thickness in mice lacking TRPV5. J. Clin. Invest. **112:** 1906–1914.

102. CHANG, Q. *et al.* 2005. The beta-glucuronidase klotho hydrolyzes and activates the TRPV5 channel. Science **310:** 490–493.

103. KURO-O, M. *et al.* 1997. Mutation of the mouse klotho gene leads to a syndrome resembling ageing. Nature **390:** 45–51.

104. HOENDEROP, J.G., B. NILIUS & R.J. BINDELS. 2005. Calcium absorption across epithelia. Physiol. Rev. **85:** 373–422.
105. TSUJIKAWA, H. *et al.* 2003. Klotho, a gene related to a syndrome resembling human premature aging, functions in a negative regulatory circuit of vitamin D endocrine system. Mol. Endocrinol. **17:** 2393–2403.
106. SIMON, D.B. *et al.* 1996. Bartter's syndrome, hypokalaemic alkalosis with hypercalciuria, is caused by mutations in the Na-K-2Cl cotransporter NKCC2. Nat. Genet. **13:** 183–188.
107. SIMON, D.B. *et al.* 1996. Genetic heterogeneity of Bartter's syndrome revealed by mutations in the K+ channel, ROMK. Nat. Genet. **14:** 152–156.
108. ZELIKOVIC, I. *et al.* 2003. A novel mutation in the chloride channel gene, CLC-NKB, as a cause of Gitelman and Bartter syndromes. Kidney Int. **63:** 24–32.
109. ESTEVEZ, R. *et al.* 2001. Barttin is a Cl- channel beta-subunit crucial for renal Cl-reabsorption and inner ear K+ secretion. Nature **414:** 558–561.
110. DANPURE, C.J. 2004. Molecular aetiology of primary hyperoxaluria type 1. Nephron. Exp. Nephrol. **98:** e39–e44.
111. CAMERON, J.S., F. MORO & H.A. SIMMONDS. 1993. Gout, uric acid and purine metabolism in paediatric nephrology. Pediatr. Nephrol. **7:** 105–118.
112. GOODYER, P. 2004. The molecular basis of cystinuria. Nephron. Exp. Nephrol. **98:** e45–e49.
113. WIEBERS, D.O. *et al.* 1979. Renal stones in Wilson's disease. Am. J. Med. **67:** 249–254.
114. VERNON, H.J. *et al.* 2005. Aprt/Opn double knockout mice: osteopontin is a modifier of kidney stone disease severity. Kidney Int. **68:** 938–947.

Correlation among Hyperphosphatemia, Type II Sodium–Phosphate Transporter Activity, and Vitamin D Metabolism in *Fgf-23* Null Mice

DESPINA SITARA

Department of Developmental Biology, Harvard School of Dental Medicine, Boston, Massachusetts, USA

ABSTRACT: Phosphate homeostasis is mostly regulated through humoral factors exerting direct or indirect effects on transporter proteins located in the intestine and kidney. Fibroblast growth factor 23 (FGF-23) is a major phosphate-regulating molecule, which can affect both renal and intestinal phosphate uptake to influence overall mineral ion homeostasis. We have found that *Fgf-23* gene knockout mice (*Fgf-23$^{-/-}$*) develop hyperphosphatemia that consequently leads to abnormal bone mineralization, and severe soft tissue calcifications. On the contrary, *FGF-23* transgenic mice develop hypophosphatemia and produce rickets-like features in the mutant bone. Further studies using our *Fgf-23$^{-/-}$* mice have identified an inverse correlation between Fgf-23, and vitamin D or NaPi2a; genomic elimination of either vitamin D or NaPi2a activities from *Fgf-23$^{-/-}$* mice could reverse severe hyperphosphatemia to hypophosphatemia, and consequently could alter skeletal mineralization, suggesting that regulation of phosphate homeostasis in *Fgf-23$^{-/-}$* mice is vitamin D- and NaPi2a-mediated process.

KEYWORDS: Fgf-23; phosphate regulation; vitamin D; bone

PHOSPHATE HOMEOSTASIS

Besides calcium, phosphate is the most abundant mineral in the human body. Maintenance of phosphate homeostasis is crucial for normal skeletal development and for preservation of bone integrity. Phosphate (Pi) is also essential for cellular function and energy metabolism as it is an important component of ATP, nucleic acids, and the lipid bilayer of the cell membranes. Calcium

Address for correspondence: Despina Sitara, Ph.D., Department of Developmental Biology, Harvard School of Dental Medicine, REB, Room 314, 188 Longwood Avenue, Boston, MA 02115. Voice: 617-432-5749; fax: 617-432-5767.

despina_sitara@hsdm.harvard.edu

Ann. N.Y. Acad. Sci. 1116: 485–493 (2007). © 2007 New York Academy of Sciences.
doi: 10.1196/annals.1402.021

and phosphate levels are well balanced in physiological conditions, and disturbances in their homeostasis can lead to diffuse precipitation of calcium phosphate in tissues, resulting in widespread organ dysfunction and damage.[1,2] Phosphate homeostasis is a hormonal process tightly controlled at the level of the intestine, kidney, and skeleton. About 85% of phosphate is stored in the skeleton, primarily complexed with calcium in hydroxyapatite crystals, which constitute the main inorganic component of the mineralized bone matrix.[2] A significant amount of ingested phosphate (roughly 70%) is absorbed in the small intestine through a sodium-dependent active transport process.[3] Pi absorption in the intestine is directly proportional to the dietary intake but is also regulated by 1,25-dihydroxyvitamin D_3 [1,25(OH)$_2$D$_3$].[2] Acute changes in serum Pi concentrations are largely controlled by the proximal tubule of the kidney. In normal physiological conditions, increased dietary phosphate intake results in a rise in serum Pi concentrations and a subsequent increase in parathyroid hormone (PTH) secretion from the parathyroid glands. PTH then downregulates the primary renal sodium phosphate transporter NaPi2a in the proximal tubules of the kidney to induce phosphate excretion, and simultaneously, induces synthesis of 1,25(OH)$_2$D$_3$ by upregulating 1α-hydroxylase (the enzyme that converts 25-hydroxy vitamin D into the active form of vitamin D, 1,25(OH)$_2$D$_3$). In turn, 1,25(OH)$_2$D$_3$ promotes a rise in serum calcium by increasing intestinal calcium absorption, calcium and phosphate mobilization from bone, and calcium reabsorption in the kidney.[4,5] This rise in serum calcium concentration decreases PTH synthesis, followed by an increase in the expression of NaPi2a.[5–7] In response to severe phosphate deprivation or wasting, bone-derived phosphate is used for the metabolic needs, causing such skeletal defects as osteomalacia. In contrast, prolonged hyperphosphatemia facilitates development of widespread tissue calcifications, and atherosclerosis.[8]

Although it is now widely accepted that phosphate directly stimulates PTH synthesis and secretion, it has become apparent that the classical PTH/vitamin D axis does not fully account for the complexity of phosphate handling.[9] A growing body of evidence suggests that circulating factors of bone origin, other than PTH, participate in the systemic regulation of Pi in response to dietary phosphate changes.[9,10] Recent studies have enhanced our understanding of the pathogenesis of human disorders of skeletal mineralization and impaired renal tubular phosphate reabsorption, and have identified key molecules, which are involved in the regulation of phosphate homeostasis. These phosphaturic hormones, known collectively as "phosphatonins,"[11] that is, factors responsible for inhibition of phosphate reabsorption,[3] include fibroblast growth factor-23 (FGF-23),[12–15] secreted frizzled-related protein-4 (sFRP-4),[16] and matrix extracellular phosphoglycoprotein (MEPE);[17] however, only FGF-23 has been implicated in the pathogenesis of various phosphate-wasting diseases.

FGF-23

FGF-23 is a 30 kDa-secreted protein that is processed by a proconvertase-type enzyme into two smaller fragments of approximately 18 kDa (amino fragment) and 12 kDa (carboxyl fragment) with unclear biological significance. Intraperitoneal (i.p.) administration of synthetic C-terminal (residues 180–251) or N-terminal fragments of FGF-23 (residues 25–179) to mice does not produce the known phosphaturic effects of full-length recombinant FGF-23 protein.[18] However, a recent study has shown that i.p. injections of various C-terminal fragments into Fgf-$23^{-/-}$ mice identified the phosphaturic bioactive domain of FGF-23 to be present in the aa 180–205 region of the protein.[19] Recent studies also suggest that the carboxyl terminal fragment of FGF-23 is required for the biological activity of FGF-23, and its association with klotho, which has been shown to be required as a cofactor of FGF-23 to induce receptor binding and subsequent intracellular signaling.[20,21]

A number of phosphate-wasting disorders have been attributed to increased FGF-23 activity, including autosomal dominant hypophosphatemic rickets (ADHR),[22] X-linked hypophosphatemia (XLH),[23–25] oncogenic osteomalacia (OOM) (also known as tumor-induced osteomalacia [TIO]),[12,26,27] and autosomal recessive hypophosphatemia (ARHP).[28] All four diseases are phenotypically similar, characterized by hypophosphatemia, decreased renal phosphate reabsorption, with normal serum concentrations of calcium and PTH, inappropriately low serum $1,25(OH)_2D_3$ levels, increased circulating FGF-23 concentrations, and defective skeletal mineralization. Marked elevation of serum FGF23 levels has been positively correlated with serum levels of phosphorus, calcium, and PTH in patients with end-stage renal disease.[29,30] In renal disease, declining vitamin D production and increased PTH secretion due to hyperparathyroidism contribute to renal osteodystrophy and vascular calcification. FGF-23 may indirectly contribute to hyperparathyroidism by downregulating $1,25(OH)_2D_3$ through inhibition of 1α-hydroxlyase, and subsequently decreasing serum calcium and enhancing PTH production.[5]

FGF-23 AND PHOSPHATE METABOLISM

FGF-23 acts in accordance with the definition of a hormone: a circulating protein that targets specific tissues distant from its site of production.[5] Bone is the tissue with the highest FGF-23 expression, followed by brain, thymus, and heart, as identified by real-time PCR.[31,32] FGF-23 is predominantly produced by osteocytes[33,34] and it regulates the amount of serum phosphate at two levels: (i) indirectly by suppression of the renal expression of 1α-hydroxylase (1α-(OH)ase) and, therefore, suppression of $1,25(OH)_2D_3$ production by the kidney,[35] and (ii) directly by inhibition of renal phosphate reabsorption by

NaPi2a.[36] In the presence of a functioning kidney in humans and animal models, PTH and $1,25(OH)_2D_3$ are increased in response to hypocalcemia. $1,25(OH)_2D_3$ activates both calcium and phosphate intestinal absorption to correct hypocalcemia and restored calcium levels suppress PTH synthesis.[5,9] FGF-23 synthesis in bone cells is also induced by $1,25(OH)_2D_3$, resulting in hypophosphatemia, phosphaturia accompanied by a reduction in NaPi2a activity, rickets, and osteomalacia.[3,10,37–39] Hypophosphatemia increases 1α-(OH)ase production and thus $1,25(OH)_2D_3$ concentrations in the blood; the latter would then stimulate Fgf-23 production, which would complete the feedback loop and downregulate 1α-(OH)ase expression to lower circulating vitamin D concentrations.[2,35,40,41] Therefore, increased production of FGF-23 sustains the reduced renal phosphate reabsorption while preventing induction of $1,25(OH)_2D_3$ synthesis and calcium mobilization.[5,9] Thus, FGF-23 and PTH have distinct actions in the regulation of $1,25(OH)_2D_3$; PTH actions would predominate when phosphate levels are high and calcium levels are low, whereas FGF-23 actions would predominate when both phosphate and calcium levels are high.[5]

The recent identification of distinct Pi transporters[42] has increased our understanding of the mechanisms and regulation of renal and intestinal Pi handling. The sodium-dependent Pi cotransporters NaPi2a and NaPi2c are located on the apical brush border membrane (BBM) of the proximal tubules. NaPi2a is the major determinant of phosphate reabsorption and urinary excretion[43,44] and it is regulated by several physiologic mediators of phosphate homeostasis, such as dietary phosphate, PTH,[45,46] $1,25(OH)_2D_3$, and FGF-23. Increased PTH levels and/or a high phosphate diet cause an endocytic internalization of the NaPi transporters, resulting in less reabsorption and an increase in urinary phosphate wasting. In contrary, low PTH levels and/or a low-phosphate diet cause recruitment of the NaPi transporters into the BBM, and thereby increase phosphate uptake.[47,48] Targeted inactivation of *NaPi2a* in mice results in hypophosphatemia and severe renal phosphate wasting, elevated serum $1,25(OH)_2D_3$ levels leading to hypercalcemia and hypercalcuria and, therefore, decreased serum PTH levels and increased serum alkaline phosphatase activity.[49] Furthermore, in *NaPi2a*$^{-/-}$ mice 1α-hydroxylase gene ablation and Pi supplementation inhibit renal calcifications, whereas high levels of vitamin D are associated with increased Ca absorption and hypercalcemia resulting in renal calcifications.[49,50]

Recently, several genetically modified animal models have confirmed the phosphaturic effects of Fgf-23. Overexpression of *FGF-23* under the control of various promoters in transgenic mice causes rickets/osteomalacia, hypophosphatemia, and renal phosphate wasting accompanied by reduced expression of NaPi2a in renal proximal tubules, possibly due to the secondary hyperparathyroidism found in these animals, but no significant changes in serum levels of calcium and $1,25(OH)_2D_3$.[36,40,51]

FGF-23 KNOCKOUT MICE

To determine the *in vivo* role and regulation of Fgf-23 in phosphate homeostasis, we generated and partially characterized $Fgf-23^{-/-}$ animals in which we replaced the entire coding region by the *lacZ* gene.[52] Ablation of *Fgf-23* results in growth retardation, severe hyperphosphatemia, increased renal phosphate reabsorption due to increased sodium-dependent phosphate (NaPi2a) uptake activity in the BBM of the kidney, hypercalcemia, highly elevated serum ALP activity, and $1,25(OH)_2D_3$ concentrations, suppressed PTH, along with abnormal bone mineralization, and soft tissue calcifications.[52] A similar phenotype has been reported in patients with familial tumor calcinosis due to inactivating mutations in the *FGF-23* gene that led to enhanced degradation of the protein.[53–56] Regarding bone biology, the $Fgf-23^{-/-}$ mice exhibit an altered growth plate structure and accumulation of unmineralized osteoid. Histomorphometry showed that osteoblast and osteoclast surface areas were reduced in $Fgf-23^{-/-}$ mice, which indicates that bone turnover was suppressed by loss of function of *Fgf-23*.[52] These results support the idea that FGF-23 may have a direct action on bone metabolism. However, it is possible that FGF-23 may be acting in a developmental manner or that the $Fgf-23^{-/-}$ mice suffer from secondary consequences due to renal failure.

Because homozygous ablation of *Fgf-23* results in increased synthesis of vitamin D, as determined by increased serum levels of $1,25(OH)_2D_3$,[1] and increased expression of the renal $1\alpha-(OH)ase$,[33] we hypothesize that hyperphosphatemia and hypercalcemia accompanied by excessive skeletal mineralization and soft tissue calcifications in $Fgf-23^{-/-}$ mice are partly due to increased vitamin D activity. Thus, hypervitaminosis D may be responsible for the phenotype of $Fgf-23^{-/-}$ mice in the presence of high phosphate levels. To test this hypothesis, we have generated and characterized $Fgf-23^{-/-}/1\alpha-(OH)ase^{-/-}$ compound mutants. Our results show that $Fgf-23^{-/-}/1\alpha-(OH)ase^{-/-}$ double mutants are larger in size than $Fgf-23^{-/-}$ mice with prolonged life span,[33] exhibit lower serum calcium and phosphate levels, which are comparable to the ones of $1\alpha-(OH)ase^{-/-}$ animals, develop secondary hyperparathyroidism, and are presented with a significant increase in urinary excretion of phosphate.[33] In agreement with impaired bone mineralization induced by *Fgf-23* deficiency, we observed distinctly reduced volumetric bone mineral density of the femoral shaft and femoral metaphysis, together with thinning of mineralized cortical bone in $Fgf-23^{-/-}$ mice. In line with the secondary hyperparathyroidism present in $1\alpha-(OH)ase^{-/-}$ mice and $Fgf-23^{-/-}/1\alpha-(OH)ase^{-/-}$ double mutants, the trabecular and cortical BMD of the hindlimbs of these animals were lower compared with wild-type control animals, and even further reduced when compared to $Fgf-23^{-/-}$ mice.[33] The bones of $Fgf-23^{-/-}/1\alpha-(OH)ase^{-/-}$ double mutants resembled the bones of the $1\alpha-(OH)ase^{-/-}$ mice with the typical features of rickets and osteomalacia, such as widening of the epiphysis and cupping of the metaphysis, decline in mineral deposition in trabecular bone, and accumulation of unmineralized osteoid. They also lack any of the ectopic bone nodules originally found in *Fgf-23*$^{-/-}$ animals. Furthermore, elimination of vitamin D activity has resulted in the rescue of all observed soft tissue calcifications.[33] These results suggest that $1,25(OH)_2D_3$ is mediating the phosphate-regulating functions of *Fgf-23*.

In our studies we have also shown that the hyperphosphatemia in *Fgf-23*$^{-/-}$ mice is accompanied by enhanced phosphate reabsorption due to a significant increase in renal NaPi2a activity in these mice. Genomic elimination of the *NaPi2a* cotransporter from *Fgf-23*$^{-/-}$ animals reverses severe hyperphosphatemia to hypophosphatemia. These findings suggest that increased NaPi2a activity in *Fgf-23*$^{-/-}$ mice is contributing to the abnormally high serum phosphate levels.

In conclusion, our observations suggest that regulation of phosphate homeostasis in *Fgf-23*$^{-/-}$ mice is vitamin D- and NaPi2a-mediated process. Therefore, our studies have provided further insights into the *in vivo* role of Fgf-23 and could form the basis to design strategies to manipulate abnormal Pi homeostasis and defective skeletal mineralization in patients suffering from diseases, such as rickets, FTC, and chronic renal failure, using FGF-23 as a therapeutic tool.

REFERENCES

1. BRINGHURST, F.R., M.B. DEMAY & H.M. KRONENBERG. 1998. *In* Williams Textbook of Endocrinology: 1155–1210. WB Saunders, Philadelphia, PA.
2. WHITE, K.E, T.E. LARSSON & M.J. ECONS. 2006. The roles of specific genes implicated as circulating factors involved in normal and disordered phosphate homeostasis: frizzled related protein-4, matrix extracellular phosphoglycoprotein, and fibroblast growth factor 23. Endocr. Rev. **27:** 221–241.
3. LANSKE, B. & M.S. RAZZAQUE. 2007. Premature aging in klotho mutant mice: Cause or consequence? Ageing Res. Rev. **6:** 73–79.
4. PORTALE, A.A., B.P. HALLORAN, R.C. MORRIS, JR. 1989. Physiologic regulation of the serum concentration of 1,25-dihydroxyvitamin D by phosphorus in normal men. J. Clin. Invest. **83:** 1494–1499.
5. SCHIAVI, S.C. 2006. Fibroblast growth factor 23: the making of a hormone. Kidney Int. **69:** 425–427.
6. GMAJ, P. & H. MURER. 1986. Cellular mechanisms of inorganic phosphate transport in kidney. Physiol. Rev. **66:** 36–70.
7. MIZGALA, C.L. & G.A. QUAMME. 1985. Renal handling of phosphate. Physiol. Rev. **65:** 431–466.
8. RAZZAQUE, M.S., D. SITARA, T. TAGUCHI, *et al.* 2006. Premature ageing-like phenotype in fibroblast growth factor 23 null mice is a vitamin-D mediated process. FASEB J. **20:** 720–722.
9. CANADILLAS, S., A. RODRIGUEZ-BENOT & M. RODRIGUEZ. 2007. More on the bone-kidney axis–lessons from hypophosphataemia. Nephrol. Dial. Transplant. **22:** 1521–1523.

10. FERRARI, S.L., J.P. BONJOUR & R. RIZZOLI. 2005. Fibroblast growth factor-23 relationship to dietary phosphate and renal phosphate handling in healthy young men. J. Clin. Endocrinol. Metab. **90:** 1519–1524.

11. BERNDT, T.J., S. SCHIAVI & R. KUMAR. 2005. "Phosphatonins" and the regulation of phosphorus homeostasis. Am. J. Physiol. Renal Physiol. **289:** F1170–F1182.

12. JONSSON, K.B. et al. 2003. Fibroblast growth factor 23 in oncogenic osteomalacia and X-linked hypophosphatemia. N. Engl. J. Med. **348:** 1656–1663.

13. SHIMADA, T. et al. 2001. Cloning and characterization of FGF23 as a causative factor of tumor-induced osteomalacia. Proc. Natl. Acad. Sci. USA **98:** 6500–6505.

14. SCHIAVI, S.C. & O.W. MOE. 2002. Phosphatonins: a new class of phosphate-regulating proteins. Curr. Opin. Nephrol. Hypertens. **11:** 423–430.

15. SCHIAVI, S.C. & R. KUMAR. 2004. The phosphatonin pathway: new insights in phosphate homeostasis. Kidney Int. **65:** 1–14.

16. BERNDT, T. et al. 2003. Secreted frizzled-related protein 4 is a potent tumor-derived phosphaturic agent. J. Clin. Invest. **112:** 785–794.

17. ARGIRO, L., M. DESBARATS, F.H. GLORIEUX & B. ECAROT. 2001. Mepe, the gene encoding a tumor-secreted protein in oncogenic hypophosphatemic osteomalacia, is expressed in bone. Genomics **74:** 342–351.

18. SHIMADA, T. et al. 2002. Mutant FGF-23 responsible for autosomal dominant hypophosphatemic rickets is resistant to proteolytic cleavage and causes hypophosphatemia in vivo. Endocrinology **143:** 3179–3182.

19. BERNDT, T.J. et al. 2007. Biological activity of FGF-23 fragments. Pflugers Arch. **454:** 615–623.

20. KURO-O, M. 2006. Klotho as a regulator of fibroblast growth factor signaling and phosphate/calcium metabolism. Curr. Opin. Nephrol. Hypertens. **15:** 437–441.

21. URAKAWA, I. et al. 2006. Klotho converts canonical FGF receptor into a specific receptor for FGF23. Nature **444:** 770–774.

22. ADHR Consortium. 2000. Autosomal dominant hypophosphataemic rickets is associated with mutations in FGF23. The ADHR Consortium. Nat. Genet. **26:** 345–358.

23. BRAME, L.A., K.E. WHITE & M.J. ECONS. 2004. Renal phosphate wasting disorders: clinical features and pathogenesis. Semin. Nephrol. **24:** 39–47.

24. ECONS, M.J. 1999. New insights into the pathogenesis of inherited phosphate wasting disorders. Bone **25:** 131–135.

25. HYP Consortium. 1995. A gene (PEX) with homologies to endopeptidases is mutated in patients with X-linked hypophosphatemic rickets. Nat. Genet. **11:** 130–136.

26. NELSON, A.E., J.J. HOGAN, I.A. HOLM, et al. 2001. Phosphate wasting in oncogenic osteomalacia: PHEX is normal and the tumor-derived factor has unique properties. Bone **28:** 430–439.

27. DE BEUR, S.M. et al. 2002. Tumors associated with oncogenic osteomalacia express genes important in bone and mineral metabolism. J. Bone Miner. Res. **17:** 1102–1110.

28. LORENZ-DEPIEREUX, B. et al. 2006. DMP1 mutations in autosomal recessive hypophosphatemia implicate a bone matrix protein in the regulation of phosphate homeostasis. Nat. Genet. **38:** 1248–1250.

29. LARSSON, T. *et al.* 2003. Circulating concentration of FGF-23 increases as renal function declines in patients with chronic kidney disease, but does not change in response to variation in phosphate intake in healthy volunteers. Kidney Int. **64:** 2272–2279.

30. IMANISHI, Y. *et al.* 2004. FGF-23 in patients with end-stage renal disease on hemodialysis. Kidney Int. **65:** 1943–1946.

31. RIMINUCCI, M. *et al.* 2003. FGF-23 in fibrous dysplasia of bone and its relationship to renal phosphate wasting. J. Clin. Invest. **112:** 683–692.

32. LIU, S. *et al.* 2003. Regulation of fibroblastic growth factor 23 expression but not degradation by PHEX. J. Biol. Chem. **278:** 37419–37426.

33. SITARA, D. *et al.* 2006. Genetic ablation of vitamin D activation pathway reverses biochemical and skeletal anomalies in Fgf-23-null animals. Am. J. Pathol. **169:** 2161–2170.

34. LIU, S. *et al.* 2006. Fibroblast growth factor 23 is a counter-regulatory phosphaturic hormone for vitamin D. J. Am. Soc. Nephrol. **17:** 1305–1315.

35. SHIMADA, T. *et al.* 2004. FGF-23 is a potent regulator of vitamin D metabolism and phosphate homeostasis. J. Bone Miner. Res. **19:** 429–435.

36. SHIMADA, T. *et al.* 2004. FGF-23 transgenic mice demonstrate hypophosphatemic rickets with reduced expression of sodium phosphate cotransporter type IIa. Biochem. Biophys. Res. Comm. **314:** 409–414.

37. QUARLES, L.D. 2003. Evidence for a bone-kidney axis regulating phosphate homeostasis. J. Clin. Invest. **112:** 642–646.

38. SAITO, H. *et al.* 2005. Circulating FGF-23 is regulated by 1alpha,25-dihydroxyvitamin D3 and phosphorus *in vivo*. J. Biol. Chem. **280:** 2543–2549.

39. SCHIAVI, S.C. & R. KUMAR. 2004. The phosphatonin pathway: new insights in phosphate homeostasis. Kidney Int. **65:** 1–14.

40. LARSSON, T. *et al.* 2004. Transgenic mice expressing Fibroblast Growth Factor 23 under the control of the {alpha}1(I) collagen promoter exhibit growth retardation, osteomalacia and disturbed phosphate homeostasis. Endocrinology **145:** 3087–3094.

41. ZHANG, M.Y. *et al.* 2002. Dietary phosphorus transcriptionally regulates 25-hydroxyvitamin D-1alpha-hydroxylase gene expression in the proximal renal tubule. Endocrinology **143:** 587–595.

42. TENENHOUSE, H.S. & Y. SABBAGH. 2002. Novel phosphate-regulating genes in the pathogenesis of renal phosphate wasting disorders. Pflugers Arch. **444:** 317–326.

43. MURER, H., N. HERNANDO, I. FORSTER & J. BIBER. 2000. Proximal tubular phosphate reabsorption: molecular mechanisms. Physiol. Rev. **80:** 1373–1409.

44. TENENHOUSE, H.S. & H. MURER. 2003. Disorders of renal tubular phosphate transport. J. Am. Soc. Nephrol. **14:** 240–248.

45. LEVI, M. *et al.* 1994. Cellular mechanisms of acute and chronic adaptation of rat renal P(i) transporter to alterations in dietary P(i). Am. J. Physiol. **267:** F900–F908.

46. QUAMME, G., J. BIBER & H. MURER. 1989. Sodium-phosphate cotransport in OK cells: inhibition by PTH and 'adaptation' to low phosphate. Am. J. Physiol. **257:** F967–F973.

47. TENENHOUSE, H.S. 2005. Regulation of phosphorus homeostasis by the type IIa Na/phosphate cotransporter. Annu. Rev. Nutr. **25:** 197–214.

48. TRAEBERT, M., H. VOLKL, J. BIBER, *et al.* 2000. Luminal and contraluminal action of 1–34 and 3–34 PTH peptides on renal type IIa Na-P(i) cotransporter. Am. J. Physiol. Renal Physiol. **278:** F792–F798.

49. BECK, L. *et al.* 1998. Targeted inactivation of Npt2 in mice leads to severe renal phosphate wasting, hypercalciuria, and skeletal abnormalities. Proc. Natl. Acad. Sci. USA **95:** 5372–5377.

50. TENENHOUSE, H.S., C. GAUTHIER, H. CHAU & R. ST-ARNAUD. 2004. 1alpha-hydroxylase gene ablation and Pi supplementation inhibit renal calcification in mice homozygous for the disrupted Npt2a gene. Am. J. Physiol. Renal Physiol. **286:** F675-F681.

51. BAI, X., D. MIAO, J. LI, *et al.* 2004. Transgenic mice overexpressing human fibroblast growth factor 23(R176Q) delineate a putative role for parathyroid hormone in renal phosphate wasting disorders. Endocrinology **145:** 5269–5279.

52. SITARA, D. *et al.* 2004. Homozygous ablation of fibroblast growth factor-23 results in hyperphosphatemia and impaired skeletogenesis, and reverses hypophosphatemia in Phex-deficient mice. Matrix Biol. **23:** 421–432.

53. BENET-PAGES, A., P. ORLIK, T.M. STROM & B. LORENZ-DEPIEREUX. 2005. An FGF23 missense mutation causes familial tumoral calcinosis with hyperphosphatemia. Hum. Mol. Genet. **14:** 385–390.

54. FRISHBERG, Y. *et al.* 2007. Hyperostosis–hyperphosphatemia syndrome: a congenital disorder of O-glycosylation associated with augmented processing of fibroblast growth factor 23. J. Bone Miner. **22:** 235–242.

55. LARSSON, T. *et al.* 2005. FGF23 mutants causing familial tumoral calcinosis are differentially processed. Endocrinology **146:** 3883–3891.

56. LARSSON, T. *et al.* 2005. A novel recessive mutation in fibroblast growth factor-23 causes familial tumoral calcinosis. J. Clin. Endocrinol. Metab. **90:** 2424–2427.

Osteoblast and Osteoclast Differentiation in Modeled Microgravity

RITU SAXENA,[a] GEORGE PAN,[b] AND JAY M. McDONALD[b,c]

[a]Department of Cell Biology, University of Alabama at Birmingham, Birmingham, Alabama, USA

[b]Department of Pathology, University of Alabama at Birmingham, Birmingham, Alabama, USA

[c]Veterans Administrations Medical Center, Birmingham, Alabama, USA

ABSTRACT: Prolonged microgravity experienced by astronauts is associated with a decrease in bone mineral density. To investigate the effect of microgravity on differentiation of osteoclasts and osteoblasts, we used a NASA-recommended, ground-based, microgravity-simulating system, the Rotary Cell Culture System (RCCS). Using the RCCS, we demonstrated that modeled microgravity (MMG) inhibited osteoblastogenesis and increased adipocyte differentiation in human mesenchymal stem cells incubated under osteogenic conditions. This transformation involves reduced RhoA activity and cofilin phosphorylation, disruption of F-actin stress fibers, and decreased integrin signaling through focal adhesion kinase. We have used the system to show that MMG also stimulates osteoclastogenesis. These systems provide the opportunity to develop pharmacological agents that will stimulate osteoblastogenesis and inhibit osteoclastogenesis.

KEYWORDS: modeled microgravity; osteoblasts; osteoclasts; bone

INTRODUCTION

Microgravity experienced by astronauts during spaceflights causes severe physiological alterations in the human body. The decrease in bone mineral density is one of the most important of these problems, the severity of which appears to correlate with flight duration.[1] It has been determined that 1–2% of bone mass is lost every month during spaceflight and this loss is site specific occurring mostly in the weight-bearing bones.[2] Several independent studies have detected decreases in osteoblastic markers of bone formation both *in vivo* and *in vitro*, which contribute substantially to the observed bone loss in microgravity.[3,4] An increase in bone resorption markers, including urinary calcium

Address for correspondence: Ritu Saxena, 514 Lyons Harrison Research Building, 701 19 Street South, Birmingham, AL 35294. Voice: 205-934-6940; fax: 205-975-9927.
   ritus@uab.edu

Ann. N.Y. Acad. Sci. 1116: 494–498 (2007). © 2007 New York Academy of Sciences.
doi: 10.1196/annals.1402.033

levels and collagen cross-links in astronauts during 6 months' spaceflight, has been reported.[5] The underlying molecular mechanisms responsible for the apparent concurrent decrease in bone formation and increase in bone resorption are unknown.

MODELED MICROGRAVITY (MMG): ROTARY CELL CULTURE SYSTEM (RCCS) SIMULATES MICROGRAVITY

Due to payload constraints, flight frequency and cost, spaceflight experiments are limited. Therefore, NASA has developed a ground-based system that is able to simulate microgravity. It is referred to as the RCCS and, the "microgravity" created by it is referred to as modeled microgravity (MMG). The rotational motion of the system prevents sedimentation, creating an optimized suspension culture capable of supporting three-dimensional cell growth on microcarrier bead scaffolds.[6] The high aspect ration vessels (HARVs) used in the RCCS provide two essential components of the optimized suspension cell culture system—first is the solid body rotation and second is the diffusion-mediated oxygenation. Solid body rotation of the media and microcarrier beads with attached cells allows minimal shear stress and mechanical damage to the cells. Oxygenation occurs as a result of a silicon membrane that allows appropriate diffusion of gases preventing hypoxic conditions and turbulence-inducing air space or bubbles.[7] We have used the RCCS to study both osteoblastogenesis and osteoclastogenesis. These studies support the hypothesis that both decreased osteoblastogenesis and increased osteoclastogenesis lead to bone loss in MMG environment.

MMG INHIBITS OSTEOBLASTOGENESIS AND STIMULATES ADIPOGENESIS IN HUMAN MESENCHYMAL STEM CELLS

To study osteoblastogenesis, human mesenchymal stem cells (hMSC) were used. The hMSC in osteogenic media *in vitro* under normal gravity, differentiate into osteoblasts capable of matrix mineralization.[8] Human MSC (10^6) were cultured on microcarrier beads for 7 days in DMEM with 10% fetal bovine serum (FBS). Cells were then transferred to HARVs for MMG or culture plates for gravity (G) control. Osteogenic differentiation was induced (media supplemented with 10 nM dexamethasone, 10 mM β-glycerophosphate, and 50 μM ascorbic acid-2-phosphate) immediately before transferring into MMG and G. HARVs were rotated to reach a state of solid body rotation (cell-beads aggregates in suspension) and medium was changed every 3 days. Cells were harvested 7 days after the initiation of MMG and processed for mRNA and protein extraction. [9]

RT-PCR analysis showed that MMG inhibited the expression of osteoblastogenesis markers, alkaline phosphatase, collagen type I, osteocalcin, and the

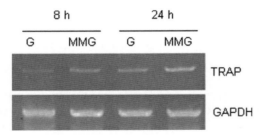

FIGURE 1. MMG enhances TRAP expression. RAW264.7 cells were incubated with microcarrier beads in MMG or G for 24 h. Cells were separated from beads and plated at normal gravity. After incubation with 20 ng/ml RANKL for 3 days RNA was extracted followed by RT-PCR for TRAP and GAPDH.

osteogenic transcription factor Runx2 and stimulated the expression of adipogenic markers adipsin, leptin, and glut 4 along with an increase in the adipogenic transcription factor PPARγ.[9] The underlying signaling mechanism involves reduced RhoA activity that leads to disruption of the cytoskeleton, which is mediated by decrease in phosphocofilin. Reduction in RhoA is associated with altered integrin signaling that leads to decreases in phospho-FAK, Ras-GTP, and phospho-ERK, ultimately resulting in the inhibition of osteoblastogenesis as determined by Western blot analysis of bone-specific markers.[7,9–11]

MMG STIMULATES OSTEOCLASTOGENESIS IN RAW264.7 CELLS

A unique system has recently been developed to study osteoclastogenesis in the MMG environment. Cells from the murine macrophage cell line RAW264.7 were used. RAW264.7 cells (2×10^6) cultured in DMEM complete media containing 10% FBS were initially incubated with 100 mg microcarrier beads at 37°C overnight. Cells were attached around the circumference of beads. Beads with cells attached were put into 10 mL fresh complete DMEM and then placed in the rotating bioreactor located in an incubator at 37°C and 5% CO_2 and allowed to rotate at 9 rpm (speed previously determined by using fluorescent beads so that they remain in suspended position). Parallel bead-attached cells were cultured in gravity. The pH, pO_2, and pCO_2 were unchanged in the RCCS; thus the apparatus did not affect the oxygenation status or pH compared to gravity (G). RAW264.7 cells attached to beads incubated in MMG or G for 24 h were separated from beads and seeded on plates at normal gravity. Cells were then incubated with 20 ng/ml RANKL for 3 days. RT-PCR (FIG. 1) and real-time PCR (data not shown) showed that TRAP expression was enhanced in 24 h MMG-treated cells as compared to the G cells after RANKL treatment. After 4 days of RANKL treatment, multinucleated TRAP-positive

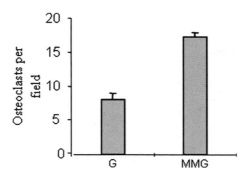

FIGURE 2. MMG enhances osteoclastogenesis. RAW264.7 cells were incubated with microcarrier beads in MMG or G for 24 h, then separated from beads and plated in 6-well plates. These cells were stimulated with 20 ng/mL RANKL. On day 4, the number of multinucleated osteoclasts from MMG and G cells were counted in several fields using the microscope.

osteoclasts derived from 24 h MMG-treated cells were increased approximately twofold compared to those from gravity-treated cells (Fig. 2). The sensitivity of MMG-treated cells to RANKL induction of osteoclastogenesis was increased (data not shown).

In summary, MMG leads to decreased osteoblastogenesis and increased adipogenesis in hMSC via disruption of stress fibers and altered integrin signaling, which are likely mediated by a decrease in RhoA. MMG also stimulates RANKL-mediated osteoclastogenesis in RAW264.7 cells as compared to gravity controls. These results are analogous to those obtained in humans on long duration spaceflights. Also, the bone loss associated with microgravity is similar to that induced on Earth by inactivity.[12] Therefore, these *in vitro* systems present a unique opportunity to investigate not only the molecular mechanisms but also test potential pharmacological agents for preventing bone loss associated with microgravity in space and disuse osteoporosis on Earth.

REFERENCES

1. BIKLE, D.D. & B.P. HALLORAN. 1999. The response of bone to unloading. J. Bone Miner. Metab. **17:** 233–244.
2. TILTON, F.E. *et al.* 1980. Long-term follow-up of Skylab bone demineralization. Aviat. Space Environ. Med. **51:** 1209–1213.
3. VICO, L. *et al.* 2000. Effects of long-term microgravity exposure on cancellous and cortical weight-bearing bones of cosmonauts. Lancet **355:** 1607–1611.
4. CARMELIET, G. *et al.* 1998. Gene expression related to the differentiation of osteoblastic cells is altered by microgravity. Bone **22**(5 Suppl): 139S–143S.
5. SMITH, S.M. *et al.* 2005. Bone markers, calcium metabolism, and calcium kinetics during extended-duration space flight on the mir space station. J. Bone Miner. Res. **20:** 208–218.

6. GOODWIN, T.J. *et al.* 1993. Reduced shear stress: a major component in the ability of mammalian tissues to form three-dimensional assemblies in simulated microgravity. J. Cell Biochem. **51:** 301–311.

7. MEYERS, V.E. *et al.* 2004. Modeled microgravity disrupts collagen I/integrin signaling during osteoblastic differentiation of human mesenchymal stem cells. J. Cell Biochem. **93:** 697–707.

8. PITTENGER, M.F. *et al.* 1999. Multilineage potential of adult human mesenchymal stem cells. Science **284:** 143–147.

9. ZAYZAFOON, M. *et al.* 2004. Modeled microgravity inhibits osteogenic differentiation of human mesenchymal stem cells and increases adipogenesis. Endocrinology **145:** 2421–2432.

10. MEYERS, V.E. *et al.* 2005. RhoA and cytoskeletal disruption mediate reduced osteoblastogenesis and enhanced adipogenesis of human mesenchymal stem cells in modeled microgravity. J. Bone Miner. Res. **20:** 1858–1866.

11. ZAYZAFOON, M. *et al.* 2005. Microgravity: the immune response and bone. Immunol. Rev. **208:** 267–280.

12. COLLET, P. *et al.* 1997. Effects of 1- and 6-month spaceflight on bone mass and biochemistry in two humans. Bone **20:** 547–551.

Evolution, Aging, and Osteoporosis

STEPHANIE TUNG[a] AND JAMEEL IQBAL[b]

[a]*Department of Ecology and Evolutionary Biology, Princeton University, Princeton, New Jersey, USA*

[b]*Mount Sinai Bone Program and Department of Medicine, Mount Sinai School of Medicine, New York, New York, USA*

ABSTRACT: Aging is a process whereby organisms lose to the capacity to effectively handle life's stresses. Associated with aging are pathophysiological processes, such as decreases in bone mass, which in the extreme form lead to significant morbidity. Evolutionary theory suggests that these pathophysiological processes are permitted to occur because an organism need only preserve its body against stress only for the amount of time needed for effective reproduction. In this review, an understanding of osteoporosis and bone loss is discussed within the context of aging theory. Specific topics covered include: (1) bone loss as an antagonistically pleiotropic physiological process, (2) age-associated stress accumulation and its negative impact on bone turnover, and (3) the mechanisms whereby gonadal failure, increases in inflammatory cytokines, and cellular bone marrow changes lead to bone loss. This review concludes by examining purported hypotheses in the context of Werner's syndrome, a disease characterized by premature aging. We suggest that future osteoporosis therapy will likely focus on prevention of aging in general as a means to prevent the development of osteoporosis.

KEYWORDS: aging; bone loss; BMD; bone; osteoclast; evolution; disposable soma; Werner's syndrome; antagonistic pleiotropy

INTRODUCTION

Bone loss invariably accompanies growing older. In contrast to popular belief, the processes that contribute to bone loss actually begin by 30 years of age and continue relentlessly until death. Extensive research has aimed to elucidate the molecular mechanisms mediating bone loss, and as a result several drugs (e.g., bisphosphonates) have been developed for clinical use to inhibit

Address for correspondence: Dr. Jameel Iqbal, Mount Sinai Bone Program and Department of Medicine, Mount Sinai School of Medicine, Box 1055, 1 Gustave Levy Place, New York, NY 10029; fax: (212) 426-8312.

jameel.iqbal@mssm.edu

Ann. N.Y. Acad. Sci. 1116: 499–506 (2007). © 2007 New York Academy of Sciences.
doi: 10.1196/annals.1402.080

the key molecules/cells involved in bone degradation. In the future, however, we will have drugs/interventions to prevent bone loss from occurring in the first place. This review will attempt to lay the groundwork for understanding where future osteoporosis research will be focused. Specifically, theories on the evolutionary significance of aging will be reviewed. These principles will then be used in examining paradigms of bone loss and osteoporosis.

UNDERSTANDING THE AGING PROCESS

Osteoporosis is defined as a decrease in bone mass with accompanying microarchitectural compromise leading to an increase in fragility and fracture risk. Osteoporosis is not separate from aging; rather, osteoporosis is considered a byproduct of the aging process. In order to understand why osteoporosis occurs, we must understand why humans age.

Aging or senescence is defined as the time-dependent loss of somatic function leading to increased vulnerability to environmental challenge and a growing risk of disease and death. Aging is usually accompanied by a decline in fertility and is thus associated with loss of Darwinian fitness.[1,2] Having osteoporotic bones, which leads to a reduced ability to withstand physical forces, also causes loss of Darwinian fitness. The bone loss that accompanies aging is therefore, at first glance, evolutionarily unexpected. After investing years in the intricate process of bone morphogenesis, it is not readily apparent why all of that effort would go to waste by allowing bones to slowly disintegrate.

A central premise of aging is that it is not a programmed event. There is no equivalent organized program for aging as there is for development. Thus, humans experience the effects of aging and bone loss *not* as the result of processes preconfigured to occur, but rather as the byproducts of the body's inability to sustain itself. Different organisms have highly variable rates of aging, suggesting that species evolved to invest different amounts of energy into self-maintenance.[3] To explain these evolutionary differences, two theories have been proposed: optimality and mutational explanations.

According to optimality explanations, constraints on the combinations of survival and fertility that an organism can achieve at each age may lead to the evolution of a single optimal genotype. This optimal genotype shows senescence since maximum fitness is achieved by increasing early performance at the expense of late performance. The two major optimality theories on aging are the antagonistic pleiotropy theory and the disposable soma theory.

The antagonistic pleiotropy theory proposes the existence of pleiotropic genes with beneficial effects early in life but adverse effects late in life.[2,4] Natural selection would favor mutations leading to such alleles since their early effects increase survival in young animals while their late effects do not impinge on the organism's reproductive fitness.[5]

In order for antagonistically pleiotropic genes to be favored in evolution, they must provide a benefit to reproductive success. Examples of such benefits include increased number or protection of eggs, and increased viability during the mating season. Despite the sound theoretical foundations of this theory, there have been only a few genes identified as antagonistically pleiotropic. One example in humans is the Huntington gene. This gene negatively impacts fitness later in life by leading to cognitive decline and impaired movement. However, individuals with Huntington's disease produce more progeny than unaffected siblings, thus providing them with a reproductive advantage.[6]

Although there is a lack of evidence to support the existence of antagonistically pleiotropic genes, there is ample evidence to suggest that antagonistically pleiotropic physiological processes govern life span and aging. Much of this evidence comes from studies in flies. Sgro and Partridge[7] selected old and young lines of *Drosophila* by breeding flies at either an old or young age. They observed that flies that mated earlier tended to die earlier and the ones that mated later tended to die later. The investigators then induced sterility in these two genetic strains using two methods: radiation and mating with a sterility mutant. They found that differences in the death rates between the flies were equalized, suggesting that parturition negatively impacted their life span. These findings support the idea of an antagonistically pleiotropic tradeoff in which early beneficial effects lead to later deleterious effects.

Studies on the effects of pregnancy and childbirth in humans show that increased reproductive success is associated with decreases in bone mineral density (BMD). Pregnancy and lactation have been shown to decrease BMD, with women losing on average 3.4% at the lumbar spine and 4.3% at the trochanter.[8] Moreover, clinical studies indicate that BMD in women decreases as the number of pregnancies increases.[9] Thus, like decreases in life span that accompany childbearing, bone mass also appears to be an antagonistically pleiotropic phenomenon; decreases in BMD serve to give vital nutrients to growing offspring, but at the expense of decreased parental bone mass.

One tenant of the concept of antagonistically pleiotropic physiology (known as the disposable soma theory) asserts that there is no need to keep the soma in good condition once it has enabled germ cells to develop into a new, independent individual that is able to reproduce.[2] An organism therefore invests only enough energy into somatic maintenance, including its bones, to keep it in sound condition for as long as it has a reasonable chance of reproduction. For humans, mating with offspring production has historically occurred in individuals less than 30 years of age.[10] Thus, evolutionarily, humans would not have had a strong selective pressure to maintain their bone strength beyond this period. This fact is reflected in population BMD studies showing that BMD decreases begin at around 30 years of age for total body and lumbar spine BMD, and at around 20 years of age for femoral neck BMD.[11]

OSTEOPOROSIS IN THE CONTEXT OF AGING

Evolutionary theory has thus far allowed for an understanding of why osteoporosis is permitted to occur when it does—because it is beyond the period of selective pressure. However, what are the key degenerative mechanisms that contribute to osteoporosis? This question has been addressed by examining the phenotypes of elderly individuals and by elucidating the mechanisms responsible for diseases characterized by accelerated aging (e.g., Werner's syndrome).

In normal individuals, aging is associated with the accumulation of oxidant stresses and the coincident rise in prooxidant cytokine levels. Studies have found that levels of the inflammatory cytokine TNF are at least twofold higher in older individuals compared to younger ones.[12] Moreover, in elderly humans levels of TNF are correlated with plasma levels of other inflammatory markers, such as IL-6, CRP, and soluble TNF receptors.[12]

Recent studies have shown that a correlation exists between oxidant stress, increased levels of inflammatory cytokines, and bone loss with increased fracture risk. Results from the Health ABC study showed that subjects with the greatest number of inflammatory markers had the highest risk of fracture.[13] Moreover, the results suggested that levels of inflammatory markers alone could be used to accurately predict the relative risk of fracture.[13] Specifically, levels of soluble TNF receptors and soluble IL-2 receptor provided the best correlation with the risk of fracture.[13] The correlation between oxidant stress and bone loss has also been demonstrated in studies showing increased BMD with long-term antioxidant use. Studies in guinea pigs have illustrated a correlation between dietary intake of the antioxidant ascorbic acid and BMD.[14] Similarly, the Third National Health and Nutrition Examination Survey found that dietary ascorbic acid intake was independently associated with BMD and self-reported fractures in premenopausal women and in men.[15] The Women's Health Initiative trial found that the users of hormone replacement therapy that had the highest BMD were those that consumed ascorbic acid.[16]

As individuals age, there are changes not only in the amount of inflammatory cytokines, but also in the composition of the cellular components of bone marrow, such that the number of granulocytes and monocytes increases. To address whether this increase impacts age-related bone loss, Perkins et al.[17] studied the propensity of bone marrow cells to form osteoclasts from 4- to 6-month-old mice compared to 24-month-old mice. They found that the elderly mice had a 2.5-fold increase in the number of osteoclast precursors/monocytes and that this translated into a 2.5-fold increase in the number of osteoclasts formed when cultured with osteoclast-inducing cytokines.

Recent work using mathematical modeling and cell cultures has confirmed that increases in the number of osteoclast precursors lead to increased osteoclast formation; importantly, the nature of this causal relationship suggests that decreases in BMD can be obtained *without* alterations in the amount of proosteoclastogenic cytokines (e.g., RANK-L or OPG) expressed per cell.[18] Increases in the number of osteoclast precursors have also been linked to increased levels of TNF.[18-20] These findings tie together several features of aging and help us understand why increased TNF levels, seen in aging, can recapitulate one of the primary mechanisms driving age-related bone loss, osteoclast precursor expansion.

Other complementary cellular mechanisms for age-related bone loss have been described in addition to an increase in the number of osteoclast precursors. One mechanism is that stromal cells and osteoblasts in the bone microenvironment produce altered levels of cytokines and this increases the propensity to form osteoclasts. Cao *et al.*[21] quantitatively analyzed the effects of aging on the production of RANK-L and M-CSF in mice. They found in co-culture experiments that stromal cells/osteoblasts from old mice resulted in a 2.2-fold increase in the number of osteoclast precursors when compared to younger mice.[21] When analyzing the mechanism for how older stromal/cells osteoblasts lead to increases in osteoclast formation, they found that older cells had increased levels of RANK-L by approximately fivefold and M-CSF by approximately threefold per cell. It should be noted, however, that when they looked at the propensity to form RANK-L/M-CSF-induced osteoclasts from the same animals' bone marrow progenitor cells, they found that old animals had a 4.4-fold increase in the number of osteoclasts formed—an effect they attributed to greater numbers of osteoclast precursors.[21] Thus, the effect of changes in the stromal/osteoblast expression profile serve to augment the effects of increases in the osteoclast precursor pool, with both leading to increases in osteoclast formation and bone loss.

Aging is also associated with decreased tissue preservation and this contributes to osteoporosis by the failure of gonadal tissue's production of sex steroids. How aging is related to decreased gonadal tissue performance is currently being elucidated, but evidence suggests the involvement of FOXO transcription factors. FOXO transcription factors are important in mediating stress resistance and are regulated by genes associated with alterations in life span.[22] Studies with Foxo3a$^{-/-}$ female mice have shown that the loss of FOXO function leads to premature ovarian failure with early depletion of functional ovarian follicles and infertility.[23,24] Recent studies in ovariectomized rats have also suggested that modifying the action of FOXO transcription factors may be able to reverse the effects of decreases in estrogen.[25] A study examining the effects of resveratrol, a compound that modifies FOXO activity, found it had equivalent bone-protective effects to hormone replacement therapy.[25]

Age-associated declines in gonadal tissue functioning contribute to osteoporosis by altering levels of hormones that normally govern the bone remodeling process. Two key hormones appear to be altered: estrogen and follicle-stimulating hormone (FSH). Estrogen (or androgen) is believed to act on osteoblasts to augment their synthetic function.[26] For unclear reasons, aging is associated with a failure of osteoblasts to adequately respond to estrogen despite an increased number of estrogen receptors.[26] The other key mechanism linking age-associated hormonal changes with bone loss is the action of FSH on osteoclast precursors. Our group recently discovered that FSH is able to directly act in the bone marrow to cause an expansion of the osteoclast precursor pool and to augment proosteoclastogenic signaling downstream of the proosteoclastogenic cytokine RANK-L.[18,27] Through these actions, FSH likely serves to increase osteoclastic functioning and leads to bone loss during the peri- and postmenopausal periods.

In addition to changes in the cellular components of the bone marrow microenvironment, *hormonal changes* have been linked to alterations in the levels of inflammatory cytokines. Work by Pacifici *et al.* first established a link between altered levels of immune cytokines, such as TNF, their increase following ovarian failure, and the resultant decrease in BMD.[28–30] We have also recently established that levels of FSH may govern bone marrow TNF production.[18] Thus, gonadal failure is likely associated with alterations in FOXO functioning, and may be intricately connected to age-related changes in the bone marrow and cytokine profile through alterations in the levels of estrogen and FSH.

WERNER'S SYNDROME: DO THE HYPOTHESES FIT?

There are several known diseases and animals models that are characterized by accelerated aging. Many of these syndromes recapitulate the features seen with normal aging, such as osteoporosis, graying of the hair, and arthritis.[31] Interestingly, all of these conditions have in common a defect in the maintenance of the genome. We will use Werner's syndrome as an example to test out the hypotheses regarding aging and bone loss discussed above.

In the context of the disposable soma theory, individuals with Werner's syndrome age because they are incapable of maintaining their body for a sufficient time. This is in line with the idea that the maintenance of an individual species is governed by multiple processes that come together to determine the amount of time needed for the preservation of the soma. Mutation of any one of these methods of preservation would thus lead to an accelerated accumulation of damage to the soma and manifest as premature aging. By having a defect in a gene necessary for the maintenance of DNA, Werner's syndrome supports this hypothesis.

Do individuals with Werner's syndrome experience similar age-related mechanisms that lead to bone loss? Similar to normal elderly individuals, those with Werner's syndrome have elevations in the levels of inflammatory cytokines, such as TNF.[32] It is currently unknown whether they have any changes in their bone marrow similar to that seen in normal aging. However, individuals with Werner's do have premature gonadal failure resulting in hypogonadism.[32]

Thus, in many ways the age-associated changes seen in Werner's syndrome appear to mesh well with the hypotheses on normal aging discussed above.

In summary, the ideas regarding aging and osteoporosis stem from the accumulation of stresses that cause increased cytokine production and gonadal tissue failure. Premature aging syndromes, such as Werner's have taught us that the effects of aging manifest as a function of the amount of self-preservation undertaken. It is likely that the future of osteoporosis therapy will be focused on treating aging itself.

ACKNOWLEDGMENT

J.I. is supported by an MSTP grant to Mount Sinai School of Medicine.

REFERENCES

1. BONDURIANSKY, R. & C.E. BRASSIL. 2005. Reproductive ageing and sexual selection on male body size in a wild population of antler flies (*Protopiophila litigata*). J. Evol. Biol. **18:** 1332–1340.
2. KIRKWOOD, T.B. 2005. Understanding the odd science of aging. Cell **120:** 437–447.
3. PARTRIDGE, L. & N.H. BARTON. 1993. Optimality, mutation and the evolution of ageing. Nature **362:** 305–311.
4. TERMAN, A. & U.T. BRUNK. 2005. Autophagy in cardiac myocyte homeostasis, aging, and pathology. Cardiovasc. Res. **68:** 355–365.
5. VIJG, J. et al. 2005. Aging and genome maintenance. Ann. N. Y. Acad. Sci. **1055:** 35–47.
6. HAYDEN, M.R. et al. 1981. Huntington's chorea on the island of Mauritius. S. Afr. Med. J. **60:** 1001–1002.
7. SGRO, C.M. & L. PARTRIDGE. 1999. A delayed wave of death from reproduction in Drosophila. Science **286:** 2521–2524.
8. ULRICH, U. et al. 2003. Bone remodeling and bone mineral density during pregnancy. Arch. Gynecol. Obstet. **268:** 309–316.
9. GUR, A. et al. 2003. Influence of number of pregnancies on bone mineral density in postmenopausal women of different age groups. J. Bone. Miner. Metab. **21:** 234–241.
10. MORBEY, Y.E. & P.A. ABRAMS. 2004. The interaction between reproductive lifespan and protandry in seasonal breeders. J. Evol. Biol. **17:** 768–778.
11. ANNAPOORNA, N. et al. 2004. An increased risk of osteoporosis during acquired immunodeficiency syndrome. Int. J. Med. Sci. **1:** 152–164.
12. BRUUNSGAARD, H. et al. 2000. TNF-alpha, leptin, and lymphocyte function in human aging. Life Sci. **67:** 2721–2731.
13. CAULEY, J.A. et al. 2007. Inflammatory markers and incident fracture risk in older men and women: the health aging and body composition study. J. Bone Miner. Res. **22:** 1088–1095.
14. KIPP, D.E. et al. 1996. Long-term low ascorbic acid intake reduces bone mass in guinea pigs. J. Nutr. **126:** 2044–2049.

15. SIMON, J.A. & E.S. HUDES. 2001. Relation of ascorbic acid to bone mineral density and self-reported fractures among US adults. Am. J. Epidemiol. **154:** 427–433.

16. WOLF, R.L. *et al.* 2005. Lack of a relation between vitamin and mineral antioxidants and bone mineral density: results from the Women's Health Initiative. Am. J. Clin. Nutr. **82:** 581–588.

17. PERKINS, S.L. *et al.* 1994. Age-related bone loss in mice is associated with an increased osteoclast progenitor pool. Bone **15:** 65–72.

18. IQBAL, J. *et al.* 2006. Follicle-stimulating hormone stimulates TNF production from immune cells to enhance osteoblast and osteoclast formation. Proc. Natl. Acad. Sci. USA **103:** 14925–14930.

19. BOYCE, B.F., E.M. SCHWARZ & L. XING. 2006. Osteoclast precursors: cytokine-stimulated immunomodulators of inflammatory bone disease. Curr. Opin. Rheumatol. **18:** 427–432.

20. LI, P. *et al.* 2004. Systemic tumor necrosis factor alpha mediates an increase in peripheral CD11bhigh osteoclast precursors in tumor necrosis factor alpha-transgenic mice. Arthritis Rheum. **50:** 265–276.

21. CAO, J.J. *et al.* 2005. Aging increases stromal/osteoblastic cell-induced osteoclastogenesis and alters the osteoclast precursor pool in the mouse. J. Bone Miner. Res. **20:** 1659–1668.

22. BRUNET, A. *et al.* 2004. Stress-dependent regulation of FOXO transcription factors by the SIRT1 deacetylase. Science **303:** 2011–2015.

23. BRENKMAN, A.B. & B.M. BURGERING. 2003. FoxO3a eggs on fertility and aging. Trends Mol. Med. **9:** 464–467.

24. CASTRILLON, D.H. *et al.* 2003. Suppression of ovarian follicle activation in mice by the transcription factor Foxo3a. Science **301:** 215–218.

25. SU, J.L. *et al.* 2007. Forkhead proteins are critical for bone morphogenetic protein-2 regulation and anti-tumor activity of resveratrol. J. Biol. Chem. **282:** 19385–19398.

26. ANKROM, M.A. *et al.* 1998. Age-related changes in human oestrogen receptor alpha function and levels in osteoblasts. Biochem. J. **333**(Pt 3): 787–794.

27. SUN, L. *et al.* 2006. FSH directly regulates bone mass. Cell **125:** 247–260.

28. KIMBLE, R.B., S. BAIN & R. PACIFICI. 1997. The functional block of TNF but not of IL-6 prevents bone loss in ovariectomized mice. J. Bone Miner. Res. **12:** 935–941.

29. PACIFICI, R. 1996. Estrogen, cytokines, and pathogenesis of postmenopausal osteoporosis. J. Bone Miner. Res. **11:** 1043–1051.

30. PACIFICI, R. *et al.* 1991. Effect of surgical menopause and estrogen replacement on cytokine release from human blood mononuclear cells. Proc. Natl. Acad. Sci. USA **88:** 5134–5138.

31. GRAY, M.D. *et al.* 1997. The Werner syndrome protein is a DNA helicase. Nat. Genet. **17:** 100–103.

32. DAVIS, T. & D. KIPLING. 2006. Werner syndrome as an example of inflamm-aging: possible therapeutic opportunities for a progeroid syndrome? Rejuv. Res. **9:** 402–407.

Index of Contributors